Regulation of Autophagy under Stress

Regulation of Autophagy under Stress

Editors

Angeles Aroca
Cecilia Gotor

MDPI

Basel • Beijing • Wuhan • Barcelona • Belgrade • Novi Sad • Cluj • Manchester

Editors
Angeles Aroca
Universidad de Sevilla
Seville, Spain

Cecilia Gotor
Instituto de Bioquı́mica,
Vegetal y Fotosíntesis, Consejo
Superior de Investigaciones,
Científicas and Universidad de Sevilla
Seville, Spain

Editorial Office
MDPI
St. Alban-Anlage 66
4052 Basel, Switzerland

This is a reprint of articles from the Special Issue published online in the open access journal *Antioxidants* (ISSN 2076-3921) (available at: https://www.mdpi.com/journal/antioxidants/special_issues/Autophagy_Stress).

For citation purposes, cite each article independently as indicated on the article page online and as indicated below:

Lastname, A.A.; Lastname, B.B. Article Title. *Journal Name* **Year**, *Volume Number*, Page Range.

ISBN 978-3-0365-8894-0 (Hbk)
ISBN 978-3-0365-8895-7 (PDF)
doi.org/10.3390/books978-3-0365-8895-7

Cover image courtesy of Angeles Aroca

Contents

Sara Laine-Menéndez, Miguel Fernández-de la Torre, Carmen Fiuza-Luces, Aitor Delmiro, Joaquín Arenas, Miguel Ángel Martín, et al.
Apoptosis-Inducing Factor Deficiency Induces Tissue-Specific Alterations in Autophagy: Insights from a Preclinical Model of Mitochondrial Disease and Exercise Training Effects
Reprinted from: *Antioxidants* **2022**, *11*, 510, doi:10.3390/antiox11030510 1

Ran Guo, Shan-Shan Wang, Xiao-You Jiang, Ye Zhang, Yang Guo, Hong-Yan Cui, Qi-Qiang Guo, et al.
CHK2 Promotes Metabolic Stress-Induced Autophagy through ULK1 Phosphorylation
Reprinted from: *Antioxidants* **2022**, *11*, 1166, doi:10.3390/antiox11061166 23

Miey Park, Anshul Sharma, Hana Baek, Jin-Young Han, Junho Yu and Hae-Jeung Lee
Stevia and Stevioside Attenuate Liver Steatosis through PPARα-Mediated Lipophagy in *db/db* Mice Hepatocytes
Reprinted from: *Antioxidants* **2022**, *11*, 2496, doi:10.3390/antiox11122496 35

Angeles Aroca and Cecilia Gotor
Hydrogen Sulfide: A Key Role in Autophagy Regulation from Plants to Mammalians
Reprinted from: *Antioxidants* **2022**, *11*, 327, doi:10.3390/antiox11020327 47

Naveed Ur Rehman, Peichun Zeng, Zulong Mo, Shaoying Guo, Yunfeng Liu, Yifeng Huang and Qingjun Xie
Conserved and Diversified Mechanism of Autophagy between Plants and Animals upon Various Stresses
Reprinted from: *Antioxidants* **2021**, *10*, 1736, doi:10.3390/antiox10111736 59

Yipin Wang, Nicholas Siu Kay Fung, Wai-Ching Lam and Amy Cheuk Yin Lo
mTOR Signalling Pathway: A Potential Therapeutic Target for Ocular Neurodegenerative Diseases
Reprinted from: *Antioxidants* **2022**, *11*, 1304, doi:10.3390/antiox11071304 83

Inmaculada Navarro-Lérida, Anna M. Aragay, Alejandro Asensio and Catalina Ribas
Gq Signaling in Autophagy Control: Between Chemical and Mechanical Cues
Reprinted from: *Antioxidants* **2022**, *11*, 1599, doi:10.3390/antiox11081599 109

Anna De Gaetano, Lara Gibellini, Giada Zanini, Milena Nasi, Andrea Cossarizza and Marcello Pinti
Mitophagy and Oxidative Stress: The Role of Aging
Reprinted from: *Antioxidants* **2021**, *10*, 794, doi:10.3390/antiox10050794 135

Yuchen Lei, Yuxiang Huang, Xin Wen, Zhangyuan Yin, Zhihai Zhang and Daniel J. Klionsky
How Cells Deal with the Fluctuating Environment: Autophagy Regulation under Stress in Yeast and Mammalian Systems
Reprinted from: *Antioxidants* **2022**, *11*, 304, doi:10.3390/antiox11020304 161

Qiuluo Liu, Yan Chen, Li Zhou, Haining Chen and Zongguang Zhou
From Intestinal Epithelial Homeostasis to Colorectal Cancer: Autophagy Regulation in Cellular Stress
Reprinted from: *Antioxidants* **2022**, *11*, 1308, doi:10.3390/antiox11071308 193

Adrián Santos-Ledo, Beatriz de Luxán-Delgado, Beatriz Caballero, Yaiza Potes, Susana Rodríguez-González, José Antonio Boga, et al.
Melatonin Ameliorates Autophagy Impairment in a Metabolic Syndrome Model
Reprinted from: *Antioxidants* **2021**, *10*, 796, doi:10.3390/antiox10050796 213

antioxidants

MDPI

Article

Apoptosis-Inducing Factor Deficiency Induces Tissue-Specific Alterations in Autophagy: Insights from a Preclinical Model of Mitochondrial Disease and Exercise Training Effects

Sara Laine-Menéndez [1], Miguel Fernández-de la Torre [1], Carmen Fiuza-Luces [1], Aitor Delmiro [1], Joaquín Arenas [1,2], Miguel Ángel Martín [1,2], Patricia Boya [3], Alejandro Lucia [4,5] and María Morán [1,2,*]

1 Mitochondrial and Neuromuscular Diseases Laboratory, Instituto de Investigación Sanitaria Hospital '12 de Octubre' ('imas12'), 28041 Madrid, Spain; sara.laine@cbm.csic.es (S.L.-M.); mftorre@ing.uc3m.es (M.F.-d.l.T.); cfiuza.imas12@h12o.es (C.F.-L.); adelmiro@h12o.es (A.D.); joaquin.arenas@salud.madrid.org (J.A.); mamcasanueva.imas12@h12o.es (M.Á.M.)

2 Spanish Network for Biomedical Research in Rare Diseases (CIBERER), U723, 28029 Madrid, Spain

3 Department of Cellular and Molecular Biology, Centro de Investigaciones Biológicas Margarita Salas, CSIC, 28040 Madrid, Spain; pboya@cib.csic.es

4 Faculty of Sports Sciences, European University of Madrid, Villaviciosa de Odón, 28670 Madrid, Spain; alejandro.lucia@universidadeuropea.es

5 Spanish Network for Biomedical Research in Fragility and Healthy Aging (CIBERFES), 28029 Madrid, Spain

* Correspondence: mmoran@h12o.es; Tel.: +34-91-779-2784; Fax: +34-91-390-8544

Citation: Laine-Menéndez, S.; Fernández-de la Torre, M.; Fiuza-Luces, C.; Delmiro, A.; Arenas, J.; Martín, M.Á.; Boya, P.; Lucia, A.; Morán, M. Apoptosis-Inducing Factor Deficiency Induces Tissue-Specific Alterations in Autophagy: Insights from a Preclinical Model of Mitochondrial Disease and Exercise Training Effects. *Antioxidants* **2022**, *11*, 510. https://doi.org/10.3390/antiox11030510

Academic Editors: Angeles Aroca and Cecilia Gotor

Received: 11 February 2022
Accepted: 3 March 2022
Published: 7 March 2022

Abstract: We analyzed the effects of apoptosis-inducing factor (AIF) deficiency, as well as those of an exercise training intervention on autophagy across tissues (heart, skeletal muscle, cerebellum and brain), that are primarily affected by mitochondrial diseases, using a preclinical model of these conditions, the Harlequin (Hq) mouse. Autophagy markers were analyzed in: (i) 2, 3 and 6 month-old male wild-type (WT) and Hq mice, and (ii) WT and Hq male mice that were allocated to an exercise training or sedentary group. The exercise training started upon onset of the first symptoms of ataxia in Hq mice and lasted for 8 weeks. Higher content of autophagy markers and free amino acids, and lower levels of sarcomeric proteins were found in the skeletal muscle and heart of Hq mice, suggesting increased protein catabolism. Leupeptin-treatment demonstrated normal autophagic flux in the Hq heart and the absence of mitophagy. In the cerebellum and brain, a lower abundance of Beclin 1 and ATG16L was detected, whereas higher levels of the autophagy substrate p62 and LAMP1 levels were observed in the cerebellum. The exercise intervention did not counteract the autophagy alterations found in any of the analyzed tissues. In conclusion, AIF deficiency induces tissue-specific alteration of autophagy in the Hq mouse, with accumulation of autophagy markers and free amino acids in the heart and skeletal muscle, but lower levels of autophagy-related proteins in the cerebellum and brain. Exercise intervention, at least if starting when muscle atrophy and neurological symptoms are already present, is not sufficient to mitigate autophagy perturbations.

Keywords: autophagy; OXPHOS; mitochondrial diseases; Harlequin; heart; skeletal muscle; cerebellum; brain

1. Introduction

Mitochondrial diseases (MD) are caused by mutations in mitochondrial or nuclear genes with subsequent impairment of the oxidative phosphorylation system (OXPHOS). Despite their low prevalence (~1 in 5000 individuals), MD represent the most frequent inborn errors of metabolism [1]. There is individual variability in symptom onset and clinical manifestations and, although virtually any tissue can be affected skeletal muscle, the heart and central nervous system (CNS) are the primary disease targets [2].

At the cellular level, OXPHOS dysfunction leads to impairments in ATP production, mitochondrial membrane potential and the assembly of OXPHOS complexes, together with an increase in reactive oxygen species (ROS) production, thereby leading to 'energy crises', oxidative stress and cellular damage [3,4]. Alterations in autophagy—the lysosomal-dependent process that recycles damaged or dysfunctional cellular components including mitochondria (i.e., 'mitophagy')—can also occur in MD; this is reflected by an increase in autophagy-related vesicles in transmitochondrial cybrids harboring pathogenic mitochondrial DNA mutations, as well as in patient-derived fibroblasts or in tissues from preclinical models [5–17]. However, the actual relevance of autophagy alterations in the pathophysiology of MD remains to be fully elucidated and, it is in fact unknown whether the different affected tissues show this type of alterations.

There is essentially no effective cure for most MD, although some strategies can help to attenuate clinical manifestations [18]. One such strategies is regular physical exercise, which has proven to mitigate exercise intolerance in affected patients [19–25]. However, besides improvements in muscle oxidative capacity, the molecular bases underlying exercise benefits are not yet fully clear. In this regard, there is growing evidence that regular exercise produces beneficial effects at the multisystemic level through numerous molecular pathways, including the promotion of autophagy—especially in those organs that are primarily affected in MD, that is, skeletal muscle, the heart and CNS [26–30]. Furthermore, exercise has been reported to attenuate dysregulation of autophagy in several conditions associated with altered autophagy, such as age-related sarcopenia, spinal muscular atrophy, hypertension or Parkinson's disease [31–34]. However, it remains to be determined whether exercise can improve autophagy in the context of MD.

Apoptosis inducing factor (AIF) is a mitochondrial protein that was first reported to induce chromatin degradation and caspase-independent cell death by migrating to the nucleus in response to an apoptogenic insult [35–37]. More recently, it has been elucidated that AIF cooperates with the mitochondrial protein MIA 40 in the import of some subunits of the respiratory complex I and of complex IV assembly factors through a disulfide relay substrate oxidation mechanism [38,39]. The deficiency of AIF results in altered assembly and low activity of the respiratory chain complex I in preclinical and clinical models [40]. The aim of the present study was to analyze the status of autophagy in the affected tissues of a well-characterized preclinical model of MD, the Harlequin (Hq) mouse. This animal exhibits a decrease of ~80% in the apoptosis inducing factor (*Aif*) gene expression due to a proviral insertion in this gene. Thus, the Hq model recapitulates important patients' alterations due to respiratory chain complex I deficiency, such as myopathy, increased risk of cardiomyopathy, and cerebellar ataxia [41–45]. In addition, we analyzed whether a physical exercise intervention combining aerobic and resistance training modalities could promote changes in autophagy in the aforementioned affected tissues. Our results showed a tissue-specific response, with increased autophagy in the heart and *biceps femoris* of the Hq mouse, and no significant training effect on autophagy in any of the analyzed tissues.

2. Materials and Methods

2.1. Mouse Model

All experimental protocols were approved by the Institutional Review Committee (*Hospital 12 de Octubre*; project numbers PROEX 111/15 and PROEX 067/18) and were conducted in accordance with Animal Research: Reporting of In Vivo Experiments (ARRIVE) guidelines and with European (European Convention for the Protection of Vertebrate Animals ETS123) and Spanish (32/2007 and R.D. 1201/2005) laws on animal protection in research. Heterozygous male (B6CBACa Aw-J/A-Aifm1Hq/J) and female Hq mice (X/Hq, B6CBACa Aw-J/A-Aifm1Hq/J), and wild-type male mice (WT, B6CBACa) of the same strain, were obtained from The Jackson Laboratory (Bar Harbor, ME, USA) were used for the present study. Mice were housed in the animal facility of *Hospital 12 de Octubre* (Madrid, Spain) under controlled conditions of temperature, humidity and ventilation, with 12-h light/dark cycles and ad libitum access to food and water.

2.2. Study Design

2.2.1. Time Course Study

To assess the impact of MD progression on autophagy, 8-week-old heterozygous female mice (X/Hq, B6CBACa Aw-J/A-Aifm1Hq/J) were crossed with WT males of the same strain (all purchased from The Jackson Laboratory), and the male mice of the F1 generation were used for tissue analyses at different ages. PCR genotyping of F1 mice was performed on tail DNA using previously described primers (Integrated DNA Technologies; Coralville, IA, USA) [43].

F1 male mice were sacrificed at 2, 3 and 6 months of age (8 WT and 8 Hq mice for each age, project number PROEX 067/18) by cervical dislocation, and the heart, *biceps femoris*, *quadriceps femoris*, brain and cerebellum were dissected, frozen in liquid nitrogen and stored at −80 °C for subsequent biochemical analysis.

2.2.2. Exercise Training Study

The study variables were analyzed in a group of mice after an 8-week exercise training program previously described by us [41]. Briefly, 44 male mice (23 WT and 21 Hq, project number PROEX 111/15) were obtained from The Jackson Laboratory from 6–8 weeks of age and were physically evaluated every week until the onset of MD symptoms. Physical evaluation included locomotion analysis (treadmill locomotion test), as well as cerebellar ataxia (rotarod) and muscular strength (handgrip) determinations. When symptoms of MD were clear, each Hq mouse was paired to a WT mouse; both mice then performed a maximal incremental test on a treadmill and were randomly allocated to a sedentary (Sed, 12 WT and 11 Hq) or an exercise training group (Ex, 11 WT and 11 Hq). In the latter, each training session lasted between 40 and 60 min, and included resistance exercises thrice a week (Monday, Wednesday and Friday), and endurance (treadmill) training five times per week (from Monday to Friday), with the duration and intensity of each session gradually increasing during the program. Sedentary animals did not perform the training program but were free to move in their cages. The mean ± SD age of the Hq mice upon symptom onset and at the end of the intervention period were 12.1 ± 0.1 weeks and 5.3 ± 0.1 months, respectively. The mean age of the mice at the end of the intervention period did not differ between the experimental groups [41].

After the 8-week intervention study, the aforementioned physical assessments were also performed in age-matched pairs of sedentary and exercise-trained mice. In order to account for a potential confounding effect caused by the short-term response to a single acute exercise stimulus—we solely intended to determine the long-term (i.e., 'training') effects of the exercise intervention—48 h elapsed after the last exercise test before mice were sacrificed by cervical dislocation. The heart, *biceps femoris*, *quadriceps femoris*, brain and cerebellum were quickly obtained and frozen in liquid nitrogen before storage at −80 °C until molecular analysis.

2.3. Cardiomyocyte Cell Culture

Cardiomyocyte isolation was performed using the PierceTM Primary Cardiomyocyte Isolation Kit (Thermo Fisher Scientific; Waltham, MA, USA) following the manufacturer's instructions. Briefly, hearts from postnatal (1–4 days old, P1–P4) mice were dissected, cut into 1–3 mm fragments, and washed twice with Hanks' Balanced Salt Solution without calcium and magnesium at 4 °C. Tissue fragments were subsequently digested by enzymatic treatment, mechanically disaggregated, and living cells were counted for cell culture seeding. Cells were cultured in Dulbecco's Modified Eagle Medium (DMEM) (Gibco; Gaithersburg, MD, USA), supplemented with 10% fetal bovine serum (Gibco), 100 IU/mL penicillin, 100 IU/mL streptomycin and the cardiomyocyte growth factor provided by Pierce in the isolation kit (1/1000), in a CO_2 incubator at 37 °C. The culture medium was replaced every 3 days, and the obtained cardiomyocyte purity was approximately 80%.

2.4. Assessment of Autophagy Flux

For the study of autophagic flux, 3-month-old male mice (5 WT and 5 Hq) were treated i.p., with 0.5 mL of sterile phosphate buffered saline (PBS) containing 40 mg/kg leupeptin hemisulfate (Merck; Darmstadt, Germany). After 30 min treatment, mice were sacrificed by cervical dislocation and tissues were dissected, frozen and stored as described above.

To assess autophagic flux in cultured cardiomyocytes, cells were treated with 30 µg/µL of hydroxychloroquine sulfate (H0915, Sigma-Aldrich; St., Louis, MI, USA) for 24 h (5–15 cells of 2–3 mice). After treatment, cells were fixed and processed for immunofluorescence as described below.

2.5. Tissue Processing

Heart and muscle tissues were processed to obtain total homogenates in ice-cold 10 mM of Tris-HCl pH 7.6, containing 150 mM of NaCl, 1 mM of EDTA, 1% Triton™ X-100 and a protease and phosphatase inhibitor cocktail (Roche Diagnostics Corporation; Indianapolis, IN, USA) 1:10 (weight: volume) in a Potter homogenizer. The brain and cerebellum were homogenized at 1:10 (weight: volume) in ice-cold RIPA buffer containing 50 mM of Tris-HCl pH 7.4, 1% NP-40, 0.5% Na-deoxycholate, 1% SDS, 150 mM of NaCl, 2 mM of EDTA, with protease and phosphatase inhibitors. Homogenates were centrifuged at 11,000× *g* for 15 min at 4 °C, and the supernatants containing the solubilized proteins were collected for analysis. Protein concentration in the homogenates was determined using the Pierce® BCA protein assay kit (Thermo Fisher Scientific; Whaltham, MA, USA).

2.6. Western Blotting

Autophagy-related protein levels were analyzed by Western blotting. Samples of tissue homogenates containing between 20 and 40 µg of denatured proteins were separated on 7.5–15% SDS-polyacrylamide gels. Resolved proteins were transferred to PVDF membranes and blocked with 5% non-fat dry milk or 5% bovine serum albumin (BSA). Blocked membranes were probed with the relevant primary antibody (Supplemental Table S1) and subsequently incubated with horseradish peroxidase (HRP)-conjugated secondary antibody. Protein signals were finally detected with ECL Prime Western blotting Detection Reagent (Amersham, GE Healthcare; Little Chalfont, UK). Protein signals were quantified with ImageJ (Rasband, W.S, ImageJ; U. S. National Institutes of Health, Bethesda, MA, USA). Data were normalized by total protein load per well according to Coomassie Blue staining of the membrane as previously described [46] for heart protein analysis, whereas glyceraldehyde 3-phosphate dehydrogenase (GAPDH) was used as the loading control for skeletal muscle analyses and γ-tubulin for brain and cerebellum studies.

2.7. Immunocytochemistry

Cardiomyocytes were seeded and fixed on glass coverslips with 4% paraformaldehyde in a CO_2 incubator. Fixed cells were washed with PBS and antigen retrieval was performed with 5% urea and 0.1 mM Tris pH 9.5 for 20 min at 95 °C. Cells were permeabilized with 0.1% Triton™ X-100, blocked with 10% goat serum and 0.01% Triton X-100 in PBS, and incubated with primary antibody in blocking solution overnight (Supplemental Table S1). After washing, cardiomyocytes were incubated with the relevant fluorophore-coupled secondary antibody in blocking solution. After secondary antibody washing, nuclei were stained with 0.5 µg/mL of 4′,6-diaminidino-2-phenylindole (DAPI) and mounted in Prolong Gold Antifade mounting medium (Thermo Scientific; Whaltham, MA, USA). Cells were observed with an LSM510 META confocal microscope (Zeiss; Oberkochen, Germany) fitted with a 40× achromatic or 63× plan-achromatic objective.

2.8. High Performance Liquid Chromatography

The amino acid content of heart and skeletal muscle homogenates was analyzed by high-performance liquid chromatography (HPLC) using a methodology certified by the Spanish National Accreditation Entity ENAC (*Entidad Nacional de Acreditación*) (ISO15189) used for the analysis of clinical samples in the Clinical Biochemistry Department of the *Hospital 12 de Octubre*. Briefly, homogenate samples containing 1.7 and 2.0 mg of total protein for heart and skeletal muscles homogenates, respectively, were deproteinized with 50% sulfosalicylic acid and analyzed by ion exchange chromatography with post-column derivatization with ninhydrin on a Biochrom 30+ amino acid analyzer (Biochrom Ltd.; Cambridge, UK). Amino acid peaks were detected and quantified with OpenLAB EZChrom Edition software A.04.10 (Siegwerk Druckfarben AG & Co. KGaA; Siegburg, Germany).

2.9. Statistical Analysis

All study variables were expressed as a median and interquartile range. All variables were tested for normality of data distribution using the D'Agostino-Pearson test. Owing to the relatively small sample size, and to the fact that most of the study variables did not follow a Gaussian (normal) distribution, the non-parametric Kruskal–Wallis test was used to determine whether a significant 'group' effect was found for the different variables, in which case pairwise comparisons were performed *post hoc* with the Dunn's test. The Mann–Whitney U test was used for comparisons between pairs of groups. Statistical significance was set at a p-value $p < 0.05$. All analyses were performed with GraphPad Prism® 7 for Windows (GraphPad Software; San Diego, CA, USA).

3. Results

3.1. Autophagy in the Heart: Time Course Study

To analyze if MD was associated with autophagy alterations depending on age, the levels of proteins representative of distinct phases of the autophagy pathway were measured at 2, 3 and 6 months of age, respectively, in WT and Hq mice of the time-course study (Figure 1). The analysis of the main regulators of autophagy showed the following results. First, the levels of the active form (phosphorylated at residue Ser2448) of the mechanistic target of rapamycin (pmTOR) tended to be higher in Hq mice than in WT mice at all ages, with the difference between both groups significant at 3 months of age (Figure 1). In turn, the levels of 5′ AMP-activated protein kinase (AMPK) phosphorylated at residue Thr172 (pAMPK), which is associated with a state of autophagy activation in response to low cellular energy availability, tended to be lower in Hq mice than in WT, with the between-group difference significant at 2 months of age. Subsequently, the levels of two proteins involved in the early stages of formation of the phagophore (i.e., the double membrane that encloses and isolates the cytoplasmic components during autophagy), Beclin 1 and autophagy-related protein 16-like (ATG16L), were quantified. We found the trend towards higher levels of both proteins in Hq mice compared to WT mice, with the difference between groups significant at 2 (for both Beclin 1 and ATG16L) and 6 months of age (for ATG16L only). The determination of microtubule-associated protein 1B light chain I (LC3B)-I and LC3B-II (which is formed after LC3B-I conjugation to phosphatidylethanolamine) showed that LC3B-I levels tended to be higher in Hq mice at all ages, with the between-group difference significant at 2 and 3 months. Next, we measured the levels of the protein p62, which tended to be higher in Hq mice than in WT mice at all ages, with between-group differences significant at 2 and 3 months. Finally, to assess the lysosomal content in Hq mice, the levels of the lysosomal protease cathepsin B (protein bands at 27 and 24 KDa) were quantified, with the results showing a trend towards higher values in Hq mice and with a significant difference vs. WT mice for the 27 KDa band at 2 months of age (Figure 1).

3.2. Autophagic Flux Analysis in the Heart

To assess whether the accumulation of autophagy proteins in the hearts of Hq mice was due to a failure in the autophagosome-lysosome flux or to an autophagy induction response, we studied the autophagic flux by treating 3-month-old WT and Hq animals with leupeptin, an inhibitor of cysteine, serine and threonine peptidases present in lysosomes. The levels of different autophagy proteins in leupeptin-treated mice are shown in Figure 2a. We found a significant group effect for the levels of Beclin 1, p62 and LC3B-II ($p = 0.0038$,

$p = 0.014$ and $p = 0.015$, respectively), with higher values of the three variables in treated mice than in untreated animals (Figure 2a). *Post hoc* analyses showed significant differences in Beclin 1 and p62 levels between leupeptin-treated Hq and non-treated WT mice as well as in p62 levels between treated and non-treated Hq, and a similar response to leupeptin in Hq and WT mice with overall higher autophagy markers for treated mice (both genotypes). These data suggest a normal autophagic flux in the heart of both Hq and WT mice and may point to an increased protein degradation response by the lysosome-autophagosome pathway in Hq mice.

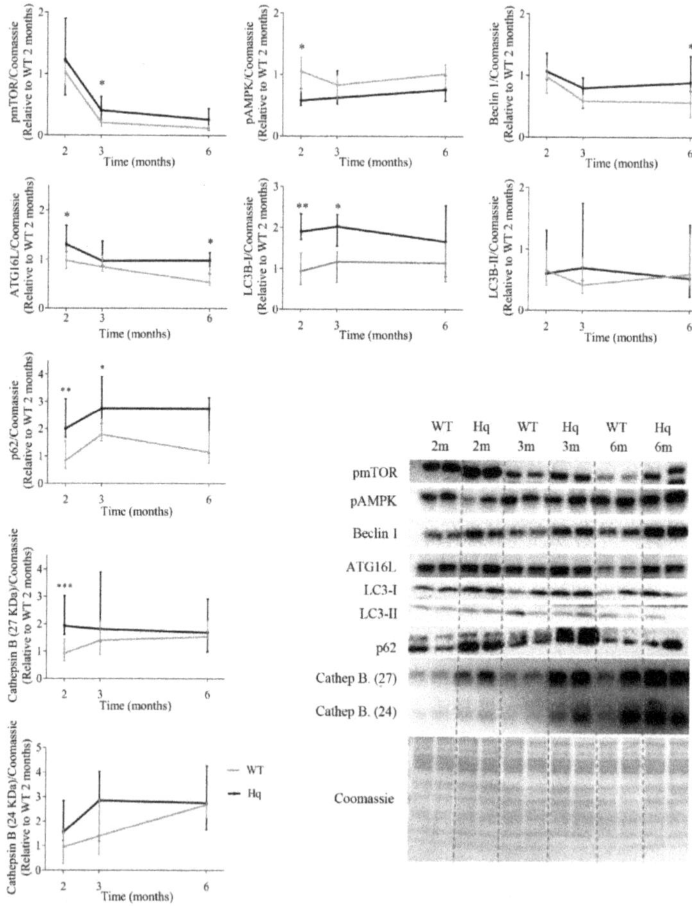

Figure 1. Autophagy proteins in the heart (time course study). Representative Western blots and quantifications of proteins involved in autophagy in total heart homogenates from wild-type (WT) and Harlequin (Hq) male mice at 2, 3 and 6 months (m) ($n = 8$ mice). Total protein content per lane estimated by Coomassie Blue staining was used as loading control. Data (median and interquartile range) are expressed relative to protein levels in the 2-month-old WT group. Black lines: Hq mice; gray lines: WT mice. Mann–Whitney U-test: * $p < 0.05$; ** $p < 0.01$; *** $p < 0.001$ significantly different from the age-matched WT group. Abbreviations: pmTOR, mTOR phosphorylated at residue Ser2448; pAMPK, AMPK phosphorylated at residue Thr172; ATG16L, autophagy-related protein 16-like; LC3B-I, microtubule-associated protein 1B light chain I; LC3B-II, microtubule-associated protein 1B light chain II; Cathep. B, cathepsin B.

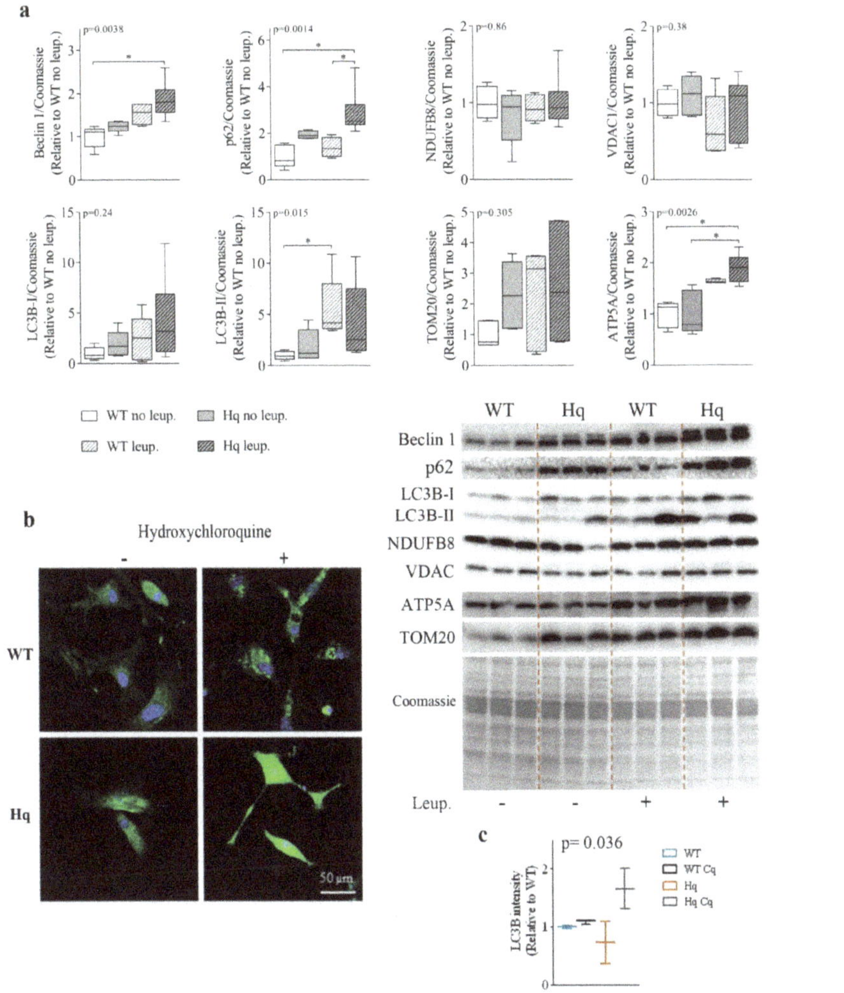

Figure 2. Autophagic flux and mitochondrial markers in the heart. (**a**) Representative Western blots and quantifications of autophagy (Beclin 1, p62, LC3B-I, LC3B-II) and mitochondrial (NDUFB8, VDAC, TOM20 and ATP5A) markers in heart homogenates of 3-month-old wild-type (WT) and Harlequin (Hq) male mice, whether untreated (no leup.) or treated with leupeptin (leup.) (n = 5–6 mice). Quantification of Western blots. Total protein content per lane estimated by Coomassie Blue staining was used as loading control. Data (median, interquartile range, minimum and maximum values) are expressed relative to the untreated WT group. (**b**) Representative images of LC3B (green) and nuclei (blue) staining in cardiomyocytes from WT and Hq male neonatal mice untreated (−) and treated (+) with hydroxychloroquine (native colors are shown). (**c**) Quantification of mean fluorescence intensity of LC3B per cell in untreated and hydroxychloroquine-treated WT and Hq neonatal cardiomyocytes (n = 5–15 cells from 2–3 mice). Data (median and interquartile range) are expressed relative to the untreated WT mean value. p-values for group effect (Kruskal–Wallis test) are shown above the graphs. Symbols for significant differences in *post hoc* (Dunn's test) pairwise comparisons: * $p < 0.05$. Abbreviations: ATP5A, ATP synthase lipid-binding protein; TOM20, mitochondrial import receptor subunit TOM20; NDUF8, mitochondrial NADH dehydrogenase (ubiquinone) 1 beta subcomplex subunit 8; VDAC, voltage-dependent anion-selective channel protein.

To assess whether the higher protein tagging for autophagic degradation in the hearts of Hq animals was due to a higher lysosomal degradation of mitochondrial proteins, or perhaps to mitophagy, we quantified the levels of several representative mitochondrial proteins in leupeptin-treated and non-treated mice. We found no significant differences in the levels of mitochondrial NADH dehydrogenase (ubiquinone) 1 beta subcomplex subunit 8 (NDUFB8), the voltage-dependent anion-selective channel protein 1 (VDAC1), or the mitochondrial import receptor subunit TOM20, respectively, between leupeptin-treated and untreated Hq mice. However, we detected a significant group effect for the ATP synthase lipid-binding protein (ATP5A) ($p = 0.0026$), which reached higher levels in leupeptin-treated animals with significant *post hoc* differences observed for leupeptin-treated Hq mice compared to both untreated Hq and WT animals (whether treated or not) mice (Figure 2). These results suggest that, overall, mitophagy is not enhanced in the heart of Hq mice and that autophagy participates in the turnover of some OXPHOS proteins.

Finally, we detected LC3B by immunofluorescence in WT and Hq cardiomyocytes isolated from neonatal mice treated with hydroxychloroquine (a lysosomal inhibitor which basifies lysosomes) for 24 h (Figure 2b,c). The analysis of LC3B fluorescence intensity per cell showed significant differences between groups ($p = 0.036$), although *post hoc* tests did not reveal differences between pairs of groups. A higher LC3B fluorescence was observed in hydroxychloroquine-treated Hq cardiomyocytes compared to untreated Hq cells, with this difference greater than that observed between treated and untreated WT cardiomyocytes. These findings may suggest a more active autophagic flux in neonatal Hq cardiomyocytes compared to WT cells (Figure 2c).

3.3. *Effects of Exercise Training on Autophagy in the Heart*

In order to determine if exercise training could induce normalization of autophagy markers in the heart tissue of Hq mice, we measured several autophagy-related proteins in sedentary and exercise-trained animals (Figure 3). We found significant differences between experimental groups for most variables (pmTOR, $p = 0.0008$; Beclin 1 and ATG16L $p < 0.0001$; LC3B-I, $p = 0.015$; LC3B-II, $p = 0.0021$; p62, $p < 0.0001$; and cathepsin B ($p = 0.002$ for 27KDa and $p = 0.0305$ for 24 KDa band)) (Figure 3). However, *post hoc* analyses revealed that the training intervention did not induce a significant effect in either Hq or WT. All the observed differences between pairs of experimental groups were due to genotype (i.e., MD) and corroborated our previous results regarding the differences in ATG16L and Beclin 1 levels between WT and sedentary Hq mice. In addition, statistical significance was also reached for *post hoc* differences in LC3B-I and II, pmTOR, p62 and cathepsin B (27 KDa) between sedentary Hq and WT mice. We found quasi-significant differences in pAMPK between sedentary Hq and WT mice ($p = 0.059$) (Figure 3).

3.4. *Sarcomere Proteins Levels in the Heart*

To analyze if autophagy might be involved in the degradation of the contractile machinery of the heart, we measured the levels of three sarcomere proteins in the hearts of mice in the intervention study. The results are shown in Figure 4. We found a significant group effect for the proteins myosin light chain 3 (MYL3), cardiac troponin I3 (TNNI3) and cardiac troponin C1 (TNNC1) (p-value = 0.0006, 0.0002 and 0.0071, respectively), with overall lower levels of these proteins in Hq mice (Figure 4). In *post hoc* pairwise comparisons, sedentary Hq mice showed lower levels of TNNI3 and TNNC1 than sedentary WT mice (Figure 4). On the other hand, the exercise training intervention did not induce any significant change (i.e., no differences between exercise-trained and sedentary Hq, respectively). However, strikingly, we found lower levels of TNNC1 in trained WT mice than in sedentary WT mice.

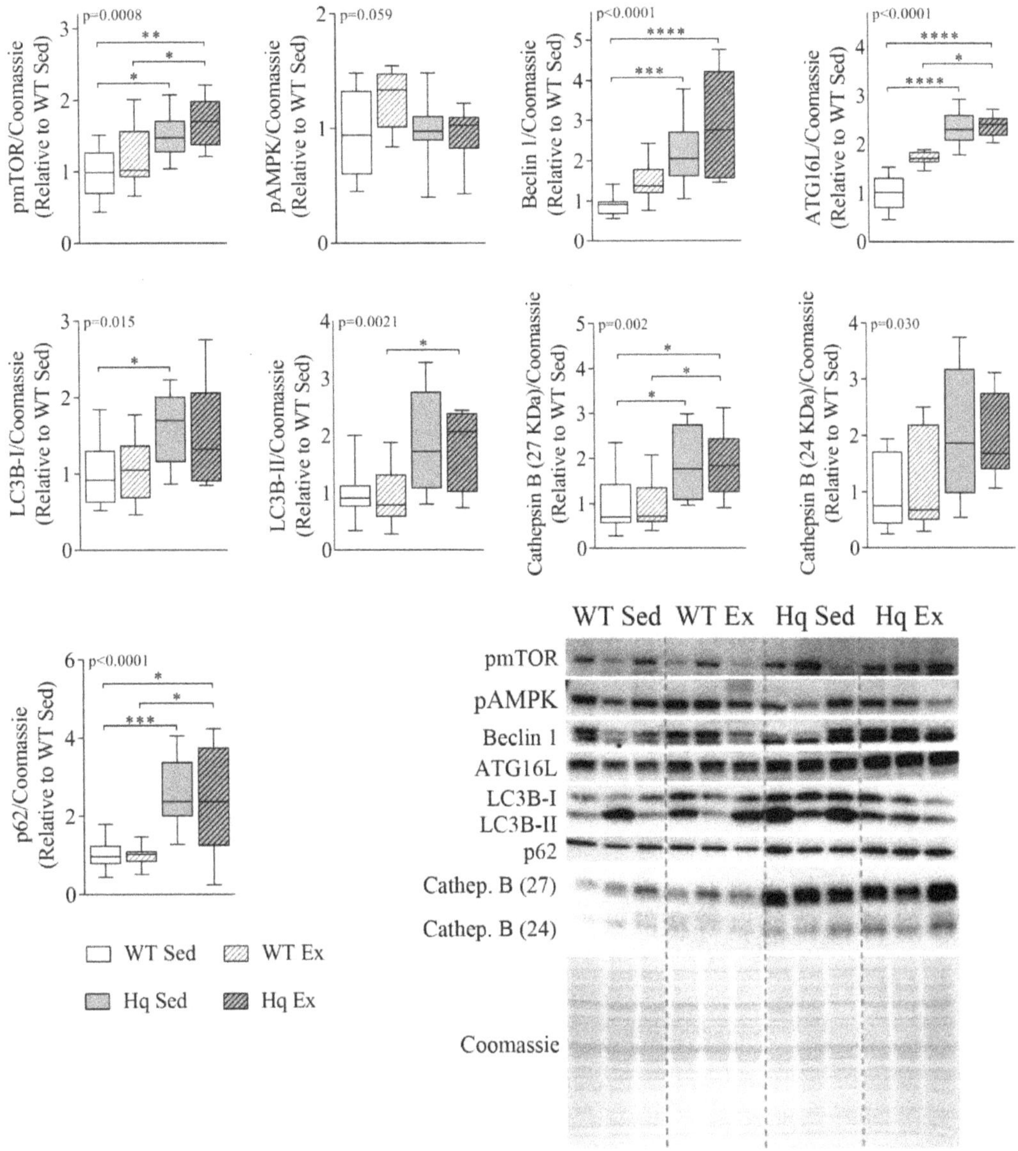

Figure 3. Effects of exercise training on autophagy in the heart tissue. Representative Western blots and quantifications of mTOR phosphorylated at residue Ser2448 (pmTOR), AMPK phosphorylated at residue Thr172 (pAMPK), Beclin 1, autophagy-related protein 16-like (ATG16L), microtubule-associated protein 1B light chain I (LC3B-I) and II (LC3B-II), p62 and cathepsin B (Cathep. B), in heart homogenates from wild-type (WT) and Harlequin (Hq), sedentary (Sed) and trained (Ex) mice of the intervention study (n = 10–12 male mice, age 5.3 months). Total protein content per lane estimated by Coomassie Blue staining was used as loading control. Data (median, interquartile range, minimum and maximum values) are relative to the control group (WT Sed). p-values for group effect (Kruskal–Wallis test) are shown above the graphs. Symbols for significant differences in *post hoc* (Dunn's test) pairwise comparisons: * $p < 0.05$; ** $p < 0.01$; *** $p < 0.001$; **** $p < 0.0001$.

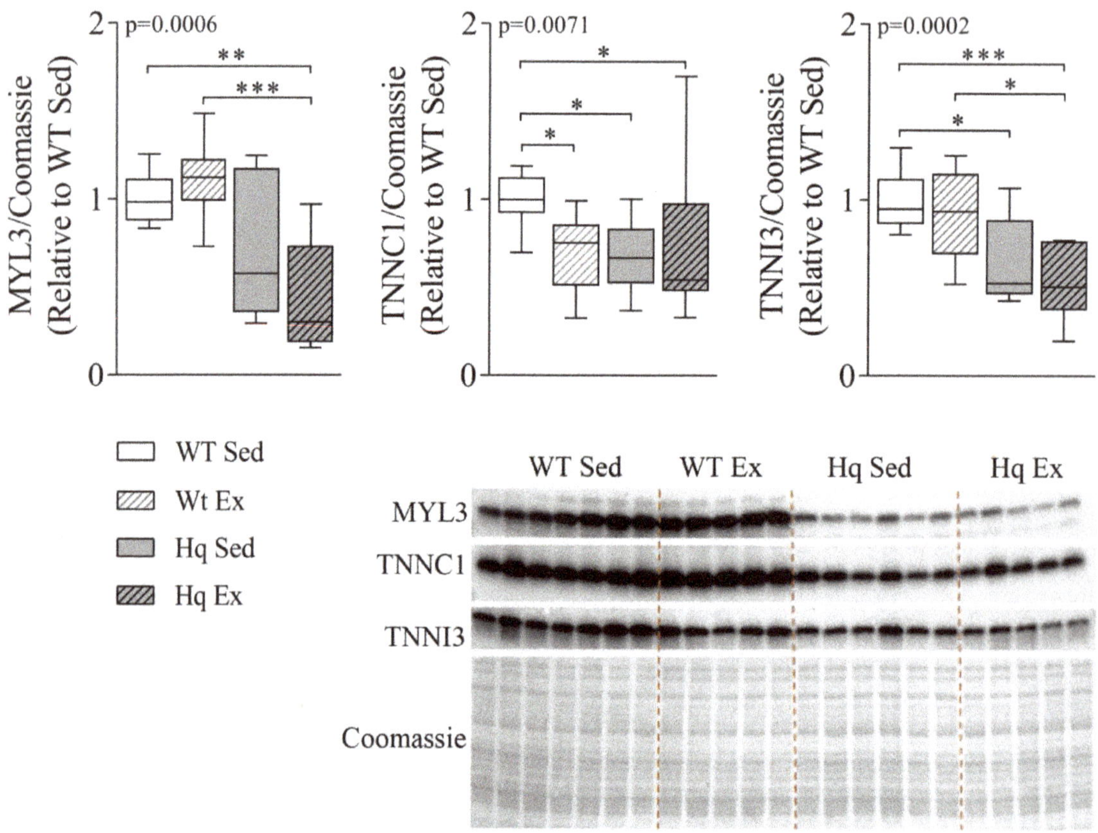

Figure 4. Sarcomere protein levels in the heart. Representative Western blots and quantifications of protein levels of myosin light chain 3 (MYL3), troponin C1 (TNNC1) and troponin I (TNNI3) in heart homogenates from wild-type (WT) and Harlequin (Hq), sedentary (Sed) and trained (Ex) mice in the intervention study (n = 10–12 male mice, age 5.3 months). Quantification of Western blots. Total protein content per lane estimated by Coomassie Blue staining was used as loading control. Significant p-values for group effect (Kruskal–Wallis test) are shown above the graphs. Symbols for significant differences in *post hoc* (Dunn's test) pairwise comparisons: * $p < 0.05$; ** $p < 0.01$; *** $p < 0.001$.

3.5. Free Amino Acid and Ammonium Levels in the Heart

To determine whether the increased autophagic flux in Hq mice might be related to increased protein catabolism, we measured the concentration of free amino acids by HPLC in total heart homogenates from sedentary WT and Hq mice of the intervention study groups. We found a trend towards higher levels of several amino acids in Hq mice compared to WT mice, with significant differences for threonine, serine, glycine and valine (Table 1). Ammonium levels were similar in both groups ($p > 0.99$).

3.6. Autophagy in Skeletal Muscle—Exercise Training Effects

To study whether autophagy was also altered in skeletal muscle, as well as the potential effects of exercise training in this process, we measured the levels of different proteins representing distinct phases of autophagy in homogenates of the *biceps femoris* of the exercise training study groups (Figure 5). We found a significant group effect for total AMPK (tAMPK) levels (p = 0.025), with *post hoc* analysis revealing higher tAMPK in both Hq mice (exercise-trained or sedentary) than in sedentary WT mice. However, the analysis of pAMPK levels showed no difference between groups, indicating that the training intervention did not produce a significant effect on total or phosphorylated AMPK levels in any of the experimental groups. We also detected a significant group effect for Beclin 1, ATG16L, LC3B-II and p62 ($p = 0.004$, $p < 0.001$, $p < 0.001$ and $p = 0.0046$, respectively), with *post hoc* comparisons showing higher levels of Beclin 1, ATG16L and LC3B-II in both Hq groups than in sedentary WT animals, as well as higher levels of these proteins in sedentary Hq mice compared to sedentary WT mice. In the case of p62, the post hoc test only showed differences between sedentary groups of mice. On the other hand, we did not observe a significant group effect for LC3B-I or lysosomal-associated membrane protein 1 (LAMP1) levels between experimental groups, for either disease or training effect (Figure 5).

Table 1. Free amino acids and ammonium in heart homogenates of wild-type (WT, $n = 5$) and Harlequin (Hq, $n = 5$) male mice (age 5.3 months).

Amino Acid	WT µmol/L	Hq µmol/L	*p*-Value
Threonine	14.3 [12.9–15.6]	22.3 [19.5–25.5]	0.008
Serine	17.0 [15.5–17.8]	22.4 [20.1–24.9]	0.016
Asparagine	7.9 [3.3–9.4]	9.2 [6.9–11.4]	0.309
Glutamate	192.6 [159.1–227.5]	211.4 [191.5–252.9]	0.309
Glutamine	206.7 [171.4–310.3]	294.9 [234.9–303.3]	0.309
Glycine	24.4 [20.6–26.2]	41.2 [36.6–42.8]	0.007
Alanine	131.5 [97.9–158.7]	170.5 [135.3–209.2]	0.095
Citrulline	7.0 [5.5–8.9]	7.1 [6.3–8.0]	0.690
Valine	4.8 [4.2–6.0]	10.4 [7.4–10.5]	0.031
Methionine	4.1 [2.9–4.6]	5.2 [3.7–5.5]	0.222
Isoleucine	9.0 [6.8–10.2]	11.6 [9.6–12.2]	0.151
Leucine	13.6 [8.2–13.9]	18.5 [14.1–19.1]	0.056
Tyrosine	4.9 [4.2–6.3]	6.5 [5.3–7.0]	0.309
Phenylalanine	5.8 [4.8–6.2]	5.2 [4.5–6.3]	0.690
Ammonium	313.9 [253.2–345.7]	336.9 [288.9–339.2]	>0.99
Lysine	24.4 [20.6–25.0]	27.9 [23.4–33.3]	0.095
Histidine	7.4 [6.7–10.2]	10.0 [8.6–11.3]	0.222
Arginine	16.2 [13.1–18.0]	17.8 [16.5–22.0]	0.309

Data are expressed as the median (interquartile range) of the amino acid concentration for each experimental group determined in homogenate aliquots containing 1.7 mg of total protein. *p*-values are shown for each comparison between WT and Hq groups (Mann–Whitney U-test).

3.7. Free Amino Acids and Ammonium Levels in Skeletal Muscle

To determine whether an increase in protein catabolism status might also occur in the skeletal muscles of Hq mice, we quantified the concentration of free amino acids and ammonium by HPLC in total quadricep homogenates from WT and Hq sedentary mice from the exercise training study. We found a trend towards higher levels of virtually all amino acids in Hq mice compared to WT mice, with significant differences for valine, ornithine, lysine and arginine (Table 2). By contrast, we detected no differences in ammonium levels between Hq and WT mice.

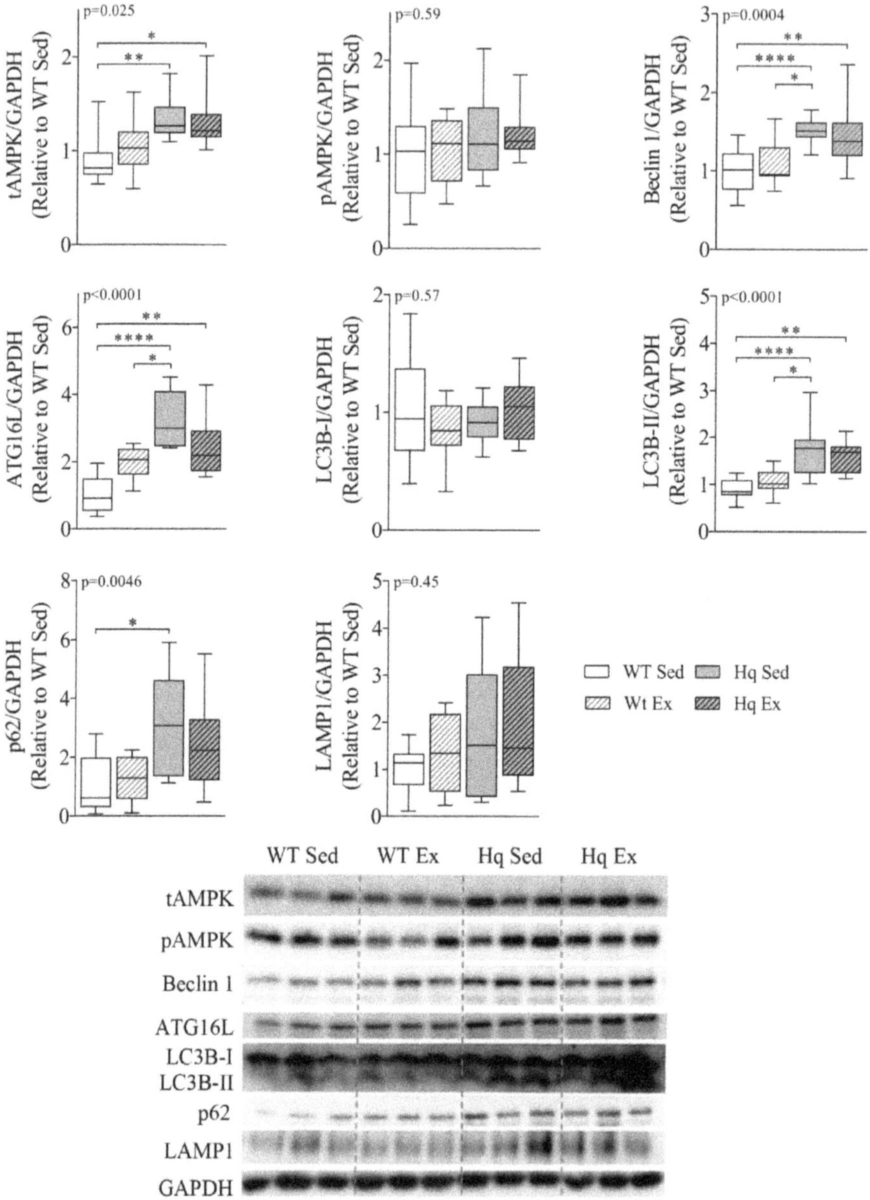

Figure 5. Autophagy proteins and effects of exercise training in skeletal muscle. Representative Western blots and quantifications of total AMPK (tAMPK), AMPK phosphorylated at residue Thr172 (pAMPK), autophagy-related protein 16-like (ATG16L), Beclin 1, microtubule-associated protein 1B light chain I (LC3B-I) and II (LC3B-II), p62 and lysosomal-associated membrane protein 1 (LAMP1) in *biceps femoris* homogenates from wild-type (WT) and Harlequin (Hq), sedentary (Sed) and exercise-trained (Ex) mice in the intervention study (n = 10–12 male mice, age 5.3 months). GAPDH was used as protein loading control. Data (median, interquartile range, minimum and maximum values) are relative to the control group (WT Sed). p-values for group effect (Kruskal–Wallis test) are shown above the graphs. Symbols for significant differences in *post hoc* (Dunn's test) pairwise comparisons: * $p < 0.05$; ** $p < 0.01$; **** $p < 0.0001$.

Table 2. Free amino acids and ammonium in *quadriceps* homogenates of wild-type (WT, $n = 5$) and Harlequin (Hq, $n = 5$) mice (age 5.3 months).

Amino Acid	WT (μM)	Hq (μM)	p-Value
Threonine	38.2 [36.5–44.7]	44.7 [44.1–62.5]	0.095
Serine	47.6 [34.9–65.5]	58.7 [47.3–82.4]	0.420
Asparagine	5.6 [4.8–7.3]	6.6 [5.8–13.1]	0.548
Glutamate	56.5 [48.4–72.9]	66.2 [57.3–131.2]	0.151
Glutamine	165.3 [143.4–202.1]	204.3 [181.8–235.5]	0.095
Glycine	205.3 [170.3–311.2]	282.1 [235.1–334.4]	0.222
Alanine	348.7 [277.4–371.1]	379.4 [354.5–475.1]	0.095
Citrulline	13.6 [12.3–16.2]	15.6 [10.1–18.2]	>0.999
Valine	24.7 [23.1–25.9]	31.1 [29.3–39.1]	0.008
Methionine	9.9 [9.4–10.4]	12.1 [8.2–13.8]	0.690
Cysteine	8.3 [7.9–8.4]	8.6 [7.8–9.4]	0.151
Isoleucine	17.9 [15.8–18.8]	21.3 [18.7–23.3]	0.055
Leucine	23.6 [22.9–25.5]	33.7 [25.7–38.3]	0.095
Tyrosine	16.2 [13.2–16.3]	16.6 [14.5–18.5]	0.222
Phenylalanine	11.5 [11.2–13.1]	15.7 [11.5–17.3]	0.095
Ammonium	575.3 [573.5–600.0]	569.0 [555.1–598.5]	0.421
Ornithine	4.4 ± [4.1–5.1]	8.3 [6.2–12.8]	0.008
Lysine	53.6 [44.9–58.3]	66.8 [63.4–84.5]	0.008
Histidine	14.8 [12.9–18.7]	18.0 [14.9–21.5]	0.309
Arginine	25.0 [18.7–25.7]	34.1 [27.9–46.4]	0.008
Hydroxyproline	11.7 [10.5–14.2]	14.6 [12.3–16.1]	0.222
Proline	34.4 [16.7–37.3]	36.7 [28.9–57.8]	0.309

Data are expressed as the median ± interquartile range of the amino acid concentration (μM) for each experimental group determined in homogenate aliquots containing 2.0 mg of total protein. *p*-value is shown for each comparison between WT and Hq groups (Mann–Whitney U-test).

3.8. Autophagy Protein Levels and Effects of Exercise Training in Central Nervous System

To assess whether autophagy was altered in the CNS, as well as the potential effects of exercise training on autophagy markers in this tissue, we measured several autophagy-related proteins in the cerebellum and brain of sedentary and exercise-trained mice (Figure 6). We found a significant effect for the levels of Beclin 1 and ATG16L in the cerebellum ($p = 0.0004$ and $p = 0.0032$, respectively) and brain ($p = 0.0007$ and $p = 0.0003$, respectively) (Figure 6a,b). The *post hoc* analysis revealed only significant differences between both groups of Hq mice (trained or sedentary) and WT sedentary mice, and no effect of the exercise training intervention in WT or Hq mice (Figure 6a,b). Regarding LC3B-I and LC3B-II, we found no differences in the cerebellum or brain, indicating the absence of autophagosome accumulation in the CNS of Hq mice and no effects of the exercise training program on autophagosome content (Figure 6a,b). By contrast, we detected a higher content of p62 and LAMP1 ($p = 0.0084$; $p = 0.0053$, respectively) in the cerebellum of Hq mice but not in the brain. *Post hoc* analyses revealed significant differences only in sedentary Hq mice in comparison with trained WT mice for p62, and differences between the exercise-trained groups of Hq and WT mice for LAMP1, but no effect of the exercise intervention in WT and Hq mice (Figure 6a).

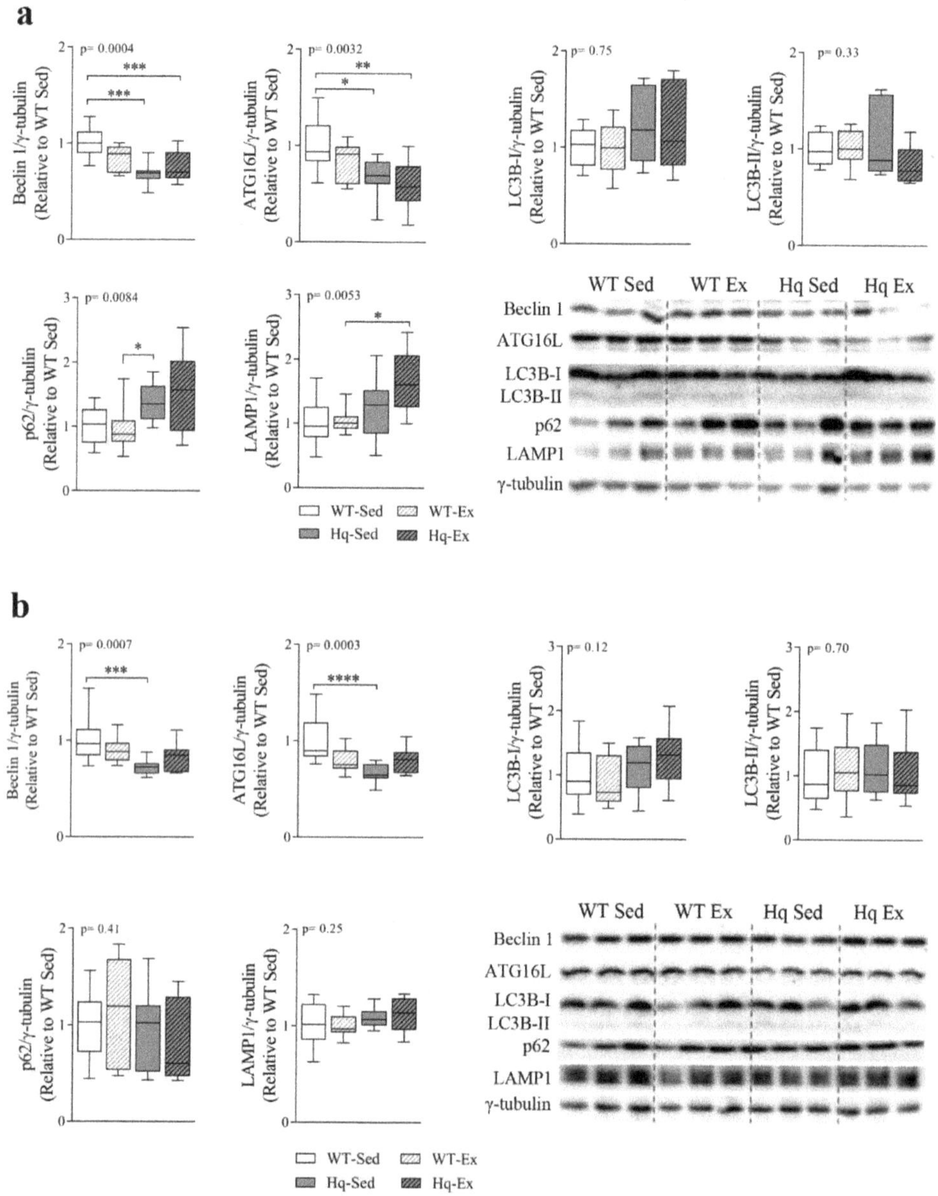

Figure 6. Autophagy proteins and effects of exercise training in central nervous system. Representative Western blot and densitometry analysis of autophagy-related protein 16-like (ATG16L), Beclin 1, microtubule-associated protein 1B light chain I (LC3B-I) and II (LC3B-II), p62 and lysosomal-associated membrane protein (LAMP1) in cerebellum (**a**) and brain (**b**) homogenates of wild-type (WT) and Harlequin (Hq), sedentary (Sed) and exercise-trained (Ex) mice of the intervention study (n = 10–12 male mice, age 5.3 months). γ-tubulin was used as protein loading control. Data (median, interquartile range, minimum and maximum values) are relative to the control group (WT Sed). p-values for group effect (Kruskal–Wallis test) are shown in the above graphs. Symbols for significant differences in *post hoc* (Dunn's test) pairwise comparisons: * $p < 0.05$; ** $p < 0.01$; *** $p < 0.001$; **** $p < 0.0001$.

Finally, the analysis of leupeptin treatment of WT and Hq mice (Figure 7) showed significant accumulation of LC3B-I and LC3B-II in both WT and Hq leupeptin-treated cerebella (p = 0.0028 and p = 0.0067, respectively) and a significant *post hoc* difference between treated Hq and non-treated WT, revealing normal autophagosome elimination in this tissue in the Hq mouse (Figure 7). Protein p62 was also apparently higher in WT and Hq mouse leupeptin-treated cerebella, supporting normal autophagic flux in the Hq cerebellum. However, we found no significant group effect, probably due to the variability in the low number of animals analyzed (n = 5–6 in each experimental group), and the time selected for the treatment, which was optimal for the heart but not for the CNS [47].

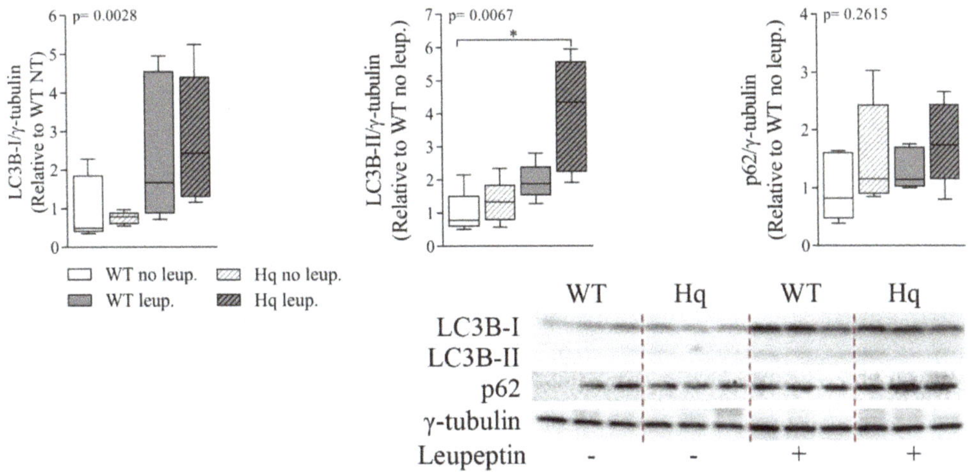

Figure 7. Autophagic flux analysis in the cerebellum. Representative Western blots and quantifications of autophagy markers p62, microtubule-associated protein 1B light chain I (LC3B-I) and II (LC3B-II), in cerebellum homogenates of 3-month-old wild-type (WT) and Harlequin (Hq) mice, untreated (no leup.) and treated with leupeptin (leup.) (n = 5–6 male mice). γ-tubulin was used as protein loading control. Data (median, interquartile range, minimum and maximum values) are expressed relative to untreated WT group. p-values for group effect (Kruskal–Wallis test) are shown in the above graphs. Symbols for significant differences in *post hoc* (Dunn's test) pairwise comparisons: * $p < 0.05$.

4. Discussion

The present study suggests that AIF deficiency causes a tissue-specific effect on autophagy, with the accumulation of autophagy-related proteins and a higher content of free amino acids in the heart and skeletal muscle tissue, as well as increased autophagy flux and lower levels of sarcomeric proteins in the heart, but, in turn, with the attenuation of autophagy in the CNS. On the other hand, an exercise intervention combining endurance and resistance training did not significantly mitigate autophagy alterations in the tissues that are primarily affected by MD (i.e., skeletal muscle, the heart and CNS).

In the present work, the finding of higher levels of proteins involved in different stages of autophagy in the heart of Hq mice suggest either an enhancement of autophagy, or an alteration of the autophagic flux due to AIF deficiency. However, the fact that treatment with leupeptin (a blocker of autophagy) in Hq mice demonstrated normal autophagic flux in the heart, points to a higher autophagy flux in this tissue as a result of AIF deficiency. Although we did not assess autophagy flux in skeletal muscles of Hq mice, the accumulation of autophagy markers observed in this tissue would suggest the existence of a similar response to that found in heart and correlates with a higher p62 and LC3II/I ratio previously reported in the Deletor mouse model of MD [48]. In contrast, the brain and the cerebellum did not show accumulation of autophagy markers in Hq mice but, in fact, a slightly lower content of Beclin 1 and ATG16L, and according to LC3B-II levels, a very low content of autophagosomes. In this regard, previous research has demonstrated a negative effect of Beclin 1 deficiency on Purkinje cell survival in mice [49] and that loss of ATG proteins in neurons leads to motor dyscoordination and progressive neurodegeneration [50,51], both of which are hallmarks of the Hq mice phenotype [42,43,52]. Considering that leupeptin treatment revealed normal autophagosome-lysosome flux in the cerebellum, our results might indicate a lower induction of autophagy in the CNS of Hq mice, which in the cerebellum could be responsible for the accumulation of p62-linked proteins. This high p62 content may also point to additional roles of p62 in the cerebellum of Hq mice, such as compensatory proteasome-mediated protein degradation [53,54], antioxidant response mediated by the Kelch-like ECH-associated protein (Keap1)/nuclear factor erythroid-derived 2-like 2 (Nrf2) pathway [55,56] or apoptosis [57]. The higher LAMP1 content found in the cerebellum indicates a higher lysosomal content in the Hq mouse, which can hypothetically be attributed to lysosomal biogenesis to promote a compensatory degradation of cellular components by direct lysosomal degradation, such as chaperone-mediated autophagy and/or microautophagy [58,59]. We analyzed if the enhanced autophagy observed in the Hq mouse heart could be related to mitophagy. In this effect, we found that the levels of markers for different mitochondrial compartments (inner and outer membrane) were not sensitive to leupeptin treatment, thereby suggesting that AIF deficiency does not promote mitochondrial degradation by mitophagy. Nevertheless, our finding of a higher content of ATP5A in the heart of leupeptin-treated mice compared with untreated mice would indicate that this mitochondrial protein is, indeed, degraded by the lysosomal pathway, which is consistent with the results of a previous study reporting that autophagy acts selectively on some mitochondrial proteins [60]. Overall, our findings indicate that the alterations in autophagy induced by AIF deficiency in the Hq mouse model are tissue-specific, with an induction of autophagy in striated muscles, but an attenuation in the CNS. Future studies are needed on the role of autophagy alterations in this mouse model of MD.

One of the mTOR-independent signals capable of inducing autophagy is oxidative stress [59], which is also involved in the physiopathology of MD. Indeed, other authors [43] as well as our group [41], have previously reported evidence of oxidative stress in the skeletal muscle of the Hq mouse together with higher protein and activity levels of antioxidant enzymes [41,43]. In addition, a higher sensitivity to oxidative stress-induced dead has been reported in the myocardium of these animals [45], and in the present study we found a higher content of peroxiredoxin 6 and a lower catalase activity in the hearts of 6-month-old Hq mice (data not shown). Therefore, we could hypothesize that oxidative stress associated to AIF deficiency could induce autophagy in the striated (skeletal and heart) muscles of Hq mice. There are also evidences of oxidative stress and enhanced ROS sensitivity in the cerebellum of the Hq mouse model [42,52]. Therefore, the aforementioned findings together with the fact that we failed to detect an increased autophagy in the CNS of Hq mice would suggest that the autophagic response to oxidative stress and mitochondrial dysfunction is also tissue-specific in this preclinical model of MD.

Autophagy is a cellular stress response that can be induced by exercise, not only in skeletal muscles but also in other tissues such as the brain, adipose tissue, pancreas or liver [27,28,30,61–64]. Interestingly, exercise intervention has been reported to mitigate autophagy alterations in several conditions such as muscle atrophy, heart failure, ischemic stroke, Parkinson's and Alzheimer's disease or ataxia [31,65–69]. In the present study, however, despite the significant exercise training-induced improvement in strength and endurance capacity previously reported by us in the same Hq mice of the present study [41], we failed to detect an attenuation of autophagy alterations in the striated muscles or the CNS of these animals with the exercise intervention. In this regard, previous research assessing the short-term effects of an acute exercise session has suggested that high-intensity exercise sessions induce autophagy to a greater extent than longer but less intense ones [63,70]. In addition, autophagy induction can be detected immediately after acute exercise in skeletal muscles, but its persistence thereafter has been questioned [71]. Thus, the mild intensity of the training program used in the present work, together with the fact that animals were sacrificed 48 h after the last physical test to prevent a potential confounding effect produced by an acute exercise stimulus, might explain, at least partly, the lack of changes in autophagy with the exercise training intervention. In addition, an earlier start of the intervention—before symptom onset—as well the use of a longer program could have also normalized the observed autophagy alterations and perhaps increased the lifespan of the animals. In fact, longer exercise interventions have been shown to attenuate proteome deregulation in the skeletal muscle and CNS of the Mutator mouse model of MD [72] or mitigate myopathy and increase the lifespan of the cytochrome C oxidase assembly factor heme A: farnesyltransferase (COX10) knockout (Cox $10^{-/-}$) mouse model [73]. The fact that our intervention started when myopathy and ataxia were already present might also explain, at least partly, the absence of attenuations in the observed autophagy alterations with exercise training. In this regard, it has been reported that the response of autophagy to exercise can be attenuated is some metabolic diseases [74]. In turn, previous research has indicated a tissue-specific effect of exercise training on autophagy response. For instance, Sprague Dawley rats that underwent a long-term, moderate intensity exercise training program showed autophagy induction in the brain cortex but not in the hippocampus, heart or skeletal muscle [61]. In other studies, acute (i.e., a single bout) as well as long-term exercise (i.e., training) in mice increased autophagy in the brain cortex but not in the cerebellum and hippocampus [30,75]. Therefore, we cannot rule out the possibility that, despite its moderate intensity, the training program used in our study may have elicited a significant autophagic response in other skeletal muscles or brain areas (e.g., brain cortex) that were not analyzed here.

The study of proteins involved in autophagy regulation in the heart showed lower levels of activated AMPK in Hq mice and a higher activation of the autophagy inhibitor mTOR [59]. The absence of activation of AMPK in the heart of Hq mice might be explained by the absence of OXPHOS deficiency previously described in this organ [44]. On the other hand, the enhanced mTOR activation in the Hq heart is in agreement with previous observations of increased mTOR activity in other mouse models of MD, such as the Aifm1 (R200 del) model of AIF deficiency and the Deletor mouse [48,76–78]. In these two pre-clinical models, mTOR overactivation was related to an enhancement of folate-mediated 1-carbon metabolism [48,76]. This metabolic route supplies glycine, serine and folate needed for methylation of transfer RNA during the translation of mitochondrial proteins and, therefore, supports the maintenance of the OXPHOS system [79]. Moreover, 1-carbon metabolism has been reported to facilitate the assembly of respiratory chain complex I also by a mechanism mediated by serine catabolism [58]. Therefore, overactivation of mTOR may act as a compensatory mechanism to promote the 1-carbon metabolism in the Hq mouse heart. In this effect, in the Deletor mouse, 1-carbon metabolism enhancement is associated with higher free amino acid content in the skeletal muscle and heart and accumulation of autophagic markers in the former [48,77]. In agreement with these findings, we observed higher serine and glycine content together with higher mTOR activation in the hearts of Hq mice than in those of the control mice. Therefore, one could hypothesize that folate-mediated 1-carbon metabolism is promoted by mTOR activation as a compensatory mechanism to enhance respiratory chain complex I and OXPHOS function in the Hq mouse heart. Under these conditions, autophagy would be driven by an mTOR-independent signal [59], as a synergistic response that induces protein catabolism and amino acid release to feed 1-carbon metabolism, despite its potential detrimental effect on myofibrils and, thus, explain the lower levels of sarcomere proteins and higher levels of free amino acids in the Hq heart.

Although the differences in amino acid content in the quadricep muscles of Hq mice were less pronounced than in the heart—where significance was only reached for valine, lysine and arginine, and a trend ($p < 0.1$) was observed for threonine, leucine, isoleucine, and phenylalanine−, overall, all amino acids tended to be more abundant in the Hq mice, suggesting active protein catabolism probably due, at least partly, to the observed enhanced autophagy. The higher content of free amino acids and autophagy markers in Hq quadriceps is in accordance with previous findings on the Deletor mouse muscle phenotype [48,77]. In addition, we have previously reported higher mTOR activation in the skeletal muscles of Hq mice in comparison with WT mice [41]. These data support the idea of an enhancement of 1-carbon metabolism and a protein catabolic response also in the skeletal muscles of the Hq mouse and might contribute to explaining the muscle atrophy and myopathy previously reported in this model [41].

The increased protein catabolism observed in the skeletal muscle tissue of the Hq mouse could also result in a higher availability of amino acids for other metabolic pathways in order to compensate for the mitochondrial defect. In this regard, computer models of human complex I deficiency in cardiomyocytes have predicted that the supply of several amino acids (glutamate, arginine, proline, valine, aspartate, lysine and glutamine) increase maximal ATP production by the mitochondrial respiratory chain [80]. Another study performed with transmitochondrial cybrids defective in mitochondrially-encoded ATP synthase membrane subunit 6 (MT-ATP6) and the $Cox10^{-/-}$ mouse model of mitochondrial myopathy has demonstrated that anaplerosis of the Krebs cycle with α-ketoglutarate, which can be obtained from glutamine and glutamate, mitigates the consequences of the mitochondrial defect [81]. Given that skeletal muscles represent the main reservoir of amino acids in the body, we can hypothesize that muscle catabolism serves as a source of amino acids for Krebs cycle anaplerosis and energy production in the Hq mouse model, particularly the CNS. Further detailed studies are needed to assess this proposed metabolic rewiring, by analyzing whether amino acids are consumed only in the muscles, or if they are also delivered to the CNS to enhance the residual energy production capacity of this tissue.

5. Conclusions

AIF deficiency induces tissue-specific alterations of autophagy in the Hq mouse model of MD, with opposite changes in striated muscle tissues (heart and skeletal muscle) and in the CNS. In striated muscles, autophagy alterations may be related to metabolic rewiring through an mTOR-dependent enhancement of 1-carbon metabolism, and subsequent amino acid-driven anaplerosis of the Krebs cycle to promote energy production in other tissues, particularly the CNS. Such metabolic rewiring could, however, have a detrimental effect on muscle structure and function due to the breakdown of contractile proteins. Further research is needed to assess the precise role of autophagy disruptions in the pathophysiology of MD.

Supplementary Materials: The following supporting information can be downloaded at: https://www.mdpi.com/article/10.3390/antiox11030510/s1, Table S1: Antibodies used; Original images; Original data.

Author Contributions: Investigation: S.L.-M., M.F.-d.l.T., C.F.-L., A.D. and M.M.; Analysis: S.L.-M., M.F.-d.l.T., C.F.-L. and M.M.; Resources: M.Á.M., A.L., P.B. and M.M.; Data curation: S.L.-M., M.F.-d.l.T., C.F.-L. and M.M.; Writing—original draft preparation: S.L.-M., M.F.-d.l.T., A.L. and M.M.; Review and editing: P.B., A.L., M.Á.M., J.A. and M.M.; Project Administration: M.M.; Funding acquisition: M.Á.M. and M.M.; Supervision: M.M.; Conceptualization: M.M., C.F.-L. and A.L. All authors have read and agreed to the published version of the manuscript.

Funding: This study was funded by the Spanish *Instituto de Salud Carlos III* (ISCIII) (grant numbers PI17/00093 and PI20/00147) and The European Regional Development Fund of the European Union.

Institutional Review Board Statement: The study was approved by the Institutional Review Committee (*Hospital 12 de Octubre*; project numbers PROEX 111/15 and PROEX 067/18) and were conducted in accordance with ARRIVE guidelines and with European (European Convention for the Protection of Vertebrate Animals ETS123) and Spanish (32/2007 and R.D. 1201/2005) laws on animal protection in research.

Informed Consent Statement: Not applicable.

Data Availability Statement: Data is contained within the article or Supplementary Materials.

Acknowledgments: We thank Ana Serrano-Puebla for sample preparation.

Conflicts of Interest: The authors declare no conflict of interest. The funders had no role in the design of the study; in the collection, analyses, or interpretation of data; in the writing of the manuscript, or in the decision to publish the results.

References

1. Schaefer, A.M.; Taylor, R.W.; Turnbull, D.M.; Chinnery, P.F. The epidemiology of mitochondrial disorders—Past, present and future. *Biochim. Biophys. Acta (BBA)-Bioenerg.* **2004**, *1659*, 115–120. [CrossRef]
2. Mcfarland, R.; Turnbull, D.M. Batteries not included: Diagnosis and management of mitochondrial disease. *J. Intern. Med.* **2009**, *265*, 210–228. [CrossRef]
3. Hüttemann, M.; Lee, I.; Pecinova, A.; Pecina, P.; Przyklenk, K.; Doan, J.W. Regulation of oxidative phosphorylation, the mitochondrial membrane potential, and their role in human disease. *J. Bioenerg. Biomembr.* **2008**, *40*, 445–456. [CrossRef]
4. Ghezzi, D.; Zeviani, M. Human diseases associated with defects in assembly of OXPHOS complexes. *Essays Biochem.* **2018**, *62*, 271–286.
5. Emperador, S.; Garrido-Pérez, N.; Amezcua-Gil, J.; Gaudó, P.; Andrés-Sanz, J.A.; Yubero, D.; Fernández-Marmiesse, A.; O'Callaghan, M.M.; Escobar, J.D.O.; Iriondo, M.; et al. Molecular Characterization of New FBXL4 Mutations in Patients with mtDNA Depletion Syndrome. *Front. Genet.* **2019**, *10*, 1300. [CrossRef]
6. Kane, M.S.; Alban, J.; Desquiret-Dumas, V.; Gueguen, N.; Ishak, L.; Ferré, M.; Amati-Bonneau, P.; Procaccio, V.; Bonneau, D.; Lenaers, G.; et al. Autophagy controls the pathogenicity of OPA1 mutations in dominant optic atrophy. *J. Cell Mol. Med.* **2017**, *21*, 2284–2297. [CrossRef]
7. De la Mata, M.; Garrido-Maraver, J.; Cotán, D.; Cordero, M.D.; Oropesa-Ávila, M.; Izquierdo, L.G.; de Miguel, M.; Lorite, J.B.; Infante, E.R.; Ybot, P.; et al. Recovery of MERRF fibroblasts and cybrids pathophysiology by coenzyme Q10. *Neurotherapeutics* **2012**, *9*, 446–463. [CrossRef]
8. Yuan, J.-H.; Sakiyama, Y.; Higuchi, I.; Inamori, Y.; Higuchi, Y.; Hashiguchi, A.; Higashi, K.; Yoshimura, A.; Takashima, H. Mitochondrial myopathy with autophagic vacuoles in patients with the m.8344A > G mutation. *J. Clin. Pathol.* **2013**, *66*, 659–664. [CrossRef]
9. Garrido-Maraver, J.; Paz, M.V.; Cordero, M.D.; Bautista-Lorite, J.; Oropesa-Ávila, M.; de la Mata, M.; Pavón, A.D.; Lavera, I.; Alcocer-Gómez, E.; Galán, F. Critical role of AMP-activated protein kinase in the balance between mitophagy and mitochondrial biogenesis in MELAS disease. *Biochim. Biophys. Acta (BBA)-Mol. Basis Dis.* **2015**, *1852*, 2535–2553. [CrossRef]
10. Lin, D.-S.; Huang, Y.-W.; Ho, C.-S.; Hung, P.-L.; Hsu, M.-H.; Wang, T.-J.; Wu, T.-Y.; Lee, T.-H.; Huang, Z.-D.; Chang, P.-C.; et al. Oxidative Insults and Mitochondrial DNA Mutation Promote Enhanced Autophagy and Mitophagy Compromising Cell Viability in Pluripotent Cell Model of Mitochondrial Disease. *Cells* **2019**, *8*, 65. [CrossRef]
11. White, K.E.; Davies, V.J.; Hogan, V.E.; Piechota, M.J.; Nichols, P.P.; Turnbull, D.M.; Votruba, M. OPA1 deficiency associated with increased autophagy in retinal ganglion cells in a murine model of dominant optic atrophy. *Investig. Ophthalmol. Vis. Sci.* **2009**, *50*, 2567–2571. [CrossRef]
12. Villanueva-Paz, M.; Povea-Cabello, S.; Villalón-García, I.; Álvarez-Córdoba, M.; Suárez-Rivero, J.M.; Talaverón-Rey, M.; Jackson, S.; Falcón-Moya, R.; Rodríguez-Moreno, A.; Sánchez-Alcázar, J.A. Parkin-mediated mitophagy and autophagy flux disruption in cellular models of MERRF syndrome. *Biochim. Biophys. Acta Mol. Basis Dis.* **2020**, *1866*, 165726. [CrossRef]
13. Morán, M.; Delmiro, A.; Blázquez, A.; Ugalde, C.; Arenas, J.; Martín, M.A. Bulk autophagy, but not mitophagy, is increased in cellular model of mitochondrial disease. *Biochim. Biophys. Acta* **2014**, *1842*, 1059–1070. [CrossRef]
14. Cotán, D.; Cordero, M.D.; Garrido-Maraver, J.; Oropesa-Avila, M.; Rodríguez-Hernández, A.; Gomez Izquierdo, L.; De la Mata, M.; De Miguel, M.; Lorite, J.B.; Infante, E.R.; et al. Secondary coenzyme Q10 deficiency triggers mitochondria degradation by mitophagy in MELAS fibroblasts. *FASEB J. Off. Publ. Fed. Am. Soc. Exp. Biol.* **2011**, *25*, 2669–2687.
15. Dombi, E.; Diot, A.; Morten, K.; Carver, J.; Lodge, T.; Fratter, C.; Ng, Y.S.; Liao, C.; Muir, R.; Blakely, E.L. The m.13051G > A mitochondrial DNA mutation results in variable neurology and activated mitophagy. *Neurology* **2016**, *86*, 1921–1923. [CrossRef]
16. Tyynismaa, H.; Mjosund, K.P.; Wanrooij, S.; Lappalainen, I.; Ylikallio, E.; Jalanko, A.; Spelbrink, J.; Paetau, A.; Suomalainen-Wartiovaara, A. Mutant mitochondrial helicase Twinkle causes multiple mtDNA deletions and a late-onset mitochondrial disease in mice. *Proc. Natl. Acad. Sci. USA* **2005**, *102*, 17687–17692. [CrossRef]
17. Deng, J.; Lu, Y.; Xie, Z.; Liu, J.; Yuan, Y.; Wang, Z. RNA-seq profiling, and impaired autophagic process in skeletal muscle of MELAS. *Biochem. Biophys. Res Commun.* **2020**, *523*, 91–97. [CrossRef]
18. Zhang, L.; Zhang, Z.; Khan, A.; Zheng, H.; Yuan, C.; Jiang, H. Advances in drug therapy for mitochondrial diseases. *Ann. Transl. Med.* **2020**, *8*, 17. [CrossRef]
19. Jeppesen, T.D.; Schwartz, M.; Olsen, D.B.; Wibrand, F.; Krag, T.; Dunø, M.; Hauerslev, S.; Vissing, J. Aerobic training is safe and improves exercise capacity in patients with mitochondrial myopathy. *Brain* **2006**, *129*, 3402–3412. [CrossRef]
20. Porcelli, S.; Marzorati, M.; Morandi, L.; Grassi, B. Home-based aerobic exercise training improves skeletal muscle oxidative metabolism in patients with metabolic myopathies. *J. Appl. Physiol.* **2016**, *121*, 699–708. [CrossRef]

21. Taivassalo, T.; Gardner, J.L.; Taylor, R.W.; Schaefer, A.; Newman, J.; Barron, M.; Haller, R.G.; Turnbull, D.M. Endurance training and detraining in mitochondrial myopathies due to single large-scale mtDNA deletions. *Brain* **2006**, *129*, 3391–3401. [CrossRef]
22. Murphy, J.L.; Blakely, E.L.; Schaefer, A.M.; He, L.; Wyrick, P.; Haller, R.G.; Taylor, R.W.; Turnbull, D.M.; Taivassalo, T. Resistance training in patients with single, large-scale deletions of mitochondrial DNA. *Brain* **2008**, *131*, 2832–2840. [CrossRef]
23. Venturelli, M.; Villa, F.; Ruzzante, F.; Tarperi, C.; Rudi, D.; Milanese, C.; Cavedon, V.; Fonte, C.; Picelli, A.; Smania, N. Neuromuscular and Muscle Metabolic Functions in MELAS before and after Resistance Training: A Case Study. *Front. Physiol.* **2019**, *10*, 503. [CrossRef]
24. Cejudo, P.; Bautista, J.; Montemayor, T.; Villagómez, R.; Jiménez, L.; Ortega, F.; Campos, Y.; Sánchez, H.; Arenas, J. Exercise training in mitochondrial myopathy: A randomized controlled trial. *Muscle Nerve* **2005**, *32*, 342–350. [CrossRef]
25. Fiuza-Luces, C.; Díez-Bermejo, J.; Fernández-De La Torre, M.; Rodríguez-Romo, G.; Sanz-Ayán, P.; Delmiro, A.; Munguía-Izquierdo, D.; Rodríguez-Gómez, I.; Ara, I.; Domínguez-González, C.; et al. Health Benefits of an Innovative Exercise Program for Mitochondrial Disorders. *Med. Sci. Sports Exerc.* **2018**, *50*, 1142–1151. [CrossRef]
26. Vainshtein, A.; Hood, D.A. The regulation of autophagy during exercise in skeletal muscle. *J. Appl. Physiol.* **2016**, *120*, 664–673. [CrossRef]
27. Campos, J.C.; Queliconi, B.B.; Bozi, L.H.M.; Bechara, L.R.G.; Dourado, P.M.M.; Andres, A.M.; Jannig, P.R.; Gomes, K.M.S.; Zambelli, V.O.; Rocha-Resende, C.; et al. Exercise reestablishes autophagic flux and mitochondrial quality control in heart failure. *Autophagy* **2017**, *13*, 1304–1317. [CrossRef]
28. Wang, L.; Wang, J.; Cretoiu, D.; Li, G.; Xiao, J. Exercise-mediated regulation of autophagy in the cardiovascular system. *J. Sport Health Sci.* **2020**, *9*, 203–210. [CrossRef]
29. Andreotti, D.Z.; Silva, J.D.N.; Matumoto, A.M.; Orellana, A.M.; de Mello, P.S.; Kawamoto, E.M. Effects of Physical Exercise on Autophagy and Apoptosis in Aged Brain: Human and Animal Studies. *Front Nutr.* **2020**, *7*, 94. [CrossRef]
30. He, C.; Sumpter, R.J.; Levine, B. Exercise induces autophagy in peripheral tissues and in the brain. *Autophagy* **2012**, *8*, 1548–1551. [CrossRef]
31. Zeng, Z.; Liang, J.; Wu, L.; Zhang, H.; Lv, J.; Chen, N. Exercise-Induced Autophagy Suppresses Sarcopenia Through Akt/mTOR and Akt/FoxO3a Signal Pathways and AMPK-Mediated Mitochondrial Quality Control. *Front. Physiol.* **2020**, *11*, 583478. [CrossRef]
32. McMillan, E.M.; Paré, M.F.; Baechler, B.L.; Graham, D.A.; Rush, J.W.E.; Quadrilatero, J. Autophagic signaling and proteolytic enzyme activity in cardiac and skeletal muscle of spontaneously hypertensive rats following chronic aerobic exercise. *PLoS ONE* **2015**, *10*, e0119382. [CrossRef]
33. Almeida, M.F.; Silva, C.M.; Chaves, R.S.; Lima, N.C.R.; Almeida, R.S.; Melo, K.P.; Demasi, M.; Fernandes, T.; Oliveira, E.M.; Netto, L.E.S.; et al. Effects of mild running on substantia nigra during early neurodegeneration. *J. Sports Sci.* **2018**, *36*, 1363–1370. [CrossRef]
34. Ng, S.Y.; Mikhail, A.; Ljubicic, V. Chronic physical activity is beneficial for SMA patients, however the cellular and molecular mechanisms of exercise biology are largely undefined in SMA. *J. Physiol.* **2019**, *597*, 4757–4778. [CrossRef]
35. Susin, S.A.; Lorenzo, H.K.; Zamzami, N.; Marzo, I.; Snow, B.E.; Brothers, G.M.; Mangion, J.; Jacotot, E.; Constantini, P.; Loeffler, M.; et al. Molecular characterization of mitochondrial apoptosis-inducing factor. *Nature* **1999**, *397*, 441–446. [CrossRef]
36. Otera, H.; Ohsakaya, S.; Nagaura, Z.I.; Lorenzo, H.K.; Zamzami, N. Export of mitochondrial AIF in response to proapoptotic stimuli depends on processing at the intermembrane space. *EMBO J.* **2005**, *24*, 1375–1386. [CrossRef]
37. Chinta, S.K.; Rane, A.; Yadava, N.; Andersen, J.K.; Nicholls, D.G.; Polster, B. Reactive oxygen species regulation by AIF- and complex I- depleted brain mitochondria. *Free Radic. Biol. Med.* **2009**, *2346*, 939–947. [CrossRef]
38. Hangen, E.; Féraud, O.; Lachkar, S.; Mou, H.; Doti, N.; Fimia, G.M.; Lam, N.-v.; Zhu, C.; Godin, I.; Muller, K.; et al. Interaction between AIF and CHCHD4 Regulates Respiratory Chain Biogenesis. *Mol. Cell* **2015**, *58*, 1001–1014. [CrossRef]
39. Meyer, K.; Buettner, S.; Ghezzi, D.; Zeviani, M.; Bano, D.; Nicotera, P. Loss of apoptosis-inducing factor critically affects MIA40 function. *Cell Death Dis.* **2015**, *6*, e1814. [CrossRef]
40. Bano, D.; Prehn, J.H.M. Apoptosis-Inducing Factor (AIF) in Physiology and Disease: The Tale of a Repented Natural Born Killer. *EBioMedicine* **2018**, *30*, 29–37. [CrossRef]
41. Fiuza-Luces, C.; Valenzuela, P.L.; Laine-Menéndez, S.; Fernández-de la Torre, M.; Bermejo-Gómez, V.; Rufián-Vázquez, L.; Arenas, J.; Martín, M.A.; Lucia, A.; Morán, M. Physical Exercise and Mitochondrial Disease: Insights from a Mouse Model. *Front. Neurol.* **2019**, *10*, 790. [CrossRef] [PubMed]
42. Klein, J.A.; Longo-Guess, C.M.; Rossmann, M.P.; Seburn, K.L.; Hurd, R.E.; Frankel, W.N.; Bronson, R.T.; Ackerman, S.L. The harlequin mouse mutation downregulates apoptosis-inducing factor. *Nature* **2002**, *419*, 367–374. [CrossRef] [PubMed]
43. Bénit, P.; Goncalves, S.; Dassa, E.P.; Brière, J.-J.; Rustin, P. The variability of the harlequin mouse phenotype resembles that of human mitochondrial-complex I-deficiency syndromes. *PLoS ONE* **2008**, *3*, e3208. [CrossRef] [PubMed]
44. Vahsen, N.; Candé, C.; Brière, J.J.; Bénit, P.; Joza, N.; Larochette, N.; Mastroberardino, P.G.; Pequignot, M.O.; Casares, N.; Lazar, V.; et al. AIF deficiency compromises oxidative phosphorylation. *EMBO J.* **2004**, *23*, 4679–4689. [CrossRef] [PubMed]
45. van Empel, V.P.M.; Bertrand, A.T.; van der Nagel, R.; Kostin, S.; Doevendans, P.A.; Crijns, H.J.; de Wit, E.; Sluiter, W.; Ackerman, S.L.; De Windt, L.J. Downregulation of apoptosis-inducing factor in harlequin mutant mice sensitizes the myocardium to oxidative stress-related cell death and pressure overload-induced decompensation. *Circ. Res.* **2005**, *96*, e92–e101. [CrossRef] [PubMed]

46. Welinder, C.; Ekblad, L. Coomassie staining as loading control in Western blot analysis. *J. Proteome Res.* **2011**, *10*, 1416–1419. [CrossRef]
47. Haspel, J.; Shaik, R.S.; Ifedigbo, E.; Nakahira, K.; Dolinay, T.; Englert, J.A.; Choi, A.M.K. Characterization of macroautophagic flux in vivo using a leupeptin-based assay. Characterization of macroautophagic flux in vivo using a leupeptin-based assay. *Autophagy* **2011**, *7*, 629–642. [CrossRef]
48. Khan, N.A.; Nikkanen, J.; Yatsuga, S.; Jackson, C.; Wang, L.; Pradhan, S.; Kivelä, R.; Pessia, A.; Velagapudi, V.; Suomalainen, A. mTORC1 Regulates Mitochondrial Integrated Stress Response and Mitochondrial Myopathy Progression. *Cell Metab.* **2017**, *26*, 419–428.e5. [CrossRef]
49. McKnight, N.C.; Zhong, Y.; Wold, M.S.; Gong, S.; Phillips, G.R.; Dou, Z.; Zhao, Y.; Heintz, N.; Zong, W.-X.; Yue, Z. Beclin 1 is required for neuron viability and regulates endosome pathways via the UVRAG-VPS34 complex. *PLoS Genet.* **2014**, *10*, e1004626. [CrossRef]
50. Hara, T.; Nakamura, K.; Matsui, M.; Yamamoto, A.; Nakahara, Y.; Suzuki-Migishima, R.; Yokoyama, M.; Mishima, K.; Saito, I.; Okano, H.; et al. Suppression of basal autophagy in neural cells causes neurodegenerative disease in mice. *Nature* **2006**, *441*, 885–889. [CrossRef]
51. Komatsu, M.; Waguri, S.; Chiba, T.; Murata, S.; Iwata, J.; Tanida, I.; Ueno, T.; Koike, M.; Uchiyama, Y.; Kominami, E.; et al. Loss of autophagy in the central nervous system causes neurodegeneration in mice. *Nature* **2006**, *441*, 880–884. [CrossRef] [PubMed]
52. Fernández-de la Torre, M.; Fiuza-Luces, C.; Valenzuela, P.L.; Laine-Menéndez, S.; Arenas, J.; Martín, M.A.; Turnbull, D.M.; Lucia, A.; Morán, M. Exercise Training and Neurodegeneration in Mitochondrial Disorders: Insights from the Harlequin Mouse. *Front. Physiol.* **2020**, *11*, 594223. [CrossRef] [PubMed]
53. Danieli, A.; Martens, S. p62-mediated phase separation at the intersection of the ubiquitin-proteasome system and autophagy. *J. Cell Sci.* **2018**, *131*, jcs214304. [CrossRef]
54. Shin, W.H.; Park, J.H.; Chung, K.C. The central regulator p62 between ubiquitin proteasome system and autophagy and its role in the mitophagy and Parkinson's disease. *BMB Rep.* **2020**, *53*, 56–63. [CrossRef] [PubMed]
55. Katsuragi, Y.; Ichimura, Y.; Komatsu, M. p62/SQSTM1 functions as a signaling hub and an autophagy adaptor. *FEBS J.* **2015**, *282*, 4672–4678. [CrossRef] [PubMed]
56. Wei, R.; Enaka, M.; Muragaki, Y. Activation of KEAP1/NRF2/P62 signaling alleviates high phosphate-induced calcification of vascular smooth muscle cells by suppressing reactive oxygen species production. *Sci. Rep.* **2019**, *9*, 10366. [CrossRef]
57. Jin, Z.; Li, Y.; Pitti, R.; Lawrence, D.; Pham, V.C.; Lill, J.R.; Ashkenazi, A. Cullin3-based polyubiquitination and p62-dependent aggregation of caspase-8 mediate extrinsic apoptosis signaling. *Cell* **2009**, *137*, 721–735. [CrossRef]
58. Lucas, S.; Chen, G.; Aras, S.; Wang, J. Serine catabolism is essential to maintain mitochondrial respiration in mammalian cells. *Life Sci. Alliance* **2018**, *1*, 1–10. [CrossRef]
59. Al-Bari, M.A.A.; Xu, P. Molecular regulation of autophagy machinery by mTOR-dependent and -independent pathways. *Ann. N. Y. Acad. Sci.* **2020**, *1467*, 3–20. [CrossRef]
60. Vincow, E.S.; Thomas, R.E.; Merrihew, G.E.; Shulman, N.J.; Bammler, T.K.; MacDonald, J.W.; MacCoss, M.J.; Pallanck, L.J. Autophagy accounts for approximately one-third of mitochondrial protein turnover and is protein selective. *Autophagy* **2019**, *15*, 1592–1605. [CrossRef]
61. Bayod, S.; Del Valle, J.; Pelegri, C.; Vilaplana, J.; Canudas, A.M.; Camins, A.; Jimenez, A.; Sanchez-Roige, S.; Lalanza, J.F.; Escorihuela, R.M.; et al. Macroautophagic process was differentially modulated by long-term moderate exercise in rat brain and peripheral tissues. *J. Physiol. Pharmacol. Off. J. Polish Physiol. Soc.* **2014**, *65*, 229–239.
62. Marques-Aleixo, I.; Santos-Alves, E.; Balça, M.M.; Rizo-Roca, D.; Moreira, P.I.; Oliveira, P.J.; Magalhães, J.; Ascensão, A. Physical exercise improves brain cortex and cerebellum mitochondrial bioenergetics and alters apoptotic, dynamic and auto(mito)phagy markers. *Neuroscience* **2015**, *301*, 480–495. [CrossRef] [PubMed]
63. Schwalm, C.; Jamart, C.; Benoit, N.; Naslain, D.; Prémont, C.; Prevet, J.; Van Thienen, R.; Deldicque, L.; Francaux, M. Activation of autophagy in human skeletal muscle is dependent on exercise intensity and AMPK activation. *FASEB J.* **2015**, *29*, 3515–3526. [CrossRef] [PubMed]
64. Rocchi, A.; He, C. Regulation of Exercise-Induced Autophagy in Skeletal Muscle. *Curr. Pathobiol. Rep.* **2017**, *5*, 177–186. [CrossRef]
65. Pan, G.; Jin, L.; Shen, W.; Zhang, J.; Pan, J.; Cheng, J.; Xie, Q.; Hu, Q.; Wu, S.; Zhang, H.; et al. Treadmill exercise improves neurological function by inhibiting autophagy and the binding of HMGB1 to Beclin1 in MCAO juvenile rats. *Life Sci.* **2020**, *243*, 117279. [CrossRef]
66. Herring, A.; Münster, Y.; Metzdorf, J.; Bolczek, B.; Krüssel, S.; Krieter, D.; Yavuz, I.; Karim, F.; Roggendorf, C.; Stang, A.; et al. Late running is not too late against Alzheimer's pathology. *Neurobiol. Dis.* **2016**, *94*, 44–54. [CrossRef]
67. Fucà, E.; Guglielmotto, M.; Boda, E.; Rossi, F.; Leto, K.; Buffo, A. Preventive motor training but not progenitor grafting ameliorates cerebellar ataxia and deregulated autophagy in tambaleante mice. *Neurobiol. Dis.* **2017**, *102*, 49–59. [CrossRef]
68. Jang, Y.; Kwon, I.; Song, W.; Cosio-Lima, L.M.; Lee, Y. Endurance Exercise Mediates Neuroprotection Against MPTP-mediated Parkinson's Disease via Enhanced Neurogenesis, Antioxidant Capacity, and Autophagy. *Neuroscience* **2018**, *379*, 292–301. [CrossRef]
69. Koo, J.-H.; Cho, J.-Y. Treadmill Exercise Attenuates α-Synuclein Levels by Promoting Mitochondrial Function and Autophagy Possibly via SIRT1 in the Chronic MPTP/P-Induced Mouse Model of Parkinson's Disease. *Neurotox. Res.* **2017**, *32*, 473–486. [CrossRef]

70. Fritzen, A.M.; Frøsig, C.; Jeppesen, J.; Jensen, T.E.; Lundsgaard, A.-M.; Serup, A.K.; Schjerling, P.; Proud, C.G.; Richter, E.A.; Kiens, B. Role of AMPK in regulation of LC3 lipidation as a marker of autophagy in skeletal muscle. *Cell. Signal.* **2016**, *28*, 663–674. [CrossRef]
71. Kim, Y.A.; Kim, Y.S.; Song, W. Autophagic response to a single bout of moderate exercise in murine skeletal muscle. *J. Physiol. Biochem.* **2012**, *68*, 229–235. [CrossRef] [PubMed]
72. Ross, J.M.; Coppotelli, G.; Branca, R.M.; Kim, K.M.; Lehtiö, J.; Sinclair, D.A.; Olson, L. Voluntary exercise normalizes the proteomic landscape in muscle and brain and improves the phenotype of progeroid mice. *Aging Cell* **2019**, *18*, e13029. [CrossRef] [PubMed]
73. Wenz, T.; Diaz, F.; Hernandez, D.; Moraes, C.T. Endurance exercise is protective for mice with mitochondrial myopathy. *J. Appl. Physiol.* **2009**, *106*, 1712–1719. [CrossRef] [PubMed]
74. Wu, N.N.; Tian, H.; Chen, P.; Wang, D.; Ren, J.; Zhang, Y. Physical Exercise and Selective Autophagy: Benefit and Risk on Cardiovascular Health. *Cells* **2019**, *8*, 1436. [CrossRef]
75. Codina-Martínez, H.; Fernández-García, B.; Díez-Planelles, C.; Fernández, Á.F.; Higarza, S.G.; Fernández-Sanjurjo, M.; Díez-Robles, S.; Iglesias-Gutiérrez, E.; Tomás-Zapico, C. Autophagy is required for performance adaptive response to resistance training and exercise-induced adult neurogenesis. *Scand. J. Med. Sci. Sports* **2020**, *30*, 238–253. [CrossRef]
76. Wischhof, L.; Gioran, A.; Sonntag-Bensch, D.; Piazzesi, A.; Stork, M.; Nicotera, P.; Bano, D. A disease-associated Aifm1 variant induces severe myopathy in knockin mice. *Mol. Metab.* **2018**, *13*, 10–23. [CrossRef]
77. Nikkanen, J.; Forsström, S.; Euro, L.; Paetau, I.; Kohnz, R.A.; Wang, L.; Chilov, D.; Viinamäki, J.; Roivainen, A.; Marjamäki, P.; et al. Mitochondrial DNA Replication Defects Disturb Cellular dNTP Pools and Remodel One-Carbon Metabolism. *Cell Metab.* **2016**, *23*, 635–648. [CrossRef]
78. Tyynismaa, H.; Carroll, C.J.; Raimundo, N.; Ahola-Erkkila, S.; Wenz, T.; Ruhanen, H.; Guse, K.; Hemminki, A.; Peltola-Mjosund, K.E.; Tulkki, V.; et al. Mitochondrial myopathy induces a starvation-like response. *Hum. Mol. Genet.* **2010**, *19*, 3948–3958. [CrossRef]
79. Morscher, R.J.; Ducker, G.S.; Li, S.H.-J.; Mayer, J.A.; Gitai, Z.; Sperl, W.; Rabinowitz, J.D. Mitochondrial translation requires folate-dependent tRNA methylation. *Nature* **2018**, *554*, 128–132. [CrossRef]
80. Zieliński, Ł.P.; Smith, A.C.; Smith, A.G.; Robinson, A.J. Metabolic flexibility of mitochondrial respiratory chain disorders predicted by computer modelling. *Mitochondrion* **2016**, *31*, 45–55. [CrossRef]
81. Chen, Q.; Kirk, K.; Shurubor, Y.I.; Zhao, D.; Arreguin, A.J.; Shahi, I.; Valsecchi, F.; Primiano, G.; Calder, E.L.; Carelli, V.; et al. Rewiring of Glutamine Metabolism Is a Bioenergetic Adaptation of Human Cells with Mitochondrial DNA Mutations. *Cell Metab.* **2018**, *27*, 1007–1025.e5. [CrossRef] [PubMed]

antioxidants

MDPI

Article

CHK2 Promotes Metabolic Stress-Induced Autophagy through ULK1 Phosphorylation

Ran Guo [1,†], Shan-Shan Wang [2,†], Xiao-You Jiang [2], Ye Zhang [2], Yang Guo [2], Hong-Yan Cui [2], Qi-Qiang Guo [2,*], Liu Cao [2,*] and Xiao-Chen Xie [3,*]

1 Department of Orthopedics, Shengjing Hospital of China Medical University, Shenyang 110004, China; sj_guoran@163.com
2 College of Basic Medical Science, Health Sciences Institute, Key Laboratory of Medical Cell Biology, Ministry of Education, China Medical University, Shenyang 110122, China; shanshanwang0128@gmail.com (S.-S.W.); 18204509380@163.com (X.-Y.J.); 18002471466@163.com (Y.Z.); cmuguoyang@163.com (Y.G.); c2447076105@163.com (H.-Y.C.)
3 Liaoning Provincial Key Laboratory of Endocrine Diseases, Department of Endocrinology and Metabolism, Institute of Endocrinology, The First Affiliated Hospital of China Medical University, Shenyang 110001, China
* Correspondence: qqguo@cmu.edu.cn (Q.-Q.G.); lcao@cmu.edu.cn (L.C.); xcxie@cmu.edu.cn (X.-C.X.); Tel.: +86-189-0091-0432 (Q.-Q.G.); +86-189-0091-1888 (L.C.); +86-139-4043-1327 (X.-C.X.)
† These authors contributed equally to this work.

Abstract: Reactive oxygen species (ROS) act as a signaling intermediate to promote cellular adaptation to maintain homeostasis by regulating autophagy during pathophysiological stress. However, the mechanism by which ROS promotes autophagy is still largely unknown. Here, we show that the ATM/CHK2/ULK1 axis initiates autophagy to maintain cellular homeostasis by sensing ROS signaling under metabolic stress. We report that ULK1 is a physiological substrate of CHK2, and that the binding of CHK2 to ULK1 depends on the ROS signal and the phosphorylation of threonine 68 of CHK2 under metabolic stress. Further, CHK2 phosphorylates ULK1 on serine 556, and this phosphorylation is dependent on the ATM/CHK2 signaling pathway. CHK2-mediated phosphorylation of ULK1 promotes autophagic flux and inhibits apoptosis induced by metabolic stress. Taken together, these results demonstrate that the ATM/CHK2/ULK1 axis initiates an autophagic adaptive response by sensing ROS, and it protects cells from metabolic stress-induced cellular damage.

Keywords: autophagy; ULK1; CHK2; oxidative stress; ROS

Citation: Guo, R.; Wang, S.-S.; Jiang, X.-Y.; Zhang, Y.; Guo, Y.; Cui, H.-Y.; Guo, Q.-Q.; Cao, L.; Xie, X.-C. CHK2 Promotes Metabolic Stress-Induced Autophagy through ULK1 Phosphorylation. *Antioxidants* **2022**, *11*, 1166. https://doi.org/10.3390/antiox11061166

Academic Editors: Angeles Aroca, Cecilia Gotor and Jiankang Liu

Received: 7 May 2022
Accepted: 9 June 2022
Published: 14 June 2022

1. Introduction

Oxidative damage to cellular biomolecules caused by excess reactive oxygen species (ROS) is the root cause of apoptosis and a potential factor leading to a range of pathologies, including neurodegenerative diseases, atherosclerosis, and aging processes [1,2]. In recent years, an increasing number of studies have shown that ROS appear to be induced in response to pathophysiological stress and act as signaling intermediates that promote adaptive cellular responses to maintain homeostasis [3,4]. Autophagy, a multistep lysosomal degradation pathway that supports nutrient recycling and metabolic adaptation, has been implicated as an essential biological behavior for maintaining homeostasis [5]. ROS have been widely recognized as being central signaling molecules that induce autophagy under various stimuli [4,6]. However, the mechanism by which ROS promote autophagy is still largely unknown.

Cell cycle checkpoint kinase 2 (CHK2) is an evolutionarily highly conserved serine/threonine-protein kinase that was initially identified as a vital transducer in the DNA damage response (DDR) [7]. When DNA damage occurs, ataxia-telangiectasia mutated (ATM) phosphorylates CHK2 at threonine 68. After the phosphorylation of CHK2 inactive monomers at T68 by ATM, dimers form and undergo subsequent autophosphorylation, and may also serve as substrates for other protein kinases. Therefore, the phosphorylation of CHK2 T68 is considered to be a prerequisite for CHK2 to be activated and to perform signaling functions, such as cell cycle arrest, DNA repair, cell survival, proliferation, and cell death [8,9]. In recent years, the ATM-CHK2 signaling pathway, which can act as a sensor of ROS and participate in autophagy to maintain cell homeostasis, has received more attention [10,11]. The oxidation of ATM (Cys2991) directly induces ATM activation (phospho-ATM Ser1981) in the absence of DNA DSBs and the MRN complex [12]. In response to elevated ROS, ATM activates TSC2 through the LKB1/AMPK signaling pathway to inhibit mTORC1, and it promotes autophagy [13]. Our previous study found that elevated levels of ROS induced by glucose deprivation and hypoxia can promote autophagy to maintain cellular homeostasis by activating the ATM/CHK2/Beclin 1 axis [10]. However, DDR signaling as a direct crosstalk mechanism between ROS and autophagy remains to be further explored.

The serine/threonine-protein kinase ULK1 is the homologous protein of yeast Atg1 in mammalian cells [14]. It forms a complex with Atg13, FIP200, and Atg101, and plays a vital role in autophagy [15–17]. Phosphorylation of ULK1 is critical in autophagy regulation [18]. Under adequate nutrient conditions, mTOR phosphorylates ULK1 and Atg13, resulting in the inactivation of the ULK1 complex. Under starvation conditions, by inhibiting the activity of mTORC1, the inhibitory effect on ULK1 is relieved, and the kinase activity of ULK1 is enhanced, which can further phosphorylate Atg13 and FIP200 to initiate the occurrence of autophagy [19]. In addition, AMPK can promote autophagy by phosphorylating multiple serine/threonine sites in the intermediate domain of ULK1 (including serines 317, 467, 555, 637, and 777, and threonine 574) [20–24]. However, the regulatory mechanism for how ROS signaling promotes autophagy through ULK1 remains unclear.

Here, we found that the autophagy-related protein ULK1 is a new interaction protein of CHK2, and that its binding depends on the accumulation of ROS caused by metabolic stress. CHK2 phosphorylates ULK1 at serine 556 to promote autophagy and inhibit cell apoptosis under metabolic stress. Our findings establish that the ATM/CHK2/ULK1 signaling pathway initiates an autophagic adaptive response by sensing ROS, and it protects cells from metabolic stress-induced cellular damage.

2. Materials and Methods

2.1. Plasmids Constructs, Cell Culture, and Viral Infection

ULK1 expression plasmids were kindly provided by Xin Pan (National Center of Biomedical Analysis, Beijing, China). ULK1 was subcloned into pGEX-5X-1. CHK2 (Addgene, no. 41901, Watertown, MA, USA) was subcloned into pcDNA3.1 with an amino-terminal 3× Flag tag. $ULK1^{Mut}$ constructs, $ULK1^{S556A}$ and $ULK1^{S556D}$, and $CHK2^{Mut}$ constructs, $CHK2^{T68A}$ and $CHK2^{T68D}$ were created using site-directed mutagenesis.

HEK293T and HEK293 cells were cultured in Dulbecco's modified Eagle's medium (DMEM) containing 10% fetal bovine serum (FBS) and supplemented with 100 μg/mL penicillin/streptomycin. H1299 and HCT116 cells were cultured in RPMI1640 medium containing 10% FBS. The cells were cultured in standard conditions (37 °C, 5% CO_2). N-acetyl cysteine (NAC) was purchased from Sigma (Beijing, China). CHK2 Inhibitor II (2-(4-(4-Chlorophenoxy)phenyl)-1H-benzimidazole-5-carboxamide hydrate) was purchased from Sigma (C3742). CHK2 inhibitor II specifically inhibits CHK2 phosphorylation at Thr68 and is specific for the inhibition of CHK2 activity. The ATM inhibitor KU-55933 (2-(4-Morpholinyl)-6-(1-thianthrenyl)-4H-Pyran-4-one) was from Sigma (SML1109).

ShRNA against CHK2, and ULK1 lenti-virus were purchased from GeneChem (Shanghai, China). The CHK2 sequence was 5′- ACAGATAAATACCGAACAT -3′ and the ULK1 sequence was 5′- CACGCCATCTCCTCAAGTT -3′. GFP-mCherry-LC3 lenti-virus was purchased from Syngentech (Beijing, China).

2.2. Western Blotting

For the Western blot analysis, cells were collected and lysed with lysis buffer (50 mM Tris, pH 7.4, 150 mM NaCl, 1 mM EDTA, 1% NP40, 1% Triton X-100, and 0.25% sodium deoxycholate) containing protease inhibitor cocktail, on ice for 30 min. Afterwards, the cells were vortexed every 10minutes and the lysates were centrifuged at 13,000× *g* for 20 min at 4 °C. Then, the quantified proteins were subjected to SDS-PAGE electrophoresis, after which the proteins were transferred to nitrocellulose, and the membranes were immunoblotted with the indicated primary antibodies. Antibody information and usage are as follows: rabbit anti-ULK1 (D9D7) mAb #6439 (1:1000, CST), rabbit anti-phospho-ULK1 (D1H4) mAb#5869 (S555 (mouse), S556 (human)) (1:1000, CST), rabbit anti-ATM (D2E2) mAb #2873 (1:1000, CST), rabbit anti-phospho-ATM (Ser1981) (D25E5) mAb #13050 (1:1000, CST), rabbit anti-CHK2 mAb #2662 (1:1000, CST), rabbit anti-phospho-CHK2 (Thr68) mAb (1:1000, CST), rabbit anti-p62/SQSTM1 P0067(1:2000, Sigma), rabbit anti-LC3A/B #4108 (1:1000, CST), mouse anti-Flag SG4110-16 (1:1000, Shanghai Genomics Technology, Shanghai, China), and mouse anti-tubulin AC012 (1:2000, ABclonal Technology, Woburn, MA, USA).

2.3. *Co-Immunoprecipitation*

HCT116 cells were transfected with the indicated plasmids using Lipofectamine 3000 (ThermoFisher, L3000008, Waltham, MA, USA). After 24 h of expression, the cells were treated with Earle's balanced salt solution (EBSS) (GIBCO, Shanghai, China), a nutrient-deprivation medium for 1 h and then lysed with IP lysis buffer. Immune complexes conjugated with primary antibodies to protein A/G beads (Santa Cruz, Dallas, TX, USA) were added to the quantified cell lysates, incubated overnight at 4 °C, and washed three times. Whole cell lysates and precipitation samples were analyzed by Western blot.

2.4. *In Vitro GST Pull-Down*

The protein of GST-ULK1 was induced by IPTG in *E. coli BL21*, and purified by glutathione sepharose4B (GE Healthcare, Chicago, IL, USA) according to the manufacturer's protocol. At the same time, FLAG-tagged CHK2 was synthesized by using a transcription and translation in vitrokit (Promega, P2221, Madison, WI, USA). The purified protein GST-ULK1 was incubated with FLAG-tagged CHK2 synthesized in vitro, and its direct binding in vitro was detected by Western blot.

2.5. *In Vitro CHK2 Kinase Assay*

Flag-tagged wild-type ULK1 or mutant ULK1 (S556A) was washed three times with kinase buffer (50 mM HEPES, pH 7.4, 10 mM MgCl2, 10 mM MnCl2, and DTT 0.2 mM), and then incubated with CHK2 recombinant human protein (PV3367, ThermoFisher SCIENTIFIC) in kinase reaction buffer (kinase buffer containing 100 μM ATP (Sigma)) at 30 °C for 45 min. Thephosphorylated proteins were subjected to SDS-PAGE electrophoresis, and afterwards, the proteins were transferred to nitrocellulose and the membranes were immunoblotted with the phospho-ULK1 (Ser556) antibody.

2.6. *Fluorescence Microscopy*

HEK293 cells stably expressing GFP-mCherry-LC3 were grown on coverslips, induced by EBSS starvation for 3 h, and then fixed with 4% paraformaldehyde for 20 min. Confocal images were obtained using a 60× oil lens objective on an inverted fluorescence microscope (Nikon, A1RHD25, Japan, Tokyo). The fluorescence assay was performed as described previously [10].

2.7. *Flow Cytometric Analysis*

Flow cytometry analysis of apoptosis was performed according to the manufacturer's instructions (KeyGENBioTECH, KGA1026, Nanjing, China). In brief, cells were treated with EBSS for 8 h or H_2O_2 for 8 h, and then harvested by trypsinization without EDTA. After two washes with PBS, the cells were stained by Annexin V-APC and 7AAD for 30 min and then resuspended in binding buffer solution for FACS analysis.

2.8. Statistical Analysis

Statistical comparisons between only two groups were carried out using two-sided *t*-tests. A one-way analysis of variance (ANOVA) was used for multiple-group comparisons. Data are presented as mean ± SEM. We tested data for normality and variance, and considered a *p* value of less than 0.05 as significant. Statistical calculations were performed using GraphPad Prism 5.0.

3. Results

3.1. ULK1 Is a Physiological Substrate of CHK2

Our previous study found that metabolic stress can promote autophagy to maintain cellular homeostasis by activating CHK2. To further explore the molecular mechanism of CHK2 regulating autophagy, we screened CHK2 binding proteins through the candidate approach. We performed a co-immunoprecipitation (Co-IP) assay and showed that endogenous CHK2 co-precipitates with endogenous ULK1 (Figure 1A,B). Next, using an in vitro glutathione S-transferase (GST) pull-down assay, we found that CHK2 can directly bind to ULK1 (Figure 1C). Furthermore, enhanced binding between ULK1 and CHK2 was observed in response to metabolic and oxidative stress (Figure 1D,E). N-acetylcysteine (NAC) is a widely used oxygen radical scavenger. The interaction was significantly reduced by using NAC (Figure 1F), indicating that ROS are signaling molecules that facilitate the binding of CHK2 and ULK1. We found that the enhanced binding between these two proteins was accompanied by the activation of CHK2 Thr68. Next, CHK2 inhibitors significantly reduced its phosphorylation at Thr68 and the interaction between CHK2 and ULK1 under metabolic stress (Figure 1G). In addition, under metabolic stress, the binding between wild-type (WT) CHK2 and ULK1 was enhanced, whereas the binding of T68A CHK2 mutant to ULK1 was not. Mimic threonine 68 phosphorylation of CHK2 (T68D) promotes its interaction with ULK1, even under non-metabolic stress (Figure 1H). Taken together, the binding of CHK2 to ULK1 is dependent on the phosphorylation of CHK2 Thr68.

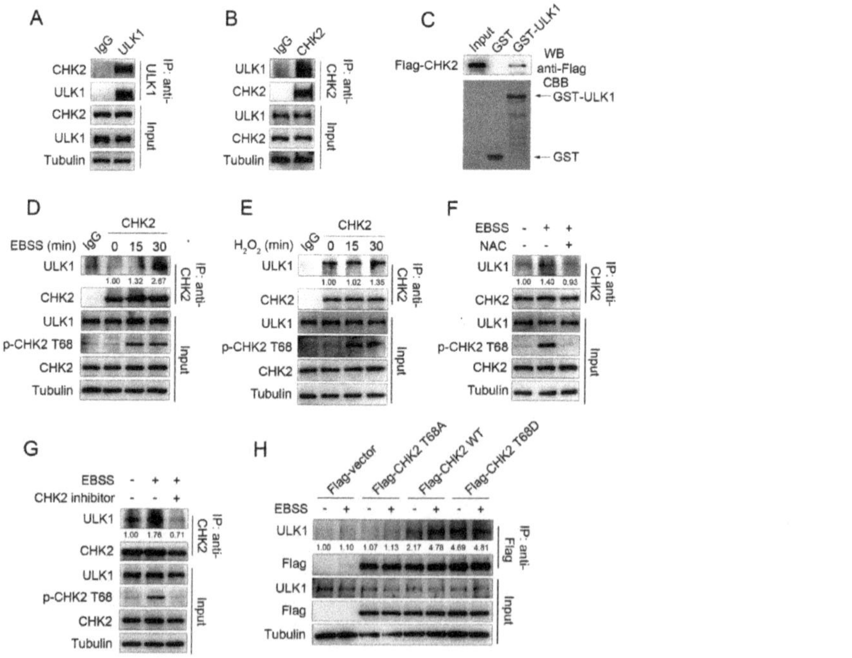

Figure 1. ULK1 is a physiological substrate of CHK2. (**A**,**B**) Immunoprecipitation assays testing the endogenous interaction between CHK2 and ULK1 in HCT116 cells. Lysates were immunoprecipitated

with ULK1 (**A**) or CHK2 (**B**) antibody. The immunoprecipitates and lysates were analyzed by Western blot. (**C**) CHK2 binds with ULK1 in vitro. GST pull-down assays were performed by incubating purified GST or GST-ULK1 with invitro-translated flag-tagged CHK2. Arrows indicate GST and GST-ULK1 bands. (**D**) Lysates from HCT116 cells treated with EBSS were immunoprecipitated with CHK2 antibody or rabbit IgG. The immunoprecipitates and lysates were analyzed by Western blot. (**E**) Lysates from HCT116 cells treated with H_2O_2 (500 μM) were immunoprecipitated with CHK2 antibody or rabbit IgG. The immunoprecipitates and lysates were analyzed by Western blot. (**F**) Lysates from HCT116 cells pretreated with NAC (SIGMA, A7250, 2 mM) for 3 h and then cultured for 30 min in EBSS starvation were immunoprecipitated with CHK2 antibody. The expression of p-CHK2 Thr68, CHK2, and ULK1 was detected by immunoblotting. (**G**) Lysates from HCT116 cells pretreated with CHK2 inhibitor (SIGMA, C3742, 20 μM) for 4 h and then treated with EBSS for 30 min were immunoprecipitated with CHK2 antibody. The expression of p-CHK2 Thr68, CHK2, and ULK1 was detected by immunoblotting. (**H**) Lysates from HCT116 cells expressing theindicated plasmids, treated or untreated with EBSS, were immunoprecipitated with FLAG antibody. The immunoprecipitates and lysates were analyzed by Western blot with the indicated antibodies. Quantitative analysis for the binding intensity of ULK1 are shown.

3.2. CHK2 Phosphorylates ULK1 at Ser556

The protein modification of ULK1, especially the phosphorylation modification, is significant in its involvement in autophagy initiation. To determine whether CHK2 could-phosphorylate ULK1, we first utilized an optimal CHK2 substrate motif to search for possible amino acids sequences containing conservative candidate target sites in the ULK1 sequence [25]. ULK1 contains a Ser556 site matching the optimal CHK2 substrate motif conserved in higher eukaryotes (Figure 2A). Next, we further performed site-directed mutagenesis combined with an in vitro kinase assay and showed that CHK2 can phosphorylate ULK1 ser556, and that site-directed mutation of Ser556 to alanine (S556A) blocked the CHK2 mediated-ULK1phosphorylation on Ser556 (Figure 2B). Further, we observed that the phosphorylation of endogenous ULK1 Ser556 was enhanced under metabolic (Figure 2C) and oxidative stress (Figure 2D). However, in the CHK2 shRNA-treated cells, the phosphorylation of ULK1 Ser556 was significantly reduced. Likewise, small-molecule inhibitors of CHK2 can block the phosphorylation of ULK1 under metabolic stress (Figure 2E). Further, the phosphorylation of CHK2 and ULK1 was also reduced in ATM shRNA-treated or pharmacological ATM inhibition cells under both metabolic (Figure 2F) and oxidative stresses (Figure 2G). Taken together, these results demonstrate that the phosphorylation of ULK1 is dependent on the ATM/CHK2 signaling pathway in response to metabolic stress. Furthermore, the antioxidant NAC was able to block the activation of the ATM/CHK2/ULK1 signaling pathway under conditions of metabolic stress (Figure 2H). These results establish a critical redox-dependent role for the ATM-CHK2 signaling pathway in ULK1 Ser 556 phosphorylation under metabolic or oxidative stress.

3.3. CHK2-Mediated ULK1 Phosphorylation Promotes Autophagy

To further explore whether CHK2-mediated phosphorylation of ULK1 was involved in the regulation of autophagy, we constructed ULK1-depleted H1299 cell lines that stably expressed different mutants of ULK1 (WT, S556A, or S556D) with or without CHK2 (Figure 3A). Increased levels of autophagy were demonstrated by a decrease in the autophagy of substrate p62, and an increase in the conversion of LC3 from the non-lipidated form (LC3-I) to the phosphatidylethanolamine-bound form (LC3-II). ULK1 WT promoted metabolic and oxidative stress-induced autophagy, whereas increased autophagy was inhibited in cells lacking CHK2 expression. Compared with the effect of ULK1 WT, the ULK1 S556A mutant was not able to promote autophagy induced by metabolic and oxidative stress. However, the ULK1 S556D mutant promoted autophagy, even in cells that did not express CHK2 in the H1299 and HEK293 cell lines (Figure 3B–E and Supplementary Figure S1A–D).

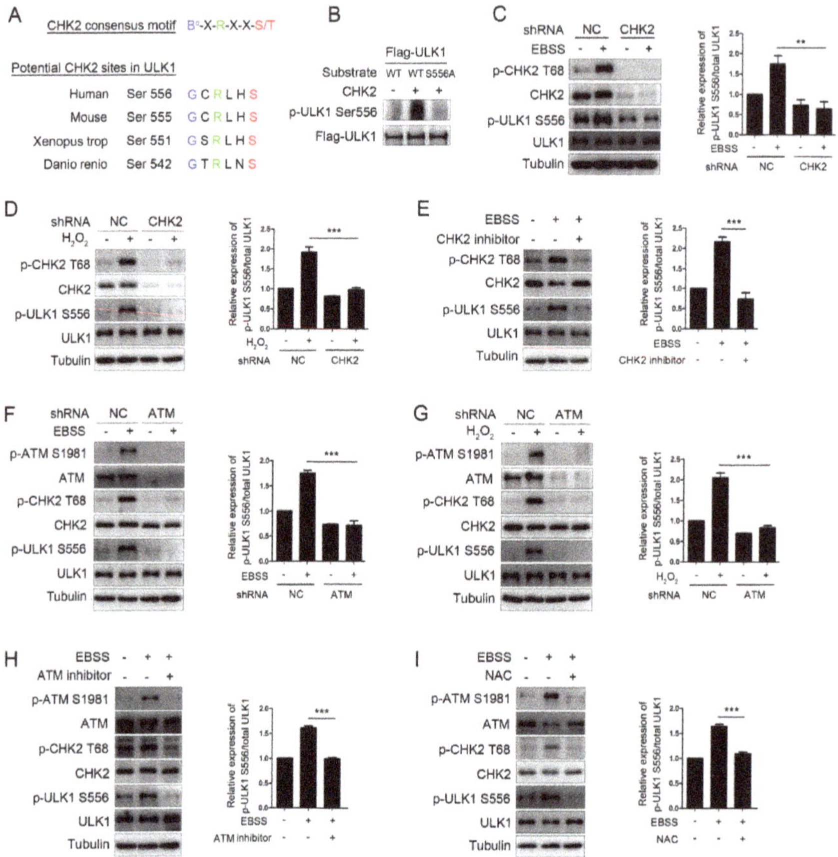

Figure 2. CHK2 phosphorylates ULK1 at Ser556. (**A**) Clustal alignment of the conserved sites in ULK1 matching the optimal CHK2 substrate motif. (**B**) In vitro CHK2 kinase assay using CHK2 recombinant human protein with FLAG-ULK1WT and FLAG-ULK1S556A as substrates, followed by immunoblotting analysis. (**C**,**D**) Western blot analysis with the indicated antibodies in H1299 cells with CHK2 knockdown. Cells were treated with EBSS (**C**) or H_2O_2 (500 μM) (**D**). (**E**) Western blot analysis with indicated antibodies in H1299 cells pretreated with CHK2 inhibitor for 3 h and treated with EBSS for 1 h. (**F**,**G**) Western blot analysis with indicated antibodies in H1299 cells with ATM knockdown. Cells were treated with EBSS (**F**) or H_2O_2 (500 μM) (**G**). (**H**) Western blot analysis with the indicated antibodies in H1299 cells pretreated with ATM inhibitor for 3 h and treated with EBSS for 1 h. (**I**) Western blot analysis with the indicated antibodies in H1299 cells pretreated with NAC for 4 h and treated with EBSS for 1 h. The results from three independent experiments are presented as mean ± SEM. ** $p < 0.05$, *** $p < 0.001$.

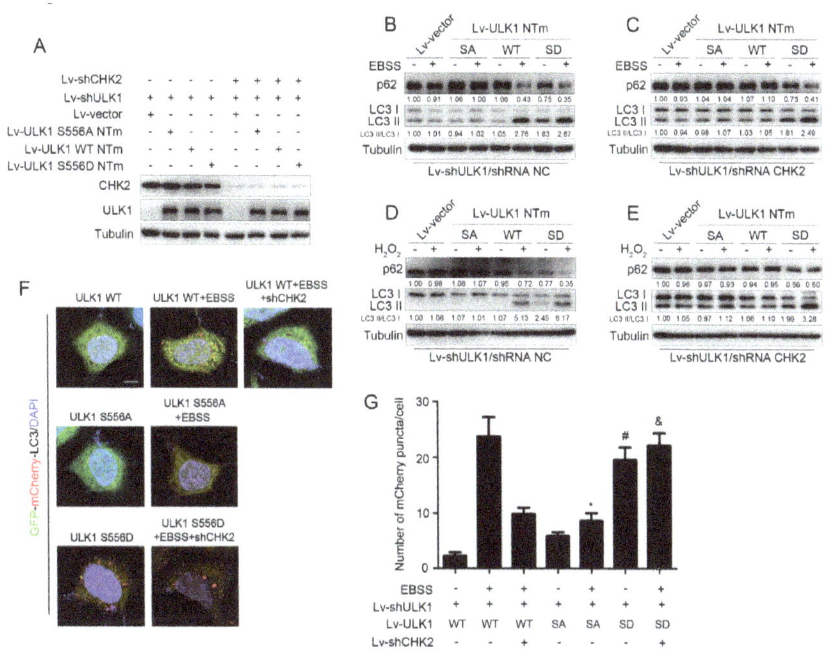

Figure 3. CHK2-mediated ULK1 phosphorylation promotes autophagy. (**A**) Western blot analysis with reconstituted expression of the indicated proteins in H1299 cells. (**B**,**C**) Western blot analysis of p62and LC3 in H1299 cells with reconstituted expression of ULK1 WT, S556A mutant, or S556D mutant treated with EBSS for 3 h with (**B**) or without (**C**) CHK2. (**D**,**E**) Western blot analysis of p62 and LC3 in H1299 cells with reconstituted expression of ULK1 WT, S556A mutant, or S556D mutant treated with H_2O_2 (500 μM) for 3 h with (**D**) or without (**E**) CHK2. (**F**) The red puncta is shown by representative confocal microscopic images in 293 cells expressing the indicated plasmids treated with EBSS for 2 h. (**G**) Data are presented as mean ± SEM. * ULK1 SA mutant treated with EBSS compared to ULK1 WT treated with EBSS, $p < 0.001$; # ULK1 SD mutant compared to ULK1 WT, $p < 0.001$ and & ULK1 SD mutant treated with EBSS without CHK2 compared to ULK1 WT treated with EBSS without CHK2, $p < 0.001$. Scale bar, 10 μm.

Furthermore, we found similar conclusions by the quantitative analysis of autophagic flux, using tandemly labeled GFP-mCherry-LC3B. Given that mCherry fluorescence can be detected in both neutral autophagosomes and acidic autolysosomes, whereas GFP fluorescence is quenched in acidic autolysosomes, we can judge the extent to which autophagic flux proceeded and represented the promotion of autophagy levels, based on the reduction in yellow puncta and the appearance of only red puncta due to GFP quenching by fusion of autophagosomes with lysosomes. The red puncta representing autophagic flux were increased after the expression of ULK1 WT under metabolic stress, but not in CHK2 knockdown cells. Compared to the effect of ULK1 WT, the ULK1 S556A mutant failed to promote metabolic stress-induced autophagic flux. However, the ULK1 S556D mutant promotes autophagic flux, even in cells that did not express CHK2 (Figure 3F,G).

3.4. CHK2-ULK1-Mediated Autophagy Protects Cells against Metabolic Stress-Induced Cell Death

To test the role of CHK2-mediated phosphorylation of ULK1 in cell fate determination under stress conditions, we examined the effects of ULK1 WT, S556A, or S556D mutants on apoptosis, with or without CHK2 in the H1299 and HEK293 cell lines, in response to metabolic and oxidative stresses. The WT and S556D mutant reduced the number of cells undergoing apoptosis compared to the control, while the S556A mutant had no effect in response to metabolic and oxidative stress (Figure 4A,C, Supplementary Figure S2A,C). In addition, compared with CHK2-expressing cells, we found that apoptosis under stress was increased in CHK2-knockdown cells, even after expressing ULK1 WT (Figure 4B,D, Supplementary Figure S2B,D).

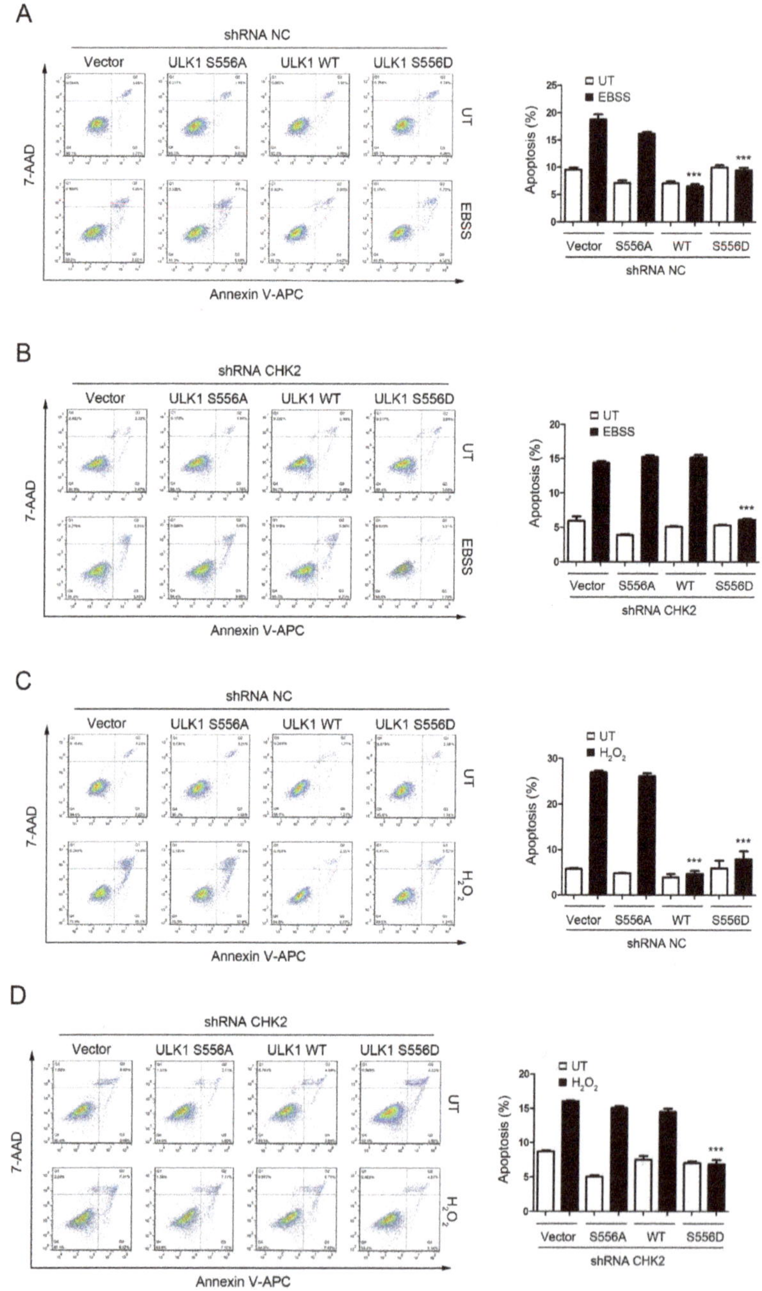

Figure 4. CHK2-ULK1 mediated autophagy protects cells against metabolic stress-induced cell death. (**A**,**B**) Flow cytometry analysis of apoptosis was performed in H1299 cells expressing the indicated

plasmids treated with EBSS for 8 h with (**A**) or without (**B**) CHK2. The results from three independent experiments are presented as mean ± SEM. *** $p < 0.001$ compared to ULK1 S556A treated with EBSS. (**C**,**D**) Flow cytometry analysis of apoptosis was performed in H1299 cells expressing the indicated plasmids treated with H_2O_2 (500 μM) for 8 h with (**C**) or without (**D**) CHK2. The results from three independent experiments are presented as mean ± SEM. *** $p < 0.001$ compared to ULK1 S556A treated with H_2O_2 (500 μM) stimulation.

4. Discussion

More and more evidence indicates that ROS are considered to be signaling molecules that trigger autophagy under various stress conditions, but their specific regulatory mechanism is still largely unknown [6]. Here, we found that the autophagy-related protein ULK1 is a novel interacting protein of CHK2, and that binding is dependent on the accumulation of ROS caused by metabolic stress. CHK2 phosphorylates serine 556 of ULK1 to promote autophagy and inhibit cell apoptosis under metabolic stress. Our findings establish that the ATM/CHK2/ULK1 signaling pathway initiates an autophagic adaptive response by sensing ROS, and it protects cells from metabolic stress-induced cellular damage.

The relationship between ROS, autophagy, and apoptosis has been well revealed [4]. However, the possible regulatory mechanism is largely unknown. Our studies have revealed that ROS act as signaling molecules and activate the ATM/CHK2/ULK1 signaling pathway under metabolic and oxidative stress, while the ROS neutralizer NAC can inhibit the activation of the ATM/CHK2/ULK1 signaling pathway under metabolic stress. We further verified that CHK2-mediated ULK1 phosphorylation could promote autophagy and inhibit apoptosis in response to metabolic and oxidative stress. Previous studies have fully demonstrated the relationship between ROS, autophagy, and apoptosis. For example, the ROS neutralizer NAC can inhibit the accumulation of excessive ROS under metabolic stress, thereby inhibiting apoptosis. Autophagy can inhibit the accumulation of ROS by promoting the clearance of damaged mitochondria in response to metabolic stress, thereby inhibit apoptosis. Conversely, the inhibition of autophagy leads to the excessive accumulation of ROS under metabolic stress, and promotes apoptosis. As the molecular mechanisms of autophagy are well understood, the focus has shifted to the initiation signals of autophagy and the specific molecular mechanisms that initiate autophagy. More and more studies have demonstrated that various stimuli inducing autophagy could lead to an increase in ROS levels [26–30]. These findings suggest that ROS may play an essential role in autophagic initiation as a variety of central signals inducing autophagy. However, the molecular mechanism by which ROS initiates autophagy remains largely unknown. In recent years, DNA damage response (DDR)-transduced ataxia-telangiectasia mutant (ATM) has played a vital role in sensing ROS signaling and autophagy initiation. ATM activates TSC2 through the LKB1/AMPK signaling pathway to inhibit mTORC1, and this promotes autophagy [13]. Our previous study found that elevated levels of ROS induced by glucose deprivation and hypoxia can promote autophagy to maintain cellular homeostasis by activating the ATM/CHK2/Beclin 1 axis [10]. In this study, we further explored other molecular mechanisms by which the ATM-CHK2 signaling pathway initiated autophagy. We found that the autophagy-related protein ULK1 is an essential substrate of CHK2 under metabolic stress conditions, and further confirmed the role of the ATM/CHK2/ULK1 axis in autophagic initiation. This study further improved the function of the ATM-CHK2 signaling pathway to sense ROS signaling molecules in the autophagy regulatory network.

The formation of autophagosomes is a critical initial event in autophagy [31,32]. The serine/threonine-protein kinase ULK1 is the core protein involved in this step, and it forms a complex with FIP200, ATG13, and ATG101 to participate in the initiation of autophagy [33,34]. The phosphorylation of ULK1 is critical in autophagy regulation. Under adequate nutrient conditions, mTOR phosphorylates ULK1, resulting in the inactivation of the ULK1 complex. AMPK can promote autophagy by phosphorylating multiple serine/threonine sites in the intermediate domain of ULK1 (including serines 317, 467, 555, 637, 777 and threonine 574). We revealed that under oxidative stress conditions, CHK2 is activated by ATM, activated CHK2 binds to ULK1 and phosphorylates serine 556 of ULK1, and activated

ULK1 promotes autophagy initiation and autophagic flux. For the first time, we revealed the specific molecular mechanism of how ROS signaling regulates ULK1 to promote autophagy, enriching the autophagy regulatory network of ULK1. ULK2 is highly homologous to ULK1, and is also particularly important in regulating autophagy. However, whether ULK2 is also involved in the autophagy process involving ROS signaling remains to be further investigated.

5. Conclusions

Our results show that the ROS–ATM–CHK2–ULK1–autophagy axis is a novel metabolic stress adaptive response pathway that protects cells from stress-induced cellular damage.

Supplementary Materials: The following supporting information can be downloaded at: https://www.mdpi.com/article/10.3390/antiox11061166/s1, Figure S1: CHK2-mediated ULK1 phosphorylation promotes autophagy; Figure S2: CHK2-ULK1 mediated autophagy protects cells against metabolic stress-induced cell death.

Author Contributions: Conceptualization, X.-C.X., L.C. and Q.-Q.G.; methodology, R.G. and S.-S.W.; software, X.-Y.J.; validation, Y.Z., Y.G. and H.-Y.C.; writing—original draft preparation, R.G.; writing—review and editing, X.-C.X. and Q.-Q.G.; funding acquisition, X.-C.X., L.C., Q.-Q.G. and R.G. All authors have read and agreed to the published version of the manuscript.

Funding: This work was supported by National Key R&D Program of China (2016YFC1302400); Key projects of the National Natural Fund of China (82030091), Natural Science Foundation of China (81702738, 81073089, 81502400), Ministry of Education Innovation Team Development Plan (IRT13101/17R107), and Scientific research project of Liaoning Provincial Department of Education (JC2019016, JC2019039).

Institutional Review Board Statement: Not applicable.

Informed Consent Statement: Not applicable.

Data Availability Statement: Data is contained within the article and Supplementary Materials.

Acknowledgments: We thank Xin Pan for providing the ULK1 plasmids.

Conflicts of Interest: The authors declare no conflict of interest.

References

1. Xie, X.; Zhang, Y.; Wang, Z.; Wang, S.; Jiang, X.; Cui, H.; Zhou, T.; He, Z.; Feng, H.; Guo, Q.; et al. ATM at the crossroads of reactive oxygen species and autophagy. *Int. J. Biol. Sci.* **2021**, *17*, 3080–3090. [CrossRef] [PubMed]
2. Forrester, S.J.; Kikuchi, D.S.; Hernandes, M.S.; Xu, Q.; Griendling, K.K. Reactive Oxygen Species in Metabolic and Inflammatory Signaling. *Circ. Res.* **2018**, *122*, 877–902. [CrossRef] [PubMed]
3. Sies, H.; Jones, D.P. Reactive oxygen species (ROS) as pleiotropic physiological signalling agents. *Nat. Rev. Mol. Cell Biol.* **2020**, *21*, 363–383. [CrossRef] [PubMed]
4. Filomeni, G.; De Zio, D.; Cecconi, F. Oxidative stress and autophagy: The clash between damage and metabolic needs. *Cell Death Differ.* **2015**, *22*, 377–388. [CrossRef]
5. Choi, A.M.; Ryter, S.W.; Levine, B. Autophagy in human health and disease. *N. Engl. J. Med.* **2013**, *368*, 651–662. [CrossRef]
6. Zhou, J.; Li, X.Y.; Liu, Y.J.; Feng, J.; Wu, Y.; Shen, H.M.; Lu, G.D. Full-coverage regulations of autophagy by ROS: From induction to maturation. *Autophagy* **2021**. [CrossRef]
7. Antoni, L.; Sodha, N.; Collins, I.; Garrett, M.D. CHK2 kinase: Cancer susceptibility and cancer therapy—Two sides of the same coin? *Nat. Rev. Cancer* **2007**, *7*, 925–936. [CrossRef]
8. Lavin, M.F. Ataxia-telangiectasia: From a rare disorder to a paradigm for cell signalling and cancer. *Nat. Rev. Mol. Cell Biol.* **2008**, *9*, 759–769. [CrossRef]
9. Lee, J.H.; Paull, T.T. Cellular functions of the protein kinase ATM and their relevance to human disease. *Nat. Rev. Mol. Cell Biol.* **2008**, *22*, 796–814. [CrossRef]
10. Guo, Q.Q.; Wang, S.S.; Zhang, S.S.; Xu, H.D.; Li, X.M.; Guan, Y.; Yi, F.; Zhou, T.T.; Jiang, B.; Bai, N.; et al. ATM-CHK2-Beclin 1 axis promotes autophagy to maintain ROS homeostasis under oxidative stress. *EMBO J.* **2020**, *39*, e103111. [CrossRef]
11. Cirotti, C.; Rizza, S.; Giglio, P.; Poerio, N.; Allega, M.F.; Claps, G.; Pecorari, C.; Lee, J.H.; Benassi, B.; Barila, D.; et al. Redox activation of ATM enhances GSNOR translation to sustain mitophagy and tolerance to oxidative stress. *EMBO Rep.* **2021**, *22*, e50500. [CrossRef] [PubMed]

12. Guo, Z.; Kozlov, S.; Lavin, M.F.; Person, M.D.; Paull, T.T. ATM activation by oxidative stress. *Science* **2010**, *330*, 517–521. [CrossRef] [PubMed]
13. Alexander, A.; Cai, S.L.; Kim, J.; Nanez, A.; Sahin, M.; MacLean, K.H.; Inoki, K.; Guan, K.L.; Shen, J.; Person, M.D.; et al. ATM signals to TSC2 in the cytoplasm to regulate mTORC1 in response to ROS. *Proc. Natl. Acad. Sci. USA* **2010**, *107*, 4153–4158. [CrossRef] [PubMed]
14. Lee, E.J.; Tournier, C. The requirement of uncoordinated 51-like kinase 1 (ULK1) and ULK2 in the regulation of autophagy. *Autophagy* **2011**, *7*, 689–695. [CrossRef]
15. Ganley, I.G.; Lam, D.H.; Wang, J.; Ding, X.; Chen, S.; Jiang, X. ULK1.ATG13.FIP200 complex mediates mTOR signaling and is essential for autophagy. *J. Biol. Chem.* **2009**, *284*, 12297–12305. [CrossRef]
16. Hosokawa, N.; Hara, T.; Kaizuka, T.; Kishi, C.; Takamura, A.; Miura, Y.; Iemura, S.; Natsume, T.; Takehana, K.; Yamada, N.; et al. Nutrient-dependent mTORC1 association with the ULK1-Atg13-FIP200 complex required for autophagy. *Mol. Biol. Cell* **2009**, *20*, 1981–1991. [CrossRef]
17. Jung, C.H.; Jun, C.B.; Ro, S.H.; Kim, Y.M.; Otto, N.M.; Cao, J.; Kundu, M.; Kim, D.H. ULK-Atg13-FIP200 complexes mediate mTOR signaling to the autophagy machinery. *Mol. Biol. Cell* **2009**, *20*, 1992–2003. [CrossRef]
18. Zachari, M.; Ganley, I.G. The mammalian ULK1 complex and autophagy initiation. *Essays Biochem.* **2017**, *61*, 585–596.
19. Puente, C.; Hendrickson, R.C.; Jiang, X. Nutrient-regulated Phosphorylation of ATG13 Inhibits Starvation-induced Autophagy. *J. Biol. Chem.* **2016**, *291*, 6026–6035. [CrossRef]
20. Kim, J.; Kundu, M.; Viollet, B.; Guan, K.L. AMPK and mTOR regulate autophagy through direct phosphorylation of Ulk1. *Nat. Cell Biol.* **2011**, *13*, 132–141. [CrossRef]
21. Egan, D.F.; Shackelford, D.B.; Mihaylova, M.M.; Gelino, S.; Kohnz, R.A.; Mair, W.; Vasquez, D.S.; Joshi, A.; Gwinn, D.M.; Taylor, R.; et al. Phosphorylation of ULK1 (hATG1) by AMP-activated protein kinase connects energy sensing to mitophagy. *Science* **2011**, *331*, 456–461. [CrossRef] [PubMed]
22. Shang, L.; Chen, S.; Du, F.; Li, S.; Zhao, L.; Wang, X. Nutrient starvation elicits an acute autophagic response mediated by Ulk1 dephosphorylation and its subsequent dissociation from AMPK. *Proc. Natl. Acad. Sci. USA* **2011**, *108*, 4788–4793. [CrossRef] [PubMed]
23. Zhao, M.; Klionsky, D.J. AMPK-dependent phosphorylation of ULK1 induces autophagy. *Cell Metab.* **2011**, *13*, 119–120. [CrossRef] [PubMed]
24. Mack, H.I.; Zheng, B.; Asara, J.M.; Thomas, S.M. AMPK-dependent phosphorylation of ULK1 regulates ATG9 localization. *Autophagy* **2012**, *8*, 1197–1214. [CrossRef]
25. Seo, G.J.; Kim, S.E.; Lee, Y.M.; Lee, J.W.; Lee, J.R.; Hahn, M.J.; Kim, S.T. Determination of substrate specificity and putative substrates of Chk2 kinase. *Biochem. Biophys. Res. Commun.* **2003**, *304*, 339–343. [CrossRef]
26. Filomeni, G.; Desideri, E.; Cardaci, S.; Rotilio, G.; Ciriolo, M.R. Under the ROS: Thiol network is the principal suspect for autophagy commitment. *Autophagy* **2010**, *6*, 999–1005. [CrossRef]
27. Chen, Y.; Azad, M.B.; Gibson, S.B. Superoxide is the major reactive oxygen species regulating autophagy. *Cell Death Differ.* **2009**, *16*, 1040–1052. [CrossRef]
28. Scherz-Shouval, R.; Shvets, E.; Elazar, Z. Oxidation as a post-translational modification that regulates autophagy. *Autophagy* **2007**, *3*, 371–373. [CrossRef]
29. Zhang, C.; Yang, L.; Wang, X.B.; Wang, J.S.; Geng, Y.D.; Yang, C.S.; Kong, L.Y. Calyxin Y induces hydrogen peroxide-dependent autophagy and apoptosis via JNK activation in human non-small cell lung cancer NCI-H460 cells. *Cancer Lett.* **2013**, *340*, 51–62. [CrossRef]
30. Levonen, A.L.; Hill, B.G.; Kansanen, E.; Zhang, J.; Darley-Usmar, V.M. Redox regulation of antioxidants, autophagy, and the response to stress: Implications for electrophile therapeutics. *Free Radic. Biol. Med.* **2014**, *71*, 196–207. [CrossRef]
31. Hurley, J.H.; Young, L.N. Mechanisms of Autophagy Initiation. *Annu. Rev. Biochem.* **2017**, *86*, 225–244. [CrossRef] [PubMed]
32. Wirth, M.; Joachim, J.; Tooze, S.A. Autophagosome formation—The role of ULK1 and Beclin1-PI3KC3 complexes in setting the stage. *Semin. Cancer Biol.* **2013**, *23*, 301–309. [CrossRef] [PubMed]
33. Lin, M.G.; Hurley, J.H. Structure and function of the ULK1 complex in autophagy. *Curr. Opin. Cell Biol.* **2016**, *39*, 61–68. [CrossRef] [PubMed]
34. Wong, P.M.; Puente, C.; Ganley, I.G.; Jiang, X. The ULK1 complex: Sensing nutrient signals for autophagy activation. *Autophagy* **2013**, *9*, 124–137. [CrossRef] [PubMed]

antioxidants

MDPI

Article

Stevia and Stevioside Attenuate Liver Steatosis through PPARα-Mediated Lipophagy in *db/db* Mice Hepatocytes

Miey Park [1,2], Anshul Sharma [1], Hana Baek [1,2], Jin-Young Han [2], Junho Yu [1,2] and Hae-Jeung Lee [1,2,*]

[1] Department of Food and Nutrition, College of BioNano Technology, Gachon University, Seongnam 13120, Gyeonggi-do, Republic of Korea

[2] Institute for Aging and Clinical Nutrition Research, Gachon University, Seongnam 13120, Gyeonggi-do, Republic of Korea

* Correspondence: skysea1010@gmail.com or skysea@gachon.ac.kr; Tel.: +82-31-750-5968; Fax: +82-31-724-4411

Abstract: Lipophagy, a type of autophagy that breaks down lipid droplets, is essential in the regulation of intracellular lipid accumulation and intracellular free fatty acid levels in numerous organisms and metabolic conditions. We investigated the effects of *Stevia rebaudiana Bertoni* (S), a low-calorie sweetener, and stevioside (SS) on hepatic steatosis and autophagy in hepatocytes, as well as in *db/db* mice. S and SS reduced the body and liver weight and levels of serum triglyceride, total cholesterol, and hepatic lipogenic proteins. In addition, S and SS increased the levels of fatty acid oxidase, peroxisome proliferator-activated receptor alpha (PPARα), and microtubule-associated protein light chain 3 B but decreased that of sequestosome 1 (p62) in the liver of *db/db* mice. Additionally, Beclin 1, lysosomal associated membrane protein 1, and phosphorylated adenosine monophosphate-activated protein kinase protein expression was augmented following S and SS treatment of *db/db* mice. Furthermore, the knockdown of PPARα blocked lipophagy in response to SS treatment in HepG2 cells. These outcomes indicate that PPARα-dependent lipophagy is involved in hepatic steatosis in the *db/db* mouse model and that SS, a PPARα agonist, represents a new therapeutic option for managing associated diseases.

Keywords: non-alcoholic fatty liver disease; stevia; stevioside; lipophagy; PPARα

Citation: Park, M.; Sharma, A.; Baek, H.; Han, J.-Y.; Yu, J.; Lee, H.-J. Stevia and Stevioside Attenuate Liver Steatosis through PPARα-Mediated Lipophagy in *db/db* Mice Hepatocytes. *Antioxidants* **2022**, *11*, 2496. https://doi.org/10.3390/antiox11122496

Academic Editors: Angeles Aroca and Cecilia Gotor

Received: 3 November 2022
Accepted: 15 December 2022
Published: 19 December 2022

1. Introduction

Non-alcoholic fatty liver disease (NAFLD) is the most common liver ailment caused by the buildup of excess fat in liver cells rather than by alcohol [1]. Hepatic steatosis, or fatty liver disease, develops when the liver's weight is more than 5% fat [2]. Fatty liver disease can be divided into two types: when the liver has fat buildup and no damage, it is called NAFLD; when the liver has fat buildup, hepatocellular injury, inflammation, and different degrees of fibrosis, it is called non-alcoholic steatohepatitis (NASH) [3]. NAFLD is significantly linked to insulin resistance and obesity and is detected in more than 76% of type 2 diabetes (T2D) patients [4]. In addition, patients with T2D are at significant risk of developing NASH [3]. Recent findings suggest that autophagy may boost the lipid metabolism. Therefore, autophagy is considered to have therapeutic potential in NAFLD [5,6].

Autophagy mediates not only the redistribution of valuable nutrients during starvation but also the treatment of excess or damaged small organs and invading microorganisms [7]. When nutrients are sufficient, cells store their energy reserves as neutral lipids, cholesteryl esters, and triglycerides in the lipid droplets (LDs). Owing to the dynamic fusion of LDs with other intracellular organelles and erroneously folded proteins [8] or infectious particles [9], these apparently inactive LDs have often been reflected as separate organelles [10].

Autophagy breaks down LDs through a process called lipophagy [10,11]. Given that starvation triggers autophagy, which results in the production of nutrients through the lysosomal decomposition of unnecessary cytoplasmic content, autophagy has been postulated to have a role in disintegrating LDs to release free fatty acids for the starving cells and aid in lipid metabolism [12].

Stevia rebaudiana Bertoni (S) is a natural sweetener that is 300-times sweeter than sucrose and has various health-promoting biological effects [13,14]. These biological effects are elicited by the plant leaf extract, which contains secondary metabolites, such as polyphenols and steviol glycosides, including stevioside (SS), rebaudioside A, and rebaudioside C [15]. SS is one of the main compounds in stevia extract and constitutes between 4% and 20% of dried leaves [16]. It exhibits non-caloric, anti-inflammatory [17], anti-tumor [14], anti-diarrheal [18], and antihypertensive [19] effects and is effective in the treatment of hyperlipidemia [20]. Studies have shown that S and SS have many health benefits; however, their effects on hepatic metabolism and autophagy remain unclear. Therefore, the purpose of this study was to investigate how S and SS affect NAFLD in HepG2 cells and *db/db* mouse models.

2. Materials and Methods

2.1. Reagents

S leaf extract was obtained from Pharminogen (Pharminogen Inc., Gyeonggi-do, Korea). The concentrate was lyophilized to remove moisture completely and stored at −20 °C. SS ($C_{38}H_{60}O_{18}$) was purchased from ChemFaces (ChemFaces Biochemical Co., Wuhan, China). Chloroquine (CQ) was procured from Sigma (Sigma-Aldrich Co., St. Louis, MO, USA), and the CYTO-ID Autophagy Detection Kit was acquired from Enzo Inc. (Enzo Life Sciences, Farmingdale, NY, USA).

2.2. HepG2 Cells Culture, Steatosis Induction, and Oil Red O Staining

HepG2 (human hepatoma) cells were acquired from the American Type Culture Collection (ATCC, Manassas, VA, USA). HepG2 cells were cultivated in Dulbecco's modified Eagle's medium (DMEM), which included 10% fetal bovine serum and 1% antibiotics (antibiotic-antimycotic, ThermoFisher, Waltham, MA, USA) at 37 °C in an environment with 5% CO_2. Free fatty acid (FFA, oleic, and palmitic acid at a molar ratio of 2:1) stock was dissolved in isopropyl alcohol.

To induce hepatic steatosis, HepG2 cells were grown in a serum-free medium and exposed to 1 mM FFA for 24 h. BSA (1%) was used as a control. HepG2 cells were fixed in formalin (10%) and stained with a working solution of Oil Red O for 30 min. After three washes, the cells were viewed under an inverted microscope, and images were taken. Following the observation, the absorbance of the lipids (in 1 mL of 100% isopropanol) was measured at 500 nm. Each experiment was conducted in triplicate.

2.3. Mouse Models and Diets

Male C57BL/6J *db/db* mice (BKS.Cg-Dock7m +/+ *Leprdb*/J, homozygote, 8-weeks-old) and negative controls (C57BL/6J mice, heterozygotes, similar age) were acquired from Jackson Laboratories (Sacramento, CA, USA). Following acclimatization for 2 weeks, the animals were randomly allocated into six groups as follows: the negative control group (C57BL/6J, N+; saline, n = 6), control group (C57BL/6J *db/db*, NC; saline, n = 6), positive control group (C57BL/6J *db/db*, PC, n = 6; saline with metformin 200 mg/kg/day), low S treatment group (C57BL/6J *db/db*, S200, n = 6; saline with 200 mg/kg/day of S extract), high S treatment group (C57BL/6J *db/db*, S500, n = 6; saline with 500 mg/kg/day of S extract), and SS treatment group (C57BL/6J *db/db*, SS, n = 6; saline with 40 mg/kg/day of SS). Mice were orally administered saline and SS (prepared in saline) during the three weeks.

Three weeks later, all 36 mice were sacrificed. All studies using the chosen mouse models were approved by Gachon University (GI-ACUC-R2020012) and performed in compliance with the guidelines of the Guide for the Care and Use of Laboratory Animals.

2.4. Liver Tissue Histological Evaluation and Oil Red O Staining

Liver tissue samples from all the experimental mice groups were fixed in formalin (10%, Sigma-Aldrich, MO, USA) for a minimum of 72 h. The tissue sections (3–4 μm) were stained with hematoxylin-eosin (H-E) and Oil Red O solutions. A competent pathologist examined the stained sections using an Olympus Provis AX70 microscope (Olympus, Tokyo, Japan). A scale bar of 100 μm is used.

2.5. Real-Time PCR Quantification of Gene Expression

The total RNA was extracted from HepG2 cells and liver tissues using a commercial kit (iNtRON Biotechnology, Gyeonggi-do, Korea) in accordance with the owner's manual. Following quantification, 0.8 μg of RNA was reverse transcribed into complementary DNA utilizing the iScript cDNA synthesis kit (Bio-Rad, Hercules, CA, USA). SYBR® Green Master Mix (TaKaRa Bio, Otsu, Japan) was used in real-time PCR performed on a QuantStudio3 machine (Thermo Fisher Scientific).

The following primers (5′–3′) were used: peroxisome proliferator-activated receptor gamma (PPARγ, NC_000072.7) forward CAGGAGAGCAGGGATTTGCA and reverse CCTACGCTCAGCCCTCTTCAT; sterol regulatory element-binding transcription factor-1c (SREBP-1c, NM_001113566.1) forward ATCGCAAACAAGCTGACCTG and reverse AGATCCAGGTTT GAGGTGGG; CCAAT/enhancer binding protein alpha (C/EBPα, NM_001287523.1) forward TTACAACAGGCCAGGTTTCC and reverse GGCTGGCGACATACAGTACA; fatty acid synthase (FAS, NC_000077.7) forward TTGCTGGCACTACAGAATGC and reverse AACAGCCTCAGAGCGACAAT; and β-actin (NM_007393.5) forward CTGTCCCTGTATGCCTCTG and reverse ATGTCACGCACGATTTCC. The expression levels of the chosen genes were compared to that of β-actin.

2.6. Western Blotting

For protein expression analysis, the total proteins were harvested from HepG2 cells and liver tissues using a lysis buffer for the extraction of proteins (iNtRON Biotechnology) comprising protease and phosphate inhibitors. Protein (30 μg) samples were electrophoretically separated on a polyacrylamide gel containing 15% sodium dodecyl sulfate and electrophoretically shifted to polyvinylidene difluoride membranes (Bio-Rad Laboratories, Hercules, CA, USA).

After blocking with 5% skim milk at room temperature (RT, 18 to 25 °C), the membranes were incubated for more than 2 h with the following primary antibodies at RT: PPARγ (1:1000 dilution), SREBP-1c (1:1000 dilution), C/EBPα (1:500 dilution), FAS (1:1000 dilution), microtubule-associated protein light chain 3 B (LC3B; 1:1000 dilution), sequestosome 1 (SQSTM1; 1:1000 dilution), TFEB (1:1000 dilution), CPT-1 (1:1000 dilution), lysosomal associated membrane protein 1 (LAMP-1; 1:1000 dilution), BECN1 (1:1000 dilution), Bcl-2 (1:1000 dilution), Bax (1:1000 dilution), PPARα (1:1000 dilution), adenosine monophosphate-activated protein kinase (AMPK; 1:1000 dilution), and p-AMPK (1:1000 dilution).

The primary antibodies were acquired from Abcam (Cambridge, UK) and Santa Cruz Biotechnology (Dallas, TX, USA). After 1 h of incubation at RT with the secondary antibody, the membranes were developed using the Miracle-Star™ Western Blot Detection System (iNtRON Biotechnology) and photographed using the ImageQuant™ LAS500 system (GE Healthcare Life Sciences, Issaquah, WA, USA). Densitometry data after western blot analysis were obtained using the Amersham Imager 680 analysis software (GE Healthcare Life Sciences, USA) and used for analysis.

2.7. Detection of Autophagy with CYTO-ID® Green

Autophagy was detected using the Cyto-ID® Autophagy Detection Kit (Enzo Life Sciences) in accordance with the owner's manual. Briefly, steatosis-induced cells were stained by incubating them with Cyto-ID® Green stain solution for 30 min at 37 °C in the dark. The samples were rinsed once, stained with Hoechst 33342 for 1 min, washed three times with PBS, and mounted onto slides. Fluorescence images were obtained using a fluorescence inverted phase-contrast microscope (KI-2000F, Korea Lab Tech, Gyeonggi, Korea) and inspected using processing software (OptiView 3.7, Korea Lab Tech).

2.8. *Small Interfering RNAs (siRNAs) and Transfection*

PPARα (Catalog No. AM51331) and negative control (NC, Catalog No. AM4635) siRNAs were obtained from Invitrogen (Waltham, MA, USA). One day before transfection, HepG2 cells were seeded in six-well plates at a density of 2×10^5 cells/well. The following day, Opti-MEM (Gibco-BRL, Thermo Fisher Scientific) was used as a culture medium substitute. Both the siRNAs were prepared in an Opti-MEM medium and mixed with Lipofectamine® RNAiMAX (7 μL) reagent (Thermo Fisher Scientific) to a final volume of 200 μL. The produced mixture was applied to each well after incubating for 20 min at room temperature (25 °C). Fresh medium was given to each well after 4 h of incubation, and cells were then allowed to grow. After 20 h, the transfected cells were treated with stevioside and harvested after 24 h.

2.9. *Statistical Analysis*

The mean ± standard error (SE) is used to express all experimental results. Each experiment was performed three times, GraphPad Prism 9.02 (GraphPad Software, USA) was used for statistical analysis using one-way analysis of variance (ANOVA), and Tukey's post-hoc testing was used to compare multiple independent groups. Statistical significance was set at $p < 0.05$.

3. Results

3.1. *S and SS Attenuated Liver Steatosis in db/db Mice*

After three weeks of oral administration of S and SS, the terminal body weights of *db/db* mice in the NC group showed significantly increased weighLinet ($p < 0.01$) from that in the S (S200 and S500) and SS groups (Figure 1a). In addition, the liver tissue weights of *db/db* mice were significantly decreased in the S200 ($p < 0.001$), S500 ($p < 0.001$), and SS ($p < 0.1$) treatment groups relative to that in the NC group (Figure 1b). The *db/db* mice had the highest plasma serum triglyceride (TG) and total serum cholesterol (TC) levels among type 2 diabetic mouse models [21].

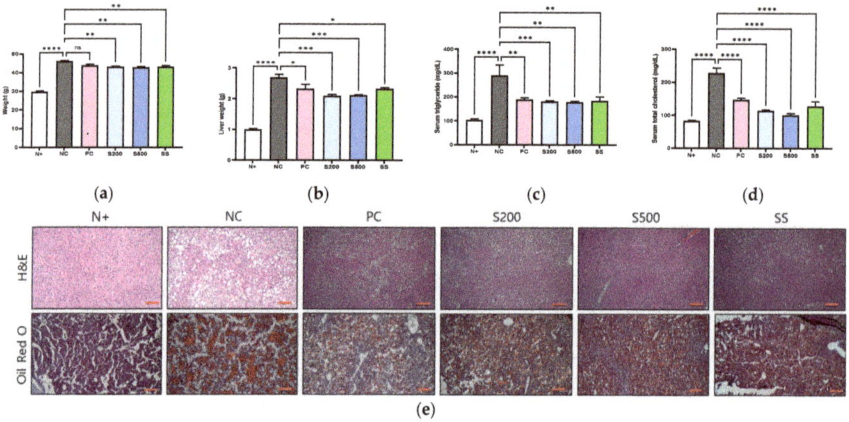

Figure 1. Stevia and stevioside attenuate lipid accumulation in *db/db* mice. Mice were orally administered saline (N+ and NC), saline with 200 mg/kg/day metformin (PC), saline with 200 or 500 mg/kg/day stevia (S200 and S500), and saline with 40 mM/kg/day stevioside (SS) for 3 weeks. (**a**) Body weight. (**b**) Liver weight. (**c**) Serum triglyceride (TG). (**d**) Serum total cholesterol (TC). (**e**) Image of liver tissues. All data are presented as the mean ± standard error (SE), $n = 6$ and represent results from three independent experiments. Scale bar, 100 μm. * $p < 0.05$, ** $p < 0.01$, *** $p < 0.001$, and **** $p < 0.0001$.

Considering these characteristics, in this study, the weight and serum TG and TC levels in the NC group were higher than in the N+ group of normal mice, and they were significantly reduced by the oral administration of S and SS (Figure 1c,d). Histological analyses of the livers from the six *db/db* mouse groups revealed that, compared to the N+ group, the NC group showed hepatocellular damage (H&E staining) and widespread LDs (Oil Red O staining), whereas the S200, S500, and SS treatment groups showed decreased accumulation of LDs (Figure 1e).

3.2. S and SS Attenuated Expressions of Hepatic Lipid Genes in db/db Mice

To examine the biological mechanisms of S in the liver, we used quantitative PCR and immunoblot analysis to examine the lipogenic markers. As shown in Figure 2, the mRNA expressions of PPARγ ($p < 0.1$), SREBP-1c ($p < 0.001$), C/EBPα ($p < 0.001$), and FAS ($p < 0.0001$) were significantly higher in the NC group relative to the N+ group (Figure 2a–d). The corresponding mRNA expression levels in the S200, S500, and SS groups were significantly lower than those in the NC group. Consistent with this, protein expression analysis also showed a significant decrease in the levels of the respective proteins in the S200-, S500-, and SS-treated groups (Figure 2e–f).

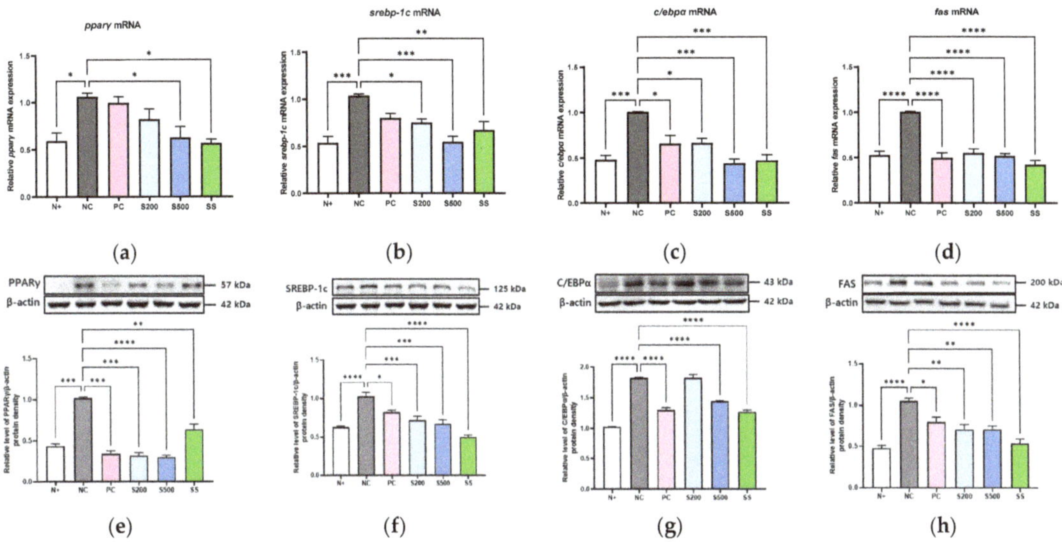

Figure 2. Stevia and stevioside attenuate adipogenic-related genes and proteins in *db/db* mice. Mice were orally administered saline (N+ and NC), saline with 200 mg/kg/day metformin (PC), saline with 200 or 500 mg/kg/day stevia (S200 and S500), and saline with 40 mM/kg/day stevioside (SS) for 3 weeks. (**a**,**e**) Peroxisome proliferator-activated receptor gamma (PPARγ). (**b**,**f**) Sterol regulatory element-binding transcription factor-1c (SREBP-1c). (**c**,**g**) CCAAT/enhancer binding protein alpha (C/EBPα). (**d**,**h**) Fatty acid synthase (FAS). All data are presented as the mean ± SE, $n = 6$ and represent results from three independent experiments. * $p < 0.05$, ** $p < 0.01$, *** $p < 0.001$, and **** $p < 0.0001$.

3.3. S and SS Activated Lipid Metabolism and AMPK Phosphorylation in db/db Mice

The expression genes linked to fatty acid oxidation, such as PPARα and CPT1, were evaluated in the livers of *db/db* mice following S and SS treatment. Immunoblot analysis revealed enhanced levels of CPT1 and PPARα protein in response to S and SS treatment compared to those in the NC group (Figure 3a,b). In addition, the NC group showed a significant reduction in AMPK phosphorylation, which showed an increasing trend following S and SS treatment and was the highest in the PC group administered metformin (Figure 3c).

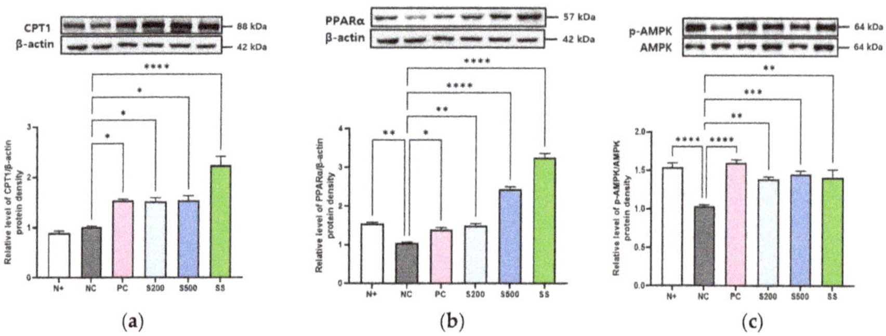

Figure 3. Stevia and stevioside activate fatty acid oxidation-related proteins in *db/db* mice. Mice were orally administered saline (N+ and NC), saline with 200 mg/kg/day metformin (PC), saline with 200 or 500 mg/kg/day stevia (S200 and S500), and saline with 40 mM/kg/day stevioside (SS) for 3 weeks. (**a**) Carnitine palmitoyl transferase-1 (CPT-1). (**b**) Peroxisome proliferator-activated receptor alpha (PPARα). (**c**) Adenosine monophosphate-activated protein kinase (AMPK). All data are presented as the mean ± SE, $n = 3$ and represent results from three independent experiments. * $p < 0.05$, ** $p < 0.01$, *** $p < 0.001$, and **** $p < 0.0001$.

3.4. S- and SS-Induced Autophagy in Liver of db/db Mice

Next, we analyzed whether S and SS regulate autophagy to mediate intracellular lipid storage in the *db/db* mice liver tissue. We found that the expression of microtubule-associated protein light chain 3 B (LC3B)-II/LC3B-I was significantly increased in the S- (S200 ($p < 0.01$) and S500 ($p < 0.1$)) and SS-treated ($p < 0.0001$) groups compared to that in the NC group, whereas SQSTM1 (p62 and sequestosome1) was significantly downregulated in the S- (S200 ($p < 0.001$) and S500 ($p < 0.0001$)) and SS-treated ($p < 0.0001$) groups (Figure 4a,b).

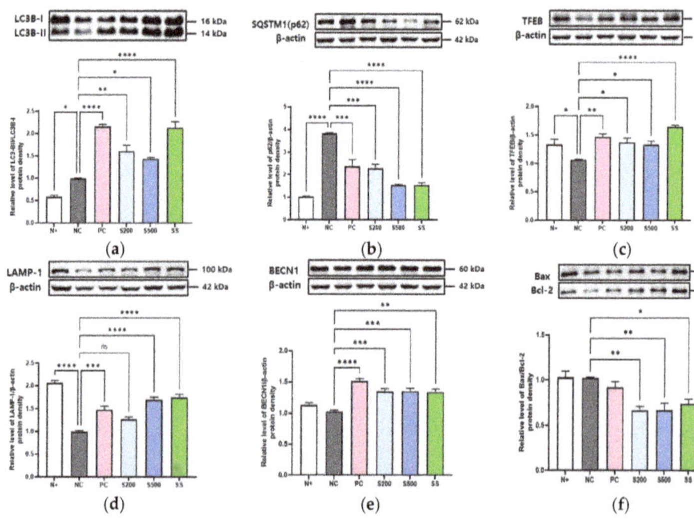

Figure 4. Stevia and stevioside induce autophagy in hepatocytes of *db/db* mice. Mice were orally administered saline (N+ and NC), saline with 200 mg/kg/day metformin (PC), saline with 200 or 500 mg/kg/day stevia (S200 and S500), and saline with 40 mM/kg/day stevioside (SS) for 3 weeks. (**a**) Microtubule-associated protein light chain 3B (LC3B)-II/LC3B-I. (**b**) Sequestosome1 (SQSTM1). (**c**) Transcription factor EB (TFEB). (**d**) Lysosomal associated membrane protein 1 (LAMP-1). (**e**) Beclin 1 (BECN1). (**f**) Bax/Bcl-2. All data are presented as the mean ± SE, $n = 3$ and represent results from three independent experiments. * $p < 0.05$, ** $p < 0.01$, *** $p < 0.001$, and **** $p < 0.0001$.

Analysis of the lysosome-related proteins, transcription factor EB (TFEB), and lysosome-associated membrane protein 1 (LAMP-1) revealed that TFEB expression was lower in the livers of *db/db* mice compared with in the normal (N+) group, although oral S and SS administration for three weeks markedly enhanced the relative levels of TFEB/β-actin protein (Figure 4c). Similar results were observed for LAMP-1 expression by immunoblot analysis (Figure 4d). The expression of Beclin 1 (BECN1) protein, which promotes crosstalk between apoptosis and autophagy, was increased in the livers of the S200, S500, and SS groups relative to that in the *db/db* mice (NC) group (Figure 4e). Furthermore, the relative Bax/Bcl-2 protein levels were lower in the S200, S500, and SS groups than those in the NC group (Figure 4f).

3.5. SS-Induced Autophagy in Steatosis-Induced Hepatocytes

Different concentrations of SS (0, 12.5, 25, 50, and 100 μM) and 1 mM FFA were applied to HepG2 cells for 24 h. SS treatment augmented LC3B-II/LC3B-I levels and reduced SQSTM1 (p62) levels in a dose-dependent manner (Figure 5a,b).

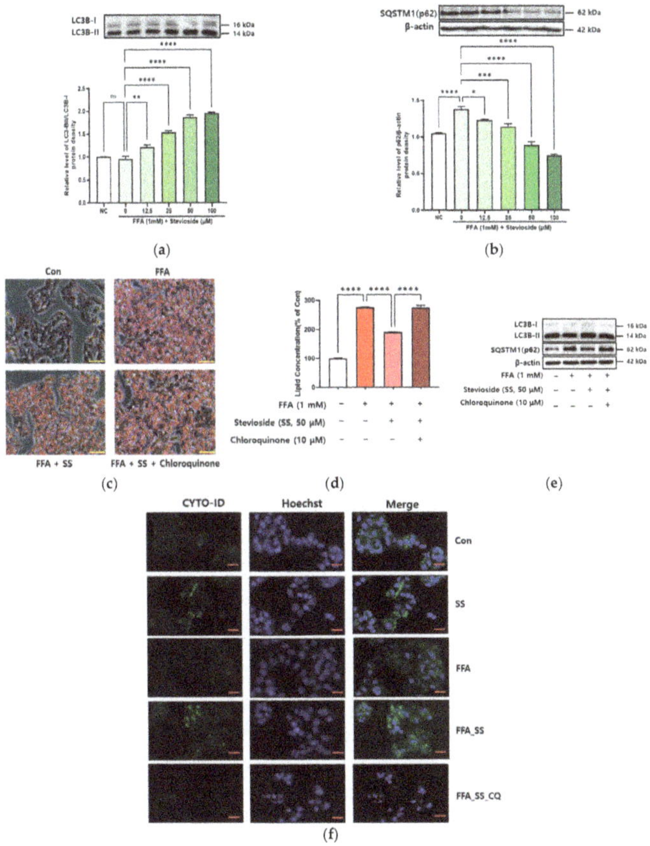

Figure 5. Stevioside induces autophagy in steatosis-induced hepatocytes. HepG2 cells were treated with 1 mM FFA and different concentrations of stevioside (SS; 0, 12.5, 25, 50, or 100 μM) in the presence or absence of 10 μM chloroquine (CQ) for 24 h. (**a**) Microtubule-associated protein light chain 3B (LC3B)-II/LC3B-I. (**b**) Sequestosome1 (SQSTM1). (**c**) HepG2 cells were stained with Oil Red O. Scale bar, 100 μm. (**d**) Intracellular lipid accumulation (the absorbance of lipids was measured at 500 nm). (**e**) Immunoblot analysis. (**f**) Cells were observed by fluorescence microscopy. Scale bar, 10 μm. All data are presented as the mean ± SE and represent results from three independent experiments. * $p < 0.05$, ** $p < 0.01$, *** $p < 0.001$, and **** $p < 0.0001$.

SS-treated HepG2 cells were treated with chloroquine (CQ), an autophagy inhibitor that prevents autophagosomes from attaching to lysosomes, and Oil Red O staining and immunoassays were used to assess increasing lipophagy (Figure 5c–e). HepG2 cells treated with FFA showed increased lipid concentration compared to the control (Con).

After FFA and SS (50 µM) treatment, the lipid concentration was decreased compared to that in the Con (Figure 5d). The lipid concentration was increased following FFA treatment after CQ treatment showed the same results with decreased LC3B-II/LC3B-I levels and increased SQSTM1 (p62) levels (Figure 5e). Next, the effect of SS on autophagy was evaluated in steatosis-induced HepG2 cells using fluorescence microscopy. The presence of autophagic green vacuoles was assessed using an autophagy detection kit in HepG2 cells treated with SS, or FFA and SS (Figure 5f). The cells were also stained with the nuclear stain Hoechst 33342. As shown in Figure 5f, SS treatment markedly increased the number of green autophagic vacuoles throughout the cells.

3.6. SS-Induced Autophagy Is Dependent on PPARα in Hepatocytes

To test our hypothesis that PPAR regulates lipophagy in hepatocytes, we used siRNAs to silence the PPAR gene in HepG2 cells. HepG2 cells transfected with the nonspecific (NS) siRNA showed significant autophagy induction following SS treatment, whereas cells with PPARα knockdown did not show any significant change in autophagy induction following SS treatment (Figure 6).

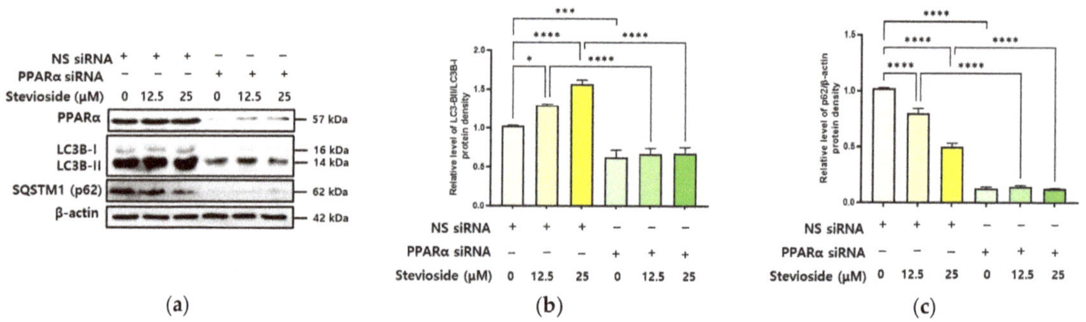

Figure 6. Stevioside-induced autophagy is dependent on PPARα in hepatocytes. HepG2 cells were transfected with the nonspecific (NS) siRNA or peroxisome proliferator-activated receptor α (PPARα) siRNA. After 24 h, the cells were treated with different concentration of stevioside (SS; 0, 12.5, and 25 µM) for another 24 h. (**a**) Immunoblot analysis of PPARα, microtubule-associated protein light chain 3B (LC3B)-II/LC3B-I, sequestosome1 (SQSTM1), and β-actin. (**b**) Relative expression of LC3B-II/LC3B-I. (**c**) Relative expression of SQSTM1(p62)/β-actin. All data are presented as the mean ± SE and represent results from three independent experiments. * $p < 0.05$, *** $p < 0.001$, and **** $p < 0.0001$.

4. Discussion

In this study, we provide in vivo and in vitro evidence that S and SS induce autophagy and that the lysosomal pathway reduces liver steatosis and hepatic lipid gene expression in *db/db* mice and HepG2 cells. Furthermore, oral S and SS treatment reduces LDs and triggers fatty acid oxidation in the livers of *db/db* mice, which is controlled by autophagy and the lysosomal pathway. We also verified that PPAR plays a critical role in lipophagy, which reduces hepatocyte steatosis.

Appropriate amounts of lipids are essential for cellular functions and cell survival [12]. The finding that macrophages can break down some of the lipids in liver cells has opened up the possibility of controlling the pathologies associated with the lipid metabolism [22]. Lipid metabolism in the liver is associated with the plasma lipid levels, lipid synthesis, and lipid exports from the liver [23]. Increased LD levels in hepatocytes do not always cause cellular dysfunction [24].

Lipid metabolism regulates various cellular processes to produce energy or structural components in the cell membrane that regulate multiple cellular processes [25]. After oxidation to FFAs, LDs are destroyed via the mitochondrial fatty acid oxidation pathway to produce ATP (energy house) and meet the energy demands of rapidly growing cells [12,26].

In mice lacking leptin receptors (*db/db*), the levels of lipogenesis markers and LDs were significantly higher following a high-calorie intake for three weeks than those in normal control mice (Figures 1 and 2). However, lipophagy and liver-specific AMPK activation were markedly lower than that in the other groups (Figures 3 and 4), which differed significantly from the results of oral administration of S or SS in this study.

The transcriptional mechanism that links autophagy and the lipid metabolism is associated with the activation of nuclear TFEB [27]. Excessive expression of TFEB promotes the decomposition of autophagy substrates and immobilizes LDs and damaged mitochondria [28]. In an obese mouse model, TFEB-knockout resulted in lipid metabolism disorders and metabolic pathway imbalances [29]. TFEB controls genes involved in lipid metabolism through peroxisome proliferator-activated receptor gamma coactivator-1 alpha (PGC-1α) and PPARα, indicating that PGC-1α regulates lipid metabolism in the liver by controlling the activity of PPARα during starvation [29].

Fatty acids and their derivatives activate transcription factors, such as PPARs, helping to regulate the expression of genes [30]. Three subtypes of PPARs (PPARβ/δ, PPARα, and PPARγ) have been identified in mammals based on their tissue-specific expression patterns. PPARα is the most prevalent subtype in hepatocytes and is implicated in various lipid metabolic pathways [31]. PPARα is profusely expressed in tissues with high capabilities for fatty acid oxidation, such as the liver, heart, and skeletal muscles. This is an essential factor in converting and utilizing energy, particularly during fasting, and serves as the primary regulator of fatty acid homeostasis [32].

SS is the principal sweetening component of the *Stevia rebaudiana* leaf and constitutes 4–20% of the leaf (dry weight basis), depending on the cultivar and growing conditions [33]. In both animals and humans, SS is absorbed and metabolized without being degraded by digestive enzymes. SS is metabolized in the colon by the intestinal flora and absorbed into the portal vein by the colon wall as S, which is then partially transported to the liver, filtered by the kidneys, and eliminated in the urine as steviol glucuronide [34]. As the human intestinal enzymes cannot cleave the steviol structure, they are released without retention in the body. Therefore, steviol glycoside metabolites rarely accumulate in humans.

Several epidemiological studies have shown that consuming plant-based foods is beneficial in alleviating fatty liver diseases, including NAFLD [35–37]. Some fruits and their bioactive compounds ameliorate fatty liver disease by promoting the inhibition of apoptosis, inflammation, and oxidative stress and alleviate hepatic steatosis by regulating AMPK and SIRT1 signaling [38–41]. Specific plant-based foods with bioactive compounds are natural sources that prevent and alleviate fatty liver disease [42]. In addition, herbal bioactive compounds and medicinal plants complement a healthy lifestyle and appear to have several advantages in improving oxidative stress, cell inflammation, and insulin resistance in NAFLD treatment [43].

In this study, SS, a natural food compound, caused PPARα-mediated autophagy, which relieved fatty acid buildup in the liver, increased β-oxidation, and alleviated hepatic steatosis through the activation of lipophagy. Though many studies have found a connection between lipophagy and NAFLD, there is still some disagreement over the precise function of (members of the autophagy/lipophagy pathways) in NAFLD [44,45]. Pharmacological treatment has been shown to activate PPARα, which reverses the normal regulation of autophagy in the fed state and triggers lipophagy or autophagic lipid destruction [46]. Therefore, our goal was to determine whether or not PPAR was involved in the process of S and SS, causing an autophagic lipid degradation state (lipophagy).

According to studies, higher TFEB expression can boost lipophagy, which, in turn, can increase hepatic lipid catabolism and the beta-oxidation of fatty acids. Lysosomal biogenesis and autophagy are both regulated by TFEB [31]. A previous human investigation discovered that human liver tissues with both simple steatosis and steatohepatitis had lower TFEB expression than did the healthy controls [47]. In contrast to the control, the expression of this master regulator was elevated in our study after S and SS treatments.

As can be observed in graphic abstract, increased expression of LC3-1 proteins can identify big lipid droplets (LD), attract them (often referred to as cargo), and help them fuse with the autophagolysosome membrane. Although p62 connects the cargo with the autophagosome, which makes it necessary for autophagy and lipophagy, increased levels typically signify accumulation or aggregation due to reduced autophagy [48]. A considerably lower expression of p62 was observed in our study after the treatment of S and SS. In the presence of LAMP-1, the newly formed autophagolysosome unites with the lysosome, playing a crucial part in this fusion process that creates the autolysosome.

The liver content of LDs, TG, and TC in *db/db* mice treated with S and SS was lowered because the lysosomal acid lipase in the autolysosome converted TG and TC into free fatty acids. Free fatty acid then underwent beta-oxidation as indicated by enhanced PPARα and CPT-1 and decreased expression of SREBP-1c, FAS, PPARγ, and C/EBPα. Furthermore, the knockdown of PPARα blocked lipophagy in response to SS treatment in HepG2 cells. Beclin 1 function has been reported to be a cross regulator between autophagy and apoptosis. According to Kang and coworkers, Beclin 1 often interacts with a number of cofactors and allows for the creation of specific complexes that cause autophagy [49].

5. Conclusions

In the current study, we showed novel beneficial effects of S and SS, which ameliorate hepatic steatosis through lipophagy activation in the liver of *db/db* mice. Furthermore, the inhibition of PPARα, which plays an imperative role in fatty acid homeostasis in the liver, blocked the effects of SS. Our findings suggest that SS, a new lipophagy enhancer, represents a viable therapeutic option for hepatic steatosis.

Author Contributions: Investigation and data arrangement, M.P. and J.-Y.H.; Original draft preparation and visualization, M.P., A.S. and H.B.; Data analysis, M.P. and J.Y.; Review and editing, H.-J.L.; Conceptualization and supervision, H.-J.L. All authors have read and agreed to the published version of the manuscript.

Funding: This work was supported by the "Cooperative Research Program of the Center for Companion Animal Research (Project No. PJ01476703)", Rural Development Administration, Republic of Korea.

Institutional Review Board Statement: All studies using the chosen mouse models were approved by Gachon University (GI-ACUC-R2020012) and performed in compliance with the guidelines of the Guide for the Care and Use of Laboratory Animals.

Informed Consent Statement: Not applicable.

Data Availability Statement: All of the data is contained within the article.

Conflicts of Interest: The authors declare no conflict of interest.

References

1. Abd El-Kader, S.M.; El-Den Ashmawy, E.M.S. Non-alcoholic fatty liver disease: The diagnosis and management. *World J. Hepatol.* **2015**, *7*, 846–858. [CrossRef]
2. Nassir, F.; Rector, R.S.; Hammoud, G.M.; Ibdah, J.A. Pathogenesis and Prevention of Hepatic Steatosis. *Gastroenterol. Hepatol.* **2015**, *11*, 167–175.
3. Benedict, M.; Zhang, X. Non-alcoholic fatty liver disease: An expanded review. *World J. Hepatol.* **2017**, *9*, 715–732. [CrossRef]
4. Portillo-Sanchez, P.; Bril, F.; Maximos, M.; Lomonaco, R.; Biernacki, D.; Orsak, B.; Subbarayan, S.; Webb, A.; Hecht, J.; Cusi, K. High Prevalence of Nonalcoholic Fatty Liver Disease in Patients With Type 2 Diabetes Mellitus and Normal Plasma Aminotransferase Levels. *J. Clin. Endocrinol. Metab.* **2015**, *100*, 2231–2238. [CrossRef]
5. Mao, Y.; Yu, F.; Wang, J.; Guo, C.; Fan, X. Autophagy: A new target for nonalcoholic fatty liver disease therapy. *Hepatic Med. Evid. Res.* **2016**, *8*, 27–37. [CrossRef]
6. Sinha, R.A.; Farah, B.L.; Singh, B.K.; Siddique, M.M.; Li, Y.; Wu, Y.; Ilkayeva, O.R.; Gooding, J.; Ching, J.; Zhou, J.; et al. Caffeine stimulates hepatic lipid metabolism by the autophagy-lysosomal pathway in mice. *Hepatology* **2014**, *59*, 1366–1380. [CrossRef]

7. Rogov, V.; Dötsch, V.; Johansen, T.; Kirkin, V. Interactions between autophagy receptors and ubiquitin-like proteins form the molecular basis for selective autophagy. *Mol. Cell* **2014**, *53*, 167–178. [CrossRef]
8. Cole, N.B.; Murphy, D.D.; Grider, T.; Rueter, S.; Brasaemle, D.; Nussbaum, R.L. Lipid droplet binding and oligomerization properties of the Parkinson's disease protein alpha-synuclein. *J. Biol. Chem.* **2002**, *277*, 6344–6352. [CrossRef]
9. Filipe, A.; McLauchlan, J. Hepatitis C virus and lipid droplets: Finding a niche. *Trends Mol. Med.* **2015**, *21*, 34–42. [CrossRef]
10. Martinez-Lopez, N.; Singh, R. Autophagy and Lipid Droplets in the Liver. *Annu. Rev. Nutr.* **2015**, *35*, 215–237. [CrossRef]
11. Kounakis, K.; Chaniotakis, M.; Markaki, M.; Tavernarakis, N. Emerging Roles of Lipophagy in Health and Disease. *Front. Cell Dev. Biol.* **2019**, *7*, 185. [CrossRef]
12. Zhang, S.; Peng, X.; Yang, S.; Li, X.; Huang, M.; Wei, S.; Liu, J.; He, G.; Zheng, H.; Yang, L.; et al. The regulation, function, and role of lipophagy, a form of selective autophagy, in metabolic disorders. *Cell Death Dis.* **2022**, *13*, 132. [CrossRef]
13. Salehi, B.; López, M.D.; Martínez-López, S.; Victoriano, M.; Sharifi-Rad, J.; Martorell, M.; Rodrigues, C.F.; Martins, N. *Stevia rebaudiana* Bertoni bioactive effects: From in vivo to clinical trials towards future therapeutic approaches. *Phytother. Res. PTR* **2019**, *33*, 2904–2917. [CrossRef]
14. Iatridis, N.; Kougioumtzi, A.; Vlataki, K.; Papadaki, S.; Magklara, A. Anti-Cancer Properties of *Stevia rebaudiana*; More than a Sweetener. *Molecules* **2022**, *27*, 1362. [CrossRef]
15. Purkayastha, S.; Markosyan, A.; Prakash, I.; Bhusari, S.; Pugh, G., Jr.; Lynch, B.; Roberts, A. Steviol glycosides in purified stevia leaf extract sharing the same metabolic fate. *Regul. Toxicol. Pharmacol. RTP* **2016**, *77*, 125–133. [CrossRef]
16. González, C.; Tapia, M.; Pérez, E.; Pallet, D.; Dornier, M. Main properties of steviol glycosides and their potential in the food industry: A review. *Fruits* **2014**, *69*, 127–141. [CrossRef]
17. Alavala, S.; Nalban, N.; Sangaraju, R.; Kuncha, M.; Jerald, M.K.; Kilari, E.K.; Sistla, R. Anti-inflammatory effect of stevioside abates Freund's complete adjuvant (FCA)-induced adjuvant arthritis in rats. *Inflammopharmacology* **2020**, *28*, 1579–1597. [CrossRef]
18. Pariwat, P.; Homvisasevongsa, S.; Muanprasat, C.; Chatsudthipong, V. A natural plant-derived dihydroisosteviol prevents cholera toxin-induced intestinal fluid secretion. *J. Pharmacol. Exp. Ther.* **2008**, *324*, 798–805. [CrossRef]
19. Ferri, L.A.; Alves-Do-Prado, W.; Yamada, S.S.; Gazola, S.; Batista, M.R.; Bazotte, R.B. Investigation of the antihypertensive effect of oral crude stevioside in patients with mild essential hypertension. *Phytother. Res. PTR* **2006**, *20*, 732–736. [CrossRef]
20. Ahmad, U.; Ahmad, R.S.; Arshad, M.S.; Mushtaq, Z.; Hussain, S.M.; Hameed, A. Antihyperlipidemic efficacy of aqueous extract of *Stevia rebaudiana* Bertoni in albino rats. *Lipids Health Dis.* **2018**, *17*, 175. [CrossRef]
21. Nishina, P.M.; Lowe, S.; Wang, J.; Paigen, B. Characterization of plasma lipids in genetically obese mice: The mutants obese, diabetes, fat, tubby, and lethal yellow. *Metabolism* **1994**, *43*, 549–553. [CrossRef]
22. Parzych, K.R.; Klionsky, D.J. An overview of autophagy: Morphology, mechanism, and regulation. *Antioxid. Redox Signal.* **2014**, *20*, 460–473. [CrossRef]
23. Nguyen, P.; Leray, V.; Diez, M.; Serisier, S.; Bloc'h, J.L.; Siliart, B.; Dumon, H. Liver lipid metabolism. *J. Anim. Physiol. Anim. Nutr.* **2008**, *92*, 272–283. [CrossRef]
24. Mashek, D.G. Hepatic lipid droplets: A balancing act between energy storage and metabolic dysfunction in NAFLD. *Mol. Metab.* **2021**, *50*, 101115. [CrossRef]
25. Xie, Y.; Li, J.; Kang, R.; Tang, D. Interplay Between Lipid Metabolism and Autophagy. *Front. Cell Dev. Biol.* **2020**, *8*, 431. [CrossRef]
26. Onal, G.; Kutlu, O.; Gozuacik, D.; Dokmeci Emre, S. Lipid Droplets in Health and Disease. *Lipids Health Dis.* **2017**, *16*, 128. [CrossRef]
27. Li, X.; Zhang, X.; Zheng, L.; Kou, J.; Zhong, Z.; Jiang, Y.; Wang, W.; Dong, Z.; Liu, Z.; Han, X.; et al. Hypericin-mediated sonodynamic therapy induces autophagy and decreases lipids in THP-1 macrophage by promoting ROS-dependent nuclear translocation of TFEB. *Cell Death Dis.* **2016**, *7*, e2527. [CrossRef]
28. Moruno-Manchon, J.F.; Uzor, N.E.; Kesler, S.R.; Wefel, J.S.; Townley, D.M.; Nagaraja, A.S.; Pradeep, S.; Mangala, L.S.; Sood, A.K.; Tsvetkov, A.S. TFEB ameliorates the impairment of the autophagy-lysosome pathway in neurons induced by doxorubicin. *Aging* **2016**, *8*, 3507–3519. [CrossRef]
29. Settembre, C.; De Cegli, R.; Mansueto, G.; Saha, P.K.; Vetrini, F.; Visvikis, O.; Huynh, T.; Carissimo, A.; Palmer, D.; Klisch, T.J.; et al. TFEB controls cellular lipid metabolism through a starvation-induced autoregulatory loop. *Nat. Cell Biol.* **2013**, *15*, 647–658. [CrossRef]
30. Montagner, A.; Polizzi, A.; Fouché, E.; Ducheix, S.; Lippi, Y.; Lasserre, F.; Barquissau, V.; Régnier, M.; Lukowicz, C.; Benhamed, F.; et al. Liver PPARα is crucial for whole-body fatty acid homeostasis and is protective against NAFLD. *Gut* **2016**, *65*, 1202–1214. [CrossRef]
31. Sinha, R.A.; Rajak, S.; Singh, B.K.; Yen, P.M. Hepatic Lipid Catabolism via PPARα-Lysosomal Crosstalk. *Int. J. Mol. Sci.* **2020**, *21*, 2391. [CrossRef]
32. Feige, J.N.; Gelman, L.; Michalik, L.; Desvergne, B.; Wahli, W. From molecular action to physiological outputs: Peroxisome proliferator-activated receptors are nuclear receptors at the crossroads of key cellular functions. *Prog. Lipid Res.* **2006**, *45*, 120–159. [CrossRef] [PubMed]
33. Geuns, J.M. Stevioside. *Phytochemistry* **2003**, *64*, 913–921. [CrossRef] [PubMed]
34. Li, Y.; Zhu, W.; Cai, J.; Liu, W.; Akihisa, T.; Li, W.; Kikuchi, T.; Xu, J.; Feng, F.; Zhang, J. The role of metabolites of steviol glycosides and their glucosylated derivatives against diabetes-related metabolic disorders. *Food Funct.* **2021**, *12*, 8248–8259. [CrossRef]

35. Xia, H.M.; Wang, J.; Xie, X.J.; Xu, L.J.; Tang, S.Q. Green tea polyphenols attenuate hepatic steatosis, and reduce insulin resistance and inflammation in high-fat diet-induced rats. *Int. J. Mol. Med.* **2019**, *44*, 1523–1530. [CrossRef] [PubMed]
36. Yang, Z.; Zhu, M.-Z.; Zhang, Y.-B.; Wen, B.-B.; An, H.-M.; Ou, X.-C.; Xiong, Y.-F.; Lin, H.-Y.; Liu, Z.-H.; Huang, J.-A. Coadministration of epigallocatechin-3-gallate (EGCG) and caffeine in low dose ameliorates obesity and nonalcoholic fatty liver disease in obese rats. *Phytother. Res.* **2019**, *33*, 1019–1026. [CrossRef] [PubMed]
37. Khoo, W.Y.; Chrisfield, B.J.; Sae-tan, S.; Lambert, J.D. Mitigation of nonalcoholic fatty liver disease in high-fat-fed mice by the combination of decaffeinated green tea extract and voluntary exercise. *J. Nutr. Biochem.* **2020**, *76*, 108262. [CrossRef]
38. Park, M.; Yoo, J.H.; Lee, Y.S.; Lee, H.J. Lonicera caerulea Extract Attenuates Non-Alcoholic Fatty Liver Disease in Free Fatty Acid-Induced HepG2 Hepatocytes and in High Fat Diet-Fed Mice. *Nutrients* **2019**, *11*, 494. [CrossRef]
39. Park, M.; Yoo, J.H.; Lee, Y.S.; Park, E.J.; Lee, H.J. Ameliorative effects of black ginseng on nonalcoholic fatty liver disease in free fatty acid-induced HepG2 cells and high-fat/high-fructose diet-fed mice. *J. Ginseng. Res.* **2020**, *44*, 350–361. [CrossRef]
40. Liu, Q.; Pan, R.; Ding, L.; Zhang, F.; Hu, L.; Ding, B.; Zhu, L.; Xia, Y.; Dou, X. Rutin exhibits hepatoprotective effects in a mouse model of non-alcoholic fatty liver disease by reducing hepatic lipid levels and mitigating lipid-induced oxidative injuries. *Int. Immunopharmacol.* **2017**, *49*, 132–141. [CrossRef]
41. Gao, J.; Chen, S.; Qiu, Z.; Fang, L.; Zhang, L.; Guo, C.; Chen, T.; Qiu, L. Myricitrin ameliorates ethanol-induced steatosis in mouse AML12 liver cells by activating AMPK, and reducing oxidative stress and expression of inflammatory cytokines. *Mol. Med. Rep.* **2018**, *17*, 7381–7387. [CrossRef] [PubMed]
42. Li, H.-Y.; Gan, R.-Y.; Shang, A.; Mao, Q.-Q.; Sun, Q.-C.; Wu, D.-T.; Geng, F.; He, X.-Q.; Li, H.-B. Plant-Based Foods and Their Bioactive Compounds on Fatty Liver Disease: Effects, Mechanisms, and Clinical Application. *Oxid. Med. Cell Longev.* **2021**, *2021*, 6621644. [CrossRef] [PubMed]
43. Bagherniya, M.; Nobili, V.; Blesso, C.N.; Sahebkar, A. Medicinal plants and bioactive natural compounds in the treatment of non-alcoholic fatty liver disease: A clinical review. *Pharmacol. Res.* **2018**, *130*, 213–240. [CrossRef]
44. Allaire, M.; Rautou, P.E.; Codogno, P.; Lotersztajn, S. Autophagy in liver diseases: Time for translation? *J. Hepatol.* **2019**, *70*, 985–998. [CrossRef] [PubMed]
45. Kwanten, W.J.; Martinet, W.; Michielsen, P.P.; Francque, S.M. Role of autophagy in the pathophysiology of nonalcoholic fatty liver disease: A controversial issue. *World J. Gastroenterol.* **2014**, *20*, 7325–7338. [CrossRef] [PubMed]
46. Lee, J.M.; Wagner, M.; Xiao, R.; Kim, K.H.; Feng, D.; Lazar, M.A.; Moore, D.D. Nutrient-sensing nuclear receptors coordinate autophagy. *Nature* **2014**, *516*, 112–115. [CrossRef]
47. Kim, S.H.; Kim, G.; Han, D.H.; Lee, M.; Kim, I.; Kim, B.; Kim, K.H.; Song, Y.M.; Yoo, J.E.; Wang, H.J.; et al. Ezetimibe ameliorates steatohepatitis via AMP activated protein kinase-TFEB-mediated activation of autophagy and NLRP3 inflammasome inhibition. *Autophagy* **2017**, *13*, 1767–1781. [CrossRef] [PubMed]
48. Rusten, T.E.; Stenmark, H. p62, an autophagy hero or culprit? *Nat. Cell Biol.* **2010**, *12*, 207–209. [CrossRef]
49. Kang, R.; Zeh, H.J.; Lotze, M.T.; Tang, D. The Beclin 1 network regulates autophagy and apoptosis. *Cell Death Differ.* **2011**, *18*, 571–580. [CrossRef]

antioxidants

MDPI

Review

Hydrogen Sulfide: A Key Role in Autophagy Regulation from Plants to Mammalians

Angeles Aroca * and Cecilia Gotor

Institute of Plant Biochemistry and Photosynthesis, University of Seville and CSIC, 41092 Seville, Spain; gotor@ibvf.csic.es
* Correspondence: aaroca@us.es

Abstract: Autophagy is a degradative conserved process in eukaryotes to recycle unwanted cellular protein aggregates and damaged organelles. Autophagy plays an important role under normal physiological conditions in multiple biological processes, but it is induced under cellular stress. Therefore, it needs to be tightly regulated to respond to different cellular stimuli. In this review, the regulation of autophagy by hydrogen sulfide is described in both animal and plant systems. The underlying mechanism of action of sulfide is deciphered as the persulfidation of specific targets, regulating the pro- or anti-autophagic role of sulfide with a cell survival outcome. This review aims to highlight the importance of sulfide and persulfidation in autophagy regulation comparing the knowledge available in mammals and plants.

Keywords: autophagy; autophagy-related genes (ATG); hydrogen sulfide; persulfidation

Citation: Aroca, A.; Gotor, C. Hydrogen Sulfide: A Key Role in Autophagy Regulation from Plants to Mammalians. *Antioxidants* **2022**, *11*, 327. https://doi.org/10.3390/antiox11020327

Academic Editor: Stanley Omaye

Received: 13 January 2022
Accepted: 6 February 2022
Published: 8 February 2022

1. Introduction

The term "autophagy", (from the Greek words *auto*, meaning "self" and *phagein*, meaning "to eat")—literally, eating one's self—was first created by Christian de Duve over 40 years ago, who discovered lysosomes and provided clear proof of their participation in this process [1]. It is an evolutionarily conserved process of degradation and recycling in eukaryotic organisms. Two common forms of autophagy have been described in mammals and plants: micro-autophagy and macro-autophagy, while they differ in a third type of autophagy described, chaperone-mediated autophagy (in mammals) and mega-autophagy (in plants) [2–6]. The differences among them have been previously described in detail elsewhere [7,8], and this review will focus on macro-autophagy (hereafter, autophagy). In this latest process, the cytoplasm and/or organelles are isolated in double membrane vesicles—named autophagosomes—and then transported to the lytic organelle (vacuole in plants and yeast, and lysosome in animals) to be degraded, resulting in the turnover of cellular components. Therefore, autophagy is a fundamental cell clearance pathway that eliminates cellular components, including nucleic acids, proteins, lipids, and organelles, to promote homeostasis, differentiation, development and cell survival.

Autophagy is a unique membrane trafficking process that involves the de novo formation of a membrane, which is generally derived from the endoplasmic reticulum (ER) by generating a double membrane structure called phagophore that elongates to sequester cytoplasmic cargo and closes to form the autophagosome [6,9].

The molecular process of autophagy was mostly unknown until 1993, when Yoshinori Oshumi described a genetic screen in yeast, leading to the discovery of AuTophagy-related Genes (ATG) [10]. 41 yeast ATG genes have been described, and many of them have orthologues in other organisms such as humans and plants.

The autophagy core process in mammals is induced in response to stress by inhibiting the mammalian kinase target of rapamycin (mTOR) or activating 5′ AMP-activated protein kinase (AMPK). In mammals, different stress stimuli can trigger autophagy, such as protein misfolding, hypoxia, nutritional and energy deficiency, ER stress, redox stress, mitochondrial damage, and pathogen infection [11]. Dysregulated autophagy plays an important role in many pathological processes, including ischemia-reperfusion injury, inflammatory and infectious diseases, obesity and type 2 diabetes, cancer and neurodegenerative diseases [12–14].

In plants, inhibition of TOR, usually induced by starvation of nutrients such as nitrogen starvation, is the main pathway that triggers autophagy. In addition, it can also be regulated by repression of glucose signaling, activating the energy sensor Snf1-related protein kinase 1 (SnRK1), which in turn inhibits TOR and activates the ATG1 autophagy initiation complex. The function of AMPK in plant autophagy remains largely unknown, although a plant ortholog of mammalian AMPK, named KIN10, was described as a positive regulator of plant autophagy [15]. In plant cells, autophagy is triggered by different biotic and abiotic stresses such as oxidative stress, salinity, hypoxia, heat and cold, nutrient starvation, ER stress and pathogen infection. Therefore, autophagy is essential for plants during reproductive and vegetative development, senescence, starvation, immune response and it is critical to cope with environmental stress [3,16,17]. Thus, autophagy must be tightly regulated to maintain cellular homeostasis.

Hydrogen sulfide (H_2S) is a colorless, flammable and highly toxic gas known for its rotten egg scent at low concentrations. It has always been considered a toxic pollutant that is found naturally in sewers, stagnant or well waters, compost pits, gas wells and volcanoes. However, it is also endogenously produced in cells by different enzymes.

H_2S is produced in animals by cystathionine β synthase (CBS, EC 4.2.1.22), cystathionine-γ-lyase (CSE, EC 4.4.1.1) and 3-mercaptopyruvate sulfurtransferase (3-MST, EC 2.8.1.2); these use cysteine or 3-mercaptopyruvate as substrates. The sulfate-reducing bacterial flora in the large intestine of animals also releases H_2S, reaching concentrations from 0.3 to 3.4 mmol L^{-1} in the colon [18,19].

In *Arabidopsis*, the plant species where the H_2S signaling has been deeply studied, H_2S is produced from cysteine by the action of L-cysteine desulfhydrases (DES1, EC 4.4.1.2; and L-CDES, EC 4.4.1.1), D-cysteine desulfhydrases (D-CDES, EC 4.4.1.15), cyanoalanine synthase (CAS, EC 4.4.1.9), cysteine synthase (CS, EC 4.2.99.8), NifS-like proteins and in the photosynthetic sulfate assimilation pathway by sulfite reductase (SiR, EC 1.8.7.1) [20].

Over the last decade, both in animal and plant systems, H_2S has been highlighted as a biological signaling molecule—namely, gasotransmitter—as important as other signal molecules such as nitric oxide (NO), carbon monoxide (CO) or hydrogen peroxide (H_2O_2) [21–23].

H_2S is already considered a physiological mediator involved in many physiological and pathological processes in animals and plants. Its regulatory function in mammals includes processes such as reducing inflammation [24], synaptic transmission [25], apoptosis [26], vascular tone [27], ischemia-reperfusion injury [28] and promoting ulcer healing [29] and protects cells from oxidative stress [30]. In plants, H_2S regulates a wide range of physiological processes, from seed germination to fruit maturation and the first description of its influence on vegetative development and disease resistance of plants dates from the late 1960s [31,32]. Today, the protective effects of H_2S against different stresses are widely known, such as drought [33], osmotic and saline stresses [34], heat [35], oxidative stress [36] and metal stresses [37]. In addition, H_2S regulates photosynthesis [38], stomatal closure/aperture [39,40] and autophagy [41–47].

2. Hydrogen Sulfide as a Regulator of Autophagy

2.1. The Anti-Autophagic Role of Sulfide in Plants

Over the last 10 years, there have been many studies on the effects of H_2S on autophagy in eukaryotic cells, but its mechanism has not been completely deciphered.

In plants, the role of H_2S in autophagy has been described as a protective effect towards a prosurvival outcome. By now, H_2S has been revealed as a negative regulator of autophagy induced by nutrient deficiency, carbon and nitrogen deprivation [41,43] and osmotic [46] and ER stress [45] (Figure 1). It was shown that only sulfide donor molecules, and no other compounds containing inorganic sulfur, are responsible for the inhibition of autophagy under nitrogen starvation in Arabidopsis roots [43]. Besides, sulfide signaling was dose-dependent, with an optimal NaHS (commonly used as sulfide generating molecule) concentration of 100–200 μM, with devastating effects at higher concentrations, inducing autophagy, probably due to its toxicity.

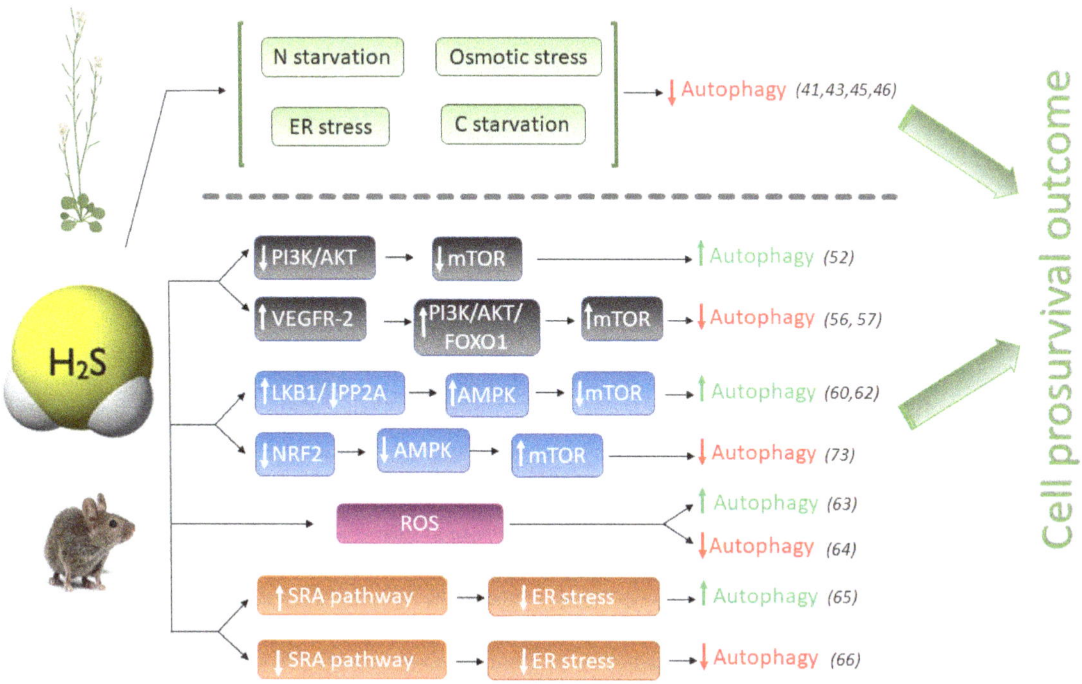

Figure 1. Schematic comparison of pro-autophagic and anti-autophagic effect of H_2S signaling in animal and plant systems. H_2S, hydrogen sulfide; LKB1, liver kinase B1; PP2A, Protein phosphatase 2; AMPK, adenosine monophosphate-activated protein kinase; mTOR, mammalian target of rapamycin; NRF2, Nuclear factor E2-related factor 2; Akt (PKB), protein kinase B; PI3K, phosphatidylinositol 3-kinase; FOXO1, Forkhead Box O1; VEGFR-2, Vascular endothelial growth factor receptor 2; SRA, Scavenger receptor A. Numbers between brackets refer to references cited.

The induction of autophagy by oxidative stress, especially during nutrient deprivation, and the ability of H_2S to activate the antioxidant response of plant cells, are well known. However, it was nicely shown that the negative regulation of autophagy through sulfide signaling was not dependent on its antioxidant activity, showing that hydrogen sulfide does not behave as an H_2O_2 or superoxide scavenger [43]. Furthermore, treatments with identical concentrations of antioxidant molecules such as glutathione and ascorbate were unable to produce the negative effect that sulfide treatment had on autophagy regulation; only NaHS treatment significantly inhibited autophagy [43].

Hydrogen sulfide has also been revealed to play a key role in autophagy during ER stress. Aggregation of misfolded proteins in the endoplasmic reticulum disrupts ER function, producing ER stress [48], which interferes with normal physiological functions of the cell. ER stress occurs when an increase of misfolded proteins accumulate in the ER which may be activated by different adverse environmental conditions such as cold or heat and pathogen infections in plants [49], or by several chemical and physiological situations such as glucose deprivation, hypoxia or genome instability in animals [50].

In a recent study, the effect of sulfide was also demonstrated to be independent from the antioxidant activity under ER stress, comparing the results observed using similar concentrations of sulfide, glutathione and ascorbate. Their results showed that when ER stress was induced with tunicamycin, no significant decrease in autophagosomes was detected upon well-established antioxidant compounds. By contrast, sulfide provoked a severe decrease of autophagosomes, indicating that the negative effect of sulfide is independent of redox conditions [45].

The anti-autophagic role of sulfide in *Arabidopsis* was also demonstrated under induced autophagy by carbon starvation, where Cys-generated sulfide in the cytosol was shown to regulate negatively autophagy and to modulate the transcriptional profile of Arabidopsis [41]. DES1, the L-Cys desulfhydrase protein located in the cytosol, catalyzes the desulfuration of L-Cys to sulfide plus ammonia and pyruvate. Consequently, the null mutant *des1-1* plants contain 30% less endogenous sulfide in leaves than WT plants. Mutant *des1-1* plants were shown to have induced autophagy under physiological conditions, and exogenous treatment with NaHS negatively regulated autophagy in this mutant background [41]. Moreover, sulfide was able to suppress autophagy induction caused by carbon starvation even in wild-type plants, whereas exogenous ammonia, also a product of DES1 activity, had no effect on carbon-induced autophagy. Therefore, it was concluded that sulfide exerts a general effect on autophagy unrelated to nutrition limitation stress.

In a different study sought to decipher the mechanism of action by which NaHS regulates autophagy, it was shown that abscisic acid (ABA) treatment induced the autophagic flux and that this induction was also repressed by NaHS [46]. One of the first plant responses to adverse environmental conditions is the increase of intracellular ABA content in order to activate downstream ABA-signaling pathway so as to help plants cope with the stress. In this situation, when plants successfully have overcome the adverse conditions and induced autophagy is not more required, NaHS repression prevents the over-activation of autophagy allowing to return back to levels in favorable growth conditions [44]. Therefore, in all studies reported up to now in plant systems, sulfide has an anti-autophagic role (Figure 1).

2.2. The Pro- or Anti-Autophagic Role of Sulfide in Mammals

However, the pro- or anti-autophagic role of H_2S in mammals has not always been completely clear, and several publications have shown that autophagy and H_2S could be a double-edged sword in cancer studies depending on the experimental settings. Hydrogen sulfide induces autophagy of hepatocellular carcinoma cells (HCC) by inhibiting the phosphatidylinositol 3-kinase (PI3K)/protein kinase B (AKT)/ mTOR (PI3K/Akt/mTOR) signaling pathway [51]. PI3Ks are a family of lipid kinases, which phosphorylate phosphoinositides that entail AKT recruitment to the cell membrane. AKT is an evolutionarily highly conserved serine/threonine protein kinase, considered one of the key downstream proteins of PI3K. mTOR is a conserved serine/threonine protein kinase and it is the catalytic core of two complexes: mTORC1 and mTORC2. Activation of the PI3K/Akt pathway further phosphorylates downstream regulators such as mTOR and the transcription factor Forkhead box O-1 (FoxO-1), upregulating the activity of mTOR complex 1 (mTORC1) that drives autophagy inhibition [52]. The PI3K/AKT signaling pathway is one of the upstream pathways that regulate mTOR. Suppression or dysfunction of PI3K can greatly block the downstream signaling pathways AKT and mTOR, and therefore the induction of autophagy (Figure 1).

NaHS treatment significantly inhibited the expression of phospho-PI3K, phospho-Akt, and mTOR proteins in HCC cells, mimicking the effect of rapamycin [51], and therefore activating autophagy. However, NaHS did not affect basal-level Akt phosphorylation in heart disease during ischemia, but further doubled myocardial Akt phosphorylation during reperfusion [53]. Zhou Y. et al., also found that NaHS enhances mTOR phosphorylation in both ischemic and reperfused hearts [53]. In another study, pretreatment of MC3T3-E1 osteoblasts with SDSS [a H_2S donor derived from β-(3,4-dihydroxyphenyl)lactic acid] stimulated Akt phosphorylation in a concentration-dependent manner [54]. H_2S also activates vascular endothelial growth factor 2 (VEGFR-2), which, in turn, activates the PI3K/AKT/FOXO-1 signaling pathway, with the opposite result of inhibition of autophagy [55,56] (Figure 1).

Thus, the role of NaHS activating the phosphorylation or dephosphorylation of the PI3K/AKT signaling pathway and the outcome of autophagy regulation has not been clearly deciphered in mammals. The different cell types, as well as the sulfide concentrations used in the experiments, were likely the consequence of different conclusions.

The adenosine monophosphate-activated protein kinase (AMPK) is involved in the regulation of metabolic energy balance, and several studies have implicated the AMPK/mTOR pathway in the regulation of autophagy. Numerous publications described the role of H_2S in activating autophagy through the AMPK/mTOR pathway, making this signaling a promising target for several diseases [18,57,58].

Another pro-autophagic effect of H_2S has been described in the regulation of the liver kinase B1 (LKB1)-AMPK signaling pathway [59]. LKB1 forms a heterotrimeric complex with the pseudokinase Ste20-related adaptor (STRAD) and the scaffolding mouse protein 25 (MO25), and this LKB1-STRAD-MO25 complex activates AMPK by phosphorylation [60]. Kundu et al. described that NaHS treatment in hyperglycemic cells increased LKB1/STRAD/MO25 complex assembly and therefore, AMPK phosphorylation, promoting autophagy [59] (Figure 1). However, a later study demonstrated that H_2S regulated AMPK phosphorylation through inhibition of protein phosphatase 2A (PP2A), and not through the LKB1/STRAD/MO25 complex [61], but sulfide still had a pro-autophagic role.

The protective effect of sulfide in several illnesses has been linked to its role promoting autophagy which may decrease ROS production. In a recent study in endothelial progenitor cells (EPCs), exogenous H_2S ameliorated the high glucose (HG)-induced injury by promoting autophagic flux and decreasing ROS production, demonstrating the protecting role of sulfide under this dysfunction [62]. Their findings demonstrated that the phosphorylation of the endothelial nitric oxide synthase (eNOS) at Thr495 determines whether this enzyme produces either NO or superoxide, and sulfide reduced the phosphorylation level of this enzyme, decreasing NO and ROS production [62].

But on the other hand, it is well known that oxidative stress may induce autophagy to protect cells from apoptosis. The effect of sulfide ameliorating oxidative stress has also been described in mice where it was demonstrated that GYY4137, a sulfide donor, attenuated the severity of lung injury by alleviating septicemia-induced ferroptosis and inhibiting the activation of autophagy in sepsis-induced acute lung injury [63]. Therefore, sulfide may play an anti-autophagic role by alleviating oxidative stress (Figure 1).

Prolonged ER stress has been associated with a wide range of diseases, including neurodegeneration, cancer, atherosclerosis, type 2 diabetes and liver disease. Autophagy is activated to remove dysfunctional proteins during ER stress. Several studies have connected the role of sulfide enhancing autophagy in reducing ER stress in mammals. In peritoneal macrophages of rats, hydrogen sulfide induces autophagy by suppressing the class A scavenger receptor (SRA) pathway (Figure 1). This cell response reduces ER-stress by inducing autophagy and protects against ischemia/reperfusion injury, maintaining liver function [64]. In other studies, NaHS treatment blocked ER stress and ER stress-associated autophagy [65].

During ER stress, H_2S has been reported to inhibit protein tyrosine phosphatase (PTP1B) [66]. PTP1B dephosphorylates PKR-like endoplasmic reticulum kinase (PERK), an ER stress sensor that autophosphorylates and induces the phosphorylation of eukaryotic initiation factor 2 alpha (eIF2α), which is necessary to mediate the induction of autophagy [67]. Therefore, it was concluded that exogenous H_2S, or induction of its endogenous synthesis, suppress the activation of PERK/eIF2α/ATF4 pathway and its subsequent effects on ER stress, which are an increased eIF2α phosphorylation [68], and the induction of autophagy. In addition, H_2S suppresses the expression of PKR-like endoplasmic reticulum kinase (PERK) [69], which induces autophagy.

Additionally, H_2S induces the activity of the transient receptor potential channel (TPRV4) and KATP channels, mediating angiogenesis and inducing vasodilation [70,71]. Through activation of TPRV4, H_2S also activates the AMPK/mTOR pathway, by this means reducing autophagy [72].

Hydrogen sulfide also exerts a cytoprotective role by upregulating cellular antioxidants by suppressing nuclear factor erythroid-2 related factor 2 (NRF2) [73]. NRF2 is a family of nuclear basic leucine zipper transcription factors that regulate the gene expression of a number of antioxidant enzymes. However, Nrf2 can also sense ROS and RNS in stressed cells, triggering the activation of AMPK, which suppresses mTOR and therefore induces autophagy [74]. Thus, NaHS could also inhibit excessive autophagy of vascular endothelial cells by the Nrf2/AMPK signaling pathway [72].

We can draw the conclusion that in mammals H_2S could play opposite effects, enhancing or decreasing autophagy induction, which may be attributed to the sulfide concentration, reaction time, cell types and/or differences among the diseases studied. The administration of exogenous H_2S in mammalian systems has also typically been performed at micromolar concentrations as in plants [75]. Higher doses of H_2S exposed in some publications lead to contradictory data.

However, in all cases, the final outcome of the role of H_2S is cell survival, which likewise has been described in plant systems. When stress is mild, in mammals H_2S often activates autophagy to protect cells, usually by reducing stress conditions, but with the progression of the disease, H_2S can act as a regulator inhibiting autophagy to avoid excess stress-induced autophagy and cell death.

3. Persulfidation as the Molecular Mechanism of Sulfide for Autophagy Regulation

During the past decade, the research of H_2S as a signaling molecule has been focused on the effect of sulfide donors on different diseases and physiological pathways, until in 2009 when Snyder's group described persulfidation or S-sulfhydration as the mechanism of H_2S signaling [76]. Since then, numerous targets have been identified to undergo persulfidation, and it has become recognized as the main mechanism by which H_2S controls several cellular functions. Persulfidation is a posttranslational modification of cysteine residues, where a thiol group (RSH) is transformed into a persulfide group (RSSH). Modified cysteines show greater reactivity than their thiol counterparts [23]. It has been proven that this new posttranslational oxidative modification can affect the subcellular localization of the modified protein [77], its activity [76,78] and stability, and it has been proposed to be a cellular mechanism to cope with oxidative stress [79,80]. Numerous studies have demonstrated that persulfidation is a widespread modification in animal and plant cells [76,81] involved in a huge range of biological processes [82], which explains the great interest of the scientific community in understanding this cell mechanism.

3.1. Regulation of Autophagy by Persulfidation in Plants

The role of persulfidation as the molecular mechanism of sulfide for autophagy regulation was first proposed in plants by Gotor's group [83] and then several autophagy-related core proteins were demonstrated to be targets for persulfidation [82,84]. Recently, the autophagy-related (ATG) proteins, ATG18a, ATG3, ATG5 and ATG7 were published to be targets for persulfidation identified by a quantitative proteomic study in *Arabidopsis* leaves [82]. Nevertheless, the role of persulfidation in those proteins was not deciphered.

In a very recent quantitative proteomic approach in Arabidopsis under nitrogen deprivation, more than 5200 proteins were identified as targets for persulfidation. In this work, authors extended the number of persulfidated proteins involved in autophagy. They found 17 proteins that play an essential role in core autophagy machinery were persulfidated; including ATG2, 3, 4, 5, 7, 11, 13, the serine/threonine kinase TARGET OF RAPAMYCIN (TOR), its effectors proteins REGULATORY-ASSOCIATED PROTEIN OF TOR 1 (RAPTOR 1) and LETHAL WITH SEC THIRTEEN PROTEIN 8 (LST8), five subunits of PP2A, the regulatory subunit of PP2A (TAP46) and the serine/threonine-protein kinase

VPS15 [84]. In addition, this study also revealed that other 58 proteins related to endocytosis and the formation of the phagophore were persulfidated, including several transporters and vacuolar sorting proteins.

The role of persulfidation as the underlying mechanism regulating autophagy through sulfide was demonstrated in the ATG4 protease, which was specifically modified by persulfidation of Cys170 residue, negatively regulating [46] (Figure 2). These authors established that persulfidation of ATG4 upon sulfide treatment inhibited its protease activity, disabling the progress of autophagy. They also revealed that an increase in the intracellular level of the plant hormone abscisic acid (ABA) triggered a decrease in the persulfidated ATG4 level; consequently, its protease activity was enhanced, activating the processing of ATG8, which was further lipidated, and, as a result, autophagy was induced.

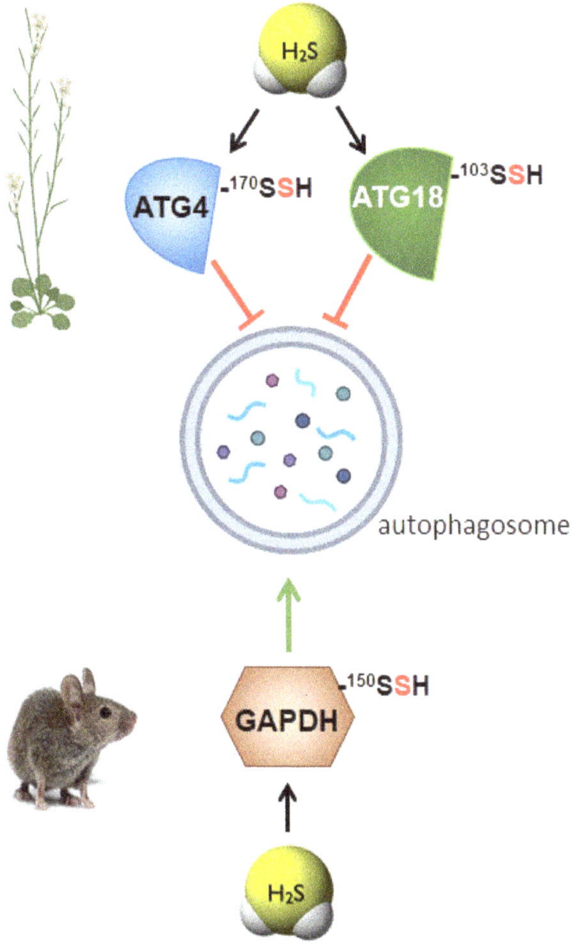

Figure 2. Persulfidation of specific proteins differently regulates autophagy in plants and mammals. ATG4, autophagy-related gene 4; ATG18, autophagy-related gene 18; GAPDH, glyceraldehyde-3-phosphate dehydrogenase.

In plants, the role of H_2S in the regulation of autophagy has been studied under stress conditions, particularly under nutrient limitation, demonstrating the negative regulation of bulk autophagy by sulfide through persulfidation of specific targets [41,46,85]. However, a recent research was published describing the role of persulfidation of ATG18a regulating autophagy under ER-stress [45], and, therefore, deciphering a new level of regulation of selective autophagy through the persulfidation of ATG18a [45] (Figure 2). ATG18a binds to phosphoinositides [86] and forms a complex with ATG2, which is involved in autophagosome biogenesis during phagophore expansion [87]. In this research, it was demonstrated that sulfide regulates ATG18a phospholipid-binding activity by reversible persulfidation at the specific residue Cys103, which reversibly activates ATG18a binding capacity to specific phospholipids. Authors proved that the mutation of Cys103 in ATG18a decreased its binding capacity to membranes and its localization time within the phagophore was shorter. In this way, the reversible persulfidation of ATG18a affects its binding to membranes, which potentially delays its release from the autophagosome, inhibiting autophagosome progression and maturation [45]. This regulation of autophagy through the persulfidation of ATG18a probably allows the plant a correct physiological response upon stress, with the final outcome of plant survival.

3.2. *Regulation of Autophagy by Persulfidation in Animals*

The molecular mechanism by which H_2S regulates autophagy in mammals has been recently established thought the study of the persulfidation of glycolytic glyceraldehyde-3-phosphate dehydrogenase (GAPDH) [88] (Figure 2). This modification of GAPDH triggers its translocation to the nucleus, which is critical to induce autophagy via deacetylation of the autophagic core protein LC3B, and the consequential autophagosome formation. Authors demonstrated that nuclear GAPDH interacts with the cell cycle activator and apoptosis regulator 2 (CCAR2/DBC1), avoiding the interaction of CCAR2 with deacetylase SIRT1, and therefore avoiding the inactivation of SIRT1. Then, activated SIRT1 deacetylates MAP1LC3B/LC3B (microtubule-associated protein 1 light chain 3 beta) inducing its translocation into the cytoplasm and activating autophagy [88]. Persulfidation of GAPDH at the same residue, Cys150 was previously described in mammals resulting in an increase of its enzymatic activity [76], although in later studies, polysulfide treatment of GAPDH showed opposite effects decreasing its enzymatic activity [89]. In plants, persulfidation of the cytosolic isoform GAPDH (GapC) was also previously described [78] and demonstrated its nuclear translocation upon the modification of the protein [77]. However, in none of these studies, the relationship between persulfidation of GAPDH and autophagy regulation was analyzed. A similar situation comes about the regulation of protein tyrosine phosphatase (PTP1B), which was described to be persulfidated at Cys215, inhibiting its enzymatic activity. This inhibition resulted in the activation of PERK alleviating ER stress [66], but authors did not relate this regulation to autophagy. Nevertheless, later was demonstrated that activation of PERK by sulfide treatments, increased eIF2 phosphorylation and induced autophagy [68], probably due to the persulfidation of PTP1B.

4. Conclusions

The aim of this review is to provide insights into the role of H_2S regulating autophagy, contrasting the available knowledge in plant and animal systems. In plants, it has been widely accepted that the beneficial effect of sulfide mitigating different stresses, as well as all the studies published, point toward an anti-autophagic role of sulfide by repressing autophagy. Accumulating experimental evidence in mammalians demonstrates the cytoprotective effect of sulfide in a wide range of physiologic and pathologic conditions and its role through regulating autophagy. However, it needs further clarification as to if sulfide exerts a pro- or anti-autophagic role in mammalians.

It is worth pointing out that the mechanisms by which hydrogen sulfide regulates autophagy are through the persulfidation of specific targets, what seems to be evolutive conserved among species.

Therefore, from all the above, we can conclude that these findings suggest that either in mammals and in plant systems, the regulation of autophagy by persulfidation of specific targets seems to be the H_2S signaling mechanism in autophagy, and, regardless, because sulfide exerts a pro-autophagic or anti-autophagic role, its beneficial effect points toward cell survival.

Author Contributions: A.A. organized, wrote and prepared the manuscript. C.G. contributed to writing and reviewing the major parts of the manuscript. All authors have read and agreed to the published version of the manuscript.

Funding: This work was supported by ERDF A way of making Europe and MCIN/AEI/10.13039/501100011033 (grant No. PID2019-109785GB-I00); Junta de Andalucía (grant No. P18-RT-3154 and US-1255781); and the Marie Skłodowska-Curie Grant Agreement (grant No. 834120 to A.A.).

Acknowledgments: We thank Marie Skłodowska-Curie grant agreement No. 834120 for funding support and Junta de Andalucía-FEDER (Project *ref. US-1255781*).

Conflicts of Interest: The authors declare that this research was conducted in the absence of any commercial or financial relationships that could be construed as potential conflicts of interest.

References

1. Deter, R.L.; De Duve, C. Influence of glucagon, an inducer of cellular autophagy, on some physical properties of rat liver lysosomes. *J. Cell Biol.* **1967**, *33*, 437–449. [CrossRef] [PubMed]
2. Marshall, R.S.; Vierstra, R.D. Autophagy: The Master of Bulk and Selective Recycling. *Annu. Rev. Plant Biol.* **2018**, *69*, 173–208. [CrossRef]
3. Avin-Wittenberg, T. Autophagy and its role in plant abiotic stress management. *Plant Cell Environ.* **2019**, *42*, 1045–1053. [CrossRef] [PubMed]
4. Gou, W.; Li, X.; Guo, S.; Liu, Y.; Li, F.; Xie, Q. Autophagy in Plant: A New Orchestrator in the Regulation of the Phytohormones Homeostasis. *Int. J. Mol. Sci.* **2019**, *20*, 2900. [CrossRef] [PubMed]
5. Feng, Y.; He, D.; Yao, Z.; Klionsky, D.J. The machinery of macroautophagy. *Cell Res.* **2014**, *24*, 24–41. [CrossRef]
6. Parzych, K.R.; Klionsky, D.J. An overview of autophagy: Morphology, mechanism, and regulation. *Antioxid. Redox Signal.* **2014**, *20*, 460–473. [CrossRef]
7. van Doorn, W.G.; Papini, A. Ultrastructure of autophagy in plant cells. *Autophagy* **2013**, *9*, 1922–1936. [CrossRef]
8. Lescat, L.; Véron, V.; Mourot, B.; Péron, S.; Chenais, N.; Dias, K.; Riera-Heredia, N.; Beaumatin, F.; Pinel, K.; Priault, M.; et al. Chaperone-Mediated Autophagy in the Light of Evolution: Insight from Fish. *J. Mol. Biol. Evol.* **2020**, *37*, 2887–2899. [CrossRef]
9. Melia, T.J.; Lystad, A.H.; Simonsen, A. Autophagosome biogenesis: From membrane growth to closure. *J. Cell Biol.* **2020**, *219*. [CrossRef]
10. Tsukada, M.; Ohsumi, Y. Isolation and characterization of autophagy-defective mutants of Saccharomyces cerevisiae. *FEBS Lett.* **1993**, *333*, 169–174. [CrossRef]
11. Ravanan, P.; Srikumar, I.F.; Talwar, P. Autophagy: The spotlight for cellular stress responses. *Life Sci.* **2017**, *188*, 53–67. [CrossRef]
12. Matsui, Y.; Takagi, H.; Qu, X.; Abdellatif, M.; Sakoda, H.; Asano, T.; Levine, B.; Sadoshima, J. Distinct Roles of Autophagy in the Heart During Ischemia and Reperfusion. *Circ. Res.* **2007**, *100*, 914–922. [CrossRef]
13. Levine, B.; Mizushima, N.; Virgin, H.W. Autophagy in immunity and inflammation. *Nature* **2011**, *469*, 323–335. [CrossRef] [PubMed]
14. Nah, J.; Yuan, J.; Jung, Y.K. Autophagy in neurodegenerative diseases: From mechanism to therapeutic approach. *Mol. Cells* **2015**, *38*, 381–389. [CrossRef] [PubMed]
15. Chen, L.; Su, Z.-Z.; Huang, L.; Xia, F.-N.; Qi, H.; Xie, L.-J.; Xiao, S.; Chen, Q.-F. The AMP-Activated Protein Kinase KIN10 Is Involved in the Regulation of Autophagy in Arabidopsis. *Front. Plant Sci.* **2017**, *8*, 1201. [CrossRef]
16. Signorelli, S.; Tarkowski, Ł.P.; Van den Ende, W.; Bassham, D.C. Linking Autophagy to Abiotic and Biotic Stress Responses. *Trends Plant Sci.* **2019**, *24*, 413–430. [CrossRef] [PubMed]
17. Üstün, S.; Hafrén, A.; Hofius, D. Autophagy as a mediator of life and death in plants. *Curr. Opin. Plant Biol.* **2017**, *40*, 122–130. [CrossRef]
18. Wu, Y.C.; Wang, X.J.; Yu, L.; Chan, F.K.L.; Cheng, A.S.L.; Yu, J.; Sung, J.J.Y.; Wu, W.K.K.; Cho, C.H. Hydrogen Sulfide Lowers Proliferation and Induces Protective Autophagy in Colon Epithelial Cells. *PLoS ONE* **2012**, *7*, e37572. [CrossRef]
19. Florin, T.; Neale, G.; Gibson, G.R.; Christl, S.U.; Cummings, J.H. Metabolism of dietary sulphate: Absorption and excretion in humans. *J. Gut.* **1991**, *32*, 766–773. [CrossRef]
20. Gotor, C.; García, I.; Aroca, Á.; Laureano-Marín, A.M.; Arenas-Alfonseca, L.; Jurado-Flores, A.; Moreno, I.; Romero, L.C. Signaling by hydrogen sulfide and cyanide through post-translational modification. *J. Exp. Bot.* **2019**, *70*, 4251–4265. [CrossRef]
21. Abe, K.; Kimura, H. The possible role of hydrogen sulfide as an endogenous neuromodulator. *J. Neurosci. Off. J. Soc. Neurosci.* **1996**, *16*, 1066–1071. [CrossRef]
22. Martelli, A.; Testai, L.; Breschi, M.C.; Blandizzi, C.; Virdis, A.; Taddei, S.; Calderone, V. Hydrogen sulphide: Novel opportunity for drug discovery. *Med. Res. Rev.* **2012**, *32*, 1093–1130. [CrossRef] [PubMed]

23. Paul, B.D.; Snyder, S.H. H2S signalling through protein sulfhydration and beyond. *Nat. Rev. Mol. Cell Biol.* **2012**, *13*, 499–507. [CrossRef]
24. Du, J.; Huang, Y.; Yan, H.; Zhang, Q.; Zhao, M.; Zhu, M.; Liu, J.; Chen, S.X.; Bu, D.; Tang, C. Hydrogen Sulfide Suppresses Oxidized Low-Density Lipoprotein (Ox-LDL)-Stimulated Monocyte Chemoattractant Protein 1 Generation from Macrophages via the Nuclear Factor B (NF-kB) Pathway. *J. Biol. Chem.* **2014**, *289*, 9741. [CrossRef] [PubMed]
25. Gerasimova, E.; Lebedeva, J.; Yakovlev, A.; Zefirov, A.; Giniatullin, R.; Sitdikova, G. Mechanisms of hydrogen sulfide (H_2S) action on synaptic transmission at the mouse neuromuscular junction. *Neuroscience* **2015**, *303*, 577–585. [CrossRef] [PubMed]
26. Guo, C.; Liang, F.; Shah Masood, W.; Yan, X. Hydrogen sulfide protected gastric epithelial cell from ischemia/reperfusion injury by Keap1 s-sulfhydration, MAPK dependent anti-apoptosis and NF-kappaB dependent anti-inflammation pathway. *Eur. J. Pharmacol.* **2014**, *725*, 70–78. [CrossRef] [PubMed]
27. Yang, G.; Wu, L.; Jiang, B.; Yang, W.; Qi, J.; Cao, K.; Meng, Q.; Mustafa, A.K.; Mu, W.; Zhang, S. H_2S as a Physiologic Vasorelaxant: Hypertension in Mice with Deletion of Cystathionine Lyase. *Science* **2008**, *322*, 587. [CrossRef]
28. Nicholson, C.K.; Calvert, J.W. Hydrogen sulfide and ischemia-reperfusion injury. *Pharmacol. Res.* **2010**, *62*, 289–297. [CrossRef] [PubMed]
29. Wallace, J.L.; Dicay, M.; McKnight, W.; Martin, G.R. Hydrogen sulfide enhances ulcer healing in rats. *FASEB J. Off. Publ. Fed. Am. Soc. Exp. Biol.* **2007**, *21*, 4070–4076. [CrossRef]
30. Wen, Y.-D.; Wang, H.; Kho, S.-H.; Rinkiko, S.; Sheng, X.; Shen, H.-M.; Zhu, Y.-Z. Hydrogen Sulfide Protects HUVECs against Hydrogen Peroxide Induced Mitochondrial Dysfunction and Oxidative Stress. *PLoS ONE* **2013**, *8*, e53147. [CrossRef]
31. Rodriguez-Kabana, R.; Jordan, J.W.; Hollis, J.P. Nematodes: Biological Control in Rice Fields: Role of Hydrogen Sulfide. *Science* **1965**, *148*, 524–526. [CrossRef] [PubMed]
32. Thompson, C.R.; Kats, G. Effects of continuous hydrogen sulfide fumigation on crop and forest plants. *Environ. Sci. Technol.* **1978**, *12*, 550–553. [CrossRef]
33. Shen, J.; Xing, T.; Yuan, H.; Liu, Z.; Jin, Z.; Zhang, L.; Pei, Y. Hydrogen Sulfide Improves Drought Tolerance in Arabidopsis thaliana by MicroRNA Expressions. *PLoS ONE* **2013**, *8*, e77047. [CrossRef] [PubMed]
34. Shi, H.; Ye, T.; Chan, Z. Exogenous application of hydrogen sulfide donor sodium hydrosulfide enhanced multiple abiotic stress tolerance in bermudagrass (*Cynodon dactylon* (L). Pers.). *Plant Physiol. Biochem. PPB/Soc. Fr. De Physiol. Veg.* **2013**, *71*, 226–234. [CrossRef]
35. Li, Z.G.; Gong, M.; Xie, H.; Yang, L.; Li, J. Hydrogen sulfide donor sodium hydrosulfide-induced heat tolerance in tobacco (*Nicotiana tabacum* L.) suspension cultured cells and involvement of $Ca^{(2+)}$ and calmodulin. *Plant Sci. Int. J. Exp. Plant Biol.* **2012**, *185–186*, 185–189. [CrossRef]
36. Zhang, H.; Hu, L.Y.; Hu, K.D.; He, Y.D.; Wang, S.H.; Luo, J.P. Hydrogen sulfide promotes wheat seed germination and alleviates oxidative damage against copper stress. *J. Integr. Plant Biol.* **2008**, *50*, 1518–1529. [CrossRef]
37. Wang, B.L.; Shi, L.; Li, Y.X.; Zhang, W.H. Boron toxicity is alleviated by hydrogen sulfide in cucumber (*Cucumis sativus* L.) seedlings. *Planta* **2010**, *231*, 1301–1309. [CrossRef]
38. Chen, J.; Wu, F.H.; Wang, W.H.; Zheng, C.J.; Lin, G.H.; Dong, X.J.; He, J.X.; Pei, Z.M.; Zheng, H.L. Hydrogen sulphide enhances photosynthesis through promoting chloroplast biogenesis, photosynthetic enzyme expression, and thiol redox modification in Spinacia oleracea seedlings. *J. Exp. Bot.* **2011**, *62*, 4481–4493. [CrossRef]
39. Garcia-Mata, C.; Lamattina, L. Hydrogen sulphide, a novel gasotransmitter involved in guard cell signalling. *N. Phytol.* **2010**, *188*, 977–984. [CrossRef]
40. Lisjak, M.; Srivastava, N.; Teklic, T.; Civale, L.; Lewandowski, K.; Wilson, I.; Wood, M.E.; Whiteman, M.; Hancock, J.T. A novel hydrogen sulfide donor causes stomatal opening and reduces nitric oxide accumulation. *Plant Physiol. Biochem. PPB/Soc. Fr. De Physiol. Veg.* **2010**, *48*, 931–935. [CrossRef]
41. Álvarez, C.; Garcia, I.; Moreno, I.; Perez-Perez, M.E.; Crespo, J.L.; Romero, L.C.; Gotor, C. Cysteine-generated sulfide in the cytosol negatively regulates autophagy and modulates the transcriptional profile in Arabidopsis. *Plant Cell* **2012**, *24*, 4621–4634. [CrossRef]
42. Romero, L.C.; Angeles Aroca, M.; Laureano-Marin, A.M.; Moreno, I.; Garcia, I.; Gotor, C. Cysteine and Cysteine-Related Signaling Pathways in Arabidopsis thaliana. *Mol. Plant* **2014**, *7*, 264–276. [CrossRef]
43. Laureano-Marin, A.M.; Moreno, I.; Romero, L.C.; Gotor, C. Negative Regulation of Autophagy by Sulfide Is Independent of Reactive Oxygen Species. *Plant Physiol.* **2016**, *171*, 1378–1391. [CrossRef]
44. Gotor, C.; Aroca, A.; Romero, L.C. Persulfidation is the mechanism underlying sulfide-signaling of autophagy. *Autophagy* **2021**, 1–3. [CrossRef] [PubMed]
45. Aroca, A.; Yruela, I.; Gotor, C.; Bassham, D.C. Persulfidation of ATG18a regulates autophagy under ER stress in Arabidopsis. *Proc. Natl. Acad. Sci. USA* **2021**, *118*, e2023604118. [CrossRef] [PubMed]
46. Laureano-Marín, A.M.; Aroca, A.; Perez-Perez, M.E.; Yruela, I.; Jurado-Flores, A.; Moreno, I.; Crespo, J.L.; Romero, L.C.; Gotor, C. Abscisic Acid-Triggered Persulfidation of the Cys Protease ATG4 Mediates Regulation of Autophagy by Sulfide. *Plant Cell* **2020**, *32*, 3902–3920. [CrossRef] [PubMed]
47. Aroca, A.; Bassham, D.C. The Role of Sulfide in Reticulophagy through the Regulation of ATG18a by Persulfidation. *FASEB J.* **2020**, *34*, 1. [CrossRef]

48. Lai, E.; Teodoro, T.; Volchuk, A. Endoplasmic Reticulum Stress: Signaling the Unfolded Protein Response. *Physiology* **2007**, *22*, 193–201. [CrossRef]
49. Liu, J.-X.; Howell, S.H. Managing the protein folding demands in the endoplasmic reticulum of plants. *New Phytol.* **2016**, *211*, 418–428. [CrossRef]
50. Corazzari, M.; Gagliardi, M.; Fimia, G.M.; Piacentini, M. Endoplasmic Reticulum Stress, Unfolded Protein Response, and Cancer Cell Fate. *Front. Oncol.* **2017**, *7*, 78. [CrossRef]
51. Wang, S.S.; Chen, Y.H.; Chen, N.; Wang, L.J.; Chen, D.X.; Weng, H.L.; Dooley, S.; Ding, H.G. Hydrogen sulfide promotes autophagy of hepatocellular carcinoma cells through the PI3K/Akt/mTOR signaling pathway. *Cell Death Dis.* **2017**, *8*, e2688. [CrossRef]
52. Fruman, D.A.; Chiu, H.; Hopkins, B.D.; Bagrodia, S.; Cantley, L.C.; Abraham, R.T. The PI3K Pathway in Human Disease. *Cell* **2017**, *170*, 605–635. [CrossRef] [PubMed]
53. Zhou, Y.; Wang, D.; Gao, X.; Lew, K.; Richards, A.M.; Wang, P. mTORC2 Phosphorylation of Akt1: A Possible Mechanism for Hydrogen Sulfide-Induced Cardioprotection. *PLoS ONE* **2014**, *9*, e99665. [CrossRef] [PubMed]
54. Yan, X.; Wu, H.; Wu, Z.; Hua, F.; Liang, D.; Sun, H.; Yang, Y.; Huang, D.; Bian, J.-S. The New Synthetic H_2S-Releasing SDSS Protects MC3T3-E1 Osteoblasts against H_2O_2-Induced Apoptosis by Suppressing Oxidative Stress, Inhibiting MAPKs, and Activating the PI3K/Akt Pathway. *Front. Pharmacol.* **2017**, *8*, 7. [CrossRef]
55. Tao, B.B.; Liu, S.Y.; Zhang, C.C.; Fu, W.; Cai, W.J.; Wang, Y.; Shen, Q.; Wang, M.J.; Chen, Y.; Zhang, L.J.; et al. VEGFR2 functions as an H_2S-targeting receptor protein kinase with its novel Cys1045-Cys1024 disulfide bond serving as a specific molecular switch for hydrogen sulfide actions in vascular endothelial cells. *Antioxid. Redox Signal.* **2013**, *19*, 448–464. [CrossRef] [PubMed]
56. Coletta, C.; Papapetropoulos, A.; Erdelyi, K.; Olah, G.; Módis, K.; Panopoulos, P.; Asimakopoulou, A.; Gerö, D.; Sharina, I.; Martin, E.; et al. Hydrogen sulfide and nitric oxide are mutually dependent in the regulation of angiogenesis and endothelium-dependent vasorelaxation. *Proc. Natl. Acad. Sci. USA* **2012**, *109*, 9161–9166. [CrossRef]
57. Xie, H.; Xu, Q.; Jia, J.; Ao, G.; Sun, Y.; Hu, L.; Alkayed, N.J.; Wang, C.; Cheng, J. Hydrogen sulfide protects against myocardial ischemia and reperfusion injury by activating AMP-activated protein kinase to restore autophagic flux. *Biochem. Biophys. Res. Commun.* **2015**, *458*, 632–638. [CrossRef]
58. Ji, L.; Li, L.; Qu, F.; Zhang, G.; Wang, Y.; Bai, X.; Pan, S.; Xue, D.; Wang, G.; Sun, B. Hydrogen sulphide exacerbates acute pancreatitis by over-activating autophagy via AMPK/mTOR pathway. *J. Cell Mol. Med.* **2016**, *20*, 2349–2361. [CrossRef]
59. Kundu, S.; Pushpakumar, S.; Khundmiri, S.J.; Sen, U. Hydrogen sulfide mitigates hyperglycemic remodeling via liver kinase B1-adenosine monophosphate-activated protein kinase signaling. *Biochim. Biophys. Acta* **2014**, *1843*, 2816–2826. [CrossRef]
60. Zeqiraj, E.; Filippi, B.M.; Deak, M.; Alessi, D.R.; Van Aalten, D.M. Structure of the LKB1-STRAD-MO25 Complex Reveals an Allosteric Mechanism of Kinase Activation. *Science* **2009**, *326*, 1707–1711. [CrossRef]
61. Wang, M.; Tang, W.; Zhu, Y.Z. An update on AMPK in hydrogen sulfide pharmacology. *Front. Pharmacol.* **2017**, *8*, 810. [CrossRef]
62. Ma, W.; Zhong, T.; Chen, J.; Ke, X.; Zuo, H.; Liu, Q. Exogenous H2S reverses high glucose-induced endothelial progenitor cells dysfunction via regulating autophagy. *Bioengineered* **2022**, *13*, 1126–1136. [CrossRef]
63. Li, J.; Li, M.; Li, L.; Ma, J.; Yao, C.; Yao, S. Hydrogen sulfide attenuates ferroptosis and stimulates autophagy by blocking mTOR signaling in sepsis-induced acute lung injury. *Mol. Immunol.* **2022**, *141*, 318–327. [CrossRef] [PubMed]
64. Ruan, Z.; Liang, M.; Deng, X.; Lai, M.; Shang, L.; Su, X. Exogenous hydrogen sulfide protects fatty liver against ischemia–reperfusion injury by regulating endoplasmic reticulum stress-induced autophagy in macrophage through mediating the class A scavenger receptor pathway in rats. *Cell Biol. Int.* **2020**, *44*, 306–316. [CrossRef] [PubMed]
65. Wang, H.; Wu, Y.; Han, W.; Li, J.; Xu, K.; Li, Z.; Wang, Q.; Xu, K.; Liu, Y.; Xie, L.; et al. Hydrogen Sulfide Ameliorates Blood-Spinal Cord Barrier Disruption and Improves Functional Recovery by Inhibiting Endoplasmic Reticulum Stress-Dependent Autophagy. *Front. Pharmacol.* **2018**, *9*. [CrossRef] [PubMed]
66. Krishnan, N.; Fu, C.; Pappin, D.J.; Tonks, N.K. H2S-Induced sulfhydration of the phosphatase PTP1B and its role in the endoplasmic reticulum stress response. *Sci. Signal.* **2011**, *4*, ra86. [CrossRef] [PubMed]
67. Py, B.F.; Boyce, M.; Yuan, J. A critical role of eEF-2K in mediating autophagy in response to multiple cellular stresses. *Autophagy* **2009**, *5*, 393–396. [CrossRef]
68. Yadav, V.; Gao, X.H.; Willard, B.; Hatzoglou, M.; Banerjee, R.; Kabil, O. Hydrogen sulfide modulates eukaryotic translation initiation factor 2α (eIF2α) phosphorylation status in the integrated stress-response pathway. *J. Biol. Chem.* **2017**, *292*, 13143–13153. [CrossRef]
69. Chen, L.; Ma, K.; Fan, H.; Wang, X.; Cao, T. Exogenous hydrogen sulfide protects against hepatic ischemia/reperfusion injury by inhibiting endoplasmic reticulum stress and cell apoptosis. *Exp. Ther. Med.* **2021**, *22*, 799. [CrossRef]
70. Jiang, B.; Tang, G.; Cao, K.; Wu, L.; Wang, R. Molecular Mechanism for H2S-Induced Activation of KATP Channels. *Antioxid. Redox Signal.* **2010**, *12*, 1167–1178. [CrossRef]
71. Naik, J.S.; Osmond, J.M.; Walker, B.R.; Kanagy, N.L. Hydrogen sulfide-induced vasodilation mediated by endothelial TRPV4 channels. *Am. J. Physiol. Heart Circ. Physiol.* **2016**, *311*, H1437–H1444. [CrossRef] [PubMed]
72. Lv, B.; Chen, S.; Tang, C.; Jin, H.; Du, J.; Huang, Y. Hydrogen sulfide and vascular regulation–An update. *J. Adv. Res.* **2021**, *27*, 85–97. [CrossRef]
73. Islam, K.N.; Polhemus, D.J.; Donnarumma, E.; Brewster, L.P.; Lefer, D.J. Hydrogen Sulfide Levels and Nuclear Factor Erythroid Related Factor 2 (NRF2) Activity Are Attenuated in the Setting of Critical Limb Ischemia (CLI). *J. Am. Heart Assoc.* **2015**, *4*, e001986. [CrossRef]

74. Kapuy, O.; Papp, D.; Vellai, T.; Bánhegyi, G.; Korcsmáros, T. Systems-Level Feedbacks of NRF2 Controlling Autophagy upon Oxidative Stress Response. *Antioxidants* **2018**, *7*, 39. [CrossRef] [PubMed]
75. Szabó, C. Hydrogen sulphide and its therapeutic potential. *Nat. Rev. Drug Discov.* **2007**, *6*, 917–935. [CrossRef] [PubMed]
76. Mustafa, A.K.; Gadalla, M.M.; Sen, N.; Kim, S.; Mu, W.; Gazi, S.K.; Barrow, R.K.; Yang, G.; Wang, R.; Snyder, S.H. H_2S Signals Through Protein S-Sulfhydration. *Sci. Signal.* **2009**, *2*, ra72. [CrossRef]
77. Aroca, A.; Schneider, M.; Scheibe, R.; Gotor, C.; Romero, L.C. Hydrogen Sulfide Regulates the Cytosolic/Nuclear Partitioning of Glyceraldehyde-3-Phosphate Dehydrogenase by Enhancing its Nuclear Localization. *Plant Cell Physiol.* **2017**, *58*, 983–992. [CrossRef] [PubMed]
78. Aroca, A.; Serna, A.; Gotor, C.; Romero, L.C. S-sulfhydration: A cysteine posttranslational modification in plant systems. *Plant Physiol.* **2015**, *168*, 334–342. [CrossRef]
79. Filipovic, M.R. Persulfidation (S-sulfhydration) and H_2S. *Handb. Exp. Pharmacol.* **2015**, *230*, 29–59. [CrossRef] [PubMed]
80. Filipovic, M.R.; Zivanovic, J.; Alvarez, B.; Banerjee, R. Chemical Biology of H_2S Signaling through Persulfidation. *Chem. Rev.* **2018**, *118*, 1253–1337. [CrossRef]
81. Ida, T.; Sawa, T.; Ihara, H.; Tsuchiya, Y.; Watanabe, Y.; Kumagai, Y.; Suematsu, M.; Motohashi, H.; Fujii, S.; Matsunaga, T.; et al. Reactive cysteine persulfides and S-polythiolation regulate oxidative stress and redox signaling. *Proc. Natl. Acad. Sci. USA* **2014**, *111*, 7606–7611. [CrossRef] [PubMed]
82. Aroca, A.; Benito, J.M.; Gotor, C.; Romero, L.C. Persulfidation proteome reveals the regulation of protein function by hydrogen sulfide in diverse biological processes in Arabidopsis. *J. Exp. Bot.* **2017**, *68*, 4915–4927. [CrossRef] [PubMed]
83. Gotor, C.; Garcia, I.; Crespo, J.L.; Romero, L.C. Sulfide as a signaling molecule in autophagy. *Autophagy* **2013**, *9*, 609–611. [CrossRef]
84. Jurado-Flores, A.; Romero, L.C.; Gotor, C. Label-Free Quantitative Proteomic Analysis of Nitrogen Starvation in Arabidopsis Root Reveals New Aspects of H2S Signaling by Protein Persulfidation. *Antioxidants* **2021**, *10*, 508. [CrossRef]
85. Laureano-Marín, A.M.; Moreno, I.; Aroca, Á.; García, I.; Romero, L.C.; Gotor, C. Regulation of Autophagy by Hydrogen Sulfide. In *Gasotransmitters in Plants: The Rise of a New Paradigm in Cell Signaling*; Lamattina, L., García-Mata, C., Eds.; Springer International Publishing: Cham, Switzerland, 2016; pp. 53–75. [CrossRef]
86. Dove, S.K.; Piper, R.C.; McEwen, R.K.; Yu, J.W.; King, M.C.; Hughes, D.C.; Thuring, J.; Holmes, A.B.; Cooke, F.T.; Michell, R.H.; et al. Svp1p defines a family of phosphatidylinositol 3,5-bisphosphate effectors. *EMBO J.* **2004**, *23*, 1922–1933. [CrossRef] [PubMed]
87. Wun, C.-L.; Quan, Y.; Zhuang, X. Recent Advances in Membrane Shaping for Plant Autophagosome Biogenesis. *Front. Plant Sci.* **2020**, *11*, 565. [CrossRef]
88. Iqbal, I.K.; Bajeli, S.; Sahu, S.; Bhat, S.A.; Kumar, A. Hydrogen sulfide-induced GAPDH sulfhydration disrupts the CCAR2-SIRT1 interaction to initiate autophagy. *Autophagy* **2021**, *17*, 3511–3529. [CrossRef]
89. Jarosz, A.P.; Wei, W.; Gauld, J.W.; Auld, J.; Özcan, F.; Aslan, M.; Mutus, B. Glyceraldehyde 3-Phosphate Dehydrogenase (GAPDH) Is Inactivated by S- Sulfuration in Vitro. *Free Radic. Biol. Med.* **2015**, *89*, 512. [CrossRef]

antioxidants

MDPI

Review

Conserved and Diversified Mechanism of Autophagy between Plants and Animals upon Various Stresses

Naveed Ur Rehman [1], Peichun Zeng [1], Zulong Mo [1], Shaoying Guo [1], Yunfeng Liu [2], Yifeng Huang [3,*] and Qingjun Xie [1,*]

[1] State Key Laboratory for Conservation and Utilization of Subtropical Agro-Bioresources, Guangdong Provincial Key Laboratory of Plant Molecular Breeding, South China Agricultural University, Guangzhou 510642, China; naveed.urrehman@yahoo.com (N.U.R.); zpc2019@stu.scau.edu.cn (P.Z.); gustaf@stu.scau.edu.cn (Z.M.); syguo6688@163.com (S.G.)

[2] State Key Laboratory for Conservation and Utilization of Subtropical Agro-Bioresources, College of Life Sciences and Technology, Guangxi University, Nanning 530004, China; yunfengliu_bio@126.com

[3] Institute of Crop and Nuclear Technology Utilization, Zhejiang Academy of Agricultural Science, Hangzhou 310001, China

* Correspondence: huangyf@zaas.ac.cn (Y.H.); qjxie@scau.edu.cn (Q.X.)

Abstract: Autophagy is a highly conserved degradation mechanism in eukaryotes, executing the breakdown of unwanted cell components and subsequent recycling of cellular material for stress relief through vacuole-dependence in plants and yeast while it is lysosome-dependent in animal manner. Upon stress, different types of autophagy are stimulated to operate certain biological processes by employing specific selective autophagy receptors (SARs), which hijack the cargo proteins or organelles to the autophagy machinery for subsequent destruction in the vacuole/lysosome. Despite recent advances in autophagy, the conserved and diversified mechanism of autophagy in response to various stresses between plants and animals still remain a mystery. In this review, we intend to summarize and discuss the characterization of the SARs and their corresponding processes, expectantly advancing the scope and perspective of the evolutionary fate of autophagy between plants and animals.

Keywords: autophagy; degradation; vacuole; autophagosomes; autophagy-related protein

Citation: Rehman, N.U.; Zeng, P.; Mo, Z.; Guo, S.; Liu, Y.; Huang, Y.; Xie, Q. Conserved and Diversified Mechanism of Autophagy between Plants and Animals upon Various Stresses. *Antioxidants* **2021**, *10*, 1736. https://doi.org/10.3390/antiox10111736

Academic Editors: Angeles Aroca and Cecilia Gotor

Received: 11 October 2021
Accepted: 27 October 2021
Published: 29 October 2021

1. Introduction

To overcome the stress challenges, eukaryotes have evolved all sorts of sophisticated mechanisms to deal with the adverse effects of stress. Among them, autophagy (meaning "self-eating") is one of the most robust mechanisms used to manage cytoplasmic material, such as nucleic acid aggregates, protein complexes, lipid bodies, and damaged organelles [1], ultimately resulting in the turnover of cellular components in the lytic organelle (vacuole in plants and yeast and lysosome in animals) [2]. Autophagy can digest certain cell components selectively or non-selectively by degrading bulk cytoplasm. In each case, the cellular components and macromolecules are encircled by a double membrane vesicle, termed an autophagosome, which merges with the vacuole for degradation and then recycles cellular components [3]. The biogenesis of the autophagosome is generally derived from the endoplasmic reticulum (ER) by generating a double membrane envelope called phagophore. However, there is still another notion that autophagosome may be produced by other membranes [4]. Owing to the discovery of *AuTophagy-related Genes* (*ATG*), the regulatory route of autophagic machinery has been well documented among various species based on the conservation of ATG proteins [5]. Briefly, initiation, nucleation, elongation, and fusion/degradation are the four phases of the autophagic process [6,7].

Autophagy is a quality control process in plants that fine-tunes the circulation of cell components. During development, it also plays a role in aging, pollen maturation, and programmed cell death (PCD) [8]. Moreover, autophagy occurs at low-intensity under normal conditions; however, it is drastically intensified when confronts with various abiotic and biotic stresses (e.g., carbon or nitrogen deficiency, salt, drought, temperature, reactive oxygen species, or infections) [9]. On the other hand, autophagy plays a crucial part in mammals' appropriate growth and development, beginning with embryogenesis [10]. It is critical for good health since its proper functioning inhibits the onset of various diseases, including cancer, liver, muscle, and heart problems, neurological disorders (such as Huntington's disease), inflammation, pathogen infections, and aging [11].

In plants, there are three types of autophagy mechanisms: microautophagy, macroautophagy, and mega-autophagy [12]. Microautophagy is a pattern in which the vacuole membrane invagination directly packages target substrates in the cytoplasm, and the bundled substrates are then degraded for cyclic use. In plants, macroautophagy is characterized by the presence of a large autophagic vacuole with a double-membrane structure that is utilized to package and transport toxic cytoplasmic components for degradation [13]. Mega-autophagy is only found in plants and occurs concomitantly with developmental programmed cell death (PCD). Throughout mega-autophagy, large amounts of hydrolases are released into the cytoplasm from the vacuole, resulting in large-scale degradation of cellular components including cytoplasm, all organelles, the plasma membrane, and part of the cell wall [12,13]. Unlike microautophagy and macroautophagy that recycle macromolecular constituents back to the cytosol from the vacuole, mega-autophagy is an extreme form of massive degradation leading to cell death [14].

In mammalian cells, the lysosomal membrane invaginations/protrusions are employed to collect cargo during microautophagy [15]. Microautophagosomes are formed close to the vacuole, while macroautophagosomes occur far from it [2,12]. Moreover, chaperone-mediated autophagy (CMA) differs from microautophagy as it requires chaperones to recognize cargo proteins where these substrates are independently unfolded and translocated via the lysosomal membrane [16]. Unlike microautophagy and CMA, macroautophagy comprises of the sequestration of cargo away from the lysosome and, subsequently, de novo synthesis of autophagosomes is employed to sequester the cargo and carry it to the lysosome [17]. In this review, we are in particular attempting to advance the current knowledge of autophagy and discuss the distinct and conserved mechanism of autophagy between plants and animals.

2. Mechanism of Autophagy in Plants and Animals

Although autophagosomes were initially discovered in mammalian cells in the 1950s [18], the molecular principles of autophagy were originally explored in yeast and subsequently expanded to animal and plant cells by the characterization of ATG proteins [13,19,20]. To date, the ATGs driving autophagy have been thoroughly understood in terms of induction, cargo recognition, phagophore generation, development, autophagosome fusion, and degradation [21]. In yeast, more than 40 *ATG* genes have been isolated, leading to the identification of many *ATG* homologs in mammals and plants (Table 1) [9]. In plants, such as Arabidopsis (*Arabidopsis thaliana*), roughly 40 ATGs have been discovered according to the protein similarity to yeast ATGs [22]. These ATG proteins are mostly clustered into four functional categories: (1) the ATG1/ATG13 kinase complex, which triggers the formation of autophagosome under nutrient deprivation; (2) the autophagy-specific class III phosphatidylinositol (PI) 3 kinase complex; (3) the ATG8/ATG12 ubiquitin-like conjugation systems that act in phagophore expansion; and (4) the ATG9 complex, which stimulates phagophore expansion [1,23].

Table 1. Identified ATGs genes in yeast, mammals, and plants.

Yeast	Mammalian	Plants	Function	Reference
ATG1	ULK1, ULK2	AtATG1a-1c,1t, OsATG1a-1d	Protein kinase; functions in the induction of autophagy	[24–26]
ATG13/APG13	ATG13	AtATG13a-13b, OsATG13a-13c	Phosphorylated by TORC1; forms complex with ATG1 to function in the induction of autophagy	[25–27]
ATG17	FIP200	Not identified	Essential for both stability and phosphorylation of ULK1	[25,28]
ATG29	Not identified	Not identified	Function in induction and regulation of autophagy	[25,29]
ATG31	Not identified	Not identified	Function in induction and regulation of autophagy	[30]
ATG9/APG9/AUT9/CVT7	ATG9A, ATG9B	AtATG9a, OsATG9a-9b	Membrane protein; deliver membrane to the forming autophagosome	[31,32]
ATG2		AtATG2a, OsATG2a	Atg18-interacting protein; function in autophagosome formation	[32,33]
ATG18/AUT10/CVT18	WIPI-1, 2, 3, 4	AtATG18a-18h, OsATG18a-18f	PI(3)P-binding protein; involved in the formation of autophagosome	[32–34]
ATG27	Not identified	Not identified	Protein required for autophagy-dependent cycling of Atg9	[35]
ATG6/VPS30/APG6	BECN1	AtATG6a, OsATG6a	Beclin1 (the core subunits), bcl2-interacting protein; functions in nucleation	[36–38]
ATG14	ATG14	AtATG14a-14b	Enhancer of autophagosome formation; function in nucleation	[37,39,40]
ATG12/APG12	ATG12	AtATG12a-12b	Ubiquitin-like, conjugates to Atg5; function in autophagosome membrane expansion	[41,42]
ATG5/APG5	ATG5	AtATG5a	Ubiquitin-like ligase, conjugated by Atg12	[42,43]
ATG16	ATG16L1	AtATG16L	interacts with Atg5; stimulate ATG8–PE conjugation reaction	[44,45]
ATG7/APG7	ATG7	AtATG7a, OsATG7	E1-like enzyme for Atg12 and Atg8/LC3 conjugation	[41,46]
ATG10/APG10	ATG10	AtATG10a	E2-like enzyme covalently conjugates Atg12 to ATG5	[41,47,48]
ATG8/APG8/AUT7	MPA1LC3B/LC3B	AtATG8a-8i, OsATG8a-8e	Ubiquitin-like conjugates to PE	[41,49,50]
ATG3/APG3	ATG3/APG3	AtATG3	Function as E2-like enzyme for Atg12 and Atg8/LC3 conjugation	[51,52]
ATG4/APG4/AUT2	ATG4A-D	AtATG4a-4b	Cytosolic cysteine protease for processing and recycling of Atg8/LC3	[32,53]

Two evolutionarily conserved protein kinase complexes, Target of Rapamycin (TOR) and Sucrose nonfermenting-1-Related protein Kinase 1 (SnRK1), compete for autophagy initiation [54]. TOR inhibits the conserved ATG1/ATG13 kinase activity, which is a negative regulator of autophagy (Figure 1) [26]. The TOR complex in Arabidopsis is made up of three main components: the TOR serine/threonine kinase [55], the regulatory-associated protein of TOR (RAPTOR) that supplies the substrates by TOR for phosphorylation [56,57], and the complex stabilizer LST8 [58]. TOR is widely expressed in actively growing tissues of Arabidopsis, such as endosperm, meristems, and embryos [55]. The reduced TOR expression, for example, results in reduced root growth, while overexpressing phenotypes show increased root growth [56]. TOR is rapidly activated under nutrient-rich conditions to accelerate development in sink tissues, in particular by Glc (glucose) after imported sucrose.

The protein kinase mechanistic target of rapamycin complex 1 (mTORC1/TORC1), which functions upstream of autophagy, includes mTOR, the regulatory associated protein of mTOR (Raptor), mammalian lethal with Sec13 protein 8 (mLst8/Lst8), proline-rich AKT substrate 40 kDa (PRAS40), and DEP domain-containing mTOR interacting protein (Deptor) [59]. Both growth factors and nutrition activate mTORC1 in the lysosome, which stimulates the translation regulating factors such as the ribosomal protein S6 kinase and the eukaryotic initiation factor 4E binding protein. Meanwhile, autophagy is suppressed by mTORC1 via phosphorylation of the ULK-complex [59]. Under glucose deficiency, AMPK directly senses the increase in the AMP:ATP ratio, leading to its activation [60]. Additionally, in response to glucose deprivation, AMPK suppresses mTORC1 by phosphorylating and activating the mTOR negative regulator tuberous sclerosis complex 2 (TSC2) (Figure 1) [61]. When nutrition levels are deprived, mTORC1 is repressed, and autophagy begins with ULK complex activation, the production of PI3KC3-mediated PI(3)P at the early autophagosomal membrane, the ATG12 complex, and the conjugation of the ATG8 family protein to the membrane lipid phosphatidylethanolamine (PE) [59].

In mammalian cells, the TOR complex suppresses ATG13–ULK1 interaction by phosphorylating ATG13, thus reducing autophagy, while AMPK stimulates autophagy by directly phosphorylating ULK1 (Figure 1) [62]. Notably, it is unclear if SnRK1/AMPK and/or TOR can directly phosphorylate ATG1 in plants, necessitating additional research [26,38]. Interestingly, even in plants overexpressing SnRK1 during hostile conditions, constitutive TOR expression inhibited autophagy, demonstrating that in both animals and plants, TOR that acts downstream of SnRK1/AMPK is crucial for autophagy induction [63]. Additionally, the overexpression of catalytic subunit of SnRK1 (KIN10) increases ATG1 phosphorylation in *Arabidopsis*, and the SnRK1–ATG1 interaction appears to exist in all plant tissues [38]. TOR is active in Arabidopsis and hyper-phosphorylates ATG13 under normal circumstances because highly phosphorylated ATG13 has a poor binding capacity for ATG1 so that the ATG1 activity is low and autophagy levels are maintained at baseline. In *Arabidopsis*, ATG1, ATG13, ATG11, and ATG101 form an active complex to stimulate autophagy (Figure 1) [26,64]. However, whether the ATG1–ATG13 complex is controlled by nutritional availability still remains a point of contention. In *Arabidopsis* membrane delivery, the nucleation, expansion, and closure of phagophores are all stimulated when the ATG1–ATG13 complex is activated [13,65]. ATG9 is involved in the development of the separation membrane at the phagophore assembly site (PAS) as well as in the supply of lipids to the growing phagophore, together with ATG2 and ATG18 (Figure 1) [4]. *Atg9* mutants in yeast and mammals do not generate autophagosomes, while in *Arabidopsis*, ATG9 deletion leads to the expansion of autophagosome-related tubules associated with the ER [4,66,67]. Furthermore, the sequence of *AtATG9* has little in common with that of yeast or humans [68], implying that *AtATG9* can work in plant-specific ways during the production of autophagosomes.

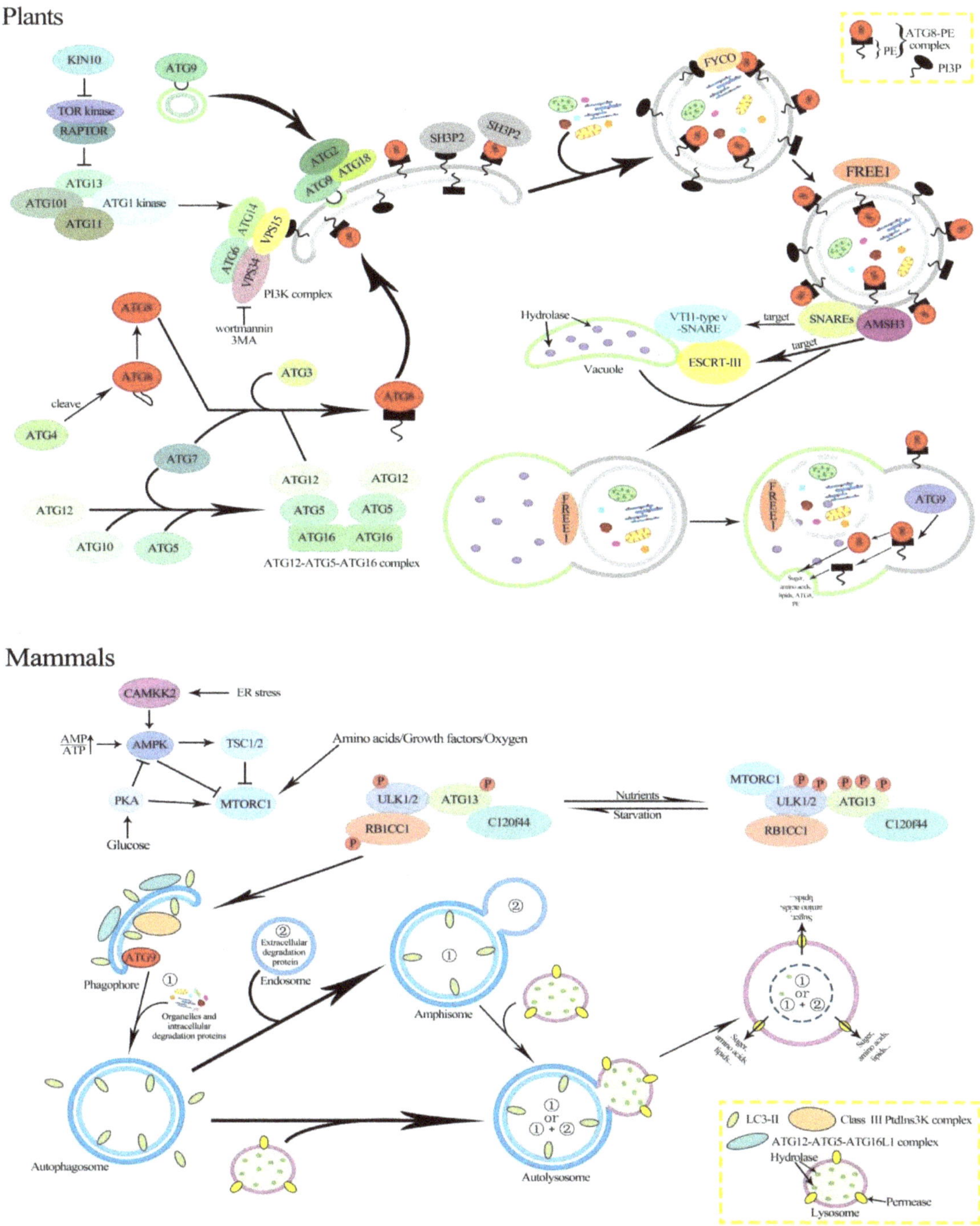

Figure 1. Schematic illustration of autophagy regulation in plants and animals. Autophagy is activated by inhibiting TOR and is blocked when TOR is overexpressed. Autophagy is triggered by the formation of an active complex between ATG13, ATG1, ATG11, and ATG101, as well as ATG11 and ATG101, which activates autophagy. Autophagosome development comprises membrane delivery, nucleation, expansion, and closure of the phagophore. ATG9 is employed in the transport of lipids to the expanding phagophore, together with ATG2 and ATG18. PI3P decoration is generated by the VPS34 lipid

kinase complex, which is followed by ATG8 conjugation to PE. Initially, ATG8 is matured by ATG4 cleaving of its C-terminal and conjugating it to PE by E2-like ATG3 and the E3-like ATG12–ATG5–ATG16 complex. For phagophore expansion, ATG8–PE binds to the autophagosomal membrane. Sealed ATG8- and PI3P-decorated autophagosomes are transported to the vacuole with the help of FYCO (FYVE and coiled-coil domain-containing) proteins that bind the autophagosome to the microtubule transport machinery. With the aid of ARP2/3 (NAP1), ESCRT (CFS1, CHMP1, FREE1, and VPS2.1), and exocyst (EXO70B1) components, SNARE-mediated fusion of autophagosomes with the tonoplast releases autophagic bodies into the vacuole. Following that, vacuolar hydrolases degrade the vesicles. Model of Ulk1 regulation by AMPK and mTORC1 in response to glucose signals. Left: when glucose is sufficient, AMPK is inactive and mTORC1 is active. The active mTORC1 phosphorylates Ulk1 on Ser 757 to prevent Ulk1 interaction with and activation by AMPK. When cellular energy level is limited, AMPK is activated and mTORC1 is inhibited by AMPK through the phosphorylation of TSC2 and Raptor. The induction complex consists of ULK1/2, ATG13, RB1CC1, and C12orf44. Under nutrient-rich conditions, MTORC1 associates with the complex and inactivates ULK1/2 and ATG13 through phosphorylation. During starvation, MTORC1 dissociates from the complex, and ATG13 and ULK1/2 become partially dephosphorylated by yet-unidentified phosphatases, allowing the complex to induce macroautophagy. RB1CC1/FIP200 and C12orf44/ATG101 are also associated with the induction complex and are essential for macroautophagy. RB1CC1/FIP200 may be the ortholog of yeast Atg17, whereas the function of C12orf44/ATG101 is not known. A signal transduction event regulated by the TOR kinase leads to the following: (1) the induction of autophagy—a membrane from an unknown source sequesters cytosol and/or organelles resulting in the formation of a double-membrane vesicle termed an autophagosome; (2) on completion—the autophagosome docks with the lysosome or vacuole. Fusion of the autophagosome outer membrane with the vacuole releases the inner vesicle into the vacuole lumen. The inner vesicle is termed an autophagic body. Breakdown within the vacuole allows the recycling of the degraded autophagic body and its hydrolyzed cargo (amino acids, fatty acids, sugars, and nucleotides).

Autophagosome expansion and vesicle closure are aided by ATG8–PE, which is found in both the inner and outer autophagosome membranes [69,70]. In *Arabidopsis*, SH3P2 (SH3 domain-containing protein 2), a membrane-associated protein, translocates the PAS (phagophore assembly site) during autophagy (Figure 1) [71]. In addition to interacting with ATG8, SH3P2 also connects with PI3P and is involved in membrane elongation and autophagosome closure via the PI3K complex [71]. To ensure the movement of autophagosomes through the microtubules' plus end, on the outer autophagosome membrane, LC3/ATG8 and PI3P bind with FYCO1 (FYVE and coiled-coil domain-containing 1) on the inner autophagosome (Figure 1) [72]. Moreover, co-sedimentation and co-localization tests in *Arabidopsis* revealed that ATG8 can bind to microtubules in vivo, implying that microtubules are involved in autophagosome migration to the vacuole [73].

In mammalian cells, autophagosomes go through a maturation process that includes PI(3)P turnover and the removal of ATG8 proteins by ATG4 proteases, as well as the recruitment of fusion machinery such as RAB7, the homotypic vacuole fusion and protein sorting (HOPS) tethering complex, and SNARES [74].

Unlike yeast, *Arabidopsis* possesses nine *ATG8* (*ATG8a–ATG8i*) homologs, two *ATG4* (*ATG4a* and *ATG4b*) homologs, and two *ATG12* (*ATG12a* and *ATG12b*) homologs [75,76]. The expression patterns of the *Arabidopsis ATG8* genes are tissue-specific, indicating that they may have diverse roles [77]. The ATG4s in Arabidopsis can cleave the C-terminus of ATG8, similar to their yeast counterparts. Furthermore, the *atg4a atg4b* double mutant exhibits autophagy defects, as shown by early senescence and lower silique synthesis, implying that ATG4s are required for plant growth [78]. The *atg12a atg12b* double mutant in *Arabidopsis* exhibits early senescence, food starvation sensitivity, and the absence of autophagic bodies, while the single mutants of *atg12a* and *atg12b* do not show, presenting functional redundancy. The ATG12–ATG5 conjugate accumulation was reduced in single mutants of *atg12a* or *atg12b* in which ATG8–PEs were not found, demonstrating that the ATG12–ATG5 binding is compulsory for ATG8–PE conjugation [79]. Mutations in plant ATG5, ATG7, or ATG10 result in hypersensitivity to nitrogen and carbon deficiency [79]. Likewise, *atg12*, *atg5*, and *atg10* mutants are unable to generate autophagic bodies in the vacuole [80].

Regarding the fusion of autophagosomes to the vacuole, several components have been implicated. For example, it was reported that SNAREs (soluble NSF attachment protein receptors) are required for accurate autophagosome targeting to the vacuole [81]. In *Arabidopsis*, the absence of VTI12, a VTI1-type v-SNARE (vesicle SNARE) on the target membrane, prevents autophagosomes from entering the vacuole under nutritional stresses, indicating that VTI12 is important for the fusion of the autophagosome [81]. AMSH3 (associated molecule with the STAM3 SH3 domain) is required for autophagosome trafficking to the vacuole in *Arabidopsis* and interacts with the ESCRT-III subunit VPS2.1 (vacuolar protein sorting 2.1) (Figure 1) [82]. Notably, in *Arabidopsis*, the plant-specific ESCRT component FREE1 (FYVE domain protein necessary for endosomal sorting 1) was discovered to interact with SH3P2 and to regulate the fusion of autophagosomes and vacuoles [71,83]. Furthermore, the interior vesicle, known as the autophagic body, is discharged into the vacuole when the autophagosome and vacuole are united and destroyed by a sequence of resident hydrolases [13]. The ATG8–PE linked to the inner autophagosome membrane is degraded into the vacuole, but ATG4 cleaves the ATG8–PE attached to the outside of autophagosome membrane, freeing ATG8 from PE and allowing it to be recycled [78].

In mammals cell, after lysosome fusion, lysosomal enzymes degrade the inner membrane of the autophagosome and its contents, and amino acids along with sugars are effluxed out of the lysosome by specific transporters, comprising of sugar efflux Spinster (SPNS), which is essential for degradation, autolysosome reformation, and the reactivation of mTORC1 [84].

3. Organelles Selective Autophagy

Organelle autophagy is essential for maintaining cellular homeostasis by preserving the integrity and quantity of organelles in changing environments and pressures. The specific selectivity of organelles by autophagy is governed by ATG8 interactions with specific autophagic receptors (termed SARs) with an ATG8-interacting motif (AIM) [85–87], resulting in different types of autophagy in regulating relevant biological processes.

3.1. Aggrephagy

Selective autophagy can also degrade nonfunctional proteins as aggregates, a process known as aggrephagy, with ubiquitin chains serving as a signal for degradation [88]. Aggrephagy receptors Cue5 in yeast and p62/SEQUESTOSOME 1 (SQSTM1) and Neighbor of BRCA 1 (NBR1) in mammals bind to ATG8 via the ubiquitin-binding domain (Figure 2) [89,90]. Plants have been shown to have a homolog of NBR1, an N-terminal PB1 (Phox and Bem1p) domain that binds to ubiquitin and ATG8 simultaneously, implying that aggrephagy mechanisms in yeast, plants, and mammals are similar (Figure 2a) [91]. NBR1 mutation causes an accumulation of ubiquitylated insoluble proteins in *Arabidopsis* during heat stress [92]. Furthermore, heat stress can drive NBR1 and ATG8 to bind with the aggregatic cytoplasmic protein, demonstrating that the plant aggrephagy receptor NBR1 is important in the regulation of proteostasis [93].

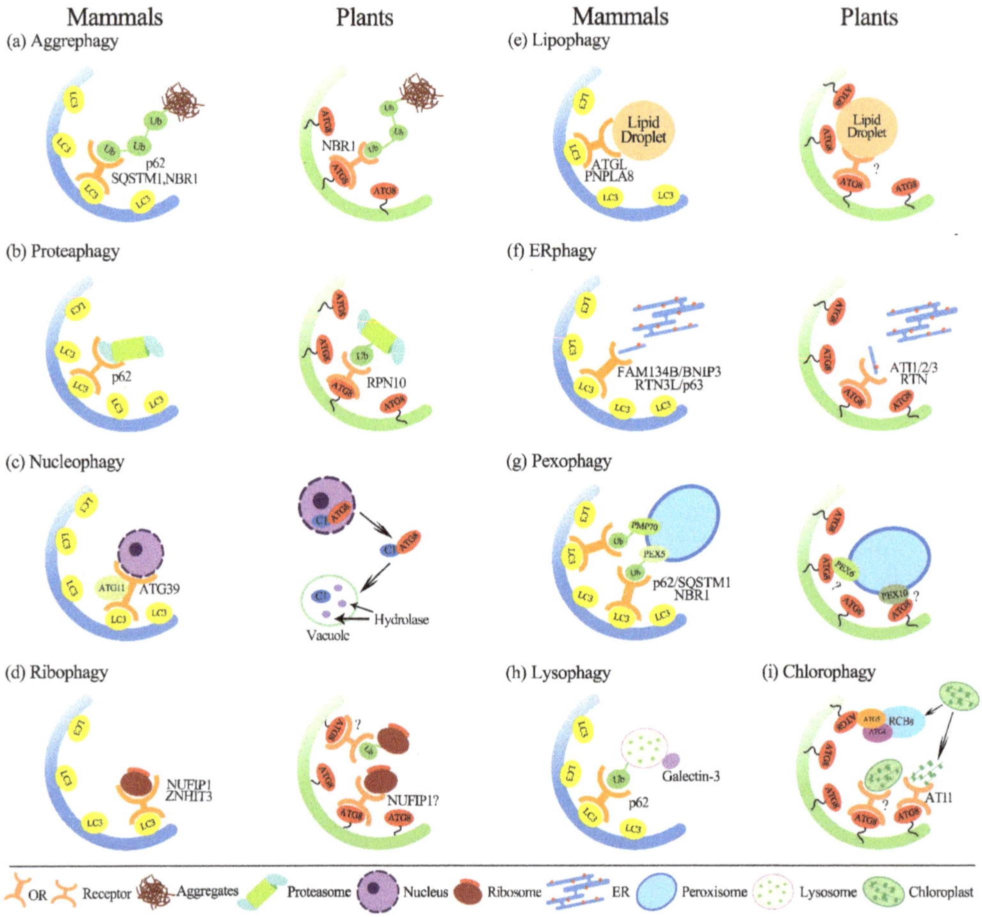

Figure 2. Schematic representation of several mechanisms of selective autophagy in plants and animals. The degradation autophagic pathways for cell organelles and aggregates are shown and distinct features of each are highlighted. (**a**) Aggrephagy. Degradation of intracellular protein aggregates that form naturally or as a result of abiotic stresses that cause protein folding. Aggrephagy is activated by aggregate ubiquitylation and autophagy-binding receptors, such as NBR1 in plants and p62/NBR1 in animals. (**b**) Proteaphagy. Degradation of proteasomes occurs in response to proteasome inactivation or nitrogen starvation. Proteaphagy is triggered by p62 in animals and RPN10 in plants and translocates it to the cytoplasm for degradation (**c**) Nucleophagy. Atg39 interacts with cargo receptor Atg11 through Atg11 binding region in animals and in plants ATG8 interacts with C1 and transports it to the cytoplasm from the nucleus. (**d**) Ribophagy. A ribophagy receptor NUFIP1 is essential for the selective degradation of ribosomes in animals and plants. (**e**) Lipophagy. PNPLA8 is required to produce autophagosomes during the lipophagy process in mammals while, in plants, no receptors have been identified so far. (**f**) Reticulophagy. The IRE1b stress sensor is required for endoplasmic reticulum degradation, which happens in response to an accumulation of unfolded proteins during ER stress. The reticulon homology domain (RTN) containing the family of reticulophagy receptors has been identified in mammals and yeast, but not in plants. ATI1 and ATI2 were the first ER-phagy receptors discovered in plants, and FAM134B, BNIP3, RTN3, and p63 have been identified as receptors in animals that translocate it to the cytoplasm for degradation. (**g**) Pexophagy. Pexophagy activates in response to ROS by phosphorylating PEX5 and PMP70 leading to ubiquitination recognized by p62, targeting peroxisomes for pexophagy. No pexophagy receptors have yet been described in plants, although the LON2 chaperone likely plays a role in peroxisome stress sensing, whereas PEX6 and PEX10 interact with ATG8. (**h**) Lysophagy. Removal of injured lysosome via concentrated recruiting of galectin-3 and LC3 onto lysosomal membranes, as these proteins are presumably recognized by p62/SQSTM1 and targeted for degradation via autophagy. (**i**) Chlorophagy. Chloroplasts are degraded in a variety of ways, including piecemeal degradation of stromal fragments in Rubisco-containing bodies (RCBs) during senescence or nutrient starvation, which may be mediated by ESCRT components such as CHMP1; the

engulfment of whole chloroplasts in response to oxidative damage, which may be mediated by PUB4-dependent ubiquitylation; and the formation of ATI1/2 bodies.

3.2. Proteaphagy

The eukaryotic proteasome contains the regulatory particle (RP), which is responsible for the recognition and unfolding of substrates, and the core particle (CP) for degradation [94]. Autophagy targets proteasomes in *Arabidopsis*, and it was previously confirmed that Arabidopsis RPN10 acts as a selective autophagy receptor and targets inactive 26S proteasomes by concurrent interactions with ubiquitylated proteasome subunits/targets and lipidated ATG8 lining, the enveloping autophagic membranes [95]. Previously, it was concluded that nitrogen deprivation induces autophagy in both proteasome subunits and is reliant on the lipidation of Atg8 via Atg7 and Atg10 [87]. Proteaphagy was increased in plants treated with the proteasome inhibitor MG132, whereas bulk autophagy remained unaltered, as determined by the lysosomal cleavage of GFP-Atg8. Notably, RPN10, a component of the RP which is essential for identifying ubiquitinated substrates, is required for proteaphagy (Figure 2b) [87,96]. RPN10 is a cytoplasmic protein that is not integrated into the proteasome, unlike other proteasomal proteins [87]. The binding motifs and sequence of RPN10 are substantially conserved among plants while neither the yeast nor human (PSMD4 in humans) homologs of Rpn10 have been confirmed to have any effect on proteaphagy or Atg8 [87,97]. Unlike yeast and plants, the animal proteasome becomes ubiquitinated upon starvation of amino acids, which is essential for its degradation by autophagy [13,87,98]. In animals, p62 can regulate the autophagosomal engagement of proteasomes by acting as a specific proteaphagy receptor (Figure 2b). Furthermore, autophagy receptor p62/SQSTM1, associated with numerous types of selective autophagy, recognizes the ubiquitin-modified proteasome [99]. These studies indicate that p62 is a key player in regulating the balance between proteasomal function and lysosomal degradation. Overall, it appears that proteaphagy occurs in a wide range of organisms, while molecular details vary that require further investigation.

3.3. Nucleophagy

The nucleus, similar to the eukaryotic cell primary organelle, is responsible for regulating gene expression and maintaining genomic integrity. When cells are stressed, they need a way to dispose of unwanted nuclear proteins and components. The mechanisms of nucleophagy are evolutionarily conserved catabolic processes which target various nuclear components such as the nuclear envelope, RNA, and DNA through a series of processes including nuclear sensing, nuclear export, and autophagic degradation in the cytoplasm [100]. For a long time, however, there was no proof that nucleophagy occurs in plants. The increase in the geminivirus nuclear protein C1 triggers autophagy, according to new research. Through the nuclear export-dependent process mediated by exportin 1 (XPO1), C1 is transported to the cytoplasm from the nucleus by interacting with ATG8h (NbATG8h or SlATG8h), one of several autophagy proteins, and when the AIM in C1 is mutated, it loses its ability to interact with ATG8 (Figure 2c) [101]. C1 degradation is prevented by inhibitors of autophagy and the removal of ATG5, ATG7, and ATG8h proteins [101]. In plants, this was the first time that autophagy was discovered to be involved in the breakdown of nuclear proteins.

Furthermore, a transcription factor BRI1-EMS Suppressor 1 (BES1), which controls brassinosteroid signaling, is ubiquitinated and interacts with DSK2A, leading to degradation in a DSK2 and core ATG-dependent way. DSK2A an adaptor of autophagy has two AIMs: a ubiquitin-like domain and a ubiquitin-associated domain. [102]. BIN2 kinase regulates DSK2 binding to ATG8, phosphorylating DSK2 around the AIM domains to improve its capacity to bind with ATG8 [102]. However, whether BES1 is damaged before or after entering the nucleus is unclear.

In mammals, nucleophagic activity is associated with genotoxic and oncogenic stress [101]. Although pathogenic conditions trigger nucleophagy in mammalian cells, the Nem1/Spo7–

Pah1 axis and the orthologous CTDNEP1/NEP1R1-lipin complex are conserved from yeast to mammals. Similarly to its counterpart, the CTDNEP1/NEP1R1-lipin complex is found in the nuclear envelope. [23]. Nucleophagy in *S. cerevisiae* is mediated by the autophagic cargo receptor Atg39, which is found on the outer nuclear membrane [103]. Atg39 interacts with Atg8 via AIM within its cytosolic N-terminal region and subsequently interacts with the cargo receptor adaptor Atg11 via an Atg11 binding region (11-BR) (Figure 2c). Both of these interactions with Atg11 and Atg8 are essential for macronucleophagy [103]. However, future research is needed to determine whether nucleophagy occurs in mammalian cells under physiological circumstances.

3.4. Ribophagy

The selective autophagy of ribosomes can be observed in plant cells in addition to the above-mentioned particular autophagy pathway: For instance, a selective autophagy mechanism relying on ATG5 has been found in *Arabidopsis* and is involved in rRNA turnover [104]. It was recently observed that NUFIP1 is a ribophagy receptor in mammals that is essential for ribosome selective degradation during starvation (Figure 2) [105]. *Arabidopsis* has a homolog of mammalian NUFIP1; however, more research is required to determine whether *Arabidopsis* NUFIP1 is likewise engaged in ribophagy (Figure 2d) [106]. A new class of ATG8 interactors with a Ubiquitin-interacting motif (UIM)-like domain interacts with ATG8 in yeast/plants and animals has just been characterized [95]. As a result, additional selective autophagy routes are likely to be uncovered soon. Plant cells can efficiently eliminate damaged or unwanted cell components through these diverse types of selective autophagy pathways, ensuring plant survival and cell viability during environmental constraints.

3.5. Lipophagy

Lipids in membrane organelles serve as energy generation substrates as well as cellular structural materials. Fatty acids are first stored as triacylglycerol (TAG) in the lipid droplets (LDs) before being used directly for β-oxidation [107–109]. Lipolysis breaks down LDs into fatty acids for the cell caused by lipophagy, a selective autophagy mechanism found in mammalian cells [110,111]. Plant lipophagy processes were less studied than yeast's and mammals' [112,113]. In rice, LDs carrying TAGs in the tapetum are required during pollen maturation as a source of lipid components [114]. LDs encased in vacuoles have been discovered in rice tapetum cells, and LD-like structures were found in greater abundance in the cytoplasm of *Osatg7* and *Osatg9* mutants than in the wild type, showing that LDs in plants may also be degraded by lipophagy [114]. Furthermore, lipidomic research revealed that these mutant anthers had impaired phosphatidylcholine (PC) editing and lipid desaturation during pollen maturation, demonstrating that autophagy is involved in regulating lipid metabolism during plant development [114,115]. Under normal and limiting conditions in Arabidopsis, for the synthesis of TAG, organelles' autophagy can offer a source of fatty acids, demonstrating that autophagy could increase TAG synthesis. A lipase sugar-dependent 1 (SDP1), responsible for the initiation of catabolism of TAG, hydrolyzes TAGs that are stored in LDs under normal conditions [116]. However, lipophagy, on the other hand, is driven by nutritional deprivation and causes the LDs to be degraded for energy production [109]. In the *atg5* mutant, for the synthesis of fatty acid and beta-oxidation, the ER and peroxisomal proteins are upregulated, and the concentrations of phospholipids, galactolipids, and sphingolipids are altered, suggesting that lipid metabolism is adversely affected in mutant autophagy, which could affect plant lipid metabolism in addition to regulating the synthesis of TAG and the degradation of LD [117].

Lipophagy in mammals activates with the autophagosomal membrane recognizing cargo by interacting with light chain 3 (LC3) [118]. Through the interaction with ATGL's LIR domain, LC3 stimulates the translocation of cytoplasmic ATGL to the LD and causes lipophagy, and by the activity of SIRT1 action, ATGL enhances lipophagy to regulate hepatic LD catabolism [119]. Lipases found in LD, such as PNPLA5 (patatin-like phospholipase domain-containing enzyme 5) have been linked to lipophagy and autophagic proteolysis [120]. In a mouse model with a high-fat diet, another lipase from the same family, PNPLA8, can similarly interact with LC3 to induce lipophagy. These lipases are vital in initiating lipophagy by promoting the recruitment of triglycerides and sterol esters, which directly contribute to the production of autophagosomes, in addition to their role in LD detection [121]. Furthermore, in deprived human hepatocytes, PNPLA3 (patatin-like phospholipase domain-containing enzyme 3) is required to produce autophagosomes during the lipophagy process (Figure 2e) [122]. Surprisingly, a forced lipophagy system based on a fusion of the LD-binding domain and p62 has been shown to diminish the number of LDs, lower the level of TG throughout embryonic development, and finally, cause developmental retardation in mouse embryos. Furthermore, lipophagy-induced embryos are resistant to lipotoxicity and indicate the elimination of excess LD [123].

3.6. ER-Phagy (Reticulophagy)

The endoplasmic (ER) reticulum is a network of membrane tubules that is significant for protein and lipid synthesis in the cytoplasm and for storing calcium. When unfolded, proteins accumulate in the ER, and the ER-associated degradation (ERAD) and the unfolded protein response (UPR) pathways are triggered [110]. UPR is a signaling pathway that aims to reduce the accumulation of misfolded proteins in organelles while enhancing their folding capacity [110]. ERAD, on the other hand, identifies misfolded proteins and translocates them to the cytoplasm for degradation by ubiquitin proteasome system (UPS) [124]. Furthermore, autophagy is triggered by ER stress, and autophagosomes generated during this time have been found to contain ER components [125]. The ER autophagy or reticulophagy helps to maintain cell homeostasis by counteracting ER enlargement during the UPR. In addition to ER stress, other stimuli have been proven to induce ER-phagy as well [125]. ER-phagy, similar to other types of selective autophagy, involves receptor proteins that play key roles in the selection of targets. In yeast *S. cerevisiae*, Atg39 and Atg40 mediate ER-phagy, where they localize to different domains of the ER and enable the production of autophagosomes by interacting with Atg8 [103]. In mammals, the family with sequence similarity 134 member B (FAM134B) protein is Atg40's functional homolog with the conserved LIR motif and positive ER fragments co-localizing with LC3B. Furthermore, whereas FAM134B downregulation causes ER enlargement, its overexpression causes ER fragmentation and lysosomal degradation [126]. Both the reticulon domain and the LIR motif of FAM134B are essential for ER-phagy (Figure 2f). The recently identified soluble members C53, CALCOCO1 (identified for homology with the xeno-phagy receptors TAXBP1 and CALCOCO2), and Sequestosome1/p62 extended the list of mammalian ER-phagy receptors [127–129]. Finally, the ER stress sensor IRE1a and two cytosolic autophagy receptors with a ubiquitin-binding domain, NBR1, and optineurin, have been involved in ER turnover and polypeptide clearance from the ER membrane [130].

The *Arabidopsis thaliana* Atg8-interacting proteins ATI1 and ATI2 were the first ER-phagy receptors reported in plants (Figure 2f) [131]. They lack homologs in yeast and higher eukaryotes and feature a single transmembrane domain and Atg8 interacting motif (AIM) in their cytosolic N-terminus and were found in the ER under favorable conditions. Carbon deprivation segregates ATI1 and ATI2 in the ER network into spherical entities that are subsequently transported to the vacuole after interacting with Atg8 [131]. The ER membrane proteins AtSec62 (*A. thaliana*), the reticulon homology domain (RHD)-containing proteins RTN1 and RTN2 (*Zea mays*), and the soluble protein Atc53 are all members of the ER-phagy receptor family in plants (*A. thaliana*) (Figure 2f) [128].

3.7. Mitophagy

Although several mechanisms mentioned above are essentially similar in plants, mitophagy regulators are considerably different in yeast/animals and plants. Mitophagy is the term for autophagic selective degradation of mitochondria. Autophagy is responsible for removing mitochondria, whether owing to injury, altered energy demands, or controlled cell maturation, as in the case of reticulocytes' loss of mitochondria. Mitophagy is induced by various stimuli that cause mitochondrial damage, including hypoxia, chemical uncouplers, and reactive oxygen species (ROS) [132]. Additionally, mitophagy can occur in mammalian reticulocytes and the enterocyte cells of the Drosophila intestinal midgut in response to developmentally controlled alterations in the cell [133]. Moreover, during *C. elegans* development, mitophagy is also required to remove paternal mitochondria from fertilized oocytes [10]. Pink1 and *Parkin* genes, linked to familial Parkinson's disease, are the most well-studied mitophagy pathways [134]. PINK1 phosphorylates a variety of targets, including ubiquitin and recruiting and activating Parkin, an E3 ubiquitin ligase [135]. Parkin then works as an amplifier of the mitophagy signal provided by PINK1. These ubiquitinating mitochondrial surface proteins can be detected by cargo receptor proteins, which transport mitochondria to autophagosomes for degradation [136]. Multiple receptors, including p62/SQSTM1, NIX/BNIP3L, BNIP3, FUNDC1, NDP52 (CALCOCO2), TAX1BP1, and optineurin (OPTN), have been involved in mitophagy in mammals [135].

Mitophagy appears to have a role in development, senescence, stress response, and programmed cell death (PCD) in plants [137]. On the other hand, plants lack many of the genes that drive mitophagy in yeast and animal cells, and no plant proteins that identify defunct mitochondria for autophagic degradation have yet been discovered. In plants, chloroplasts co-exist in energy generation alongside mitochondria, and chloroplasts are targeted by autophagy via a process called chlorophagy [138]. An early study indicated that mitochondria were encased with a double membrane structure similar to ER in mung bean (*Vigna radiata*) during autophagy [139]. Notably, these mitochondrial autophagous structures have been found to combine with lytic vacuoles. Recently, mitochondrial proteins and vesicles were shown to be degraded by autophagy in Arabidopsis during senescence [25]. A homolog of yeast ATG11 (and mammalian FIP200/RB1CC1) has recently been discovered in Arabidopsis. It is involved in mitophagy in nitrogen-depleted circumstances [25,64]. During senescence-induced mitophagy, ATG7, an E1-like enzyme, is also important as it facilitates the conjugation of ATG8 with phosphatidylethanolamine (PE) and ATG12 with ATG5, resulting in ATG8–PE and ATG5–ATG12 complexes, respectively [25,76].

3.8. Pexophagy

Peroxisomes are small round organelles surrounded by a single lipid bilayer present in most eukaryotes [140]. Despite their morphological similarity and conserved functions in all eukaryotes, major differences in peroxisomes have been found between plants and animals [141]. The selective autophagy pathways in eukaryotes require specific cargo receptor(s) and/or adaptors. Two kinds of pexophagy cargo receptors have been described in yeast and mammals, which differ in their capacities to bind ubiquitylated cargos [142]. In yeast, two AIM-containing pexophagy receptors (Atg30 and Atg36) become attached to peroxisome surface proteins such as Pex3, Pex5, or Pex14 (33), and then the Atg30 and Atg36 recruit the autophagic machinery by interacting with Atg8 and Atg11 [143]. Mammals do not have Atg30 or Atg36 but instead use p62/SQSTM1 or NBR1 as pexophagy adaptors that bind ubiquitylated forms of PEX5 or PMP70 (Figure 2g) [144].

Although plant peroxisome proteins are targets of ubiquitylation, plants do not have obvious orthologs of either Atg30 or Atg36, and there is no direct evidence that plant NBR1 is the pexophagy receptor, even though the co-localization of NBR1 and ATG8 in electron-dense peroxisomal cores in Arabidopsis plants exposed to Cd has recently been reported [145]. However, the peroxisome proteins PEX6 and PEX10 in *Arabidopsis* were recently shown to interact with ATG8 via their AIMs, suggesting that they may be the potential receptor for driving pexophagy in plants (Figure 2g) [13]. In addition, by using forward genetic screening, the peroxisomal matrix protease LON 2 was identified, mutation of which consistently recovers Arabidopsis *atg* mutants. Notably, the autophagy of *lon2* peroxisomes does not require NBR1, but NBR1 may play an important role in LON2-independent pexophagy [146]. Collectively, it is still worthy to explore if the ubiquitination of PEXs, such as PEX3, PEX5, and PEX14 reported in yeast and mammals, is also involved in plant pexophagy [147].

3.9. Lysophagy

The lysosome, a membrane-bound acidic organelle is required for eliminating unwanted intracellular compounds. The lysosome contains a significant number of hydrolytic enzymes that are involved in degradation. The lysosome's destabilization and the leakage of these hydrolytic enzymes are detrimental to the cell [148]. Furthermore, if damaged lysosomes are not removed, the intracellular lysosome's number remains unchanged, and cells are incapable of sustaining cellular homeostasis. As a result, to keep cellular homeostasis, the cell uses an autophagic mechanism called lysophagy [149,150]. Lysophagy can be triggered by several factors that cause lysosomal degradation (Figure 2h). Photochemical internalization is a method that enhances gene transport by light-induced lysosome breakdown that has been used to activate lysophagy in the lab [149]. Mineral crystals such as monosodium urate and silica, viral or bacterial toxins, lysosomotropic chemicals, lipids, and proteases have all been shown to disrupt lysosomal membranes in vivo [148,150]. In the case of lysosomal damage, ubiquitination coincides with the vigorous employment of the autophagy receptor SQSTM1/p62 that is essential for efficient lysophagy (Figure 2h). Although research into the mechanisms that regulate lysophagy is incomplete, it has been revealed that, following lysosomal injury, LC3 and galectin-3 are employed to the wounded lysosome [150]. In mouse embryonic fibroblasts (MEFs), damaged lysosomal membranes are galectin-3 positive, ubiquitinated, and co-localize with p62 [150]. Furthermore, in HeLa cells, a similar connection between p62 and ubiquitin has been linked with damaged lysosomes [149]. These findings point out to ubiquitination and subsequent recruitment of the cargo adapter protein p62 in this mechanism. There are still many unanswered questions about the molecular aspects of lysophagy. However, biochemical and functional studies of ubiquitin, ubiquitin receptors, and the factors that affect their activity could help us better understand this process.

4. Chlorophagy

Plants and photoautotrophs have chloroplasts that are responsible for photosynthesis and are essential for the metabolism. Despite the fact that plants are sessile, cellular components must be used and recycled to survive and thrive in a variety of conditions. After the degradation of cellular macromolecules, their components are mobilized and reused during plant senescence. For instance, the degradation of chloroplast proteins such as Rubisco (ribulose-1, 5-bisphosphate carboxylase/oxygenase) is a significant source of nitrogen.

In senescing leaves, the primary method for chloroplast protein degradation was sequential breakdown within the vacuole known as chlorophagy [151]. The chloroplast, similar to the nucleus, can be digested piecemeal or through complete organelle autophagy. Senescence-associated vacuoles (SAVs) and Rubisco-containing bodies (RCBs) can cause piecemeal chloroplast degradation (Figure 2i). RCBs are Rubisco and Gln synthetase-containing double-membrane entities produced from the chloroplast envelope. Furthermore, the RCB is then encircled by various membrane structures, including the isolation membrane in the cytoplasm, after pinching off from the chloroplast.

Damaged chloroplasts are ubiquitylated by PUB4 (Plant U-BOX Protein 4), the cytosolic ubiquitin ligase as part of the whole chloroplast process. The ubiquitylated chloroplasts are then encapsulated and transported to the vacuole via ATG8-decorated autophagic vesicles [152]. Autophagy mediated by RCB has been demonstrated to be dependent on Atg4 and Atg5, and is essential for the recycling of protein during abiotic stresses [153,154]. The RCBs are formed by the fission of stroma-filled tubules protruding from the chloroplast, a process in which the ESCRT component CHMP1 (Charged Multivesicular Body 1) may play a key role, and they are degraded in the vacuole by autophagy [13]. SAVs have also been linked to chloroplast autophagy on a piecemeal basis. These tiny and lytic vacuoles are only present in senescing tissues and contain stromal proteins, including Rubisco and Gln synthetase, similar to RCBs. Unlike RCBs, they do contain chlorophyll *a* and do not appear to employ any autophagic machinery. The decline in chloroplast number is hampered in *atg4* mutants [155], suggesting that there may be an *ATGs*-dependent selective autophagy of the chloroplast. Recently, plant-specific proteins ATG8-interacting protein 1 and 2 (ATI1 and ATI2), which are found in the plastid-derived bodies and ER, are proposed to be involved in *ATGs*-mediated chlorophagy (Figure 2i) [156]. Plastid proteins from the outer envelope are translocated to the vacuole via the ATI1-decorated plastid structures (ATI1-PS)-mediated autophagy [156]. Notably, incomplete degraded chloroplasts in the vacuole were also found in plastid protein Tic40 (*ppi40*) mutants of *Arabidopsis* [157]. Because these defects were observed in a starvation-independent way, it is assumed that plants could use the autophagy-independent pathway to eliminate chloroplasts for quality control. This possibility was recently validated by the characterization of the Chloroplast Vesiculation protein (CV)-containing vesicles (CCVs) pathway [158].

In summary, the mechanisms underpinning chlorophagy are still poorly understood, leaving many open questions. For instance, it is unclear when and how cargo receptor proteins target the entire chloroplast to trigger chlorophagy. Although plants have putative cargo receptors of autophagy, such as ATI1 and ATI2 [131], as well as RCB receptors, it is still unclear how these proteins are specifically assigned in operating chlorophagy, and thus, novel approaches and/or components are needed.

5. Advances of Selective Autophagy in Plant

Plant selective autophagy research is progressing at a rapid pace. The majority of the previously documented selective autophagy pathways in metazoans have lately been discovered to work in plants as well [159,160].

Plants have an NBR1-like protein that is necessary for autophagosome degradation of ubiquitinated peroxisomes [161]. Peroxisomes are ubiquitous organelles in plants that control numerous metabolic events such as photorespiration, fatty acid β-oxidation, and the glyoxylate cycle [162]. Reactive oxygen species (ROS) are primarily produced by peroxisomes, making them prone to oxidative damage. Multiple antioxidative enzymes are found in the peroxisomes to eliminate excessive ROS, such as catalase, which degrades H_2O_2 exclusively [163]. High glucose levels of wild-type plant roots cause the accumulation of ROS. On one side, reactive oxygen species (ROS) oxidize active IAA (indole3-acetic acid) and constitutive pexophagy, on the other hand, it is aided by the increased level of ROS, which reduces root meristem activity by inhibiting IAA production [164]. However, autophagy deficit in *atg5* and *atg7* affects the transmission of the high glucose signal to the peroxisomes, increasing IAA and root meristem activity and resulting in increased primary roots compared to wild type under enhanced glucose conditions [164]. Recently, Arabidopsis plants exposed to Cd were found to have NBR1 and ATG8 co-localized in electron-dense peroxisomal cores [145]. Peroxisome oxidation and pexophagy were induced by Cd exposure, however, the *Arabidopsis* mutant *rbohC* (NADPH oxidase C) and *gox2* (glycolate oxidase 2) inhibit this process significantly by reducing ROS generation in Arabidopsis. Pexophagy is a key component of quick plant responses to Cd (cadmium), as it protects peroxisomal populations and the cell redox homeostasis [145].

In the case of chlorophagy, ESCRT III component CHMP1 proteins are involved in the efficient recycling of fragmented chlorophagy vesicles containing stromal proteins, according to an exquisite cell biology investigation [165]. Unlike chloroplasts related chlorophagy, little is known mechanistically about plant mitophagy even though accumulating genetic and cytological evidences suggest that mitochondria are recycled through autophagy [25], because some of the known mitophagy receptors and regulators are absent in plants [137]. It is also indistinct whether chlorophagy and mitophagy have any similarities, such as if they would share the same autophagy receptor. As a case, the ATI1/2 may service as receptor for both chlorophagy and ER-phagy [134,156]. Additionally, if plants, such as metazoans, utilize piecemeal mitophagy mechanisms is yet to be identified.

Moreover, mitophagy has been discovered in plants, where it is involved in development, stress response, senescence, and PCD [137]. The relationship between mitophagy and senescence is well-known, although its mechanistic understanding is lacking despite the fact that the core ATG proteins are well conserved in plants and are necessary for the senescence-induced degradation of mitochondrial vesicles and mitochondria-resident proteins [64]. Arabidopsis has a number of mitochondrial membrane proteins with ATG8-interacting motifs, according to a bioinformatic investigation [166], which are thought to be mitophagy receptors [137]. *Arabidopsis* mutants lacking key autophagy components are more vulnerable to UVB exposure, resulting in increased leaf chlorosis [152], demonstrating that autophagy is critical for cellular homeostasis in response of UVB. The number of mitochondria in wild-type leaves falls in response to UVB exposure but increases in *atg* (*atg2*, *atg5*, and *atg7*) mutants [138]. Notably, following UVB damage, confocal and electron microscopy observations reveal that *atg5* and *atg7* mutant plants accumulate fragmented and tiny mitochondria in the cytoplasm [138]. A substantial percentage of mitochondria in UVB-damaged *atg* leaves fail to collect tetramethylrhodamine ethyl ester (TMRE) [138], suggesting that damaged mitochondria stay in the mutant cytoplasm and their membranes are depolarized in response to UVB damage. Additionally, a variety of evolutionary conserved mitochondrion-associated proteins are also involved in the quality control of mitochondria [167]. Notably, the clustered mitochondria protein (CLU) is essential for mitochondria distribution and function in yeast, plants, and animals [168]. Mitochondrial membrane potential is abolished when uncouplers such as 2,4-dinitrophenol (DNP), carbonyl cyanide, and p-trifluoro-methoxyphenylhydrazone (FCCP) are applied. More recently, in Arabidopsis roots, mitophagy eliminates depolarized mitochondria in response to uncoupler treatment [169]. These findings support the idea that plant mitophagy plays a critical role in mitochondrial quality control. Upon uncoupler treatment, friendly mitochondria (FMT) labeled with YFP is recruited to mitochondria and co-localizes with mCherry ATG8 [169]. Moreover, in *fmt* mutants, the uncoupler-induced mitochondrial degradation was reduced [169], suggesting that FMT has a direct role in mitophagy activation. In terrestrial plants, the shape and volume of mitochondria vary dramatically throughout reproductive development [170]. The tapetum is the anther's innermost layer, which supplies nutrition to pollen grains as they mature, and later undergoes programmed cell death (PCD) [171]. Mitochondrial fragmentation and a reduction in overall mitochondrial volume occur before PCD in Arabidopsis tapetal cells [170]. Previously, it was observed that autophagy is necessary for the regulated PCD of tapetal cells in rice [114]. As a result, autophagy could be involved in mitochondrial degradation during PCD of tapetal cells.

6. Conclusions and Future Perspective

Over the last few decades, autophagy research has progressed to unprecedented depths, with studies on the interaction and communication between autophagosomes and other organelles. Many of the protein components and molecular mechanisms involved in autophagy have been identified and key regulatory factors have also been discovered, such as the TOR complex [172] and SnRK1 [173]. Studies on the roles of autophagy upon various stress have enabled us to understand the potential contributions of autophagy to crop breeding. However, several important open questions about the underlying molecular mechanisms of autophagy still remain to be further investigated, including the identification of specific SARs for certain types of organelles' selective autophagy, possible crosstalk between autophagy and other regulatory pathways (ubiquitin-proteasome etc.), and the manipulation of *ATGs* or autophagic machinery for robust improvement of crop yield and therapy of human diseases.

It is no doubt that the rapidly expanding collection of SARs and cargo proteins by high throughput screening of ATG8-interacting proteins would extend our knowledge of the multiple roles of autophagy in organism development and growth, as well as their response to stress. However, it is still difficult to precisely determine the specific SARs and/or cargo proteins due to the greater diversity of gene families and functions in both plants and animals. Moreover, as mentioned above, the evolutionary divergence of certain SARs derived from animals or plants may also impose restrictions on the identification of these SAR homologs in plants or animals. For example, detailed studies on the detection of ubiquitinated cargo by mammalian p62 have been published, while similar attempts in plants have not been as successful. On the other hand, the employed experimental approaches are currently limited, and thus specific genetic screening is still desirable, such as suppressors or enhancers screening of *atg* mutants or subcellular localization of fluorescence-labled ATGs.

To summarize, despite the fact that there is still much work to be done, autophagy investigations are nonetheless exciting and relevant since they have the potential to target virtually all organelles for degradation, thereby facilitating the quality control of organelles upon various stresses. With further research and the application of new methodologies, we will undoubtedly obtain a better knowledge of the autophagy interaction network, as well as the extensive insights into the conserved and distinct mechanisms of autophagy between plants and animals.

Author Contributions: Writing—original draft preparation, N.U.R., Y.H., and Q.X.; writing—review and editing, N.U.R., P.Z., Z.M., Y.L., Y.H., and Q.X.; visualization, P.Z., Z.M., and S.G.; supervision, Q.X.; project administration, S.G., Y.L., and Q.X.; funding acquisition, N.U.R., Y.L., Y.H., and Q.X. All authors have read and agreed to the published version of the manuscript.

Funding: This work was supported by the Major Program of Guangdong Basic and Applied Research (grant 2019B030302006 to Q.X.), the National Natural Science Foundation of China (grant 31971920 to Q.X. and Y.H.), and the Guangdong Basic and Applied Basic Research Foundation (2021A1515012053 to Q.X.), as well as the "Top Young Scientist of the Pearl River Talent Plan" (No. 20170104 to Q.X). and the State Key Laboratory for Conservation and Utilization of Subtropical Agro-Bioresources (grant SKLCUSA-a202008 to Q.X. and Y.L.).

Institutional Review Board Statement: Not applicable.

Informed Consent Statement: Not applicable.

Data Availability Statement: The data presented in this study are available in review.

Acknowledgments: We are grateful for the support of the experimental platform and funding from the Guangdong Provincial Key Laboratory of Plant Molecular Breeding (GPKLPMB201804 to Y.H. and Q.X). We apologize in advance to colleagues whose valuable work was not cited due to article length considerations.

Conflicts of Interest: The authors declare no conflict of interest.

Abbreviations

AIM	Autophagy-interacting motif
AMPK	AMP-activated protein kinase
ATG	autophagy-related
CAMKK2/CaMKKβ	calcium/calmodulin-dependent protein kinase kinase 2, beta
CHMP1	Charged Multivesicular Body 1
CMA	chaperone-mediated autophagy
Deptor	DEP domain-containing mTOR interacting protein
ERAD	endoplasmic reticulum associated degradation
FREE1	FYVE domain protein necessary for endosomal sorting 1
FYCO1	FYVE and coiled-coil domain containing 1
HOPS	homotypic vacuole fusion and protein sorting
LC3	light chain 3
PAS	phagophore assembly site
PC	phosphatidylcholine
PCD	programmed cell death
PE	phosphatidylethanolamine

PUB4	Plant U-BOX Protein 4
RAPTOR	regulatory-associated protein of TOR
RCBs	Rubisco-containing bodies
RHD	reticulon homology domain
ROS	reactive oxygen species
SAV	Senescence-associated vacuoles
SnRK1	Sucrose nonfermenting-1-Related protein Kinase 1
TMRE	tetramethylrhodamine ethyl ester
TOR	target of Rapamycin
ULK	unc-51-like kinase
UPR	unfolded protein response
UPS	ubiquitin proteasome system
XPO1	export-dependent process mediated by exportin 1

References

1. Wang, P.; Mugume, Y.; Bassham, D.C. New advances in autophagy in plants: Regulation, selectivity and function. *Semin. Cell Dev. Biol.* **2018**, *80*, 113–122. [CrossRef] [PubMed]
2. Bassham, D.C.; Laporte, M.; Marty, F.; Moriyasu, Y.; Ohsumi, Y.; Olsen, L.J.; Yoshimoto, K. Autophagy in development and stress responses of plants. *Autophagy* **2006**, *2*, 2–11. [CrossRef]
3. Bassham, D.C. Plant autophagy-more than a starvation response. *Curr. Opin. Plant Biol.* **2007**, *10*, 587–593. [CrossRef]
4. Zhuang, X.; Chung, K.P.; Cui, Y.; Lin, W.; Gao, C.; Kang, B.H.; Jiang, L.; Bassham, D.C. ATG9 regulates autophagosome progression from the endoplasmic reticulum in Arabidopsis. *Proc. Natl. Acad. Sci. USA* **2017**, *114*, E426–E435. [CrossRef] [PubMed]
5. Hayward, A.P.; Tsao, J.; Dinesh-Kumar, S.P. Autophagy and plant innate immunity: Defense through degradation. *Semin. Cell Dev. Biol.* **2009**, *20*, 1041–1047. [CrossRef] [PubMed]
6. Codogno, P.; Mehrpour, M.; Proikas-Cezanne, T. Canonical and non-canonical autophagy: Variations on a common theme of self-eating? *Nat. Rev. Mol. Cell Biol.* **2012**, *13*, 7–12. [CrossRef]
7. Xing, H.; Tan, J.; Miao, Y.; Lv, Y.; Zhang, Q. Crosstalk between exosomes and autophagy: A review of molecular mechanisms and therapies. *J. Cell. Mol. Med.* **2021**, *25*, 2297–2308. [CrossRef]
8. Khan, M.M.M.S.; Hemalatha, S. Autophagy: Molecular Insight and Role in Plant Programmed Cell Death and Defense Mechanism. *Int. Res. J. Biol. Sci.* **2015**, *4*, 78–83.
9. Tang, J.; Bassham, D.C. Autophagy in crop plants: What's new beyond Arabidopsis? *Open Biol.* **2018**, *8*, 180162. [CrossRef]
10. Sato, M.; Sato, K. Dynamic Regulation of Autophagy and Endocytosis for Cell Remodeling During Early Development. *Traffic* **2013**, *14*, 479–486. [CrossRef]
11. Yang, Y.; Klionsky, D.J. Autophagy and disease: Unanswered questions. *Cell Death Differ.* **2020**, *27*, 858–871. [CrossRef]
12. Van Doorn, W.G.; Papini, A. Ultrastructure of autophagy in plant cells: A review. *Autophagy* **2013**, *9*, 1922–1936. [CrossRef] [PubMed]
13. Marshall, R.S.; Vierstra, R.D. Autophagy: The Master of Bulk and Selective Recycling. *Annu. Rev. Plant Biol.* **2018**, *69*, 173–208. [CrossRef]
14. Su, T.; Li, X.; Yang, M.; Shao, Q.; Zhao, Y.; Ma, C.; Wang, P. Autophagy: An Intracellular Degradation Pathway Regulating Plant Survival and Stress Response. *Front. Plant Sci.* **2020**, *11*, 1–16. [CrossRef]
15. Mijaljica, D.; Prescott, M.; Devenish, R.J. Microautophagy in mammalian cells: Revisiting a 40-year-old conundrum. *Autophagy* **2011**, *7*, 673–682. [CrossRef]
16. Massey, A.; Kiffin, R.; Cuervo, A.M. Pathophysiology of chaperone-mediated autophagy. *Int. J. Biochem. Cell Biol.* **2004**, *36*, 2420–2434. [CrossRef]
17. Yorimitsu, T.; Klionsky, D.J. Atg11 Links Cargo to the Vesicle-forming Machinery in the Cytoplasm to Vacuole Targeting Pathway. *Mol. Biol. Cell* **2005**, *16*, 1593–1605. [CrossRef] [PubMed]
18. De duve, C.; Pressman, B.C.; Gianetto, R.; Wattiaux, R.; Appelmans, F. Tissue fractionation studies. 6. Intracellular distribution patterns of enzymes in rat-liver tissue. *Biochem. J.* **1955**, *60*, 604–617. [CrossRef] [PubMed]
19. Ohsumi, Y. Molecular dissection of autophagy: Two ubiquitin-like systems. *Nat. Rev. Mol. Cell Biol.* **2001**, *2*, 211–216. [CrossRef]
20. Tsukada, M.; Ohsumi, Y. Isolation and characterization of autophagy-defective mutants of Saccharomyces cerevisiae. *FEBS Lett.* **1993**, *333*, 169–174. [CrossRef]
21. Masclaux-Daubresse, C.; Chen, Q.; Havé, M. Regulation of nutrient recycling via autophagy. *Curr. Opin. Plant Biol.* **2017**, *39*, 8–17. [CrossRef]
22. Chung, T. How phosphoinositides shape autophagy in plant cells. *Plant Sci.* **2019**, *281*, 146–158. [CrossRef] [PubMed]
23. Kim, S.H.; Kwon, C.; Lee, J.H.; Chung, T. Genes for plant autophagy: Functions and interactions. *Mol. Cells* **2012**, *34*, 413–423. [CrossRef]
24. Kuma, A.; Mizushima, N. Physiological role of autophagy as an intracellular recycling system: With an emphasis on nutrient metabolism. *Semin. Cell Dev. Biol.* **2010**, *21*, 683–690. [CrossRef]

25. Li, F.; Chung, T.; Vierstra, R.D. AUTOPHAGY-RELATED11 plays a critical role in general autophagy- and senescence-induced mitophagy in Arabidopsis. *Plant Cell* **2014**, *26*, 788–807. [CrossRef] [PubMed]
26. Suttangkakul, A.; Li, F.; Chung, T.; Vierstra, R.D. The ATG1/ATG13 protein kinase complex is both a regulator and a target of autophagic recycling in Arabidopsis. *Plant Cell* **2011**, *23*, 3761–3779. [CrossRef] [PubMed]
27. Hosokawa, N.; Hara, T.; Kaizuka, T.; Kishi, C.; Takamura, A.; Miura, Y.; Iemura, S.; Natsume, T.; Takehana, K.; Yamada, N.; et al. Nutrient-dependent mTORC1 association with the ULK1-Atg13-FIP200 complex required for autophagy. *Mol. Biol. Cell* **2009**, *20*, 1981–1991. [CrossRef]
28. Hara, T.; Takamura, A.; Kishi, C.; Iemura, S.-I.; Natsume, T.; Guan, J.-L.; Mizushima, N. FIP200, a ULK-interacting protein, is required for autophagosome formation in mammalian cells. *J. Cell Biol.* **2008**, *181*, 497–510. [CrossRef] [PubMed]
29. Kawamata, T.; Kamada, Y.; Suzuki, K.; Kuboshima, N.; Akimatsu, H.; Ota, S.; Ohsumi, M.; Ohsumi, Y. Characterization of a novel autophagy-specific gene, ATG29. *Biochem. Biophys. Res. Commun.* **2005**, *338*, 1884–1889. [CrossRef]
30. Kabeya, Y.; Kawamata, T.; Suzuki, K.; Ohsumi, Y. Cis1/Atg31 is required for autophagosome formation in Saccharomyces cerevisiae. *Biochem. Biophys. Res. Commun.* **2007**, *356*, 405–410. [CrossRef] [PubMed]
31. Young, A.R.J.; Chan, E.Y.W.; Hu, X.W.; Köchl, R.; Crawshaw, S.G.; High, S.; Hailey, D.W.; Lippincott-Schwartz, J.; Tooze, S.A. Starvation and ULK1-dependent cycling of mammalian Atg9 between the TGN and endosomes. *J. Cell Sci.* **2006**, *119*, 3888–3900. [CrossRef] [PubMed]
32. Sawa-Makarska, J.; Baumann, V.; Coudevylle, N.; von Bülow, S.; Nogellova, V.; Abert, C.; Schuschnig, M.; Graef, M.; Hummer, G.; Martens, S. Reconstitution of autophagosome nucleation defines Atg9 vesicles as seeds for membrane formation. *Science* **2020**, 369. [CrossRef] [PubMed]
33. Shintani, T.; Suzuki, K.; Kamada, Y.; Noda, T.; Ohsumi, Y. Apg2p Functions in Autophagosome Formation on the Perivacuolar Structure. *J. Biol. Chem.* **2001**, *276*, 30452–30460. [CrossRef] [PubMed]
34. Proikas-Cezanne, T.; Waddell, S.; Gaugel, A.; Frickey, T.; Lupas, A.; Nordheim, A. WIPI-1alpha (WIPI49), a member of the novel 7-bladed WIPI protein family, is aberrantly expressed in human cancer and is linked to starvation-induced autophagy. *Oncogene* **2004**, *23*, 9314–9325. [CrossRef]
35. Yen, W.; Legakis, J.E.; Nair, U.; Klionsky, D.J. Atg27 Is Required for Autophagy-dependent Cycling of Atg9. *Mol. Biol. Cell* **2007**, *18*, 581–593. [CrossRef]
36. Liang, X.H.; Jackson, S.; Seaman, M.; Brown, K.; Kempkes, B.; Hibshoosh, H.; Levine, B. Induction of autophagy and inhibition of tumorigenesis by beclin 1. *Nature* **1999**, *402*, 672–676. [CrossRef] [PubMed]
37. Kametaka, S.; Okano, T.; Ohsumi, M.; Ohsumi, Y. Apg14p and Apg6/Vps30p form a protein complex essential for autophagy in the yeast, Saccharomyces cerevisiae. *J. Biol. Chem.* **1998**, *273*, 22284–22291. [CrossRef] [PubMed]
38. Chen, L.; Su, Z.Z.; Huang, L.; Xia, F.N.; Qi, H.; Xie, L.J.; Xiao, S.; Chen, Q.F. The AMP-activated protein kinase kin10 is involved in the regulation of autophagy in arabidopsis. *Front. Plant Sci.* **2017**, *8*, 1–11. [CrossRef] [PubMed]
39. Liu, F.; Hu, W.; Li, F.; Marshall, R.S.; Zarza, X.; Munnik, T.; Vierstra, R.D. AUTOPHAGY-RELATED14 and Its Associated Phosphatidylinositol 3-Kinase Complex Promote Autophagy in Arabidopsis. *Plant Cell* **2020**, *32*, 3939–3960. [CrossRef] [PubMed]
40. Itakura, E.; Kishi-Itakura, C.; Mizushima, N. The hairpin-type tail-anchored SNARE syntaxin 17 targets to autophagosomes for fusion with endosomes/lysosomes. *Cell* **2012**, *151*, 1256–1269. [CrossRef]
41. Yamauchi, S.; Mano, S.; Oikawa, K.; Hikino, K.; Teshima, K.M.; Kimori, Y.; Nishimura, M.; Shimazaki, K.; Takemiya, A. Autophagy controls reactive oxygen species homeostasis in guard cells that is essential for stomatal opening. *Proc. Natl. Acad. Sci. USA* **2019**, *116*, 19187–19192. [CrossRef]
42. Mizushima, N.; Sugita, H.; Yoshimori, T.; Ohsumi, Y. A new protein conjugation system in human. The counterpart of the yeast Apg12p conjugation system essential for autophagy. *J. Biol. Chem.* **1998**, *273*, 33889–33892. [CrossRef]
43. Minina, E.A.; Moschou, P.N.; Vetukuri, R.R.; Sanchez-Vera, V.; Cardoso, C.; Liu, Q.; Elander, P.H.; Dalman, K.; Beganovic, M.; Lindberg Yilmaz, J.; et al. Transcriptional stimulation of rate-limiting components of the autophagic pathway improves plant fitness. *J. Exp. Bot.* **2018**, *69*, 1415–1432. [CrossRef] [PubMed]
44. Mizushima, N.; Kuma, A.; Kobayashi, Y.; Yamamoto, A.; Matsubae, M.; Takao, T.; Natsume, T.; Ohsumi, Y.; Yoshimori, T. Mouse Apg16L, a novel WD-repeat protein, targets to the autophagic isolation membrane with the Apg12-Apg5 conjugate. *J. Cell Sci.* **2003**, *116*, 1679–1688. [CrossRef]
45. Harada, K.; Kotani, T.; Kirisako, H.; Sakoh-Nakatogawa, M.; Oikawa, Y.; Kimura, Y.; Hirano, H.; Yamamoto, H.; Ohsumi, Y.; Nakatogawa, H. Two distinct mechanisms target the autophagy-related E3 complex to the pre-autophagosomal structure. *Elife* **2019**, *8*, e43088. [CrossRef] [PubMed]
46. Tanida, I. Autophagosome formation and molecular mechanism of autophagy. *Antioxid. Redox Signal.* **2011**, *14*, 2201–2214. [CrossRef] [PubMed]
47. Shin, J.-H.; Yoshimoto, K.; Ohsumi, Y.; Jeon, J.-S.; An, G. OsATG10b, an autophagosome component, is needed for cell survival against oxidative stresses in rice. *Mol. Cells* **2009**, *27*, 67–74. [CrossRef] [PubMed]
48. Nemoto, T.; Tanida, I.; Tanida-Miyake, E.; Minematsu-Ikeguchi, N.; Yokota, M.; Ohsumi, M.; Ueno, T.; Kominami, E. The Mouse APG10 Homologue, an E2-like Enzyme for Apg12p Conjugation, Facilitates MAP-LC3 Modification. *J. Biol. Chem.* **2003**, *278*, 39517–39526. [CrossRef]

49. Kabeya, Y.; Mizushima, N.; Ueno, T.; Yamamoto, A.; Kirisako, T.; Noda, T.; Kominami, E.; Ohsumi, Y.; Yoshimori, T. LC3, a mammalian homologue of yeast Apg8p, is localized in autophagosome membranes after processing. *EMBO J.* **2000**, *19*, 5720–5728. [CrossRef]
50. Bu, F.; Yang, M.; Guo, X.; Huang, W.; Chen, L. Multiple Functions of ATG8 Family Proteins in Plant Autophagy. *Front. Cell Dev. Biol.* **2020**, *8*, 466. [CrossRef] [PubMed]
51. Tanida, I.; Tanida-Miyake, E.; Komatsu, M.; Ueno, T.; Kominami, E. Human Apg3p/Aut1p Homologue Is an Authentic E2 Enzyme for Multiple Substrates, GATE-16, GABARAP, and MAP-LC3, and Facilitates the Conjugation of hApg12p to hApg5p. *J. Biol. Chem.* **2002**, *277*, 13739–13744. [CrossRef]
52. Wang, S.; Li, Y.; Ma, C. Atg3 promotes Atg8 lipidation via altering lipid diffusion and rearrangement. *Protein Sci.* **2020**, *29*, 1511–1523. [CrossRef]
53. Tanida, I.; Sou, Y.; Ezaki, J.; Minematsu-Ikeguchi, N.; Ueno, T.; Kominami, E. HsAtg4B/HsApg4B/Autophagin-1 Cleaves the Carboxyl Termini of Three Human Atg8 Homologues and Delipidates Microtubule-associated Protein Light Chain 3- and GABAA Receptor-associated Protein-Phospholipid Conjugates. *J. Biol. Chem.* **2004**, *279*, 36268–36276. [CrossRef]
54. Rodrigues, A.; Adamo, M.; Crozet, P.; Margalha, L.; Confraria, A.; Martinho, C.; Elias, A.; Rabissi, A.; Lumbreras, V.; González-Guzmán, M.; et al. ABI1 and PP2CA phosphatases are negative regulators of Snf1-related protein kinase1 signaling in Arabidopsis. *Plant Cell* **2013**, *25*, 3871–3884. [CrossRef]
55. Menand, B.; Desnos, T.; Nussaume, L.; Bergert, F.; Bouchez, D.; Meyer, C.; Robaglia, C. Expression and disruption of the Arabidopsis TOR (target of rapamycin) gene. *Proc. Natl. Acad. Sci. USA* **2002**, *99*, 6422–6427. [CrossRef] [PubMed]
56. Deprost, D.; Truong, H.N.; Robaglia, C.; Meyer, C. An Arabidopsis homolog of RAPTOR/KOG1 is essential for early embryo development. *Biochem. Biophys. Res. Commun.* **2005**, *326*, 844–850. [CrossRef]
57. Anderson, G.H.; Veit, B.; Hanson, M.R. The Arabidopsis AtRaptor genes are essential for post-embryonic plant growth. *BMC Biol.* **2005**, *3*, 1–11. [CrossRef] [PubMed]
58. Moreau, M.; Azzopardi, M.; Clément, G.; Dobrenel, T.; Marchive, C.; Renne, C.; Martin-Magniette, M.-L.; Taconnat, L.; Renou, J.-P.; Robaglia, C.; et al. Mutations in the Arabidopsis Homolog of LST8/GβL, a Partner of the Target of Rapamycin Kinase, Impair Plant Growth, Flowering, and Metabolic Adaptation to Long Days. *Plant Cell* **2012**, *24*, 463–481. [CrossRef]
59. Saxton, R.A.; Sabatini, D.M. mTOR Signaling in Growth, Metabolism, and Disease. *Cell* **2017**, *169*, 361–371. [CrossRef]
60. Liu, Y.; Nguyen, P.T.; Wang, X.; Zhao, Y.; Meacham, C.E.; Zou, Z.; Bordieanu, B.; Johanns, M.; Vertommen, D.; Wijshake, T.; et al. TLR9 and beclin 1 crosstalk regulates muscle AMPK activation in exercise. *Nature* **2020**, *578*, 605–609. [CrossRef] [PubMed]
61. Inoki, K.; Zhu, T.; Guan, K.-L. TSC2 Mediates Cellular Energy Response to Control Cell Growth and Survival. *Cell* **2003**, *115*, 577–590. [CrossRef]
62. Kim, J.; Kundu, M.; Viollet, B.; Guan, K.-L. AMPK and mTOR regulate autophagy through direct phosphorylation of Ulk1. *Nat. Cell Biol.* **2011**, *13*, 132–141. [CrossRef] [PubMed]
63. Pu, Y.; Luo, X.; Bassham, D.C. Tor-dependent and -independent pathways regulate autophagy in arabidopsis thaliana. *Front. Plant Sci.* **2017**, *8*, 1–13. [CrossRef] [PubMed]
64. Li, F.; Vierstra, R.D. Arabidopsis ATG11, a scaffold that links the ATG1-ATG13 kinase complex to general autophagy and selective mitophagy. *Autophagy* **2014**, *10*, 1466–1467. [CrossRef]
65. Li, F.; Vierstra, R.D. Autophagy: A multifaceted intracellular system for bulk and selective recycling. *Trends Plant Sci.* **2012**, *17*, 526–537. [CrossRef]
66. Yamamoto, H.; Kakuta, S.; Watanabe, T.M.; Kitamura, A.; Sekito, T.; Kondo-Kakuta, C.; Ichikawa, R.; Kinjo, M.; Ohsumi, Y. Atg9 vesicles are an important membrane source during early steps of autophagosome formation. *J. Cell Biol.* **2012**, *198*, 219–233. [CrossRef]
67. Orsi, A.; Razi, M.; Dooley, H.C.; Robinson, D.; Weston, A.E.; Collinson, L.M.; Tooze, S.A. Dynamic and transient interactions of Atg9 with autophagosomes, but not membrane integration, are required for autophagy. *Mol. Biol. Cell* **2012**, *23*, 1860–1873. [CrossRef]
68. Zhuang, X.; Chung, K.P.; Luo, M.; Jiang, L. Autophagosome Biogenesis and the Endoplasmic Reticulum: A Plant Perspective. *Trends Plant Sci.* **2018**, *23*, 677–692. [CrossRef]
69. Johansen, T.; Lamark, T. Selective autophagy mediated by autophagic adapter proteins. *Autophagy* **2011**, *7*, 279–296. [CrossRef] [PubMed]
70. Liu, Y.; Bassham, D.C. Autophagy: Pathways for self-eating in plant cells. *Annu. Rev. Plant Biol.* **2012**, *63*, 215–237. [CrossRef]
71. Zhuang, X.; Wang, H.; Lam, S.K.; Gao, C.; Wang, X.; Cai, Y.; Jiang, L. A BAR-domain protein SH3P2, which binds to phosphatidylinositol 3-phosphate and ATG8, regulates autophagosome formation in Arabidopsis. *Plant Cell* **2013**, *25*, 4596–4615. [CrossRef] [PubMed]
72. Pankiv, S.; Alemu, E.A.; Brech, A.; Bruun, J.A.; Lamark, T.; Øvervatn, A.; Bjørkøy, G.; Johansen, T. FYCO1 is a Rab7 effector that binds to LC3 and PI3P to mediate microtubule plus end—Directed vesicle transport. *J. Cell Biol.* **2010**, *188*, 253–269. [CrossRef] [PubMed]
73. Ketelaar, T.; Voss, C.; Dimmock, S.A.; Thumm, M.; Hussey, P.J. Arabidopsis homologues of the autophagy protein Atg8 are a novel family of microtubule binding proteins. *FEBS Lett.* **2004**, *567*, 302–306. [CrossRef] [PubMed]
74. Reggiori, F.; Ungermann, C. Autophagosome Maturation and Fusion. *J. Mol. Biol.* **2017**, *429*, 486–496. [CrossRef] [PubMed]

75. Hanaoka, H.; Noda, T.; Shirano, Y.; Kato, T.; Hayashi, H.; Shibata, D.; Tabata, S.; Ohsumi, Y. Leaf senescence and starvation-induced chlorosis are accelerated by the disruption of an Arabidopsis autophagy gene. *Plant Physiol.* **2002**, *129*, 1181–1193. [CrossRef]
76. Doelling, J.H.; Walker, J.M.; Friedman, E.M.; Thompson, A.R.; Vierstra, R.D. The APG8/12-activating enzyme APG7 is required for proper nutrient recycling and senescence in Arabidopsis thaliana. *J. Biol. Chem.* **2002**, *277*, 33105–33114. [CrossRef]
77. Slávikovà, S.; Shy, G.; Yao, Y.; Glozman, R.; Levanony, H.; Pietrokovski, S.; Elazar, Z.; Galili, G. The autophagy-associated Atg8 gene family operates both under favourable growth conditions and under starvation stresses in Arabidopsis plants. *J. Exp. Bot.* **2005**, *56*, 2839–2849. [CrossRef] [PubMed]
78. Yoshimoto, K.; Hanaoka, H.; Sato, S.; Kato, T.; Tabata, S. Processing of ATG8s, Ubiquitin-Like Proteins, and Their Deconjugation by ATG4s Are Essential for Plant Autophagy Published by : American Society of Plant Biologists (ASPB) Linked references are available on JSTOR for this article. *Plant Cell* **2004**, *16*, 2967–2983. [CrossRef]
79. Chung, T.; Phillips, A.R.; Vierstra, R.D. ATG8 lipidation and ATG8-mediated autophagy in Arabidopsis require ATG12 expressed from the differentially controlled ATG12A and ATG12B loci. *Plant J.* **2010**, *62*, 483–493. [CrossRef]
80. Phillips, A.R.; Suttangkakul, A.; Vierstra, R.D. The ATG12-conjugating enzyme ATG10 is essential for autophagic vesicle formation in Arabidopsis thaliana. *Genetics* **2008**, *178*, 1339–1353. [CrossRef]
81. Surpin, M.; Zheng, H.; Morita, M.T.; Saito, C.; Avila, E.; Blakeslee, J.J.; Bandyopadhyay, A.; Kovaleva, V.; Carter, D.; Murphy, A.; et al. The VTI Family of SNARE Proteins Is Necessary for Plant Viability and Mediates Different Protein Transport Pathways. *Plant Cell* **2003**, *15*, 2885–2899. [CrossRef]
82. Katsiarimpa, A.; Kalinowska, K.; Anzenberger, F.; Weis, C.; Ostertag, M.; Tsutsumi, C.; Schwechheimer, C.; Brunner, F.; Hückelhoven, R.; Isono, E. The deubiquitinating enzyme AMSH1 and the ESCRT-III subunit VPS2.1 are required for autophagic degradation in Arabidopsis. *Plant Cell* **2013**, *25*, 2236–2252. [CrossRef] [PubMed]
83. Gao, C.; Zhuang, X.; Cui, Y.; Fu, X.; He, Y.; Zhao, Q.; Zeng, Y.; Shen, J.; Luo, M.; Jiang, L. Dual roles of an Arabidopsis ESCRT component FREE1 in regulating vacuolar protein transport and autophagic degradation. *Proc. Natl. Acad. Sci. USA* **2015**, *112*, 1886–1891. [CrossRef] [PubMed]
84. Yu, L.; Chen, Y.; Tooze, S.A. Autophagy pathway: Cellular and molecular mechanisms. *Autophagy* **2018**, *14*, 207–215. [CrossRef] [PubMed]
85. Farré, J.C.; Subramani, S. Mechanistic insights into selective autophagy pathways: Lessons from yeast. *Nat. Rev. Mol. Cell Biol.* **2016**, *17*, 537–552. [CrossRef] [PubMed]
86. Maqbool, A.; Hughes, R.K.; Dagdas, Y.F.; Tregidgo, N.; Zess, E.; Belhaj, K.; Round, A.; Bozkurt, T.O.; Kamoun, S.; Banfield, M.J. Structural basis of host autophagy-related protein 8 (ATG8) Binding by the irish potato famine pathogen effector protein PexRD54. *J. Biol. Chem.* **2016**, *291*, 20270–20282. [CrossRef] [PubMed]
87. Marshall, R.S.; Li, F.; Gemperline, D.C.; Book, A.J.; Vierstra, R.D. Autophagic Degradation of the 26S Proteasome Is Mediated by the Dual ATG8/Ubiquitin Receptor RPN10 in Arabidopsis. *Physiol. Behav.* **2015**, *58*, 1053–1066. [CrossRef]
88. Yoon, S.H.; Chung, T. Protein and RNA Quality Control by Autophagy in Plant Cells. *Mol. Cells* **2019**, *42*, 285–291.
89. Lu, K.; Psakhye, I.; Jentsch, S. Autophagic clearance of PolyQ proteins mediated by ubiquitin-Atg8 adaptors of the conserved CUET protein family. *Cell* **2014**, *158*, 549–563. [CrossRef]
90. Pankiv, S.; Clausen, T.H.; Lamark, T.; Brech, A.; Bruun, J.A.; Outzen, H.; Øvervatn, A.; Bjørkøy, G.; Johansen, T. p62/SQSTM1 binds directly to Atg8/LC3 to facilitate degradation of ubiquitinated protein aggregates by autophagy. *J. Biol. Chem.* **2007**, *282*, 24131–24145. [CrossRef] [PubMed]
91. Svenning, S.; Lamark, T.; Krause, K.; Johansen, T. Plant NBR1 is a selective autophagy substrate and a functional hybrid of the mammalian autophagic adapters NBR1 and p62/SQSTM1. *Autophagy* **2011**, *7*, 993–1010. [CrossRef]
92. Zhou, J.; Wang, J.; Yu, J.Q.; Chen, Z. Role and regulation of autophagy in heat stress responses of tomato plants. *Front. Plant Sci.* **2014**, *5*, 1–12. [CrossRef]
93. Jung, H.; Lee, H.N.; Marshall, R.S.; Lomax, A.W.; Yoon, M.J.; Kim, J.; Kim, J.H.; Vierstra, R.D.; Chung, T. Arabidopsis cargo receptor NBR1 mediates selective autophagy of defective proteins. *J. Exp. Bot.* **2020**, *71*, 73–89. [CrossRef]
94. Livneh, I.; Cohen-Kaplan, V.; Cohen-Rosenzweig, C.; Avni, N.; Ciechanover, A. The life cycle of the 26S proteasome: From birth, through regulation and function, and onto its death. *Cell Res.* **2016**, *26*, 869–885. [CrossRef] [PubMed]
95. Marshall, R.S.; Hua, Z.; Mali, S.; McLoughlin, F.; Vierstra, R.D. ATG8-binding UIM Proteins Define a New Class of Autophagy Adaptors and Receptors. *Cell* **2019**, *177*, 766–781e24. [CrossRef] [PubMed]
96. Bard, J.A.; Goodall, E.A.; Greene, E.R.; Jonsson, E.; Dong, K.C.; Martin, A. Structure and Function of the 26S Proteasome Jared. *Physiol. Behav.* **2019**, *176*, 139–148.
97. Demishtein, A.; Fraiberg, M.; Berko, D.; Tirosh, B.; Elazar, Z.; Navon, A. SQSTM1/p62-mediated autophagy compensates for loss of proteasome polyubiquitin recruiting capacity. *Autophagy* **2017**, *13*, 1697–1708. [CrossRef]
98. Cohen-Kaplan, V.; Livneh, I.; Avni, N.; Fabre, B.; Ziv, T.; Kwon, Y.T.; Ciechanover, A. p62- and ubiquitin-dependent stress-induced autophagy of the mammalian 26S proteasome. *Proc. Natl. Acad. Sci. USA* **2016**, *113*, E7490–E7499. [CrossRef]
99. Kirkin, V.; Rogov, V.V. A Diversity of Selective Autophagy Receptors Determines the Specificity of the Autophagy Pathway. *Mol. Cell* **2019**, *76*, 268–285. [CrossRef] [PubMed]
100. Papandreou, M.E.; Tavernarakis, N. Nucleophagy: From homeostasis to disease. *Cell Death Differ.* **2019**, *26*, 630–639. [CrossRef] [PubMed]

101. Li, F.; Zhang, M.; Zhang, C.; Zhou, X. Nuclear autophagy degrades a geminivirus nuclear protein to restrict viral infection in solanaceous plants. *New Phytol.* **2020**, *225*, 1746–1761. [CrossRef]
102. Nolan, T.M.; Brennan, B.; Yang, M.; Chen, J.; Zhang, M.; Li, Z.; Wang, X.; Bassham, D.C.; Walley, J.; Yin, Y. Selective Autophagy of BES1 Mediated by DSK2 Balances Plant Growth and Survival. *Dev. Cell* **2017**, *41*, 33–46e7. [CrossRef] [PubMed]
103. Mochida, K.; Oikawa, Y.; Kimura, Y.; Kirisako, H.; Hirano, H.; Ohsumi, Y.; Nakatogawa, H. Receptor-mediated selective autophagy degrades the endoplasmic reticulum and the nucleus. *Nature* **2015**, *522*, 359–362. [CrossRef] [PubMed]
104. Floyd, B.E.; Morriss, S.C.; MacIntosh, G.C.; Bassham, D.C. Evidence for autophagy-dependent pathways of rRNA turnover in Arabidopsis. *Autophagy* **2015**, *11*, 2199–2212. [CrossRef]
105. Wyant, G.A.; Abu-Remaileh, M.; Frenkel, E.M.; Laqtom, N.N.; Dharamdasani, V.; Lewis, C.A.; Chan, S.H.; Heinze, I.; Ori, A.; Sabatini, D.M. NUFIP1 is a ribosome receptor for starvation-induced ribophagy. *Science* **2018**, *360*, 751–758. [CrossRef]
106. Rodor, J.; Jobet, E.; Bizarro, J.; Vignols, F.; Carles, C.; Suzuki, T.; Nakamura, K.; Echeverría, M. AtNUFIP, an essential protein for plant development, reveals the impact of snoRNA gene organisation on the assembly of snoRNPs and rRNA methylation in Arabidopsis thaliana. *Plant J.* **2011**, *65*, 807–819. [CrossRef]
107. Fan, J.; Yan, C.; Roston, R.; Shanklin, J.; Xu, C. Arabidopsis lipins, PDAT1 ACYLTRANSFERASE, and SDP1 triacylglycerol lipase synergistically direct fatty acids toward β-oxidation, thereby maintaining membrane lipid homeostasis. *Plant Cell* **2014**, *26*, 4119–4134. [CrossRef]
108. Fan, J.; Yu, L.; Xu, C. A central role for triacylglycerol in membrane lipid breakdown, fatty acid β-oxidation, and plant survival under extended darkness. *Plant Physiol.* **2017**, *174*, 1517–1530. [CrossRef]
109. Fan, J.; Yu, L.; Xu, C. Dual role for autophagy in lipid metabolismin arabidopsis. *Plant Cell* **2019**, *31*, 1598–1613. [CrossRef]
110. Wang, M.; Kaufman, R.J. Protein misfolding in the endoplasmic reticulum as a conduit to human disease. *Nature* **2016**, *529*, 326–335. [CrossRef] [PubMed]
111. Zechner, R.; Madeo, F.; Kratky, D. Cytosolic lipolysis and lipophagy: Two sides of the same coin. *Nat. Rev. Mol. Cell Biol.* **2017**, *18*, 671–684. [CrossRef]
112. Jaishy, B.; Abel, E.D. Thematic review series: Lipotoxicity: Many roads to cell dysfunction and cell death lipids, lysosomes, and autophagy. *J. Lipid Res.* **2016**, *57*, 1619–1635. [CrossRef] [PubMed]
113. Van Zutphen, T.; Todde, V.; De Boer, R.; Kreim, M.; Hofbauer, H.F.; Wolinski, H.; Veenhuis, M.; Van Der Klei, I.J.; Kohlwein, S.D. Lipid droplet autophagy in the yeast Saccharomyces cerevisiae. *Mol. Biol. Cell* **2014**, *25*, 290–301. [CrossRef] [PubMed]
114. Kurusu, T.; Koyano, T.; Hanamata, S.; Kubo, T.; Noguchi, Y.; Yagi, C.; Nagata, N.; Yamamoto, T.; Ohnishi, T.; Okazaki, Y.; et al. OsATG7 is required for autophagy-dependent lipid metabolism in rice postmeiotic anther development. *Autophagy* **2014**, *10*, 878–888. [CrossRef]
115. Hanamata, S.; Kurusu, T.; Kuchitsu, K. Roles of autophagy in male reproductive development in plants. *Front. Plant Sci.* **2014**, *5*, 1–6. [CrossRef] [PubMed]
116. Eastmond, P.J. Sugar-dependent1 encodes a patatin domain triacylglycerol lipase that initiates storage oil breakdown in germinating Arabidopsis seeds. *Plant Cell* **2006**, *18*, 665–675. [CrossRef]
117. Havé, M.; Luo, J.; Tellier, F.; Balliau, T.; Cueff, G.; Chardon, F.; Zivy, M.; Rajjou, L.; Cacas, J.L.; Masclaux-Daubresse, C. Proteomic and lipidomic analyses of the Arabidopsis atg5 autophagy mutant reveal major changes in endoplasmic reticulum and peroxisome metabolisms and in lipid composition. *New Phytol.* **2019**, *223*, 1461–1477. [CrossRef]
118. Wang, C.W. Lipid droplets, lipophagy, and beyond. *Biochim. Biophys. Acta Mol. Cell Biol. Lipids* **2016**, *1861*, 793–805. [CrossRef]
119. Sathyanarayan, A.; Mashek, M.T.; Mashek, D.G. ATGL Promotes Autophagy/Lipophagy via SIRT1 to Control Hepatic Lipid Droplet Catabolism. *Cell Rep.* **2017**, *19*, 1–9. [CrossRef]
120. Dupont, N.; Chauhan, S.; Arko-Mensah, J.; Castillo, E.F.; Masedunskas, A.; Weigert, R.; Robenek, H.; Proikas-Cezanne, T.; Deretic, V. Neutral lipid stores and lipase PNPLA5 contribute to autophagosome biogenesis. *Curr. Biol.* **2014**, *24*, 609–620. [CrossRef]
121. Ward, C.; Martinez-Lopez, N.; Otten, E.G.; Carroll, B.; Maetzel, D.; Singh, R.; Sarkar, S.; Korolchuk, V.I. Autophagy, lipophagy and lysosomal lipid storage disorders. *Biochim. Biophys. Acta Mol. Cell Biol. Lipids* **2016**, *1861*, 269–284. [CrossRef]
122. Negoita, F.; Blomdahl, J.; Wasserstrom, S.; Winberg, M.E.; Osmark, P.; Larsson, S.; Stenkula, K.G.; Ekstedt, M.; Kechagias, S.; Holm, C.; et al. PNPLA3 variant M148 causes resistance to starvation-mediated lipid droplet autophagy in human hepatocytes. *J. Cell. Biochem.* **2019**, *120*, 343–356. [CrossRef]
123. Tatsumi, T.; Takayama, K.; Ishii, S.; Yamamoto, A.; Hara, T.; Minami, N.; Miyasaka, N.; Kubota, T.; Matsuura, A.; Itakura, E.; et al. Forced lipophagy reveals that lipid droplets are required for early embryonic development in mouse. *Development* **2018**, *145*, dev161893. [CrossRef] [PubMed]
124. Stein, A.; Ruggiano, A.; Carvalho, P.; Rapoport, T.A. Key steps in ERAD of luminal ER proteins reconstituted with purified components. *Cell* **2014**, *158*, 1375–1388. [CrossRef]
125. Schuck, S.; Gallagher, C.M.; Walter, P. ER-phagy mediates selective degradation of endoplasmic reticulum independently of the core autophagy machinery. *J. Cell Sci.* **2014**, *127*, 4078–4088. [CrossRef] [PubMed]
126. Khaminets, A.; Heinrich, T.; Mari, M.; Grumati, P.; Huebner, A.K.; Akutsu, M.; Liebmann, L.; Stolz, A.; Nietzsche, S.; Koch, N.; et al. Regulation of endoplasmic reticulum turnover by selective autophagy. *Nature* **2015**, *522*, 354–358. [CrossRef]
127. Nthiga, T.M.; Kumar Shrestha, B.; Sjøttem, E.; Bruun, J.; Bowitz Larsen, K.; Bhujabal, Z.; Lamark, T.; Johansen, T. CALCOCO 1 acts with VAMP -associated proteins to mediate ER -phagy. *EMBO J.* **2020**, *39*, 1–24. [CrossRef]

128. Stephani, M.; Picchianti, L.; Gajic, A.; Beveridge, R.; Skarwan, E.; de Hernandez, V.S.M.; Mohseni, A.; Clavel, M.; Zeng, Y.; Naumann, C.; et al. A cross-kingdom conserved er-phagy receptor maintains endoplasmic reticulum homeostasis during stress. *Elife* **2020**, *9*, 1–105. [CrossRef]
129. Stefely, J.A.; Zhang, Y.; Freiberger, E.C.; Kwiecien, N.W.; Thomas, H.E.; Davis, A.M.; Lowry, N.D.; Vincent, C.E.; Shishkova, E.; Clark, N.A.; et al. Mass spectrometry proteomics reveals a function for mammalian CALCOCO1 in MTOR-regulated selective autophagy. *Autophagy* **2020**, *16*, 2219–2237. [CrossRef]
130. Tschurtschenthaler, M.; Adolph, T.E.; Ashcroft, J.W.; Niederreiter, L.; Bharti, R.; Saveljeva, S.; Bhattacharyya, J.; Flak, M.B.; Shih, D.Q.; Fuhler, G.M.; et al. Defective ATG16L1-mediated removal of IRE1α drives Crohn's disease-like ileitis. *J. Exp. Med.* **2017**, *214*, 401–422. [CrossRef] [PubMed]
131. Honig, A.; Avin-Wittenberg, T.; Ufaz, S.; Galili, G. A new type of compartment, defined by plant-specific Atg8-interacting proteins, is induced upon exposure of Arabidopsis plants to carbon starvation. *Plant Cell* **2012**, *24*, 288–303. [CrossRef]
132. Frank, M.; Duvezin-Caubet, S.; Koob, S.; Occhipinti, A.; Jagasia, R.; Petcherski, A.; Ruonala, M.O.; Priault, M.; Salin, B.; Reichert, A.S. Mitophagy is triggered by mild oxidative stress in a mitochondrial fission dependent manner. *Biochim. Biophys. Acta Mol. Cell Res.* **2012**, *1823*, 2297–2310. [CrossRef] [PubMed]
133. Chang, T.K.; Shravage, B.V.; Hayes, S.D.; Powers, C.M.; Simin, R.T.; Harper, J.W.; Baehrecke, E.H. Uba1 functions in Atg7- and Atg3-independent autophagy. *Nat. Cell Biol.* **2016**, *33*, 839–841. [CrossRef]
134. Matsuda, N.; Sato, S.; Shiba, K.; Okatsu, K.; Saisho, K.; Gautier, C.A.; Sou, Y.S.; Saiki, S.; Kawajiri, S.; Sato, F.; et al. PINK1 stabilized by mitochondrial depolarization recruits Parkin to damaged mitochondria and activates latent Parkin for mitophagy. *J. Cell Biol.* **2010**, *189*, 211–221. [CrossRef]
135. Lazarou, M.; Sliter, D.A.; Kane, L.A.; Sarraf, S.A.; Burman, J.L.; Sideris, D.P.; Fogel, A.I.; Youle, R.J.; Branch, S.N. The ubiquitin kinase PINK1 recruits autophagy receptors to induce mitophagy. *Nature.* **2016**, *524*, 309–314. [CrossRef]
136. Sarraf, S.A.; Raman, M.; Guarani-Pereira, V.; Sowa, M.E.; Huttlin, E.L.; Gygi, S.P.; Harper, J.W. Landscape of the PARKIN-dependent ubiquitylome in response to mitochondrial depolarization. *Nature* **2013**, *496*, 372–376. [CrossRef] [PubMed]
137. Broda, M.; Millar, A.H.; Van Aken, O. Mitophagy: A Mechanism for Plant Growth and Survival. *Trends Plant Sci.* **2018**, *23*, 434–450. [CrossRef] [PubMed]
138. Nakamura, S.; Hagihara, S.; Izumi, M. Mitophagy in plants. *Biochim. Biophys. Acta Gen. Subj.* **2021**, *1865*, 129916. [CrossRef] [PubMed]
139. Toyooka, K.; Okamoto, T.; Minamikawa, T. Cotyledon cells of Vigna mungo seedlings use at least two distinct autophagic machineries for degradation of starch granules and cellular components. *J. Cell Biol.* **2001**, *154*, 973–982. [CrossRef]
140. Kao, Y.-T.; Gonzalez, K.L.; Bartel, B. Peroxisome Function, Biogenesis, and Dynamics in Plants. *Plant Physiol.* **2018**, *176*, 162–177. [CrossRef]
141. Kamada, T.; Nito, K.; Hayashi, H.; Mano, S.; Hayashi, M.; Nishimura, M. Functional Differentiation of Peroxisomes Revealed by Expression Profiles of Peroxisomal Genes in Arabidopsis thaliana. *Plant Cell Physiol.* **2003**, *44*, 1275–1289. [CrossRef]
142. Zientara-Rytter, K.; Sirko, A. To deliver or to degrade—An interplay of the ubiquitin-proteasome system, autophagy and vesicular transport in plants. *FEBS J.* **2016**, *283*, 3534–3555. [CrossRef]
143. Farré, J.-C.; Burkenroad, A.; Burnett, S.F.; Subramani, S. Phosphorylation of mitophagy and pexophagy receptors coordinates their interaction with Atg8 and Atg11. *EMBO Rep.* **2013**, *14*, 441–449. [CrossRef]
144. Sargent, G.; van Zutphen, T.; Shatseva, T.; Zhang, L.; Di Giovanni, V.; Bandsma, R.; Kim, P.K. PEX2 is the E3 ubiquitin ligase required for pexophagy during starvation. *J. Cell Biol.* **2016**, *214*, 677–690. [CrossRef] [PubMed]
145. Calero-Muñoz, N.; Exposito-Rodriguez, M.; Collado-Arenal, A.M.; Rodríguez-Serrano, M.; Laureano-Marín, A.M.; Santamaría, M.E.; Gotor, C.; Díaz, I.; Mullineaux, P.M.; Romero-Puertas, M.C.; et al. Cadmium induces reactive oxygen species-dependent pexophagy in Arabidopsis leaves. *Plant. Cell Environ.* **2019**, *42*, 2696–2714. [CrossRef]
146. Young, P.G.; Passalacqua, M.J.; Chappell, K.; Llinas, R.J.; Bartel, B. A facile forward-genetic screen for Arabidopsis autophagy mutants reveals twenty-one loss-of-function mutations disrupting six ATG genes. *Autophagy* **2019**, *15*, 941–959. [CrossRef] [PubMed]
147. Michaeli, S.; Galili, G.; Genschik, P.; Fernie, A.R.; Avin-Wittenberg, T. Autophagy in Plants—What's New on the Menu? *Trends Plant Sci.* **2016**, *21*, 134–144. [CrossRef] [PubMed]
148. Aits, S.; Jäättelä, M. Lysosomal cell death at a glance. *J. Cell Sci.* **2013**, *126*, 1905–1912. [CrossRef]
149. Hung, Y.H.; Chen, L.M.W.; Yang, J.Y.; Yuan Yang, W. Spatiotemporally controlled induction of autophagy-mediated lysosome turnover. *Nat. Commun.* **2013**, *4*, 2111. [CrossRef]
150. Maejima, I.; Takahashi, A.; Omori, H.; Kimura, T.; Takabatake, Y.; Saitoh, T.; Yamamoto, A.; Hamasaki, M.; Noda, T.; Isaka, Y.; et al. Autophagy sequesters damaged lysosomes to control lysosomal biogenesis and kidney injury. *EMBO J.* **2013**, *32*, 2336–2347. [CrossRef]
151. Wittenbach, V.A.; Lin, W.; Hebert, R.R. Vacuolar Localization of Proteases and Degradation of Chloroplasts in Mesophyll Protoplasts from Senescing Primary Wheat Leaves. *Plant Physiol.* **1982**, *69*, 98–102. [CrossRef]
152. Izumi, M.; Ishida, H.; Nakamura, S.; Hidema, J. Entire photodamaged chloroplasts are transported to the central vacuole by autophagy. *Plant Cell* **2017**, *29*, 377–394. [CrossRef] [PubMed]

153. Ishida, H.; Yoshimoto, K.; Izumi, M.; Reisen, D.; Yano, Y.; Makino, A.; Ohsumi, Y.; Hanson, M.R.; Mae, T. Mobilization of Rubisco and stroma-localized fluorescent proteins of chloroplasts to the vacuole by an ATG gene-dependent autophagic process. *Plant Physiol.* **2008**, *148*, 142–155. [CrossRef]
154. Wada, S.; Ishida, H.; Izumi, M.; Yoshimoto, K.; Ohsumi, Y.; Mae, T.; Makino, A. Autophagy plays a role in chloroplast degradation during senescence in individually darkened leaves. *Plant Physiol.* **2009**, *149*, 885–893. [CrossRef] [PubMed]
155. Wada, S.; Hayashida, Y.; Izumi, M.; Kurusu, T.; Hanamata, S.; Kanno, K.; Kojima, S.; Yamaya, T.; Kuchitsu, K.; Makino, A.; et al. Autophagy supports biomass production and nitrogen use efficiency at the vegetative stage in rice. *Plant Physiol.* **2015**, *168*, 60–73. [CrossRef]
156. Michaeli, S.; Honig, A.; Levanony, H.; Peled-Zehavi, H.; Galili, G. Arabidopsis ATG8-INTERACTING PROTEIN1 is involved in autophagy-dependent vesicular trafficking of plastid proteins to the vacuole. *Plant Cell* **2014**, *26*, 4084–4101. [CrossRef] [PubMed]
157. Niwa, Y.; Kato, T.; Tabata, S.; Seki, M.; Kobayashi, M.; Shinozaki, K.; Moriyasu, Y. Disposal of chloroplasts with abnormal function into the vacuole in Arabidopsis thaliana cotyledon cells. *Protoplasma* **2004**, *223*, 229–232. [CrossRef] [PubMed]
158. Wang, S.; Blumwald, E. Stress-Induced Chloroplast Degradation in Arabidopsis Is Regulated via a Process Independent of Autophagy and Senescence-Associated Vacuoles. *Plant Cell* **2014**, *26*, 4875–4888. [CrossRef]
159. Michaeli, S.; Galili, G. Degradation of organelles or specific organelle components via selective autophagy in plant cells. *Int. J. Mol. Sci.* **2014**, *15*, 7624–7638. [CrossRef] [PubMed]
160. Avin-Wittenberg, T.; Fernie, A.R. At long last: Evidence for pexophagy in plants. *Mol. Plant* **2014**, *7*, 1257–1260. [CrossRef]
161. Deosaran, E.; Larsen, K.B.; Hua, R.; Sargent, G.; Wang, Y.; Kim, S.; Lamark, T.; Jauregui, M.; Law, K.; Lippincott-Schwartz, J.; et al. NBR1 acts as an autophagy receptor for peroxisomes. *J. Cell Sci.* **2013**, *126*, 939–952. [CrossRef]
162. Su, T.; Li, W.; Wang, P.; Ma, C. Dynamics of Peroxisome Homeostasis and Its Role in Stress Response and Signaling in Plants. *Front. Plant Sci.* **2019**, *10*, 705. [CrossRef]
163. Su, T.; Wang, P.; Li, H.; Zhao, Y.; Lu, Y.; Dai, P.; Ren, T.; Wang, X.; Li, X.; Shao, Q.; et al. The Arabidopsis catalase triple mutant reveals important roles of catalases and peroxisome-derived signaling in plant development. *J. Integr. Plant Biol.* **2018**, *60*, 591–607. [CrossRef] [PubMed]
164. Huang, L.; Yu, L.J.; Zhang, X.; Fan, B.; Wang, F.Z.; Dai, Y.S.; Qi, H.; Zhou, Y.; Xie, L.J.; Xiao, S. Autophagy regulates glucose-mediated root meristem activity by modulating ROS production in Arabidopsis. *Autophagy* **2019**, *15*, 407–422. [CrossRef] [PubMed]
165. Takahashi, Y.; He, H.; Tang, Z.; Hattori, T.; Liu, Y.; Young, M.M.; Serfass, J.M.; Chen, L.; Gebru, M.; Chen, C.; et al. An autophagy assay reveals the ESCRT-III component CHMP2A as a regulator of phagophore closure. *Nat. Commun.* **2018**, *9*, 2855. [CrossRef] [PubMed]
166. Xie, Q.; Michaeli, S.; Peled-Zehavi, H.; Galili, G. Chloroplast degradation: One organelle, multiple degradation pathways. *Trends Plant Sci.* **2015**, *20*, 264–265. [CrossRef] [PubMed]
167. Arimura, S. Fission and Fusion of Plant Mitochondria, and Genome Maintenance. *Plant Physiol.* **2018**, *176*, 152–161. [CrossRef] [PubMed]
168. Wakim, J.; Goudenege, D.; Perrot, R.; Gueguen, N.; Desquiret-Dumas, V.; de la Barca, J.M.; Dalla Rosa, I.; Manero, F.; Le Mao, M.; Chupin, S.; et al. CLUH couples mitochondrial distribution to the energetic and metabolic status. *J. Cell Sci.* **2017**, *130*, 1940–1951. [CrossRef]
169. Ma, J.; Liang, Z.; Zhao, J.; Wang, P.; Ma, W.; Mai, K.K.; Fernandez Andrade, J.A.; Zeng, Y.; Grujic, N.; Jiang, L.; et al. Friendly mediates membrane depolarization-induced mitophagy in Arabidopsis. *Curr. Biol.* **2021**, *31*, 1931–1944e4. [CrossRef]
170. Chen, P.-Y.; Wu, C.-C.; Lin, C.-C.; Jane, W.-N.; Suen, D.-F. 3D Imaging of Tapetal Mitochondria Suggests the Importance of Mitochondrial Fission in Pollen Growth. *Plant Physiol.* **2019**, *180*, 813–826. [CrossRef] [PubMed]
171. Hanamata, S.; Sawada, J.; Bunki, T.; Ono, S.; Ogawa, K.; Fukunaga, T.; Nonomura, K.I.; Kurusu, T.; Kuchitsu, K. Monitoring autophagy in rice tapetal cells during pollen maturation. *Plant Biotechnol.* **2019**, *36*, 99–105. [CrossRef] [PubMed]
172. Liu, Y.; Bassham, D.C. TOR is a negative regulator of autophagy in Arabidopsis thaliana. *PLoS ONE* **2010**, *5*, e11883. [CrossRef] [PubMed]
173. Soto-Burgos, J.; Bassham, D.C. SnRK1 activates autophagy via the TOR signaling pathway in Arabidopsis thaliana. *PLoS ONE* **2017**, *12*, e0182591. [CrossRef] [PubMed]

antioxidants

MDPI

Review

mTOR Signalling Pathway: A Potential Therapeutic Target for Ocular Neurodegenerative Diseases

Yipin Wang, Nicholas Siu Kay Fung, Wai-Ching Lam * and Amy Cheuk Yin Lo *

Department of Ophthalmology, School of Clinical Medicine, Li Ka Shing Faculty of Medicine, The University of Hong Kong, Hong Kong 999077, China; u3005138@connect.hku.hk (Y.W.); nfung@hku.hk (N.S.K.F.)
* Correspondence: waichlam@hku.hk (W.-C.L.); amylo@hku.hk (A.C.Y.L.)

Abstract: Recent advances in the research of the mammalian target of the rapamycin (mTOR) signalling pathway demonstrated that mTOR is a robust therapeutic target for ocular degenerative diseases, including age-related macular degeneration (AMD), diabetic retinopathy (DR), and glaucoma. Although the exact mechanisms of individual ocular degenerative diseases are unclear, they share several common pathological processes, increased and prolonged oxidative stress in particular, which leads to functional and morphological impairment in photoreceptors, retinal ganglion cells (RGCs), or retinal pigment epithelium (RPE). mTOR not only modulates oxidative stress but is also affected by reactive oxygen species (ROS) activation. It is essential to understand the complicated relationship between the mTOR pathway and oxidative stress before its application in the treatment of retinal degeneration. Indeed, the substantial role of mTOR-mediated autophagy in the pathogenies of ocular degenerative diseases should be noted. In reviewing the latest studies, this article summarised the application of rapamycin, an mTOR signalling pathway inhibitor, in different retinal disease models, providing insight into the mechanism of rapamycin in the treatment of retinal neurodegeneration under oxidative stress. Besides basic research, this review also summarised and updated the results of the latest clinical trials of rapamycin in ocular neurodegenerative diseases. In combining the current basic and clinical research results, we provided a more complete picture of mTOR as a potential therapeutic target for ocular neurodegenerative diseases.

Keywords: mTOR; AMD; DR; glaucoma; oxidative stress; hypoxia; inflammation; ROS; rapamycin; clinical trial

Citation: Wang, Y.; Fung, N.S.K.; Lam, W.-C.; Lo, A.C.Y. mTOR Signalling Pathway: A Potential Therapeutic Target for Ocular Neurodegenerative Diseases. *Antioxidants* **2022**, *11*, 1304. https://doi.org/10.3390/antiox11071304

Academic Editors: Angeles Aroca and Cecilia Gotor

Received: 31 May 2022
Accepted: 20 June 2022
Published: 29 June 2022

1. Introduction

The mammalian target of rapamycin (mTOR), also known as the mechanistic target of rapamycin, is a 289-kDa serine/threonine kinase that belongs to the phosphoinositide 3-kinase (PI3K)-related kinase family and is highly conserved in evolution. It plays a central role in cell growth, cell survival autophagy, and metabolism via two distinct protein complexes of mTOR complex 1 (mTORC1) and mTOR complex 2 (mTORC2) [1]. mTOR also integrates with other signalling pathways, including PI3K/AKT, tuberous sclerosis complex subunit 1 (TSC1)/tuberous sclerosis complex subunit 2 (TSC2)/Rheb, LKBL/adenosine 5′-monophosphate-activated protein kinase (AMPK), and VAM6/Rag GTPases, by affecting transcription and protein synthesis [2]. Dysregulation of the mTOR signalling pathway is involved in many diseases, such as cancers, neurodegenerative diseases, and diabetes mellitus [3–6]. Clinical data show that the mTOR signal is abnormally overactivated in nearly 30% of cancers and is one of the most frequently altered cascades in human cancers [7].

Recently, more and more studies are focused on the potential therapeutic effects of mTOR inhibitors in neurodegenerative diseases that are linked with oxidative stress. In a rat model of Alzheimer's disease (AD) induced by zinc injection, inhibition of mTOR by rapamycin attenuated zinc-induced tau phosphorylation and elevated levels of oxidative stress, as well as the synaptic impairment and decrease in cognitive function [8]. In a Parkinson's disease (PD) model induced by 6-hydroxydopamine (6-OHDA), pathogenic oxidative stress increased the negative mTOR regulator tuberous sclerosis complex 2 (TSC2) and increased autophagy in dopaminergic neurons, implicating that mTOR is a potential intervention target for oxidative-stress-induced dysfunctional autophagy in PD [9]. The retina is a part of the central nervous system; it contains complex neural circuitry and transduces the converted electrical potentials to the brain [10]. The neuroprotective roles of rapamycin may be a novel therapeutic pathway in ocular neurodegenerative diseases, such as diabetic retinopathy (DR), age-related macular degeneration (AMD), and glaucoma, which share common pathophysiological mechanisms, especially increased and prolonged oxidative stress, which would ultimately result in retinal neuronal death [11–14]. Recently, a large number of studies have been conducted to elucidate the neuroprotective role of rapamycin and its underlying mechanism(s) in the treatment of ocular degenerative diseases [15–27]. For instance, a study has shown that rapamycin ameliorated high-glucose-induced ROS formation and inflammatory injury in retinal pigment epithelial (RPE) cells [28].

In this review, we introduced the mTOR signalling pathway and its role in ocular neurodegenerative diseases under oxidative stress, trying to highlight and summarise the current understanding of the mechanisms of mTOR inhibitors, especially rapamycin and its analogues, in different retinal models, including DR, AMD, and glaucoma. In updating the latest basic research findings and clinical trial results, we attempted to shed light on the novel therapeutic strategies of mTOR in ocular degenerative diseases.

2. mTOR Signalling Pathway

mTOR is part of the catalytic subunit in two structurally and functionally distinct complexes, known as mTOR complex 1 (mTORC1) and mTOR complex 2 (mTORC2) [1]. mTORC1 consists of five components: mTOR; regulatory-associated protein of mTOR (Raptor); mammalian lethal with Sec13 protein 8 (mLST8, also referred as GβL); proline-rich AKT substrate 40 kDa (PRAS40); and DEP-domain-containing mTOR-interacting protein (Deptor) [29]. Raptor has been reported to function as a scaffolding protein that controls mTORC1 activity by regulating assembly of the complex and by recruiting substrates for mTOR [30]. The role of mLST8 in mTORC1 function is still unclear; however, it is established that mLST8 is indispensable for mTORC2 integrity and kinase activity [31]. PRAS40 and Deptor have been identified as distinct negative regulators of mTORC1. When the activity of mTORC1 decreases, PRAS40 and Deptor are recruited to the complex and promote the inhibition of mTORC1. It has been reported that PRAS40 regulates mTORC1 kinase activity by direct inhibition of substrate binding [32]. On the other hand, mTORC1 once activated would directly phosphorylate PRAS40 and Deptor, which reduce their physical interaction with mTORC1 and further activate mTORC1.

Similar to mTORC1, mTORC2 also comprises six different proteins, among which three are identical to those in mTORC1, including mTOR, mLST8, and Deptor [29]. The remaining three components include rapamycin-insensitive companion of mTOR (Rictor), mammalian stress-activated protein kinase interacting protein (mSIN1), and protein observed with Rictor-1 (Protor-1). It is proposed that Rictor and mSIN1 stabilise each other and establish the structural foundation of mTORC2. Rictor also interacts with Protor-1, but the physiological function of this interaction is not yet clear [33].

The mTOR signalling pathway integrates both intracellular and extracellular signals and transduces divergent signalling cascades. Apart from intracellular signals, such as cellular energy status and hypoxia stress, extracellular signals that include growth factors, amino acids, and hormones play essential roles in regulating the activity of mTORC1 [34]. Insulin, a pivotal hormone serving to maintain energy balance and glucose homeostasis, initiates a signalling cascade through the insulin receptor (IR), insulin receptor substrate (IRS), class I phosphoinositide 3-kinases (PI3Ks), phosphoinositide-dependent protein kinase 1 (PDK1), and AKT (also known as protein kinase B) [35]. AKT activates mTORC1 through phosphorylation of TSC1/TSC2 and PRAS40 (Figure 1). It has been reported that there is a feedback loop between AKT and mTOCR2 induced by insulin [36].

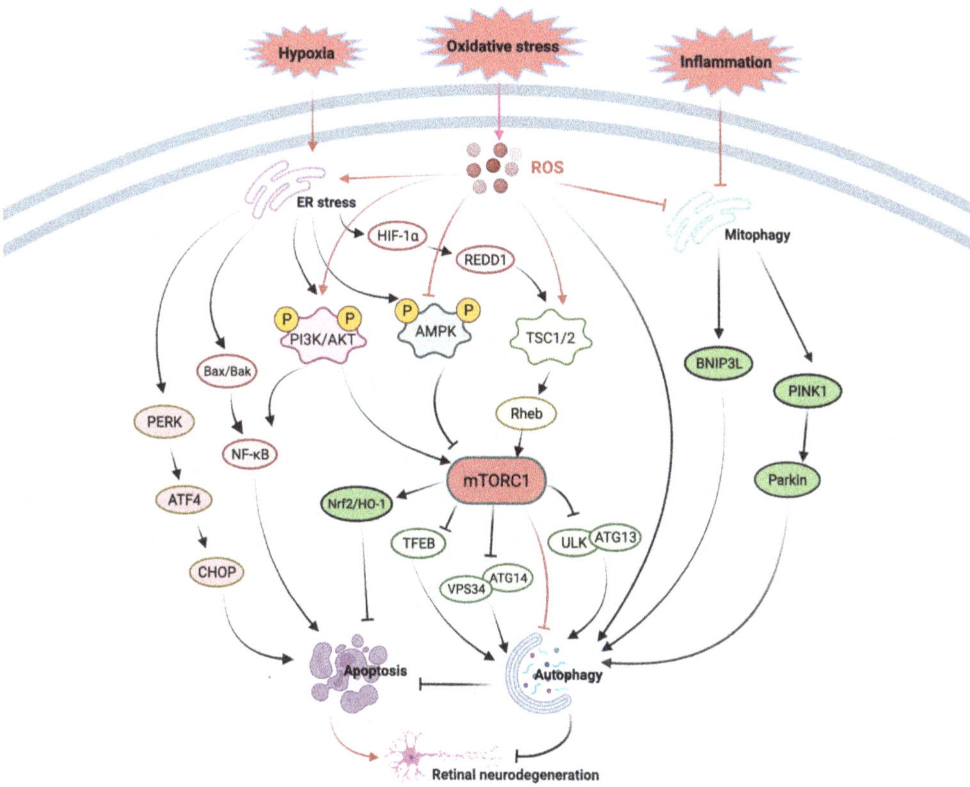

Figure 1. The common mTOR signalling pathway in ocular neurodegenerative diseases. Abbreviations: AMP′, 5′ adenosine monophosphate-activated protein kinase; ATF4, activating transcription factor 4; ATG, autophagy-regulating protease; BNIP3L, adenovirus E1B 19 kDa protein-interacting protein 3-like; CHOP, C/EBP homologous protein; ER, endoplasmic reticulum; HIF-1α, hypoxia-inducible factor 1-alpha; HO-1, heme oxygenase-1; mTORC1, mammalian target of rapamycin complex 1; Nrf2, nuclear factor-erythroid factor 2-related factor 2; PERK, protein kinase R (PKR)-like endoplasmic reticulum kinase; PI3K, phosphoinositide 3-kinase; PINK1, PTEN-induced kinase 1; REDD1, regulated in development and DNA damage responses 1; Rheb, ras homolog enriched in brain; ROS, reactive oxygen species; TFEB, transcription factor EB; TSC, tuberous sclerosis complex; ULK1, unc-51-like kinase 1; VPS34, vacuolar protein sorting 34.

In addition to AKT, mitogen-activated protein kinase (MAPK) and proinflammatory cytokine TNFα also promote the activation of mTORC1 through inhibiting TSC1/TSC2 [37]. Wnt/β-catenin signalling plays a role in regulating mTORC1 by suppressing glycogen synthase kinase 3β (GSK3-β), which phosphorylates and increases TSC1/2 activity [38]. AMPK is one of the signalling pathways that inhibits the activation of mTORC1 by direct phosphorylation of TSC2, promoting TSC1/2 activity [39]. AMPK can be activated by DNA damage via p53-dependent transcription of Sestrin1/2 [40].

Oxidative stress reflects an imbalance in the production of reactive oxygen species (ROS) and the antioxidative capacity in the cell. The retina is susceptible to ROS due to its high energy consumption and exposure to light. ROS is not only a by-product generated in the retinal cells but also a signal transducer involved in the PI3K/AKT/mTOR signalling pathway (Figure 1). The increased level of intracellular ROS activates the kinases, including PKC, MARK, and PI3K, which leads to the amplification of their downstream signalling, respectively [41,42]. ROS can also inhibit the activation of phosphatase and tensin homolog (PTEN), which negatively regulates the synthesis of PIP3, a signalling molecule in the plasma membrane, which plays a role in the activation of AKT [43].

In regard to the downstream signalling of mTOR, there are many substrates for mTORC1, such as p70 ribosomal protein S6 kinase (p70S6K) and eukaryotic initiation factor 4E (eIF4E)-binding protein (4EBP), that are responsible for translation control [44]. Other mTORC1 substrates, such as Unc-51-like autophagy activating kinase 1 (ULK1) and ATG13, play essential roles in the regulation of autophagy (Figure 1). Studies have shown that AKT, which regulates mTORC1 activity, is also an mTORC2 substrate. Protein kinase C (PKC), which regulates diverse cellular functions, is an mTORC2 substrate as well [30,45].

The mTOR signalling pathway plays a central role in the regulation of autophagy, whereas autophagy maintains retinal homeostasis by removing dysfunctional organelles and unfolded proteins [46]. Hence, mTOR modulates oxidative stress by both direct and indirect mechanisms. A study has shown that autophagy played a neuroprotective role in DR [47]. The autophagic genes beclin-1 and LC3 were moderately up-regulated, which was accompanied with increased phosphorylated AMPK and decreased phosphorylated mTOR in the diabetic retinas [48]. After the inhibition of autophagy by 3-MA in STZ-induced diabetic rats, RGC apoptosis increased when compared with the vehicle-treated group [48]. Studies have also demonstrated that autophagy-deficient cells lacking BECN1, ATG5, or ATG7 caused the accumulation of impaired organelles [49,50]. There is a feedback loop in which autophagy modulates oxidative stress through activating transcription factors, such as NRF2 and p53 [40,51].

A recent study quantified the expression of mTORC1- and mTORC2-specific partner proteins in normal adult rat retina, brain, and liver, and further localised these components in normal adult human and mouse retina [52]. They found a relatively higher content of mTORC1, mTORC2, and their components included higher Raptor (mTORC1) and Rictor (mTORC2) in the retina than the brain and liver. The two mTOR complexes may serve distinct purposes within the retina, while the mTORC1 complex is predominantly expressed in retinal ganglion cells (RGCs) and their axons with lower expression in the inner plexiform layer (IPL) and inner nuclear layer (INL); mTORC2 complex proteins, such as Rictor, were mainly found in astrocytes and Müller cells. However, no detectable immunoreactivity of mTORC1- and mTORC2-specific components was found in photoreceptor cells, RPE, and vascular endothelial cells. The composition and topology of mTOR components closely parallel the physiological characteristics of the retina as a tissue of high energy and oxygen consumption and the role of the mTOR signalling pathway in retinal metabolism and homeostasis in response to glucose.

3. mTOR Inhibitors

mTOR inhibitors are a class of drugs that inhibit the activity of the serine/threonine-specific protein kinase coded by the *MTOR* gene. This protein kinase belongs to the family of phosphatidylinositol-3 kinase (PI3K)-related kinases (PIKKs) and serves as the catalytic subunit of two multi-protein complexes (mTORC1 and mTORC2) [53] (Table 1).

Table 1. The potency, specificity, and adverse effects of mTOR inhibitors.

Drug	Target	Potency (IC_{50}, nM)	Pros/Cons	Development Status	Adverse Effects
1st generation mTOR inhibitors					
Rapamycin	mTOR/FKBP12	0.1	1st FDA-approved mTOR inhibitor/ low biological utilisation due to its poor water solubility and stability	FDA-approved	Hyperglycaemia, fatigue, nausea/vomiting, anaemia, stomatitis, mucositis, pulmonary and metabolic toxicities [54–56]
Temsirolimus	mTOR/FKBP12	1.76	Relatively high water solubility and stability, intravenous administration only	FDA-approved	
Everolimus	mTOR/FKBP12	1.6–2.4	Relatively high water solubility and stability, low toxicity and high efficacy for some types of tumours	FDA-approved	
Ridaforolimus	mTOR/FKBP12	0.2–5.6	Latest developed rapalogs, well-tolerated in children	FDA-approved	
2nd generation mTOR inhibitors					
Torin1	mTORC1/mTORC2	0.29 (mTORC1)/5 (mTOR)	Strong anti-proliferation activity/poor stability and low oral bioavailability	Preclinical	Hyperglycaemia, fatigue, nausea/vomiting, stomatitis, mucositis, diarrhoea, decreased appetite, liver dysfunction, pneumonia [57–59]
PP242	mTORC1/mTORC2	8 (mTOR)	Relatively strong selectivity to mTOR	Preclinical	
AZD8055	mTORC1/mTORC2	10 (mTORC1)/2.8 (mTOR)	Potent anti-proliferation and apoptosis induction activity/relatively high liver toxicity	Phase I	
OSI-027	mTORC1/mTORC2	4 (mTORC1)/22.6 (mTOR)	Strong inhibitory effects on mTOR, dose-dependent manner in patients with some types of tumours	Phase I	
PI-103	mTOR/PI3K	3–3.6 (PI3K)	1st developed mTOR/PI3K dual inhibitor/poor drug properties	Preclinical	Hyperglycaemia, fatigue, nausea/vomiting, mucositis, diarrhoea, decreased appetite, rash [60–62]
GSK2126458	mTOR/PI3K	0.18 (mTORC1)/ 0.019–0.13 (PI3K)	Confirmed target engagement in blood and lungs/affect insulin release and blood glucose level	Preclinical	
NVP-BEZ235	mTOR/PI3K	mTOR (20.7)/4–75 (PI3K)	Potent PI3K inhibitory effects on PI3K	Phase I	

The best-established mTOR inhibitors are rapamycin and its analogues (rapalogs). Rapamycin is a good example of the first-generation mTOR inhibitors, which were initially introduced as immunosuppressive drugs and approved by the FDA in 1997 in use for transplant surgery to prevent allograft rejection [63]. Other therapeutic effects, such as anti-cancer activity, angiogenesis, and neuroprotection, were discovered and exploited in the following years [64–66]. Rapamycin is a macrolide compound containing two binding moieties for mTOR and FKBP12, respectively. This engagement of the binding moieties has limited the modification of the ATP binding pocket. Thus, further drug development for rapamycin mainly focused on improving its pharmacokinetics and stability due to its low aqueous solubility [67]. Rapalogs were derived and evolved from rapamycin; they have a more favourable pharmacokinetic profile when compared to their parent drug.

However, they still have the same binding sites for mTOR and FKBP12 as in rapamycin. Their mechanism of action is also identical; they bind with FKBP12 to form a complex, which then binds to the FRB domain of mTOR. Through modulating mTOR formation, both rapamycin and rapalogs inhibit the kinase activity of mTORC1 (Figure 2). Rapalogs include temsirolimus (CCI-779), everolimus (RAD001), ridaforolimus (AP23573), umirolimus, and zotarolimus (ABT-578) [68–71].

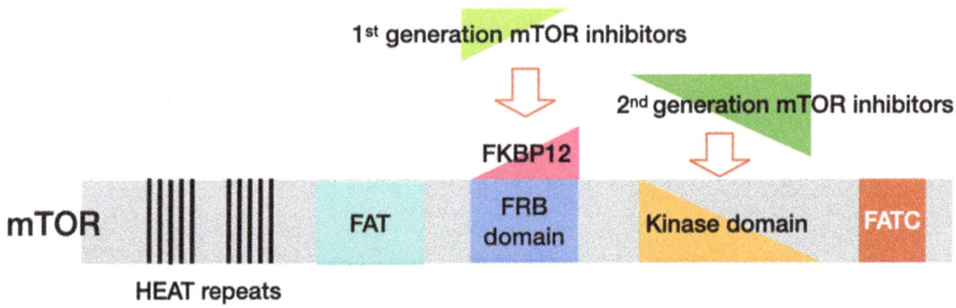

Figure 2. Domains of the mTOR protein and two generations of mTOR inhibitors.

mTOR protein is composed of several structural domains, including HEAT repeats and FAT, FRB, and FATC domains. The first-generation mTOR inhibitors bind to FKBP12 and then interact with the FRB domain of mTOR to inhibit mTOR activity. The second-generation mTOR inhibitors are ATP-competitive mTOR inhibitors that act as ATP analogues and bind to the kinase domain of mTOR. Abbreviations: FAT domain, FKBP12-rapamycin-associated protein, ataxia-telangiectasia and transactivation/transformation domain; FATC domain, FAT carboxyterminal domain; FRB domain, FKBP12-rapamycin-binding domain; HEAT, Huntingtin, elongation factor 3 (EF3), protein phosphatase 2A (PP2A), and the yeast kinase TOR1.

The second-generation mTOR inhibitors are ATP-competitive mTOR kinase inhibitors, which have been developed as two types. They include mTORC1/mTORC2 dual inhibitors, such as Torin1, PP242, and AZD8055, and dual mTOR/PI3K inhibitors, such as PI-103, OSI-027, GSK2126458, and NVP-BEZ235 [72–74]. They are developed to compete with ATP in the catalytic domain of mTOR (Figure 2). Compared with rapalogs, which only inhibit mTORC1 activity, the second-generation mTOR inhibitors are designed to target both mTORC1 and mTORC2 and inhibit all the catalytic isoforms of PI3K (Figure 3). As a result of the blockade in the feedback activation of PI3K/AKT signalling in mTORC1, they impose more potent inhibition on the mTOR pathway and stronger induction of autophagy than rapalogs (Table 1) [75]. This new generation mTOR inhibitors has been introduced into clinical trials mainly for the treatment of various cancers [71,76].

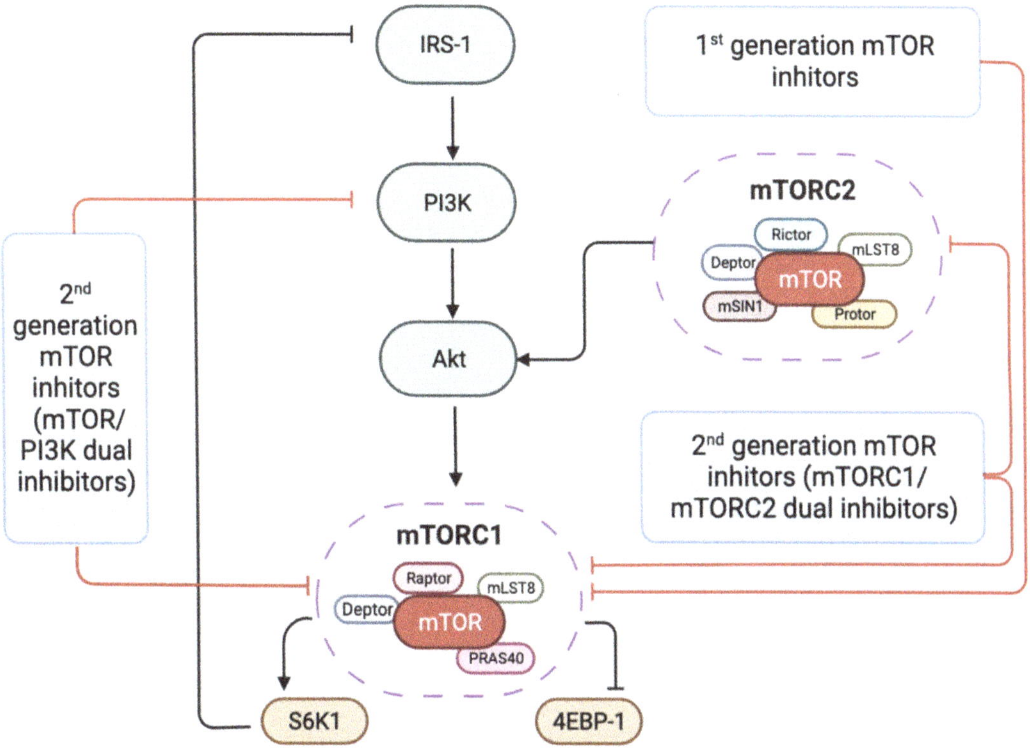

Figure 3. The signalling pathway that is modulated by different generation mTOR inhibitors. Inhibition of mTORC1 results in the suppression of 4E-BP1 and S6K1 phosphorylation. Inhibited S6K1 reduces protein synthesis through phosphorylation of the 40S ribosomal subunit, which has been suggested to decrease the translational efficiency of a class of mRNA transcripts with a 5′-terminal oligopolypirymidine. There is a negative feedback loop in which mTORC1 activation can inhibit the PI3K pathway by S6K1-mediated phosphorylation and degradation of IRS-1, and it fills an important gap in our understanding the underlying mechanisms by which mTORC1 inhibits PI3K-Akt signalling. Abbreviations: IRS-1, insulin receptor substrate 1; PI3K, phosphatidylinositol-3-kinase; AKT, protein kinase B; PRAS40, proline-rich AKT substrate of 40 kDa; mTOR, mammalian target of rapamycin; S6K1, S6 kinase 1; 4E-BP1.

4. mTOR in Ocular Neurodegenerative Diseases

4.1. mTOR and Diabetic Retinopathy (DR)

The pathogenesis of DR is complex and multifactorial. Many biochemical mechanisms are involved in the development and progression of DR. For instance, polyol pathway activation, increased advanced glycation end-products (AGEs) formation, activation of PKC, and induction of the hexosamine pathway are related to the pathogenesis of DR [77]. These pathways induce inflammation, oxidative stress, and vascular dysfunction in the retina. Chronic hyperglycaemia in diabetes mellitus favours the production and accumulation of ROS, which leads to oxidative-stress-induced impairment in different cells of the retina, especially the retinal neurons [78]. The classical view is that microvascular alteration is the primary event in the pathogenesis of DR. However, there is growing evidence showing that neurodegeneration occurs in the early stage of DR, which could also be related to the development of microvascular abnormalities [79–82]. Diabetic retinal neurodegeneration is characterised by reactive gliosis and retinal neuron apoptosis. While RGC and amacrine cell death induced by diabetes-induced apoptosis occur in the early phase of DR, other retinal neurons, such as photoreceptors, also have an increased apoptotic rate [83,84]. Reactive gliosis (glial activation) may also be involved in retinal neurons apoptosis and may be associated with the neurodegenerative process with microvascular disease [85]. In addition to macroglial cells, activated microglia, the main resident immune cells of the retina, and infiltrating monocytes also mediate diabetes-induced inflammation [86].

Recently, more and more studies have emphasised the key roles of inflammation, oxidative stress, and autophagy in the pathogenesis of DR. As described above, the mTOR pathway plays a substantial role in regulating autophagy. The activated mTORC1 inhibits autophagy through various steps, including the inhibitory phosphorylation of ULK1 and transcription factor EB (TFEB), which initiates autophagy and promotes lysosomal biogenesis required to degrade the contents of autophagosomes, respectively. LC3B and Beclin-1 are cellular autophagy markers involved in the initial stage of autophagosome formation. Park et al. observed a slight increase in Beclin-1 and the ratio of LC3B-II-to-LC3B-I after 1, 4, and 8 weeks of hyperglycaemia in STZ-induced diabetic rats compared with the non-diabetic rat [48]. Of note, the level of Beclin-1 decreased dramatically, even lower than the basal level, after prolonged hyperglycaemia over 8 weeks. Significant upregulation of phosphorylated AMPK but downregulation of phosphorylated mTOR was also observed in the early stage of DR before 8 weeks of hyperglycaemia. After the inhibition of autophagy using 3-methyladenine (3-MA), apoptosis of RGCs was significantly increased in the diabetic retinas. These results indicated that autophagy induced by hyperglycaemia may act as a survival attempt to rescue RGC apoptosis; however, the insufficient activation of autophagy failed to maintain retinal homeostasis. It suggested that AMPK-activation-induced autophagy may play a neuroprotective role in DR.

A similar result was also reported in a recent study. Fang et al. found that LC3-II and p62 levels, as well as the phosphorylated proteins in the PI3K/Akt/mTOR signalling pathway in STZ-induced diabetic rats, were increased when compared with the normal control group [87]. This suggested that retinal autophagy was initiated in DR but was inadequate to protect retinal neurons due to the excessive activation of the PI3K/Akt/mTOR signalling in DR. After the administration of a traditional Chinese medicine, Mingmu Xiaomeng tablets (MMXM), which have been proven as an effective mTOR inhibitor, LC3-II, p62, p-PI3K, p-Akt, and p-mTOR protein levels were significantly decreased in retinal tissue compared with that of the untreated diabetic rats. Glial fibrillary acidic protein (GFAP) is a specific glial cell marker that is used to reflect the response of glial cells under pathological conditions. The therapeutic effects of MMXM were also observed in retinal Müller cells (RMC), including the inhibition of GFAP overreaction and the restraint of local inflammation. Since RMC is a key player in inducing the expression of acute-phase response proteins and other inflammation-related genes in DR, the expressions of downstream inflammatory cytokines, such as IL-1β, IL-4, IL-6, TNF-α, and VEGF, were also significantly reduced.

RMCs provide structural and neurotrophic support to the retina via uptake and regulation of neurotransmitters. RMC hyperplasia is one of the early pathological changes in DR [88], and upregulated GFAP expression in RMCs was found in animal models and tissue from diabetic patients. A recent study showed that mTORC1 may play an important role in RMC dysfunction during DR [89] (Table 2). Besides the upstream kinases, mTOR senses multiple stimuli, including growth factors, amino acids, nutrients, energy status, and cellular stress. Guo et al. found that high glucose (HG) treatment increased glutamine synthetase activity in the cultured RMC, which promoted the biosynthesis of Gln, leading to the activation of mTORC1 through ADP-ribosylation factor 1 (Arf 1) in Rag GTPases-dependent and Rag-independent manners [89]. RMC proliferation and activation in high-fat diet and STZ-induced mouse diabetic models were inhibited by rapamycin. Lopes de Faria et al. investigated the correlation between the autophagy machinery and ER stress in RMCs in the HG condition and found that lysosome-mediated autophagy was impaired in RMCs. However, autophagy was activated due to the sustained but insufficient ER stress response inducing the misfolded/unfolded protein under oxidative stress condition [90]. There were higher amounts of autophagosomes in the cytosol and accumulating p62/SQTSM1 cargo than those in the normal control (NC) group, which usually occurred when autophagic flux was compromised. Furthermore, the lysosome proteolytic activity decreased due to the malfunction of cathepsin L. After the administration of rapamycin, the recovery of cathepsin L activity improved the autophagic flux and reduced p62/SQTSM1 cargo accumulation leading to the amelioration of ER stress.

Table 2. Involvement of mTOR in DR: in vitro and animal studies.

Target Cells or Tissue	Disease Model	mTOR Regulator	Autophagy-Related Markers	Related Pathways	Effects of Regulated mTOR	References
R28 cells	Hypoxia-induced AMD model	Insulin	LC3A↓	PI3K/AKT/mTOR↑	Oxidative stress↓ VEGF↓	[49]
rMC-1	HG	Rapamycin	Beclin1↑ p62↓	mTOR↓	Apoptosis↓ VEGF↓	[90]
ARPE-19	HG	Curcumin	-	PI3K/AKT/mTOR↓	TNF-α / IL-1β /IL-6↓	
661W cells	HG	3-MA	LC3B2↓ p62↑	PI3K/AKT/mTOR↑	ROS↑ Mitophagy↓ Apoptosis↑	[91]
Ex vivo mouse retinal explants	HG	Octreotide	LC3-II↑ LC3-II net flux↑	mTOR/S6K1↓	Apoptosis↑	[92]
RGCs	STZ-induced diabetic rats	3-MA	LC3B↓ Beclin-1↑	AMPK↓/mTOR↑	Apoptosis↑	[28]
RGCs	STZ-induced diabetic mice	Rapamycin	-	mTOR/S6K1↓	GLUT1↓ GFAP↓	[93]
RMCs	STZ-induced diabetic rats/HG	PPP1CA	-	YAP/GS/Gln/ mTORC1↑	RMCs activation/ proliferation↑	[89]
RMCs	STZ-induced diabetic rats	MMXM	LC3-II↑ p62↓	PI3K/AKT/mTOR↓	IL-1β / IL-6↓ VEGF↓ GFAP↓	[87]
Retina tissue	STZ-induced diabetic rats	Rapamycin	-	mTORC1/S6K1↓	VEGF↓ PEDF↓ HRCECs proliferation/migration↓	[94]
Retina tissue	STZ-induced diabetic rats	Phosphatidic acid	-	mTOR/S6K1↑	Apoptosis↓	[95]
Retina tissue	STZ-induced diabetic rats/Ins2^{Akita} mice	Insulin/phloridzin	-	AKT/mTORC2↑ mTORC1/S6K1/ 4E-BP1 ↔	Retinal protein synthesis↑	[96]

Ribosomal protein S6 kinase beta-1 (S6K1) is a downstream target of the mTORC1 pathway. It is phosphorylated and activated by mTORC1; therefore; the level of p-S6 is often used as an indicator of the activation degree of the mTORC1 pathway. In the STZ-induced diabetic model, the expressions of p-S6 and VEGF were upregulated in the retina [94]. After the administration of rapamycin in HG-induced human retinal capillary endothelial cells (HRCECs), the expression of p-S6 was decreased, and the proliferation and migration of HRCECs were restrained. It indicated that mTORC1 is involved in the development of DR, targeting different cells in the retina.

mTOR also serves as a regulator to maintain the balance between retinal neuronal death and survival based on the equilibrium between apoptosis and autophagy [97]. Due to microvascular alteration in DR, retinal neurons are subjected to ischaemia/reperfusion (I/R) injuries. Tang et al. compared autophagy level in hyperglycaemic and normoglycaemic states following I/R injury and found an elevated autophagy after two hours of ischaemia induced by middle cerebral artery occlusion (MCAO) surgery in the retinae of Akita mice [98]. After two hours of reperfusion, the expression levels of LC3B and LAMP1 were still higher in the inner retinae of Akita MCAO mice compared with sham-treated Akita mice. In contrast, the expression level of LCB3 was higher in WT sham-treated mice than WT MCAO-injured mice. It indicated that the upregulated autophagy was pre-existing in the chronic hyperglycaemic condition and sustained at a higher level in the retinal I/R episode. Moreover, the autophagy upregulation exhausted and returned to basal levels after longer time reperfusion of 22 h. It is possible that upregulated autophagy has a beneficial effect in the I/R-injured retina under hyperglycaemia. Amato et al. also found high glucose treatment significantly increased apoptosis and decreased the autophagic flux by the up-regulation of mTOR in ex vivo mouse retinal explants [92]. Compared with untreated explants, LC3 immunolabeling was dramatically reduced in different retinal layers, including GCL, INL, and OPL. After the administration of octreotide, a well-known inhibitor of the PI3K/AKT/mTOR pathway, in HG-treated retinal explants, apoptosis was reduced below control levels, and LC3 expression was increased in different types of retinal neurons, especially the bipolar cells and ganglion cells. It suggested that mTOR may play a significant role in the crosstalk between apoptosis and autophagy in DR.

More importantly, rapamycin may have an antioxidative effect and plays a role in the amelioration of diabetic oxidative stress. Özdemir et al. found that oral rapamycin treatment reduced nitrotyrosine and malondialdehyde (MDA) levels, both of which are oxidative stress markers, in STZ-induced diabetic rat retina [99]. It is possible that rapamycin prohibits the induction of inducible NO synthase (iNOS) in the retina by reducing the expression of inflammatory mediators, such as VEGF, TNF-α, and IL-1β [100]. The inhibition for the secretions of inflammatory mediators through the AKT/mTOR pathway was further confirmed by Ran et al., who found that curcumin had comparable effects with rapamycin to inhibit the phosphorylation of AKT and mTOR, as well as reducing the HG-induced ROS in RPEC [28]. Semaglutide and rosiglitazone are two commonly used antidiabetic drugs. Yang et al. found the combined treatment of these two drugs inhibited the PI3K/Akt/MTOR signalling pathway and the inhibition of mTOR reduced oxidative stress in STZ-induced diabetic rat retina [101]. The combination administration also downregulated GFAP expression in Müller cells; however, the relationship between mTOR signalling inhibition and the alleviation of Müller cells activation is unknown.

4.2. mTOR and Age-Related Macular Degeneration (AMD)

AMD is the leading cause of central vision impairment in the industrialised world [102]. There are two basic types of AMD, wet (exudative) AMD and dry (nonexudative) AMD. Wet AMD is a serious type of AMD featured with neovascularisation and subretinal haemorrhage. Most AMD patients have dry AMD, which accounts for 90% of AMD cases [103]. Dry AMD is characterised by RPE dysfunction, drusen formation, and progressive loss of neurons [104]. Although the precise mechanism of AMD has not been delineated, many studies have shown that oxidative stress acts as an initial trigger for the pathogenesis of AMD [105–108] and plays a central role in the progression of AMD [109–111]. As an early sign of AMD, drusen-like deposits have been found in SOD (superoxide dismutase) knockout mice [112,113]. The decreased autophagy in RPE cells exposed to oxidative stress reduces the removal of aggregated proteins and damaged organelles, leading to the formation and accumulation of those subretinal deposits [114,115]. Phospholipid decosahexaenoic acid (DHA) from the shredded photoreceptor outer segments is one of the main sources of ROS after lipid peroxidation in RPE cells [116]. The end-products of lipid peroxidation activate the nuclear factor kappa-light-chain-enhancer of activated B cells' (NF-κB) signalling pathway, triggering a proinflammatory cascade, which could lead to choroidal neovascularization (CNV) formation [110].

mTOR, the key member of the PI3K/AKT/mTOR signalling pathway, plays a fundamental role in cellular nutrient, oxygen, and energy sensing [117]. A previous study has shown the strong association between hypoxia and RPE-associated neovascularisation in dry AMD [118]. The long-lasting hypoxia activates PI3K/mTOR, which increases the expression of hypoxia-inducible factor-1a (HIF-1a). The accumulation of HIF-1a protein significantly induces apoptosis and the secretion of angiogenic factors. Lin et al. found that Silibinin, a traditional medicine extract, inhibited the PI3K/mTOR signalling pathway, leading to the reduction in HIF-1a subunit accumulation, suppressing RPE apoptosis and secretion of VEGF in a rat model of VEGF-induced AMD [119]. Interestingly, Silibinin reversed hypoxia-initiated autophagy induction in hypoxia-conditioned ARPE-19 cells, although the mTOR pathway had been inhibited. This is probably because of the interaction between autophagy and oxidative stress since oxidative stress activates autophagy and elevated autophagy, in turn, reduces oxidative stress. Mitter et al. found that there was a dynamic alteration in autophagic flux in cultured RPE cells based on the time exposed to oxidative stress [115]. The autophagy activity increased significantly when exposed to short-term (4 hrs to 24 hrs) oxidative stress and decreased when exposed to long-term (1 d to 14 d) oxidative stress. Rapamycin not only protected ARPE-19 cells from an acute lethal dose of H_2O_2 but also rescued the autophagy activity, leading to a reduction in ROS generation and lipofuscin-like granule accumulation upon long-term oxidative stress. Besides the in vitro model, rapamycin also played a key role in attenuating an inflammatory response and oxidative stress in sodium-iodate ($NaIO_3$)-induced retinal degeneration in mice as well [21].

mTOR is an essential upstream regulator of autophagy, which inhibits the ULK1-ATG13-RB1CC1/FIP200 complex. To investigate the relationship between autophagy and

AMD, Yao et al. used the Cre-loxP system to knock out the Rb1cc1 gene in mice [114]. After the deletion of Rb1cc1, significant autophagy defects were observed in the RPE, including decreased conversion of LC3-I to LC3-II, accumulation of autophagy-targeted precursors, and increased numbers of mitochondria, accompanied by the deposition of inflammatory and oxidatively damaged proteins and subretinal drusenoid deposits. In contrast, Cai et al. enhanced autophagy by overexpressing miR-29, a key precursor molecule that post-transcriptionally repressed LAMPTOR1/p18 and reduced the recruitment of mTORC1 to lysosomal membranes [120]. Upon inhibition of mTORC1 activity, the elevated autophagy enhanced the removal of protein aggregates. Similar results were reported using rAAV-mTOR shRNA to block the activity of both mTOR complex 1 and 2 in the mouse laser-induced CNV model [121]. Besides the removal of protein aggregates, Ebeling et al. found that rapamycin improved the clearance of damaged mitochondria in donated human RPE cells with AMD [122]. It also showed that rapamycin increased basal respiration and attenuated mitochondrial function in RPE cells [122]. Table 3 is a summary of the in vitro and in vivo studies investigating the involvement of mTOR in AMD (Table 3).

Table 3. Involvement of mTOR in AMD: in vitro and animal studies.

Target Cells or Tissue	Disease Model	mTOR Regulator	Autophagy-related Markers	Related Pathways	Effects of Regulated mTOR	Reference
ARPE-19	H_2O_2-induced RPE cell injury model	Silibinin	LC3A↓	PI3K/AKT/mTOR↓	Oxidative stress↓ VEGF↓	[46]
ARPE-19/hRPE	H_2O_2-induced RPE cell injury model	a-MSH	-	PI3K/AKT/mTOR↑	Oxidative stress↓ Apoptosis↓	[123]
hRPE/HUVEC	Hypoxia induced RPE cell injury model	Temsirolimus	-	mTOR↓	VEGF↓ PEDF↓	[19]
hRPE	Human AMD patient	Rapamycin	LC3-II/I↑	mTOR↓	Mt function↑ Mitophagy↑	[122]
ARPE-19/hRPE	H_2O_2-induced RPE cell injury model (acute/chronic)	Rapamycin	LC3 puncta↑	mTOR↓	Oxidative stress↓ ROS↓ Lipofuscin-like deposit↓	[115]
ARPE-19	H_2O_2-induced RPE cell injury model	Resveratrol	LC3-II/I↑ p62↓	mTOR↓	Apoptosis↓ VEGFA↓ IL-6/ IL-8↓	[124]
ARPE-19	Lipid-peroxidation-induced RPE injury model	Glucosamine	LC3-II/I↑ p62 ↗↘	AMPK↑/mTOR↓	Lipofuscin-like deposit↓	[125]
ARPE-19/hRPE	αB-crystallin R120G-mutation-induced protein aggregation model	miR-29	LC3-II/I↑ p62↓	mTOR↓	Protein aggregation↓	[120]
Retina tissue	Laser-induced model of CNV	GSK2126458	-	PI3K/mTOR↓	Vascular leakage↓ CNV lesions↓ Apoptosis↓ Serum glucose level↑	[126]
Retina tissue	Laser-induced model of CNV	rAAV-mTOR shRNA	LC3B↑ ATG7↑	PI3K/mTOR↓	Vascular leakage↓ CNV lesions↓ Apoptosis↓	[127]
Retina tissue	Laser-induced model of CNV	rAAV2-shmTOR-SD	-	mTOR↓	CNV lesions↓ Apoptosis↓	[127]
Retina tissue	$NaIO_3$-induced retinal degeneration	Rapamycin	-	mTOR↓	Oxidative stress↓ Apoptosis↓ GFAP↓ IL-6/ MCP-1/TNF-α↓	[21]

4.3. *mTOR and Retinitis Pigmentosa (RP)*

In contrast to AMD, retinitis pigmentosa (RP) is a genetic disorder with early onset and characterised by diffuse progressive degeneration of predominantly rod photoreceptors with subsequent dysfunction of cone photoreceptors. Although this inherited retinal degeneration does not share the same common pathological processes with other ocular neurodegenerative diseases, such as AMD, DR, and glaucoma, that are induced by oxidative stress, hypoxia, and inflammation (Figure 1), mTOR plays a critical role in the pathogenesis of RP. In a rat model for retinitis pigmentosa, D'Cruz et al. found that the mutation of receptor tyrosine kinase gene Mertk caused an RPE phagocytosis defect, which led to the accumulation of rod outer segment debris [128]. More importantly, the mutation of this autophagy-related gene was found in RP patients as well [129]. As a master regulator of the autophagic signalling pathway, mTOR may also be involved in the regulation of Mertk. *MERTK* expression was shown to be regulated by rapamycin in a time-course-dependent manner [130]. In an *rd1* mouse model of retinitis pigmentosa, mTOR was upregulated in photoreceptors. Furthermore, the progression of retinal degeneration in *rd1* mice was alleviated after rapamycin treatment [131].

4.4. *mTOR and Glaucoma*

Glaucoma, a leading cause of irreversible blindness in the world, is one of the most common ocular neurodegenerative diseases. It is characterised by a progressive death of RGCs and structural damage to the optic nerve (ON) [132]. Elevated intraocular ocular pressure (IOP) has always been thought to be the major risk factor of this disease; however, RGC and nerve fibre loss may also occur in a person with normal IOP [133,134]. A great deal of studies have provided evidence showing the involvement of the mTOR signalling pathway in the pathogenesis of glaucoma. A recent study found that AMPK, a critical regulator of mTORC1, was highly expressed in RGCs from both mice with high IOP and patients with primary open-angle glaucoma [135]. Ocular hypertension-induced AMPK overexpression strongly inhibited mTORC1, leading to RGC dendrite retraction and synapse elimination in the early stage. The restoration of mTORC1 activity by knocking down AMPK rescued dendrites and synaptic contacts and promoted RGC survival. It indicated that activated mTORC1 is essential for RGC dendritic maintenance and regeneration, and the inhibition of mTORC1 may diminish its neuroprotective effects in hypertension-induced RGC injury. Similar results were found by Park et al. They observed a significant decrease in mTOR in rat glaucomatous retinas [48]. Furthermore, when autophagy was inhibited by 3-MA, apoptosis of RGCs was significantly decreased in glaucomatous retina.

Not only in RGCs, mTOR-mediated autophagy was also activated in Müller cells in an ischaemic injury model induced by $CoCl_2$ [136]. After the treatment of lutein, a potent anti-oxidant, autophagosome formation induced by rapamycin was suppressed [137]. Moreover, the rMC-1 cell viability and survival rate significantly increased when autophagy was inhibited by lutein [137]. Autophagy is generally considered as a neuroprotective mechanism in the early onset of stress condition. However, over-upregulated autophagy may exacerbate hypoxia-induced cell damage to retinal neurons.

Owing to the versatile roles of the mTOR signalling pathway in multiple cellular functions, it could induce off-target effects [138]. The non-specific effects are also time-dependent; sustained daily rapamycin treatment may promote neuroprotection through activation of multiple pathways downstream or crosstalk with mTOR. Su et al. found that rapamycin promoted RGC survival in a rat chronic hypertensive glaucoma model via inhibition of neurotoxic mediators release and suppression of RGC apoptosis [15]. Moreover, the anti-apoptotic effects were induced directly by rapamycin instead of acting through the PI3K/AKT cell survival pathway. Rapamycin also played a role in inhibiting the activation of microglia in the glaucomatous retinas, preventing the release of pro-inflammatory factors [139]. Topical administration of rapamycin has also shown robust neuroprotective effects in a rat glaucoma model [18]. Strikingly, rapamycin eye drops could reduce IOP by inhibiting RhoA protein activation that regulates actin cytoskeleton in trabecular meshwork (TM) cells.

The trabecular meshwork, which controls the outflow of aqueous humour (AH), plays a critical role in the regulation of IOP. TM cells in the AH pathway are constantly subjected to oxidative stress, which increases the generation of intracellular reactive oxygen species (ROS), leading to mitochondrial dysfunction and apoptosis [140,141]. Besides, the TM cell is one type of post-dividing cell that does not have the capacity to remove excess harmful substances, such as damaged DNA and lipids and collagen deposits [142]. A study also showed that the autophagy homeostasis of TM cells was disrupted in glaucoma patients [143]. Decreased autophagy activity can be considered as an indication of progressive dysregulation of TM function. Studies on promoting autophagy activity in TM cells by blocking the mTOR signalling pathway have been conducted. Zhu et al. found that rapamycin treatment decreased α-actin and myocilin expression in the TM cells of a glucocorticoid-induced glaucoma (GIG) mouse, which was responsible for the extracellular matrix deposition in the TM cells [20]. Rapamycin also recovered the TM ultrastructural and morphological changes in a glaucomatous mouse model, including mitochondrial and collagen fibre arrangement and basement membrane integrity. As a result, the elevated IOP was alleviated after treatment. He et al. found that rapamycin dramatically cleared the damaged mitochondria and accumulated ROS in the TM-1 cells that were exposed to rotenone-induced oxidative stress [16]. Rapamycin also promoted mitochondrial function and prevented TM cell death. Moreover, Igarashi et al. found that topical rapamycin treatment ameliorated TM fibrosis and suppressed collagen deposition in rabbit eyes after trabeculectomy [22]. Table 4 is a summary of the in vitro and in vivo studies investigating the involvement of mTOR in glaucoma (Table 4).

Table 4. Involvement of mTOR in glaucoma: in vitro and animal studies.

Target Cells or Tissue	Disease Model	mTOR Regulator	Autophagy-Related Markers	Related Pathways	Effects of Regulated mTOR	References
NSC-34 /661W cells	2bpIns-OPTN-induced cell death	Rapamycin	LC LC3-II/I↑↑ LC3↑ ATG5↑	mTOR↓	Apoptosis↓ ER stress↓	[144]
TM-1 cells	Rotenone-induced oxidative stress model	Rapamycin	LC3-II/I↑ p62↓	PI3K/AKT/mTOR↑	Apoptosis↓ Oxidative stress↓Mitophagy↑	[16]
RGC-5	E50K-OPTN-induced RGC death	Rapamycin	-	mTOR↓	Apoptosis↓	[145]
Retina tissue/RGC-5	Rat CoCl2-induced hypoxia model	Rapamycin	-	mTOR/RhoA/ROCK↓	IOP↓ RGCs loss↓ Microglial activation↓Mitophagy↑	[119]
HCF cells/TM cells	TGFβ1 induced fibrosis/rabbit model of glaucoma filtration surgery	Rapamycin/Torin-1	-	AKT/mTOR↓	HCF proliferation/migration↓ TM fibrosis↓	[22]
RGCs/TM cells	Mouse glucocorticoid-induced glaucoma model	Rapamycin	LC3-II/I↑ Beclin-1↑ p62↓	mTOR↓	IOP↓ RGCs loss↓ TM fibrosis↓ Mitophagy↑	[20]
RGCs	Mouse chronic hypertensive glaucoma model	Rac1 cKO	LC3-II/I↑ Beclin-1↑ p62↓	mTOR↓	Apoptosis↓ RGCs loss↓	[146]
RGCs	Rat hypertensive glaucoma model	3-MA	LC3B↓ Beclin-1↓	AMPK↓/mTOR↑	Apoptosis↑	[48]
RGCs	Rat hypertensive glaucoma model	Rapamycin	LC3-II↑ p62↓	mTOR↓	Axon loss↓	[147]
RGCs	E50K-OPTN-induced normal tension glaucoma model	Rapamycin	LC3↑ p62↓	mTOR↓	Apoptosis↓ Axon loss↓	[148]
RGCs	DBA 2J mouse model for experimental glaucoma	Rapamycin	-	mTOR↓	Apoptosis↓ Axon loss↓	[26]
RGCs	Rat microbead occlusion model/ex vivo rat glaucoma model	Rapamycin	LC3-II/I↑ p62↓	mTOR↓	Apoptosis↓ RGCs loss↓	[18]
RGCs	E50K-OPTN-induced RGC death	Rapamycin	LC3-II↑ p62↓	mTOR↓	Apoptosis↓ RGCs loss↓ TDP-43 aggregation↓	[43]
RGCs	Rat laser-induced glaucoma model	Rapamycin	-	mTORC1/S6K1↓	Apoptosis↓ RGCs loss↓ VEGFR-2↓	[24]
RGCs	Mouse microbead occlusion model	Rapamycin	-	AMPK↑/mTOR↓	RGCs loss↓	[135]
RGCs	Circumlimbal-suture-induced OHT rat model	Rapamycin	LC3-II/I↑ p62↓	AMPK↑/mTOR↓	Apoptosis↓ RGCs loss↓	[149]
BV2 microglia/primary RGCs/retina tissue	Rat chronic hypertensive glaucoma model	Rapamycin	-	AKT↔/mTOR↓	Apoptosis↓ iNOS/TNF-a/NF-kB↓ Microglial activation↓	[15]
Retina tissue	Ndufs4 KO mouse model of mitochondrial optic neuropathy	Rapamycin	-	mTOR↓	Apoptosis↓ Microglial activation↓ Inflammation↓	[88]

5. Clinical Trials of mTOR Inhibitors in Ocular Neurodegenerative Diseases

As mentioned above, rapamycin, the most established mTOR inhibitor, exhibited potent anti-angiogenic and neuroprotective effects in animal models of DR. Up to now, there have been three clinical trials that evaluated the safety and tolerability of rapamycin in patients with diabetic macular oedema (DMO). Krishnadev et al. conducted a phase I/II study that included five adult participants with diabetic macular oedema (DMO) [150]. The participants received subconjunctival sirolimus injection (440 μg) every 2 months for 12 months with the fellow eye as control. There were no significant drug-related adverse events and repeated subconjunctival injections were well-tolerated. Limited efficacy results were observed, including a 2-line improvement in visual acuity (VA) and 2 log unit decrease in retinal thickness in one participant and improvement in central retinal thickness in three participants; however, one participant had a 2-line worsening of VA and a 1 log unit increase in retinal thickness. Dugel et al. conducted a phase I study to evaluate the safety and tolerability of different dosages of sirolimus in DMO patients with two administration routes, single subconjunctival (SCJ), and intravitreal (IVT) injection, respectively (220, 440, 880, 1320, or 1760 μg vs. 44, 110, 176, 264, or 352 μg) [151]. Twenty-five DMO patients were assigned into each treatment group with the fellow eye as a control. During 90 days of observation, there were no significant drug-related adverse events and dose-limiting toxicities. For the SCJ group, a median increase in BCVA (5.0, 3.0, 4.0, and 4.0 letters) was observed at day 7, 14, 45, and 90, respectively. At day 45, the median decrease in retinal thickness was −23.7 μm. In comparison, the median increase in BCVA of IVT (2.0, 4.0, 4.0, and 4.0 letters) was observed at day 7, 14, 45, and 90, respectively. At day 45, the median decrease in retinal thickness was −52.0 μm. These clinical data provided support for prospective larger randomised trials of rapamycin in the treatment of DR.

To date, there have been three clinical trials that evaluated the safety and efficacy of rapamycin in the treatment of AMD-associated GA. Wong et al. conducted a phase II trial that included 11 participants with bilateral GA [152]. The participants received subconjunctival sirolimus injection (440 μg) every 3 months for 24 months with the fellow eye as a control. Although the treatment was safe and well-tolerated, no significant beneficial effects of sirolimus were observed in the prevention of GA progression; a drug-associated VA decrease was instead found when compared with untreated eyes. In the phase I/II trial conducted by Petrou et al., ocular adverse events, including accelerated retinal thinning and abnormal perilesional changes, on fundus autofluorescence (FAF) were found in two of six participants besides drug-related endophthalmitis [153]. Later, a larger phase II trial was conducted, which included 52 participants with GA treated with monthly intravitreal injection of sirolimus (440 μg) [154]. The trial was suspended because of the observed sterile endophthalmitis in three participants treated with sirolimus. No significant structural or functional benefits were observed after sirolimus injection when compared with the sham group.

There are several potential reasons for the unsatisfactory efficacy in the clinical trial of rapamycin for GA. Firstly, the neuroprotective effects of rapamycin that are efficient in the experimental models of AMD may not be potent enough to prevent and slow down the progression of GA alone, especially in its later stages. Secondly, the protective effect induced by upregulated autophagy in early AMD may exacerbate the apoptosis of retinal neurons in the late stage of AMD. Furthermore, due to the complexity of the mTOR signalling pathway, its effects to modulate the pathological process of disease is unpredictable, and off-target effects are inevitable, especially in the long-term treatment with rapamycin. A systemic kinome-wide approach is required to profile the selectivity and potency of mTOR inhibitors. Recently, Liu et al used chemical proteomics and assays to study the enzymatic activity, protein binding, and disruption of cellular signalling of some mTOR inhibitors, including Torin 1, PP242, Ku-0063794, and WYE354 [155]. The mTOR pathway also has a different contribution in the retinal neurodegeneration in different pathological contexts. Although chronic ischaemic changes are the common pathological pathway of glaucoma and DR, and RGCs are under energetic stress due to ischaemia, rapamycin-induced autophagy played a positive role in promoting RGCs' survival in the diabetic retinas, whereas increased RGC apoptosis was found in the glaucomatous retinas with rapamycin. AMD is a complex disease with a distinct pathological context in different stages of the disease. The modulation of rapamycin may result in paradoxical outcomes in different types of AMD as well. Indeed, clinical trials of mTOR inhibitors in CNV have also been performed with more favourable results.

A series of pilot clinical trials have been conducted to evaluate the safety and efficacy of mTOR inhibitors for CNV both as a single drug and co-treatment with anti-VEGF therapy. Nussenblatt et al. performed a phase I/II clinical trial that included three CNV patients receiving an oral dose of rapamycin (2 mg daily) combined with intravitreal anti–VEGF injection [156]. The treatment was safe and well-tolerated and there were no systematic drug-related adverse events during the six-month observation. Compared with other immunosuppressive drugs, including daclizumab and infliximab, there was no significant difference in the reduction in anti-VEGF injection frequency nor VA improvement and retinal structure amelioration. Furthermore, a recent phase II trial was performed in 2021 to evaluate the safety of the monotherapy with intravitreal sirolimus, and its efficacy was compared with conventional anti-VEGF treatment in exudative AMD [157]. Twenty participants with CNV were assigned to each treatment group with the fellow eye as control. No obvious adverse events were observed, and the treatment is safe and tolerable. VA improvement (6 letters) was observed for both treatment groups; however, there was no significant difference between each other. Most importantly, significant anatomic improvement was found after sirolimus treatment. The mean central subfield thickness (CST) was decreased by 40 μm in the sirolimus group compared with the 20 μm CST increase in the anti-VEGF group. The second-generation mTOR inhibitor, Palmoid 529, was also tested in a clinical trial for the treatment of CNV. Dalal et al. conducted a phase I trial that included 13 CNV patients to assess the safety and efficacy of Palmoid 529 subconjunctival injection (1.9 mg, every four weeks) for 12 weeks in a short period [158]. There were no drug-related adverse events and no ocular or systemic safety concerns for the treatment. Probably due to the limited sample size, no treatment effects were found in those anti-VEGF refractory patients. Larger-scale randomised studies are, therefore, required to test the efficacy of the dual inhibitor in the treatment of CNV. Table 5 is a summary of the clinical trials of mTOR inhibitors in ocular degenerative diseases including AMD and DR (Table 5).

Table 5. Summary of clinical trials of mTOR inhibitors in ocular neurodegenerative diseases.

Study (NCT Number)	Design	Subjects	Intervention	Treatment Regimen	Results	Reference
Phase II trial Naor et al. 2010 (NCT00656643)	Four-arm study in US; placebo injection as control	131 with diabetic macular oedema	Sirolimus subconjunctival injection	Two subconjunctival injections of 220, 440, 880 µg, or placebo (1:1:1:1) observation through day 180	Awaiting results	[159]
Phase I/II trial Krishnadev et al. 2011 (NCT00711490)	Single-arm study in US; fellow eye as control	5 with diabetic macular oedema	Sirolimus subconjunctival injection	440 µg injection every 2 months for 12 months follow-up period	Safe and well-tolerated; efficacy trials required	[150]
Phase I trial Dugel et al. 2012 (NCT00401115)	Two-arm study in US; fellow eye as control	50 with diabetic macular oedema (n = 25 for SCJ and IVT, respectively)	Sirolimus single subconjunctival (SCJ)/intravitreal injection (IVT)	SCJ (220, 440, 880, 1320, or 1760 µg)/IVT (44, 110, 176, 264, or 352 µg); observation through day 90	Safe and well-tolerated (no dose-limiting toxicities); efficacy trials required	[151]
Phase I/II trial Naor et al. 2010 (NCT00712491)	Two-arm study in US; fellow eye as control	20 with AMD (CNV); n = 10 for each arm	Rapamycin intravitreal injection	Three injections of 352 or 1320 µg observation through 12 months	Awaiting results	[160]
Phase II trial Nussenblatt et al. 2010 (NCT00304954)	Four-arm study in US; fellow eye as control	13 with AMD (CNV)	Intravenous daclizumab/intravenous infliximab/oral rapamycin/observation with anti-VEGF therapy	Daily 2 mg oral tablet (n = 3) vs. daclizumab, vs. infliximab vs. no immunosuppression plus intraocular anti-VEGF therapy for 6 months fellow up	Safe and well-tolerated; no benefit	[156]
Phase II trial Abraham et al. 2010 (NCT00766337)	Three-arm study in US; placebo comparator as control	62 with AMD (CNV)	Sirolimus in combination with ranibizumab subconjunctival injection	440 or 1320 µg both with 500 µg ranibizumab every 2 months for 24 months fellow up	Awaiting results	[161]
Phase II trial Wong et al. 2013 (NCT00766649)	Single-arm study in US; fellow eye as control	11 with AMD (GA)	Rapamycin subconjunctival injection	440 µg injection every three months for 24 months follow-up	Safe and well-tolerated; no benefit	[152]
Phase I trial Dalal et al. 2013 (NCT01271270)	Single-arm study in US; fellow eye as control	13 with AMD (CNV)	Palomid 529 subconjunctival injection	1.9 mg injection every 4 weeks for 12 weeks follow-up	Safe and well-tolerated; efficacy trials required	[158]
Phase I/II trial Petrou et al. 2014 (NCT01445548)	Single-arm study in US; fellow eye as control	6 with AMD (GA)	Rapamycin intravitreal injection	440 µg injection every two months for 12 months follow-up	Ocular adverse events appeared; no benefit	[153]
Phase II trial Gensler et al. 2017 (NCT01675947)	Two-arm study in US; sham treatment as control	52 with AMD (GA); n = 27 for rapamycin	Rapamycin intravitreal injection	440 µg injection monthly for 24 months follow-up	Safe and well-tolerated; no benefit	[154]
Phase II trial Minturn et al. 2021 (NCT02357342)	Two-arm study in US; fellow eye as control	40 with AMD (CNV); n = 20 for each arm	Sirolimus intravitreal injection/anti-VEGF therapy	440 µg injection every two months for 6 months follow-up	Safe and well-tolerated; CST decreased by 40 µm in sirolimus group (p = 0.03)	[157]

6. Discussion

In line with the potent antioxidant and anti-inflammatory effects of mTOR inhibitors in different ocular neurodegenerative disease models shown above, rapamycin, a lead mTOR inhibitor, presents an attractive treatment option in the clinical trials of DR and AMD with a favourable safety profile and sustained ocular pharmacokinetics (Table 5). However, some studies reported that mTOR activation may have beneficial effects on the survival of cellular components in the retina. Cao et al showed that NGF (nerve growth factor) protected RPE cells against H_2O_2-induced cell apoptosis through the PI3K/Akt/mTOR and ERK/MARK signalling pathway [23]. Co-treatment with rapamycin diminished NGF-induced S6 phosphorylation and protective effects against oxidative stress in ARPE-19 cells. Since the modulation of mTOR was conducted in an RPE cell line, it is possible that mTOR upregulation may induce distinctive effects in vivo. mTOR may also have different contributions to different cellular components in the retina and to different disease conditions, respectively. Park et al. reported that the upregulation of mTOR decreased RGC apoptosis in glaucomatous retinas, which was instead increased in the diabetic retinas [48]. mTOR-mediated autophagy may, therefore, play different roles in RGCs' survival in different disease conditions. It is unclear whether prolonged and/or overregulated autophagy may have detrimental effects to the retinal cell survival. Especially, the mTOR signalling pathway may also affect the metabolism in the retina. Fang et al. found that short-term rapamycin treatment (6 weeks) induced metabolic impairment in mice, but prolonged rapamycin treatment (20 weeks) reversed the detrimental effects, with better metabolic profiles, increased oxygen consumption and ketogenesis, and markedly enhanced insulin sensitivity [162]. Yet, conflicting results were reported; short-term hyperglycaemia (1 month) upregulated mTORC1 activity, inhibited autophagy, and prevented RGCs death, whereas prolonged hyperglycaemia (6 months) downregulated mTORC1 activity, promoted autophagy, and induced RGCs damage [93]. In light of this, it is important to address how mTOR signalling contributes to retinal neurodegeneration. A systematic profiling of mTOR signalling has been conducted in the foetal fibroblasts [163], but there is no publication covering the mTOR genetic profile in the disease model of retinal neurodegeneration. With more evidence on how mTOR modulates autophagy, cell proliferation, apoptosis, and metabolism in the retina, precise treatment using new drug delivery techniques and gene therapy may avoid adverse effects and provide higher therapeutic effectiveness. Indeed, rAAV-mTOR shRNA (recombinant adeno-associated virus-delivered mTOR inhibiting short hairpin RNA) treatment significantly reduced CNV lesions and decreased local inflammation in a laser-induced mouse model [121]. Another gene therapy study using rAAV2-shmTOR-SD achieved similar results [127]. Furthermore, the development of newer compounds that selectively induce or target autophagy may have a more promising therapeutic perspective in ocular neurodegenerative diseases (Figure 4). Wen et al. found that inhibition of mTORC2 alone resulted in blood–optic nerve barrier disruption, but co-treatment with rapamycin and mTORC2 activator SC79 improved RGC survival [164].

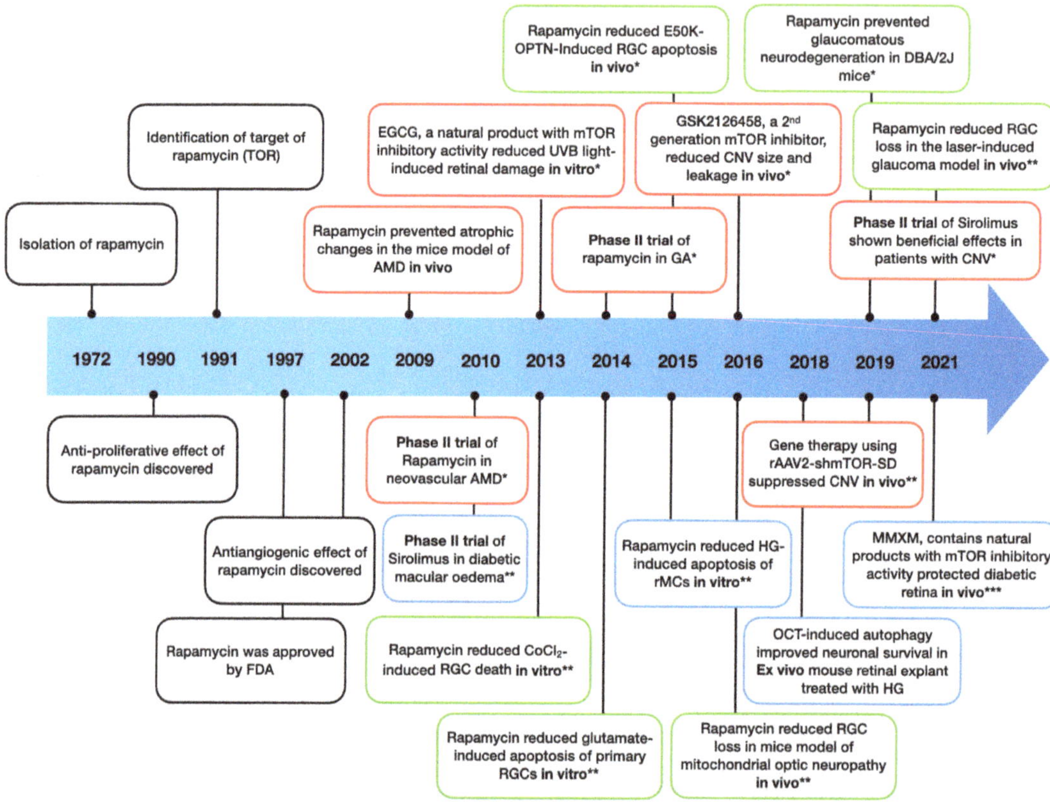

Figure 4. Timeline of mTOR inhibitors from discovery to the clinic for the treatment of ocular neurodegenerative diseases. Abbreviations: AMD, age-related macular degeneration; CNV, choroidal neovascularization; EGCG, epigallocatechin gallate; HG, high glucose; GA, geographic atrophy; MMXM, Mingmu Xiaomeng; OCT, octreotide; OPTN, optineurin; RGC, retinal ganglion cell; rMCs, retinal Müller cells. References in the figure: 1992 [55], 1990 [60], 1991 [1], 1997 [63], 2002 [61], 2009 [159], 2010 * [162], 2010 ** [161], 2013 * [160], 2013 ** [123], 2014 * [146], 2014 ** [16], 2015 * [44], 2015 ** [94], 2016 * [120], 2016 ** [83], 2018 [89], 2019 * [19], 2019 ** [136], 2021 * [153], 2021 ** [28], 2021 *** [91].

7. Conclusions

Autophagy is an essential catabolic process critical for stress responses and the maintenance of cellular homeostasis. Autophagy promotes cell survival by eliminating damaged cellular components in response to oxidative stresses. As one of the key regulators of autophagy, the involvement of the mTOR signalling pathway in the pathophysiology of major ocular neurogenerative diseases, including DR (Table 2), AMD (Table 3), glaucoma (Table 4), and RP, was summarised in this review, which focused on the common pathological processes, including mitochondrial dysfunction, elevated ROS level, and increased ER stress induced by oxidative stress, hypoxia, and inflammation (Figure 1). Each of these processes plays a substantial role in the regulation of mTOR by modulating the upstream signalling pathways, such as PI3K/Akt, AMPK, and TSC1/2. Although rapamycin may be an attractive treatment option in DR and AMD, more clinical trials are still needed. It is also essential to understand how mTOR modulates autophagy, cell proliferation, apoptosis, and metabolism in the retina. New drug delivery techniques and gene therapy, as well as selective regulators in the mTOR pathway, may help to avoid the adverse effects and provide more precise treatment, yielding higher therapeutic efficacy.

Author Contributions: Y.W. Conceptualization; data curation; formal analysis; investigation; methodology; validation; visualization; writing—original draft; writing—review and editing. N.S.K.F. Supervision; writing—review and editing. W.-C.L. Supervision; writing—review and editing. A.C.Y.L. Conceptualization; formal analysis; funding acquisition; investigation; methodology; project administration; resources; supervision; validation; visualization; writing—original draft; writing—review and editing. All authors have read and agreed to the published version of the manuscript.

Funding: This research was funded by General Research Fund, Research Grants Council, The Government of the Hong Kong Special Administrative Region (17112919) to A.C.Y.L.

Conflicts of Interest: The authors declare no conflict of interest. The funders had no role in the design of the study; in the collection, analyses, or interpretation of data; in the writing of the manuscript; or in the decision to publish the results.

References

1. Laplante, M.; Sabatini, D.M. MTOR Signaling at a Glance. *J. Cell Sci.* **2009**, *122*, 3589–3594. [CrossRef] [PubMed]
2. Dowling, R.J.O.; Topisirovic, I.; Fonseca, B.D.; Sonenberg, N. Dissecting the Role of MTOR: Lessons from MTOR Inhibitors. *Biochim. Biophys. Acta—Proteins Proteom.* **2010**, *1804*, 433–439. [CrossRef]
3. Mossmann, D.; Park, S.; Hall, M.N. MTOR Signalling and Cellular Metabolism Are Mutual Determinants in Cancer. *Nat. Rev. Cancer* **2018**, *18*, 744–757. [CrossRef] [PubMed]
4. Masui, K.; Tanaka, K.; Akhavan, D.; Babic, I.; Gini, B.; Matsutani, T.; Iwanami, A.; Liu, F.; Villa, G.R.; Gu, Y.; et al. MTOR Complex 2 Controls Glycolytic Metabolism in Glioblastoma through FoxO Acetylation and Upregulation of C-Myc. *Cell Metab.* **2013**, *18*, 726–739. [CrossRef] [PubMed]
5. Shigihara, N.; Fukunaka, A.; Hara, A.; Komiya, K.; Honda, A.; Uchida, T.; Abe, H.; Toyofuku, Y.; Tamaki, M.; Ogihara, T.; et al. Human IAPP–Induced Pancreatic β Cell Toxicity and Its Regulation by Autophagy. *J. Clin. Investig.* **2014**, *124*, 3634–3644. [CrossRef] [PubMed]
6. Griffin, R.J.; Moloney, A.; Kelliher, M.; Johnston, J.A.; Ravid, R.; Dockery, P.; O'Connor, R.; O'Neill, C. Activation of Akt/PKB, Increased Phosphorylation of Akt Substrates and Loss and Altered Distribution of Akt and PTEN Are Features of Alzheimer's Disease Pathology. *J. Neurochem.* **2005**, *93*, 105–117. [CrossRef]
7. Fruman, D.A.; Rommel, C. PI3K and Cancer: Lessons, Challenges and Opportunities. *Nat. Rev. Drug Discov.* **2014**, *13*, 140–156. [CrossRef]
8. Lai, C.; Chen, Z.; Ding, Y.; Chen, Q.; Su, S.; Liu, H.; Ni, R.; Tang, Z. Rapamycin Attenuated Zinc-Induced Tau Phosphorylation and Oxidative Stress in Rats: Involvement of Dual MTOR/P70S6K and Nrf2/HO-1 Pathways. *Front. Immunol.* **2022**, *13*, 218. [CrossRef]
9. Ha, J.-Y.; Kim, J.-S.; Kang, Y.-H.; Bok, E.; Kim, Y.-S.; Son, J.H. Tnfaip8 L1/Oxi-β Binds to FBXW5, Increasing Autophagy through Activation of TSC2 in a Parkinson's Disease Model. *J. Neurochem.* **2014**, *129*, 527–538. [CrossRef]
10. Hodgson, J.G.; Agopyan, N.; Gutekunst, C.-A.; Leavitt, B.R.; LePiane, F.; Singaraja, R.; Smith, D.J.; Bissada, N.; McCutcheon, K.; Nasir, J.; et al. A YAC Mouse Model for Huntington's Disease with Full-Length Mutant Huntingtin, Cytoplasmic Toxicity, and Selective Striatal Neurodegeneration. *Neuron* **1999**, *23*, 181–192. [CrossRef]
11. Pardue, M.T.; Allen, R.S. Neuroprotective Strategies for Retinal Disease. *Prog. Retin. Eye Res.* **2018**, *65*, 50–76. [CrossRef] [PubMed]
12. Chalke, S.D.; Kale, P.P. Combinational Approaches Targeting Neurodegeneration, Oxidative Stress, and Inflammation in the Treatment of Diabetic Retinopathy. *Curr. Drug Targets* **2021**, *22*, 1810–1824. [CrossRef]
13. IZZOTTI, A.; BAGNIS, A.; SACCA, S. The Role of Oxidative Stress in Glaucoma. *Mutat. Res. Mutat. Res.* **2006**, *612*, 105–114. [CrossRef] [PubMed]
14. Armstrong, A.; Mandala, A.; Malhotra, M.; Gnana-Prakasam, J.P. Canonical Wnt Signaling in the Pathology of Iron Overload-Induced Oxidative Stress and Age-Related Diseases. *Oxid. Med. Cell. Longev.* **2022**, *2022*, 7163326. [CrossRef]
15. Su, W.; Li, Z.; Jia, Y.; Zhuo, Y. Rapamycin Is Neuroprotective in a Rat Chronic Hypertensive Glaucoma Model. *PLoS ONE* **2014**, *9*, e99719. [CrossRef] [PubMed]
16. He, J.N.; Zhang, S.D.; Qu, Y.; Wang, H.L.; Tham, C.C.; Pang, C.P.; Chu, W.K. Rapamycin Removes Damaged Mitochondria and Protects Human Trabecular Meshwork (TM-1) Cells from Chronic Oxidative Stress. *Mol. Neurobiol.* **2019**, *56*, 6586–6593. [CrossRef]
17. Heras-Sandoval, D.; Pérez-Rojas, J.M.; Hernández-Damián, J.; Pedraza-Chaverri, J. The Role of PI3K/AKT/MTOR Pathway in the Modulation of Autophagy and the Clearance of Protein Aggregates in Neurodegeneration. *Cell. Signal.* **2014**, *26*, 2694–2701. [CrossRef]
18. Ishikawa, M.; Takaseki, S.; Yoshitomi, T.; Covey, D.F.; Zorumski, C.F.; Izumi, Y. The Neurosteroid Allopregnanolone Protects Retinal Neurons by Effects on Autophagy and GABRs/GABAA Receptors in Rat Glaucoma Models. *Autophagy* **2020**, *17*, 743–760. [CrossRef]
19. Liegl, R.; Koenig, S.; Siedlecki, J.; Haritoglou, C.; Kampik, A.; Kernt, M. Temsirolimus Inhibits Proliferation and Migration in Retinal Pigment Epithelial and Endothelial Cells via MTOR Inhibition and Decreases VEGF and PDGF Expression. *PLoS ONE* **2014**, *9*, e88203. [CrossRef]

20. Zhu, X.; Wu, S.; Zeng, W.; Chen, X.; Zheng, T.; Ren, J.; Ke, M. Protective Effects of Rapamycin on Trabecular Meshwork Cells in Glucocorticoid-Induced Glaucoma Mice. *Front. Pharmacol.* **2020**, *11*, 1006. [CrossRef]
21. Niu, Z.; Shi, Y.; Li, J.; Qiao, S.; Du, S.; Chen, L.; Tian, H.; Wei, L.; Cao, H.; Wang, J.; et al. Protective Effect of Rapamycin in Models of Retinal Degeneration. *Exp. Eye Res.* **2021**, *210*, 108700. [CrossRef] [PubMed]
22. Igarashi, N.; Honjo, M.; Aihara, M. Effects of Mammalian Target of Rapamycin Inhibitors on Fibrosis after Trabeculectomy. *Exp. Eye Res.* **2021**, *203*, 108421. [CrossRef] [PubMed]
23. Cao, G.F.; Liu, Y.; Yang, W.; Wan, J.; Yao, J.; Wan, Y.; Jiang, Q. Rapamycin Sensitive MTOR Activation Mediates Nerve Growth Factor (NGF) Induced Cell Migration and pro-Survival Effects against Hydrogen Peroxide in Retinal Pigment Epithelial Cells. *Biochem. Biophys. Res. Commun.* **2011**, *414*, 499–505. [CrossRef] [PubMed]
24. Zhou, J.; Chen, F.; Yan, A.; Xia, X. Role of Mammalian Target of Rapamycin in Regulating HIF-1α and Vascular Endothelial Growth Factor Signals in Glaucoma. *Arch. Physiol. Biochem.* **2021**, *127*, 44–50. [CrossRef] [PubMed]
25. Shang, P.; Valapala, M.; Grebe, R.; Hose, S.; Ghosh, S.; Bhutto, I.A.; Handa, J.T.; Lutty, G.A.; Lu, L.; Wan, J.; et al. The Amino Acid Transporter SLC36A4 Regulates the Amino Acid Pool in Retinal Pigmented Epithelial Cells and Mediates the Mechanistic Target of Rapamycin, Complex 1 Signaling. *Aging Cell* **2017**, *16*, 349–359. [CrossRef]
26. Harder, J.M.; Guymer, C.; Wood, J.P.M.; Daskalaki, E.; Chidlow, G.; Zhang, C.; Balasubramanian, R.; Cardozo, B.H.; Foxworth, N.E.; Deering, K.E.; et al. Disturbed Glucose and Pyruvate Metabolism in Glaucoma with Neuroprotection by Pyruvate or Rapamycin. *Proc. Natl. Acad. Sci. USA* **2020**, *117*, 33619–33627. [CrossRef]
27. Schreiber, K.H.; Arriola Apelo, S.I.; Yu, D.; Brinkman, J.A.; Velarde, M.C.; Syed, F.A.; Liao, C.Y.; Baar, E.L.; Carbajal, K.A.; Sherman, D.S.; et al. A Novel Rapamycin Analog Is Highly Selective for MTORC1 in Vivo. *Nat. Commun.* **2019**, *10*, 3194. [CrossRef]
28. Ran, Z.; Zhang, Y.; Wen, X.; Ma, J. Curcumin Inhibits High Glucose-Induced Inflammatory Injury in Human Retinal Pigment Epithelial Cells through the ROS-PI3K/AKT/MTOR Signaling Pathway. *Mol. Med. Rep.* **2019**, *19*, 1024–1031. [CrossRef]
29. Saxton, R.A.; Sabatini, D.M. MTOR Signaling in Growth, Metabolism, and Disease. *Cell* **2017**, *168*, 960–976. [CrossRef]
30. Nojima, H.; Tokunaga, C.; Eguchi, S.; Oshiro, N.; Hidayat, S.; Yoshino, K.; Hara, K.; Tanaka, N.; Avruch, J.; Yonezawa, K. The Mammalian Target of Rapamycin (MTOR) Partner, Raptor, Binds the MTOR Substrates P70 S6 Kinase and 4E-BP1 through Their TOR Signaling (TOS) Motif. *J. Biol. Chem.* **2003**, *278*, 15461–15464. [CrossRef]
31. Hwang, Y.; Kim, L.C.; Song, W.; Edwards, D.N.; Cook, R.S.; Chen, J. Disruption of the Scaffolding Function of MLST8 Selectively Inhibits MTORC2 Assembly and Function and Suppresses MTORC2-Dependent Tumor Growth In Vivo. *Cancer Res.* **2019**, *79*, 3178–3184. [CrossRef] [PubMed]
32. Wang, L.; Harris, T.E.; Roth, R.A.; Lawrence, J.C. PRAS40 Regulates MTORC1 Kinase Activity by Functioning as a Direct Inhibitor of Substrate Binding. *J. Biol. Chem.* **2007**, *282*, 20036–20044. [CrossRef] [PubMed]
33. Jacinto, E.; Facchinetti, V.; Liu, D.; Soto, N.; Wei, S.; Jung, S.Y.; Huang, Q.; Qin, J.; Su, B. SIN1/MIP1 Maintains Rictor-MTOR Complex Integrity and Regulates Akt Phosphorylation and Substrate Specificity. *Cell* **2006**, *127*, 125–137. [CrossRef] [PubMed]
34. Kroemer, G.; Mariño, G.; Levine, B. Autophagy and the Integrated Stress Response. *Mol. Cell* **2010**, *40*, 280–293. [CrossRef] [PubMed]
35. Schultze, S.M.; Hemmings, B.A.; Niessen, M.; Tschopp, O. PI3K/AKT, MAPK and AMPK Signalling: Protein Kinases in Glucose Homeostasis. *Expert Rev. Mol. Med.* **2012**, *14*, e1. [CrossRef]
36. Yoon, M.-S. The Role of Mammalian Target of Rapamycin (MTOR) in Insulin Signaling. *Nutrients* **2017**, *9*, 1176. [CrossRef]
37. Rehbein, U.; Prentzell, M.T.; Cadena Sandoval, M.; Heberle, A.M.; Henske, E.P.; Opitz, C.A.; Thedieck, K. The TSC Complex-MTORC1 Axis: From Lysosomes to Stress Granules and Back. *Front. Cell Dev. Biol.* **2021**, *9*, 751892. [CrossRef]
38. Valvezan, A.J.; Klein, P.S. GSK-3 and Wnt Signaling in Neurogenesis and Bipolar Disorder. *Front. Mol. Neurosci.* **2012**, *5*, 1. [CrossRef]
39. Inoki, K.; Zhu, T.; Guan, K.-L. TSC2 Mediates Cellular Energy Response to Control Cell Growth and Survival. *Cell* **2003**, *115*, 577–590. [CrossRef]
40. Budanov, A.V.; Karin, M. P53 Target Genes Sestrin1 and Sestrin2 Connect Genotoxic Stress and MTOR Signaling. *Cell* **2008**, *134*, 451–460. [CrossRef]
41. Nguyen, T.; Sherratt, P.J.; Pickett, C.B. Regulatory Mechanisms Controlling Gene Expression Mediated by the Antioxidant Response Element. *Annu. Rev. Pharmacol. Toxicol.* **2003**, *43*, 233–260. [CrossRef] [PubMed]
42. Kim, K.C.; Kang, K.A.; Zhang, R.; Piao, M.J.; Kim, G.Y.; Kang, M.Y.; Lee, S.J.; Lee, N.H.; Surh, Y.-J.; Hyun, J.W. Up-Regulation of Nrf2-Mediated Heme Oxygenase-1 Expression by Eckol, a Phlorotannin Compound, through Activation of Erk and PI3K/Akt. *Int. J. Biochem. Cell Biol.* **2010**, *42*, 297–305. [CrossRef] [PubMed]
43. Zhang, X.-S.; Zhang, X.; Wu, Q.; Li, W.; Zhang, Q.-R.; Wang, C.-X.; Zhou, X.-M.; Li, H.; Shi, J.-X.; Zhou, M.-L. Astaxanthin Alleviates Early Brain Injury Following Subarachnoid Hemorrhage in Rats: Possible Involvement of Akt/Bad Signaling. *Mar. Drugs* **2014**, *12*, 4291–4310. [CrossRef] [PubMed]
44. Oshiro, N.; Takahashi, R.; Yoshino, K.; Tanimura, K.; Nakashima, A.; Eguchi, S.; Miyamoto, T.; Hara, K.; Takehana, K.; Avruch, J.; et al. The Proline-Rich Akt Substrate of 40 KDa (PRAS40) Is a Physiological Substrate of Mammalian Target of Rapamycin Complex 1. *J. Biol. Chem.* **2007**, *282*, 20329–20339. [CrossRef] [PubMed]
45. Schalm, S.S.; Blenis, J. Identification of a Conserved Motif Required for MTOR Signaling. *Curr. Biol.* **2002**, *12*, 632–639. [CrossRef]
46. Lin, W.; Xu, G. Autophagy: A Role in the Apoptosis, Survival, Inflammation, and Development of the Retina. *Ophthalmic Res.* **2019**, *61*, 65–72. [CrossRef]

47. Cherra 3rd, S.J.; Chu, C.T. Autophagy in Neuroprotection and Neurodegeneration: A Question of Balance. *Future Neurol.* **2008**, *3*, 309–323. [CrossRef]
48. Park, H.Y.L.; Kim, J.H.; Park, C.K. Different Contributions of Autophagy to Retinal Ganglion Cell Death in the Diabetic and Glaucomatous Retinas. *Sci. Rep.* **2018**, *8*, 13321. [CrossRef]
49. Wu, J.J.; Quijano, C.; Chen, E.; Liu, H.; Cao, L.; Fergusson, M.M.; Rovira, I.I.; Gutkind, S.; Daniels, M.P.; Komatsu, M.; et al. Mitochondrial Dysfunction and Oxidative Stress Mediate the Physiological Impairment Induced by the Disruption of Autophagy. *Aging* **2009**, *1*, 425–437. [CrossRef]
50. Tal, M.C.; Sasai, M.; Lee, H.K.; Yordy, B.; Shadel, G.S.; Iwasaki, A. Absence of Autophagy Results in Reactive Oxygen Species-Dependent Amplification of RLR Signaling. *Proc. Natl. Acad. Sci. USA* **2009**, *106*, 2770–2775. [CrossRef]
51. Murata, H.; Takamatsu, H.; Liu, S.; Kataoka, K.; Huh, N.; Sakaguchi, M. NRF2 Regulates PINK1 Expression under Oxidative Stress Conditions. *PLoS ONE* **2015**, *10*, e0142438. [CrossRef] [PubMed]
52. Losiewicz, M.K.; Elghazi, L.; Fingar, D.C.; Rajala, R.V.S.; Lentz, S.I.; Fort, P.E.; Abcouwer, S.F.; Gardner, T.W. MTORC1 and MTORC2 Expression in Inner Retinal Neurons and Glial Cells. *Exp. Eye Res.* **2020**, *197*, 108131. [CrossRef] [PubMed]
53. Ballou, L.M.; Lin, R.Z. Rapamycin and MTOR Kinase Inhibitors. *J. Chem. Biol.* **2008**, *1*, 27–36. [CrossRef] [PubMed]
54. Bellmunt, J.; Szczylik, C.; Feingold, J.; Strahs, A.; Berkenblit, A. Temsirolimus Safety Profile and Management of Toxic Effects in Patients with Advanced Renal Cell Carcinoma and Poor Prognostic Features. *Ann. Oncol.* **2008**, *19*, 1387–1392. [CrossRef]
55. Sellami, D.B.; Urva, S.R.; Grosch, K.; Cheung, W.K.; Anak, O. Meta-Analysis on the Relationship between Everolimus Exposure and Safety and Efficacy. *J. Clin. Oncol.* **2012**, *30*, 3099. [CrossRef]
56. Hartford, C.M.; Desai, A.A.; Janisch, L.; Karrison, T.; Rivera, V.M.; Berk, L.; Loewy, J.W.; Kindler, H.; Stadler, W.M.; Knowles, H.L.; et al. A Phase I Trial to Determine the Safety, Tolerability, and Maximum Tolerated Dose of Deforolimus in Patients with Advanced Malignancies. *Clin. Cancer Res.* **2009**, *15*, 1428–1434. [CrossRef]
57. Banerji, U.; Dean, E.J.; Gonzalez, M.; Greystoke, A.P.; Basu, B.; Krebs, M.; Puglisi, M.; Grinsted, L.; Oelmann, E.; Burke, W.; et al. First-in-Human Phase I Trial of the Dual MTORC1 and MTORC2 Inhibitor AZD2014 in Solid Tumors. *J. Clin. Oncol.* **2012**, *30*, 3004. [CrossRef]
58. Tan, D.S.; Dumez, H.; Olmos, D.; Sandhu, S.K.; Hoeben, A.; Stephens, A.W.; Poondru, S.; Gedrich, R.; Kaye, S.B.; Schoffski, P. First-in-Human Phase I Study Exploring Three Schedules of OSI-027, a Novel Small Molecule TORC1/TORC2 Inhibitor, in Patients with Advanced Solid Tumors and Lymphoma. *J. Clin. Oncol.* **2010**, *28*, 3006. [CrossRef]
59. Tabernero, J.; Cervantes, A.; Gordon, M.S.; Chiorean, E.G.; Burris, H.A.; Macarulla, T.; Perez-Fidalgo, A.; Martin, M.; Jessen, K.; Liu, Y.; et al. Abstract CT-02: A Phase I, Open Label, Dose Escalation Study of Oral Mammalian Target of Rapamycin Inhibitor INK128 Administered by Intermittent Dosing Regimens in Patients with Advanced Malignancies. In *Proceedings of the Clinical Trials*; American Association for Cancer Research: Philadelphia, PA, USA, 2012; p. CT-02-CT-02.
60. Banerji, U.; Ranson, M.; Schellens, J.H.; Esaki, T.; Dean, E.; Zivi, A.; van der Noll, R.; Stockman, P.K.; Marotti, M.; Garrett, M.D.; et al. Abstract LB-66: Results of Two Phase I Multicenter Trials of AZD5363, an Inhibitor of AKT1, 2 and 3: Biomarker and Early Clinical Evaluation in Western and Japanese Patients with Advanced Solid Tumors. In *Proceedings of the Clinical Trials*; American Association for Cancer Research: Philadelphia, PA, USA, 2013; p. LB-66-LB-66.
61. Spencer, A.; Yoon, S.-S.; Harrison, S.J.; Morris, S.; Smith, D.; Freedman, S.J.; Brigandi, R.; Oliff, A.; Opalinska, J.B.; Chen, C. Novel AKT Inhibitor GSK2110183 Shows Favorable Safety, Pharmacokinetics, and Clinical Activity in Multiple Myeloma. Preliminary Results From a Phase I First-Time-In-Human Study. *Blood* **2011**, *118*, 1856. [CrossRef]
62. Tabernero, J.; Saura, C.; Roda Perez, D.; Dienstmann, R.; Rosello, S.; Prudkin, L.; Perez-Fidalgo, J.A.; Graña, B.; Jones, C.; Musib, L.; et al. First-in-Human Phase I Study Evaluating the Safety, Pharmacokinetics (PK), and Intratumor Pharmacodynamics (PD) of the Novel, Oral, ATP-Competitive Akt Inhibitor GDC-0068. *J. Clin. Oncol.* **2011**, *29*, 3022. [CrossRef]
63. Ochiai, T.; Gunji, Y.; Nagata, M.; Komori, A.; Asano, T.; Isono, K. Effects of Rapamycin in Experimental Organ Allografting. *Transplantation* **1993**, *56*, 15–19. [CrossRef] [PubMed]
64. Bové, J.; Martínez-Vicente, M.; Vila, M. Fighting Neurodegeneration with Rapamycin: Mechanistic Insights. *Nat. Rev. Neurosci.* **2011**, *12*, 437–452. [CrossRef] [PubMed]
65. Kong, D.; Banerjee, S.; Huang, W.; Li, Y.; Wang, Z.; Kim, H.-R.C.; Sarkar, F.H. Mammalian Target of Rapamycin Repression by 3,3′-Diindolylmethane Inhibits Invasion and Angiogenesis in Platelet-Derived Growth Factor-D–Overexpressing PC3 Cells. *Cancer Res.* **2008**, *68*, 1927–1934. [CrossRef]
66. Huang, S.; Houghton, P.J. Resistance to Rapamycin: A Novel Anticancer Drug. *Cancer Metastasis Rev.* **2001**, *20*, 69–78. [CrossRef]
67. Simamora, P.; Alvarez, J.M.; Yalkowsky, S.H. Solubilization of Rapamycin. *Int. J. Pharm.* **2001**, *213*, 25–29. [CrossRef]
68. Elit, L. CCI-779 Wyeth. *Curr. Opin. Investig. Drugs* **2002**, *3*, 1249–1253.
69. Dumont, F.J. Everolimus. Novartis. *Curr. Opin. Investig. Drugs* **2001**, *2*, 1220–1234.
70. Mita, M.; Sankhala, K.; Abdel-Karim, I.; Mita, A.; Giles, F. Deforolimus (AP23573) a Novel MTOR Inhibitor in Clinical Development. *Expert Opin. Investig. Drugs* **2008**, *17*, 1947–1954. [CrossRef]
71. Kwitkowski, V.E.; Prowell, T.M.; Ibrahim, A.; Farrell, A.T.; Justice, R.; Mitchell, S.S.; Sridhara, R.; Pazdur, R. FDA Approval Summary: Temsirolimus as Treatment for Advanced Renal Cell Carcinoma. *Oncologist* **2010**, *15*, 428–435. [CrossRef] [PubMed]
72. Chresta, C.M.; Davies, B.R.; Hickson, I.; Harding, T.; Cosulich, S.; Critchlow, S.E.; Vincent, J.P.; Ellston, R.; Jones, D.; Sini, P.; et al. AZD8055 Is a Potent, Selective, and Orally Bioavailable ATP-Competitive Mammalian Target of Rapamycin Kinase Inhibitor with in Vitro and in Vivo Antitumor Activity. *Cancer Res.* **2010**, *70*, 288–298. [CrossRef]

73. Xue, Q.; Hopkins, B.; Perruzzi, C.; Udayakumar, D.; Sherris, D.; Benjamin, L.E. Palomid 529, a Novel Small-Molecule Drug, Is a TORC1/TORC2 Inhibitor That Reduces Tumor Growth, Tumor Angiogenesis, and Vascular Permeability. *Cancer Res.* **2008**, *68*, 9551–9557. [CrossRef] [PubMed]
74. Thoreen, C.C.; Kang, S.A.; Chang, J.W.; Liu, Q.; Zhang, J.; Gao, Y.; Reichling, L.J.; Sim, T.; Sabatini, D.M.; Gray, N.S. An ATP-Competitive Mammalian Target of Rapamycin Inhibitor Reveals Rapamycin-Resistant Functions of MTORC1. *J. Biol. Chem.* **2009**, *284*, 8023–8032. [CrossRef] [PubMed]
75. Rodrik-Outmezguine, V.S.; Chandarlapaty, S.; Pagano, N.C.; Poulikakos, P.I.; Scaltriti, M.; Moskatel, E.; Baselga, J.; Guichard, S.; Rosen, N. MTOR Kinase Inhibition Causes Feedback-Dependent Biphasic Regulation of AKT Signaling. *Cancer Discov.* **2011**, *1*, 248–259. [CrossRef] [PubMed]
76. Carew, J.S.; Kelly, K.R.; Nawrocki, S.T. Mechanisms of MTOR Inhibitor Resistance in Cancer Therapy. *Target. Oncol.* **2011**, *6*, 17–27. [CrossRef]
77. Brownlee, M. Biochemistry and Molecular Cell Biology of Diabetic Complications. *Nature* **2001**, *414*, 813–820. [CrossRef]
78. Sifuentes-Franco, S.; Pacheco-Moisés, F.P.; Rodríguez-Carrizalez, A.D.; Miranda-Díaz, A.G. The Role of Oxidative Stress, Mitochondrial Function, and Autophagy in Diabetic Polyneuropathy. *J. Diabetes Res.* **2017**, *2017*, 2760716. [CrossRef]
79. Zafar, S.; Sachdeva, M.; Frankfort, B.J.; Channa, R. Retinal Neurodegeneration as an Early Manifestation of Diabetic Eye Disease and Potential Neuroprotective Therapies. *Curr. Diab. Rep.* **2019**, *19*, 17. [CrossRef]
80. Simó, R.; Stitt, A.W.; Gardner, T.W. Neurodegeneration in Diabetic Retinopathy: Does It Really Matter? *Diabetologia* **2018**, *61*, 1902–1912. [CrossRef]
81. Hernández, C.; Simó, R. Neurodegeneration in Diabetic Retinopathy: Current Concepts and Therapeutic Implications. *Av. Diabetol.* **2014**, *30*, 72–79. [CrossRef]
82. Wang, W.; Tam, K.C.; Ng, T.C.; Goit, R.K.; Chan, K.L.S.; Lo, A.C.Y. Long-Term Lutein Administration Attenuates Retinal Inflammation and Functional Deficits in Early Diabetic Retinopathy Using the Ins2 Akita/+ Mice. *BMJ Open Diabetes Res. Care* **2020**, *8*, e001519. [CrossRef]
83. Sergeys, J.; Etienne, I.; Van Hove, I.; Lefevere, E.; Stalmans, I.; Feyen, J.H.M.; Moons, L.; Van Bergen, T. Longitudinal In Vivo Characterization of the Streptozotocin-Induced Diabetic Mouse Model: Focus on Early Inner Retinal Responses. *Investig. Ophthalmol. Vis. Sci.* **2019**, *60*, 807–822. [CrossRef] [PubMed]
84. Kern, T.S.; Berkowitz, B.A. Photoreceptors in Diabetic Retinopathy. *J. Diabetes Investig.* **2015**, *6*, 371–380. [CrossRef]
85. de Hoz, R.; Rojas, B.; Ramírez, A.I.; Salazar, J.J.; Gallego, B.I.; Triviño, A.; Ramírez, J.M. Retinal Macroglial Responses in Health and Disease. *Biomed Res. Int.* **2016**, *2016*, 2954721. [CrossRef] [PubMed]
86. Altmann, C.; Schmidt, M.H.H. The Role of Microglia in Diabetic Retinopathy: Inflammation, Microvasculature Defects and Neurodegeneration. *Int. J. Mol. Sci.* **2018**, *19*, 110. [CrossRef] [PubMed]
87. Fang, Y.; Shi, K.; Lu, H.; Lu, L.; Qiu, B. Mingmu Xiaomeng Tablets Restore Autophagy and Alleviate Diabetic Retinopathy by Inhibiting PI3K/Akt/MTOR Signaling. *Front. Pharmacol.* **2021**, *12*, 632040. [CrossRef]
88. Yu, Y.; Chen, H.; Su, S.B. Neuroinflammatory Responses in Diabetic Retinopathy. *J. Neuroinflam.* **2015**, *12*, 141. [CrossRef]
89. Guo, Y.; Cang, X.; Zhu, L.; Zhu, M.; Li, A.; Wang, Z.; Zhang, Y.; Wang, X.; Song, E. PPP1CA/YAP/GS/Gln/MTORC1 Pathway Activates Retinal Müller Cells during Diabetic Retinopathy. *Exp. Eye Res.* **2021**, *210*, 108703. [CrossRef]
90. De Faria, J.M.L.; Duarte, D.A.; Montemurro, C.; Papadimitriou, A.; Consonni, S.R.; De Faria, J.B.L. Defective Autophagy in Diabetic Retinopathy. *Investig. Ophthalmol. Vis. Sci.* **2016**, *57*, 4356–4366. [CrossRef]
91. Taki, K.; Horie, T.; Kida, T.; Mimura, M.; Ikeda, T.; Oku, H. Impairment of Autophagy Causes Superoxide Formation and Caspase Activation in 661 w Cells, a Cell Line for Cone Photoreceptors, under Hyperglycemic Conditions. *Int. J. Mol. Sci.* **2020**, *21*, 4240. [CrossRef]
92. Amato, R.; Catalani, E.; Dal Monte, M.; Cammalleri, M.; Di Renzo, I.; Perrotta, C.; Cervia, D.; Casini, G. Autophagy-Mediated Neuroprotection Induced by Octreotide in an Ex Vivo Model of Early Diabetic Retinopathy. *Pharmacol. Res.* **2018**, *128*, 167–178. [CrossRef]
93. Madrakhimov, S.B.; Yang, J.Y.; Kim, J.H.; Han, J.W.; Park, T.K. MTOR-Dependent Dysregulation of Autophagy Contributes to the Retinal Ganglion Cell Loss in Streptozotocin-Induced Diabetic Retinopathy. *Cell Commun. Signal.* **2021**, *19*, 29. [CrossRef]
94. Liu, Y.; Zheng, Y.; Zhou, Y.; Liu, Y.; Xie, M.; Meng, W.; An, M. The Expression and Significance of MTORC1 in Diabetic Retinopathy. *BMC Ophthalmol.* **2020**, *20*, 297. [CrossRef] [PubMed]
95. Fox, T.E.; Young, M.M.; Pedersen, M.M.; Han, X.; Gardner, T.W.; Kester, M. Diabetes Diminishes Phosphatidic Acid in the Retina: A Putative Mediator for Reduced MTOR Signaling and Increased Neuronal Cell Death. *Investig. Ophthalmol. Vis. Sci.* **2012**, *53*, 7257–7267. [CrossRef] [PubMed]
96. Fort, P.E.; Losiewicz, M.K.; Pennathur, S.; Jefferson, L.S.; Kimball, S.R.; Abcouwer, S.F.; Gardner, T.W. MTORC1-Independent Reduction of Retinal Protein Synthesis in Type 1 Diabetes. *Diabetes* **2014**, *63*, 3077–3090. [CrossRef] [PubMed]
97. Zhu, Z.; Yang, C.; Iyaswamy, A.; Krishnamoorthi, S.; Sreenivasmurthy, S.G.; Liu, J.; Wang, Z.; Tong, B.C.-K.; Song, J.; Lu, J.; et al. Balancing MTOR Signaling and Autophagy in the Treatment of Parkinson's Disease. *Int. J. Mol. Sci.* **2019**, *20*, 728. [CrossRef]
98. Tang, L.H.C.; Fung, F.K.C.; Lai, A.K.W.; Wong, I.Y.H.; Shih, K.C.; Lo, A.C.Y. Autophagic Upregulation Is Cytoprotective in Ischemia/Reperfusion-Injured Retina and Retinal Progenitor Cells. *Int. J. Mol. Sci.* **2021**, *22*, 8446. [CrossRef]
99. Ozdemir, G.; Kılınç, M.; Ergün, Y.; Sahin, E. Rapamycin Inhibits Oxidative and Angiogenic Mediators in Diabetic Retinopathy. *Can. J. Ophthalmol.* **2014**, *49*, 443–449. [CrossRef]

100. Chauhan, A.; Sharma, U.; Jagannathan, N.R.; Reeta, K.H.; Gupta, Y.K. Rapamycin Protects against Middle Cerebral Artery Occlusion Induced Focal Cerebral Ischemia in Rats. *Behav. Brain Res.* **2011**, *225*, 603–609. [CrossRef]
101. Yang, X.; Wu, S.; Feng, Z.; Yi, G.; Zheng, Y.; Xia, Z. Combination Therapy with Semaglutide and Rosiglitazone as a Synergistic Treatment for Diabetic Retinopathy in Rodent Animals. *Life Sci.* **2021**, *269*, 119013. [CrossRef]
102. Steinmetz, J.D.; Bourne, R.R.A.; Briant, P.S.; Flaxman, S.R.; Taylor, H.R.B.; Jonas, J.B.; Abdoli, A.A.; Abrha, W.A.; Abualhasan, A.; Abu-Gharbieh, E.G.; et al. Causes of Blindness and Vision Impairment in 2020 and Trends over 30 Years, and Prevalence of Avoidable Blindness in Relation to VISION 2020: The Right to Sight: An Analysis for the Global Burden of Disease Study. *Lancet Glob. Health* **2021**, *9*, e144–e160. [CrossRef]
103. Evans, J.R. 28 000 Cases of Age Related Macular Degeneration Causing Visual Loss in People Aged 75 Years and above in the United Kingdom May Be Attributable to Smoking. *Br. J. Ophthalmol.* **2005**, *89*, 550–553. [CrossRef] [PubMed]
104. Bowes Rickman, C.; Farsiu, S.; Toth, C.A.; Klingeborn, M. Dry Age-Related Macular Degeneration: Mechanisms, Therapeutic Targets, and Imaging. *Investig. Opthalmol. Vis. Sci.* **2013**, *54*, ORSF68. [CrossRef] [PubMed]
105. Hollyfield, J.G.; Bonilha, V.L.; Rayborn, M.E.; Yang, X.; Shadrach, K.G.; Lu, L.; Ufret, R.L.; Salomon, R.G.; Perez, V.L. Oxidative Damage–Induced Inflammation Initiates Age-Related Macular Degeneration. *Nat. Med.* **2008**, *14*, 194–198. [CrossRef] [PubMed]
106. Bellezza, I. Oxidative Stress in Age-Related Macular Degeneration: Nrf2 as Therapeutic Target. *Front. Pharmacol.* **2018**, *9*, 1280. [CrossRef] [PubMed]
107. Mulfaul, K.; Ozaki, E.; Fernando, N.; Brennan, K.; Chirco, K.R.; Connolly, E.; Greene, C.; Maminishkis, A.; Salomon, R.G.; Linetsky, M.; et al. Toll-like Receptor 2 Facilitates Oxidative Damage-Induced Retinal Degeneration. *Cell Rep.* **2020**, *30*, 2209–2224.e5. [CrossRef] [PubMed]
108. Nita, M.; Grzybowski, A. The Role of the Reactive Oxygen Species and Oxidative Stress in the Pathomechanism of the Age-Related Ocular Diseases and Other Pathologies of the Anterior and Posterior Eye Segments in Adults. *Oxid. Med. Cell. Longev.* **2016**, *2016*, 3164734. [CrossRef]
109. Piippo, N.; Korhonen, E.; Hytti, M.; Kinnunen, K.; Kaarniranta, K.; Kauppinen, A. Oxidative Stress Is the Principal Contributor to Inflammasome Activation in Retinal Pigment Epithelium Cells with Defunct Proteasomes and Autophagy. *Cell. Physiol. Biochem.* **2018**, *49*, 359–367. [CrossRef]
110. Kim, S.-Y.; Kambhampati, S.P.; Bhutto, I.A.; McLeod, D.S.; Lutty, G.A.; Kannan, R.M. Evolution of Oxidative Stress, Inflammation and Neovascularization in the Choroid and Retina in a Subretinal Lipid Induced Age-Related Macular Degeneration Model. *Exp. Eye Res.* **2021**, *203*, 108391. [CrossRef]
111. Abokyi, S.; To, C.-H.; Lam, T.T.; Tse, D.Y. Central Role of Oxidative Stress in Age-Related Macular Degeneration: Evidence from a Review of the Molecular Mechanisms and Animal Models. *Oxid. Med. Cell. Longev.* **2020**, *2020*, 7901270. [CrossRef]
112. Imamura, Y.; Noda, S.; Hashizume, K.; Shinoda, K.; Yamaguchi, M.; Uchiyama, S.; Shimizu, T.; Mizushima, Y.; Shirasawa, T.; Tsubota, K. Drusen, Choroidal Neovascularization, and Retinal Pigment Epithelium Dysfunction in SOD1-Deficient Mice: A Model of Age-Related Macular Degeneration. *Proc. Natl. Acad. Sci. USA* **2006**, *103*, 11282–11287. [CrossRef]
113. Justilien, V.; Pang, J.-J.; Renganathan, K.; Zhan, X.; Crabb, J.W.; Kim, S.R.; Sparrow, J.R.; Hauswirth, W.W.; Lewin, A.S. SOD2 Knockdown Mouse Model of Early AMD. *Investig. Opthalmol. Vis. Sci.* **2007**, *48*, 4407. [CrossRef] [PubMed]
114. Yao, J.; Jia, L.; Khan, N.; Lin, C.; Mitter, S.K.; Boulton, M.E.; Dunaief, J.L.; Klionsky, D.J.; Guan, J.L.; Thompson, D.A.; et al. Deletion of Autophagy Inducer RB1CC1 Results in Degeneration of the Retinal Pigment Epithelium. *Autophagy* **2015**, *11*, 939–953. [CrossRef] [PubMed]
115. Mitter, S.K.; Song, C.; Qi, X.; Mao, H.; Rao, H.; Akin, D.; Lewin, A.; Grant, M.; Dunn, W.; Ding, J.; et al. Dysregulated Autophagy in the RPE Is Associated with Increased Susceptibility to Oxidative Stress and AMD. *Autophagy* **2014**, *10*, 1989–2005. [CrossRef]
116. Stone, W.L.; Farnsworth, C.C.; Dratz, E.A. A Reinvestigation of the Fatty Acid Content of Bovine, Rat and Frog Retinal Rod Outer Segments. *Exp. Eye Res.* **1979**, *28*, 387–397. [CrossRef]
117. Fernandes, S.A.; Demetriades, C. The Multifaceted Role of Nutrient Sensing and MTORC1 Signaling in Physiology and Aging. *Front. Aging* **2021**, *2*, 38. [CrossRef]
118. Arjamaa, O.; Nikinmaa, M.; Salminen, A.; Kaarniranta, K. Regulatory Role of HIF-1α in the Pathogenesis of Age-Related Macular Degeneration (AMD). *Ageing Res. Rev.* **2009**, *8*, 349–358. [CrossRef]
119. Wang, M.; Li, Y.J.; Ding, Y.; Zhang, H.N.; Sun, T.; Zhang, K.; Yang, L.; Guo, Y.Y.; Liu, S.B.; Zhao, M.G.; et al. Silibinin Prevents Autophagic Cell Death upon Oxidative Stress in Cortical Neurons and Cerebral Ischemia-Reperfusion Injury. *Mol. Neurobiol.* **2016**, *53*, 932–943. [CrossRef]
120. Cai, J.; Zhang, H.; Zhang, Y.F.; Zhou, Z.; Wu, S. MicroRNA-29 Enhances Autophagy and Cleanses Exogenous Mutant AB-Crystallin in Retinal Pigment Epithelial Cells. *Exp. Cell Res.* **2019**, *374*, 231–248. [CrossRef]
121. Park, T.K.; Lee, S.H.; Choi, J.S.; Nah, S.K.; Kim, H.J.; Park, H.Y.; Lee, H.; Lee, S.H.S.; Park, K. Adeno-Associated Viral Vector-Mediated MTOR Inhibition by Short Hairpin RNA Suppresses Laser-Induced Choroidal Neovascularization. *Mol. Ther.—Nucleic Acids* **2017**, *8*, 26–35. [CrossRef]
122. Ebeling, M.C.; Polanco, J.R.; Qu, J.; Tu, C.; Montezuma, S.R.; Ferrington, D.A. Improving Retinal Mitochondrial Function as a Treatment for Age-Related Macular Degeneration. *Redox Biol.* **2020**, *34*, 101552. [CrossRef]
123. Cheng, L.; Cheng, L.; Bi, H.; Zhang, Z.; Yao, J.; Zhou, X.; Jiang, Q. Alpha-Melanocyte Stimulating Hormone Protects Retinal Pigment Epithelium Cells from Oxidative Stress through Activation of Melanocortin 1 Receptor-Akt-MTOR Signaling. *Biochem. Biophys. Res. Commun.* **2014**, *443*, 447–452. [CrossRef] [PubMed]

124. Josifovska, N.; Albert, R.; Nagymihály, R.; Lytvynchuk, L.; Moe, M.C.; Kaarniranta, K.; Veréb, Z.J.; Petrovski, G. Resveratrol as Inducer of Autophagy, pro-Survival, and Anti-Inflammatory Stimuli in Cultured Human RPE Cells. *Int. J. Mol. Sci.* **2020**, *21*, 813. [CrossRef] [PubMed]
125. Chen, C.L.; Chen, Y.H.; Liang, C.M.; Tai, M.C.; Lu, D.W.; Chen, J.T. Glucosamine-Induced Autophagy through AMPK–MTOR Pathway Attenuates Lipofuscin-like Autofluorescence in Human Retinal Pigment Epithelial Cells in Vitro. *Int. J. Mol. Sci.* **2018**, *19*, 1416. [CrossRef] [PubMed]
126. Ma, J.; Sun, Y.; López, F.J.; Adamson, P.; Kurali, E.; Lashkari, K. Blockage of PI3K/MTOR Pathways Inhibits Laser-Induced Choroidal Neovascularization and Improves Outcomes Relative to VEGF-A Suppression Alone. *Investig. Ophthalmol. Vis. Sci.* **2016**, *57*, 3138–3144. [CrossRef] [PubMed]
127. Lee, S.H.S.; Chang, H.S.; Kim, H.J.; Choi, J.S.; Kim, J.; Kim, J.H.; Woo, H.N.; Nah, S.K.; Jung, S.J.; Lee, J.Y.; et al. Effects of Stuffer DNA on the Suppression of Choroidal Neovascularization by a RAAV Expressing a MTOR-Inhibiting ShRNA. *Mol. Ther.—Methods Clin. Dev.* **2019**, *14*, 171–179. [CrossRef] [PubMed]
128. D'Cruz, P.M. Mutation of the Receptor Tyrosine Kinase Gene Mertk in the Retinal Dystrophic RCS Rat. *Hum. Mol. Genet.* **2000**, *9*, 645–651. [CrossRef] [PubMed]
129. Gal, A.; Li, Y.; Thompson, D.A.; Weir, J.; Orth, U.; Jacobson, S.G.; Apfelstedt-Sylla, E.; Vollrath, D. Mutations in MERTK, the Human Orthologue of the RCS Rat Retinal Dystrophy Gene, Cause Retinitis Pigmentosa. *Nat. Genet.* **2000**, *26*, 270–271. [CrossRef]
130. Xue, G.; Kohler, R.; Tang, F.; Hynx, D.; Wang, Y.; Orso, F.; Prêtre, V.; Ritschard, R.; Hirschmann, P.; Cron, P.; et al. MTORC1/Autophagy-Regulated MerTK in Mutant BRAFV600 Melanoma with Acquired Resistance to BRAF Inhibition. *Oncotarget* **2017**, *8*, 69204–69218. [CrossRef]
131. Yang, J.-L.; Zou, T.-D.; Yang, F.; Yang, Z.-L.; Zhang, H.-B. Inhibition of MTOR Signaling by Rapamycin Protects Photoreceptors from Degeneration in Rd1 Mice. *Zool. Res.* **2021**, *42*, 482–486. [CrossRef]
132. Resnikoff, S.; Pascolini, D.; Etya'ale, D.; Kocur, I.; Pararajasegaram, R.; Pokharel, G.P.; Mariotti, S.P. Global Data on Visual Impairment in the Year 2002. *Bull. World Health Organ.* **2004**, *82*, 844–851.
133. Ventura, L.M.; Sorokac, N.; Santos, R.D.L.; Feuer, W.J.; Porciatti, V. The Relationship between Retinal Ganglion Cell Function and Retinal Nerve Fiber Thickness in Early Glaucoma. *Investig. Opthalmol. Vis. Sci.* **2006**, *47*, 3904. [CrossRef] [PubMed]
134. Davis, B.M.; Crawley, L.; Pahlitzsch, M.; Javaid, F.; Cordeiro, M.F. Glaucoma: The Retina and Beyond. *Acta Neuropathol.* **2016**, *132*, 807–826. [CrossRef] [PubMed]
135. Belforte, N.; Agostinone, J.; Alarcon-Martinez, L.; Villafranca-Baughman, D.; Dotigny, F.; Cueva Vargas, J.L.; Di Polo, A. AMPK Hyperactivation Promotes Dendrite Retraction, Synaptic Loss, and Neuronal Dysfunction in Glaucoma. *Mol. Neurodegener.* **2021**, *16*, 43. [CrossRef] [PubMed]
136. Del Olmo-Aguado, S.; Núñez-Álvarez, C.; Ji, D.; Manso, A.G.; Osborne, N.N. RTP801 Immunoreactivity in Retinal Ganglion Cells and Its Down-Regulation in Cultured Cells Protect Them from Light and Cobalt Chloride. *Brain Res. Bull.* **2013**, *98*, 132–144. [CrossRef]
137. Fung, F.K.C.; Law, B.Y.K.; Lo, A.C.Y. Lutein Attenuates Both Apoptosis and Autophagy upon Cobalt (II) Chloride-Induced Hypoxia in Rat Muller Cells. *PLoS ONE* **2016**, *11*, e0167828. [CrossRef]
138. Yao, A.; van Wijngaarden, P. Metabolic Pathways in Context: MTOR Signalling in the Retina and Optic Nerve—A Review. *Clin. Exp. Ophthalmol.* **2020**, *48*, 1072–1084. [CrossRef]
139. Yu, A.K.; Datta, S.; McMackin, M.Z.; Cortopassi, G.A. Rescue of Cell Death and Inflammation of a Mouse Model of Complex 1-Mediated Vision Loss by Repurposed Drug Molecules. *Hum. Mol. Genet.* **2017**, *26*, 4929–4936. [CrossRef]
140. Schieber, M.; Chandel, N.S. ROS Function in Redox Signaling and Oxidative Stress. *Curr. Biol.* **2014**, *24*, R453–R462. [CrossRef]
141. Liton, P.B. The Autophagic Lysosomal System in Outflow Pathway Physiology and Pathophysiology. *Exp. Eye Res.* **2016**, *144*, 29–37. [CrossRef]
142. Abu-Hassan, D.W.; Acott, T.S.; Kelley, M.J. The Trabecular Meshwork: A Basic Review of Form and Function. *J. Ocul. Biol.* **2014**, *2*. [CrossRef]
143. Porter, K.; Hirt, J.; Stamer, W.D.; Liton, P.B. Autophagic Dysregulation in Glaucomatous Trabecular Meshwork Cells. *Biochim. Biophys. Acta* **2015**, *1852*, 379–385. [CrossRef] [PubMed]
144. Medchalmi, S.; Tare, P.; Sayyad, Z.; Swarup, G. A Glaucoma- and ALS-Associated Mutant of OPTN Induces Neuronal Cell Death Dependent on Tbk1 Activity, Autophagy and ER Stress. *FEBS J.* **2021**, *288*, 4576–4595. [CrossRef] [PubMed]
145. Chalasani, M.L.S.; Kumari, A.; Radha, V.; Swarup, G. E50K-OPTN-Induced Retinal Cell Death Involves the Rab GTPase-Activating Protein, TBC1D17 Mediated Block in Autophagy. *PLoS ONE* **2014**, *9*, e95758. [CrossRef] [PubMed]
146. Zhang, M.L.; Zhao, G.L.; Hou, Y.; Zhong, S.M.; Xu, L.J.; Li, F.; Niu, W.R.; Yuan, F.; Yang, X.L.; Wang, Z.; et al. Rac1 Conditional Deletion Attenuates Retinal Ganglion Cell Apoptosis by Accelerating Autophagic Flux in a Mouse Model of Chronic Ocular Hypertension. *Cell Death Dis.* **2020**, *11*, 734. [CrossRef] [PubMed]
147. Kitaoka, Y.; Munemasa, Y.; Kojima, K.; Hirano, A.; Ueno, S.; Takagi, H. Axonal Protection by Nmnat3 Overexpression with Involvement of Autophagy in Optic Nerve Degeneration. *Cell Death Dis.* **2013**, *4*, e860. [CrossRef] [PubMed]
148. Ying, H.H.; Turturro, S.S.; Nguyen, T.T.; Shen, X.X.; Zelkha, R.R.; Johnson, E.E.C.; Morrison, J.J.C.; Yue, B.B.Y.J.T. Induction of Autophagy in Rats upon Overexpression of Wild-Type and Mutant Optineurin Gene. *BMC Cell Biol.* **2015**, *16*, 14. [CrossRef]
149. Lee, S.H.; Shim, K.S.; Kim, C.Y.; Park, T.K. Characterization of the Role of Autophagy in Retinal Ganglion Cell Survival over Time Using a Rat Model of Chronic Ocular Hypertension. *Sci. Rep.* **2021**, *11*, 5767. [CrossRef]

150. Krishnadev, N.; Forooghian, F.; Cukras, C.; Wong, W.; Saligan, L.; Chew, E.Y.; Nussenblatt, R.; Ferris, F.; Meyerle, C. Subconjunctival Sirolimus in the Treatment of Diabetic Macular Edema. *Graefe's Arch. Clin. Exp. Ophthalmol.* **2011**, *249*, 1627. [CrossRef]
151. Dugel, P.U.; Blumenkranz, M.S.; Haller, J.A.; Williams, G.A.; Solley, W.A.; Kleinman, D.M.; Naor, J. A Randomized, Dose-Escalation Study of Subconjunctival and Intravitreal Injections of Sirolimus in Patients with Diabetic Macular Edema. *Ophthalmology* **2012**, *119*, 124–131. [CrossRef]
152. Wong, W.T.; Dresner, S.; Forooghian, F.; Glaser, T.; Doss, L.; Zhou, M.; Cunningham, D.; Shimel, K.; Harrington, M.; Hammel, K.; et al. Treatment of Geographic Atrophy With Subconjunctival Sirolimus: Results of a Phase I/II Clinical Trial. *Investig. Opthalmol. Vis. Sci.* **2013**, *54*, 2941. [CrossRef]
153. Petrou, P.A.; Cunningham, D.; Shimel, K.; Harrington, M.; Hammel, K.; Cukras, C.A.; Ferris, F.L.; Chew, E.Y.; Wong, W.T. Intravitreal Sirolimus for the Treatment of Geographic Atrophy: Results of a Phase I/II Clinical Trial. *Investig. Ophthalmol. Vis. Sci.* **2014**, *56*, 330–338. [CrossRef] [PubMed]
154. Gensler, G.; Clemons, T.E.; Domalpally, A.; Danis, R.P.; Blodi, B.; Wells, J.; Rauser, M.; Hoskins, J.; Hubbard, G.B.; Elman, M.J.; et al. Treatment of Geographic Atrophy with Intravitreal Sirolimus: The Age-Related Eye Disease Study 2 Ancillary Study. *Ophthalmol. Retin.* **2018**, *2*, 441–450. [CrossRef] [PubMed]
155. Liu, Q.; Kirubakaran, S.; Hur, W.; Niepel, M.; Westover, K.; Thoreen, C.C.; Wang, J.; Ni, J.; Patricelli, M.P.; Vogel, K.; et al. Kinome-Wide Selectivity Profiling of ATP-Competitive Mammalian Target of Rapamycin (MTOR) Inhibitors and Characterization of Their Binding Kinetics. *J. Biol. Chem.* **2012**, *287*, 9742–9752. [CrossRef] [PubMed]
156. Nussenblatt, R.B.; Byrnes, G.; Sen, H.N.; Yeh, S.; Faia, L.; Meyerle, C.; Wroblewski, K.; Li, Z.; Liu, B.; Chew, E.; et al. A Randomized Pilot Study of Systemic Immunosuppression in the Treatment of Age-Related Macular Degeneration with Choroidal Neovascularization. *Retina* **2010**, *30*, 1579–1587. [CrossRef] [PubMed]
157. Minturn, R.J.; Bracha, P.; Klein, M.J.; Chhablani, J.; Harless, A.M.; Maturi, R.K. Intravitreal Sirolimus for Persistent, Exudative Age-Related Macular Degeneration: A Pilot Study. *Int. J. Retin. Vitr.* **2021**, *7*, 11. [CrossRef] [PubMed]
158. Dalal, M.; Jacobs-El, N.; Nicholson, B.; Tuo, J.; Chew, E.; Chan, C.C.; Nussenblatt, R.; Ferris, F.; Meyerle, C. Subconjunctival Palomid 529 in the Treatment of Neovascular Age-Related Macular Degeneration. *Graefe's Arch. Clin. Exp. Ophthalmol.* **2013**, *251*, 2705–2709. [CrossRef]
159. Pravin, N.J.D. A Multicenter, Randomized, Double-Masked, Dose-Ranging, Placebo-Controlled Phase 2 Study to Assess Sirolimus in the Treatment of Patients With Diabetic Macular Edema (DIAMOND Study). *Investig. Ophthalmol. Vis. Sci* **2010**, *51*, 4244.
160. *Phase 1/2 Study of an Ocular Sirolimus (Rapamycin) Formulation in Patients with Age-Related Macular Degeneration*; 2010. Available online: https://www.clinicaltrials.gov/ct2/show/NCT00712491 (accessed on 30 May 2022).
161. Abraham, P.; Yue, H.; Wilson, L. Randomized, Double-Masked, Sham-Controlled Trial of Ranibizumab for Neovascular Age-Related Macular Degeneration: PIER Study Year 2. *Am. J. Ophthalmol.* **2010**, *150*, 315–324.e1. [CrossRef]
162. Fang, Y.; Westbrook, R.; Hill, C.; Boparai, R.K.; Arum, O.; Spong, A.; Wang, F.; Javors, M.A.; Chen, J.; Sun, L.Y.; et al. Duration of Rapamycin Treatment Has Differential Effects on Metabolism in Mice. *Cell Metab.* **2013**, *17*, 456–462. [CrossRef]
163. Fu, Y.; Zheng, X.; Jia, X.; Binderiya, U.; Wang, Y.; Bao, W.; Bao, L.; Zhao, K.; Fu, Y.; Hao, H.; et al. A Quantitative Transcriptomic Analysis of the Physiological Significance of MTOR Signaling in Goat Fetal Fibroblasts. *BMC Genomics* **2016**, *17*, 879. [CrossRef]
164. Wen, Y.T.; Zhang, J.R.; Kapupara, K.; Tsai, R.K. MTORC2 Activation Protects Retinal Ganglion Cells via Akt Signaling after Autophagy Induction in Traumatic Optic Nerve Injury. *Exp. Mol. Med.* **2019**, *51*, 1–11. [CrossRef] [PubMed]

antioxidants

Review

Gq Signaling in Autophagy Control: Between Chemical and Mechanical Cues

Inmaculada Navarro-Lérida [1,2,3,4], Anna M. Aragay [5], Alejandro Asensio [1,2,3,4] and Catalina Ribas [1,2,3,4,*]

1 Molecular Biology Department and Center of Molecular Biology "Severo Ochoa", CSIC-UAM, 28049 Madrid, Spain
2 Health Research Institute "La Princesa", 28006 Madrid, Spain
3 Center for Biomedical Research in Cardiovascular Diseases Network (CIBERCV), ISCIII, 28029 Madrid, Spain
4 Connexion Cancer-CSIC, 28006 Madrid, Spain
5 Department of Biology, Molecular Biology Institute of Barcelona (IBMB), Spanish National Research Council (CSIC), 08028 Barcelona, Spain
* Correspondence: cribas@cbm.csic.es; Tel.: +34-91-196-4728; Fax: +34-91-1964420

Abstract: All processes in human physiology relies on homeostatic mechanisms which require the activation of specific control circuits to adapt the changes imposed by external stimuli. One of the critical modulators of homeostatic balance is autophagy, a catabolic process that is responsible of the destruction of long-lived proteins and organelles through a lysosome degradative pathway. Identification of the mechanism underlying autophagic flux is considered of great importance as both protective and detrimental functions are linked with deregulated autophagy. At the mechanistic and regulatory levels, autophagy is activated in response to diverse stress conditions (food deprivation, hyperthermia and hypoxia), even a novel perspective highlight the potential role of physical forces in autophagy modulation. To understand the crosstalk between all these controlling mechanisms could give us new clues about the specific contribution of autophagy in a wide range of diseases including vascular disorders, inflammation and cancer. Of note, any homeostatic control critically depends in at least two additional and poorly studied interdependent components: a receptor and its downstream effectors. Addressing the selective receptors involved in autophagy regulation is an open question and represents a new area of research in this field. G-protein coupled receptors (GPCRs) represent one of the largest and druggable targets membrane receptor protein superfamily. By exerting their action through G proteins, GPCRs play fundamental roles in the control of cellular homeostasis. Novel studies have shown Gαq, a subunit of heterotrimeric G proteins, as a core modulator of mTORC1 and autophagy, suggesting a fundamental contribution of Gαq-coupled GPCRs mechanisms in the control of this homeostatic feedback loop. To address how GPCR-G proteins machinery integrates the response to different stresses including oxidative conditions and mechanical stimuli, could provide deeper insight into new signaling pathways and open potential and novel therapeutic strategies in the modulation of different pathological conditions.

Keywords: GPCR; Gq; autophagy; oxidative stress; mechanotransduction; extracellular matrix

Citation: Navarro-Lérida, I.; Aragay, A.M.; Asensio, A.; Ribas, C. Gq Signaling in Autophagy Control: Between Chemical and Mechanical Cues. *Antioxidants* **2022**, *11*, 1599. https://doi.org/10.3390/antiox11081599

Academic Editors: Angeles Aroca and Cecilia Gotor

Received: 22 July 2022
Accepted: 15 August 2022
Published: 18 August 2022

1. Introduction

Cells are constantly exposed to a huge number of intra-and extra-cellular signals which requires a proper cell response for adaptation and homeostasis control [1]. Among the different sensors and integrators for such varying signals, the family of G-protein coupled receptors (GPCRs) arise as one of the most important transducers [2,3]. With over 800 GPCRs encoded in the human genome, this family of transmembrane receptors can bind a plethora of stimuli which include hormones, metabolites, and inflammatory mediators, influencing a diverse network of signaling pathways [4].

Apart from the classical chemical inputs, recent studies have begun to unravel the potential of mechanical and architectural properties of the environment as a new and alternative way to dynamically modulate cellular homeostasis [5]. Extracellular Matrix (ECM) is considered as an intricate meshwork of proteins and carbohydrates organized in a specific manner which acts not only as a reservoir of growth factor and bioactive molecules but also as a highly dynamic entity which provide mechanical rigidity and structural support to the cells [6].

All GPCRs contain seven transmembrane domains embedded within the cell membrane with several intracellular domains that trigger guanine nucleotide binding proteins downstream signaling pathways and, interestingly, extracellular domains in GPCRs have been reported as potentially critical elements in the interaction with components of the ECM and as a force sensing mechanism [7,8]. This opens the questions about the potential of GPCRs as linkers in the integration of all these signals (both chemo- and mechanical stimuli) raising the possibility that specific GPCRs and its downstream effectors can mediate the crosstalk between both types of triggers during the control of homeostatic feedback loops. Indeed, the ECM has the potential to significantly impact virtually on every physiological cellular mechanism [9,10]. Excessive deposition and increased stiffness of ECM has been directly linked with the progression of many different pathologies and has the potential to regulate cellular metabolism [11].

An important downstream process in the crossroad between chemo- and mechanosensing regulatory responses is autophagy [12]. The catabolic activity of autophagy is a fundamental cellular process that eliminates molecules and subcellular elements (including proteins, lipids, nucleic acids and even organelles) through a lysosomal-mediated degradative pathway for providing energy sources for ATP production or building blocks for protein synthesis [12]. Activated by different types of stress including those related with DNA damage, hypoxia, oxidative stress, inflammation, and food deprivation and by different challenges arising from mechanical (stretching, shear stress or hypotonic shock), autophagy can have both beneficial and deleterious effects [13–15]. Indeed, vascular and heart diseases, infectious diseases, neurodegenerative pathologies, and cancer have all been related to autophagy dysfunctions. Thus, autophagy represents a double-edge sword and for this reason the possibility to regulate autophagy in a time- and local-dependent manner represents a novel and valid therapeutic approach to control most of these disorders.

Several studies show the potential role of GPCRs as autophagy modulators through their downstream heterotrimeric G proteins [16–19] or through β-arrestin1 [20]. Interestingly, recent data from our laboratory demonstrate that Gαq can act as a general and relevant modulator of mTORC1 signaling over autophagy in response to fluctuations in different types of nutrients [21]; however, the contribution of mechanical forces and stromal remodeling impact in this regulation remains elusive. In this review, we will try to summarize the most recent advances in GPCRs, its connection with autophagy and the mechanisms underlying autophagic flux control. A better understanding of the interplay between autophagy and GPCR signaling networks will be very helpful to develop pharmacological strategies based on specific GPCR modulation with potential application to a great number of pathological situations, ranging from vascular and cancer to neurological diseases.

2. GPCRs Regulation and Functions, beyond the Classical Modulation

GPCRs share a common structural characteristic, the transmembrane region constituted by seven transmembrane spanning α-helices linked by three intracellular and three extracellular loops, together with an intracellular C-terminus and an extracellular N-terminus domain [7,22].

The GPCR superfamily has been subdivided into six classes based on how their ligand binds, or on their physiological function and structure: A, B, C, D, E and F. The classification considers amino acid sequences and functional correlation between species, with classes D and E missing in the mammalian system. Another analysis is based on the phylogenetic tree groups classifying GPCRs in five families: (G) Glutamate, (R) Rhodopsin, (A) Adhesion, (F) Frizzled/taste 2 and (S) secretin [23].

In all cases, upon activation by specific ligands, GPCRs undergo specific conformational changes allowing them to bind to heterotrimeric G proteins (α, β and γ subunits). This results in the activation of Galpha subunits by sequentially promoting the exchange of guanosine diphosphate (GDP) for guanosine triphosphate (GTP) to the heterotrimeric G proteins [7]. Then, GTP-bound G protein α-subunit dissociates from the $\beta\gamma$ dimer, and then both of which bind to their respective downstream effector molecules. Recent findings further delineate complex receptor state transitions as transformations catalyzed both by G proteins and effectors that bind to the ligand-bound receptors [24,25]. This GPCR signaling can be terminated by the phosphorylation of the active receptor by specific kinases (GPCR kinases, or GRKs), followed by the binding of arrestin proteins which leads to GPCRs desensitization and internalization via clathrin-coated vesicle-mediated endocytosis [26–28]. Association of the GPCR with clathrin-coated pits induces its internalization and degradation through lysosomes [29] or alternatively, GPCR can be recycled back to the plasma membrane [30].

An additional aspect to be considered is that apart from the canonical signaling route from cell surface GPCRs and their downstream signaling partners, recent studies demonstrate that intracellular GPCRs can signal from internal cell compartments. They have been found at lysosomes, endosomes, endoplasmic reticulum, nuclei and mitochondria, displaying diverse cellular responses from their signaling at the cell surface [30]. This raises the concept of the existence of multiple signaling platforms that can be specifically activated by different stimuli. Thus, for some GPCRs, receptor activation and/or inhibition may occur at the cell surface; while for others, the fact that a ligand can get across the cellular membrane may change its functional response. From a pharmacological point of view, this opens the possibility of a new way to selectively target specific pools of GPCRs.

Common ligands for GPCRs are strikingly diverse: spanning ions, small molecules, lipids, peptides and proteins. Apart from its chemo-sensory function, recent studies have unveiled the participation of GPCRs in mechano-transduction [31–34]. Experimental evidence strongly supports the critical role of mechanical forces in the direct activation of these receptors. Mechanical stimuli can activate GPCRs without the involvement of their cognate agonists [35,36]. Supporting these observations, stimuli such as shear stress, hypotonic conditions and cell stretching, that alter membrane organization, have been reported as inducers of conformational transitions of GPCRs between an inactive to an activated state [35,37,38]. However, there are many aspects to be explored to further clarify how GPCRs might themselves be mechano-sensors and the mechanisms and functions behind this novel regulatory pathway.

Recent studies have shown that extracellular N-terminus within adhesive GPCRs can act as an anchor mechanism to the extracellular matrix (ECM), playing key roles in response to mechanical tension and in the control of their activity [37,38]. A recent revision approaches the different GPCRs which can be directly modulated by mechanical forces, highlighting the critical role of specific GPCRs in mechanotransduction such as adhesion GPCRs, APJ/apelin, AT_1R, B_2AR, B_2R, ET_1AR, GPR68, H_1R, M_5R PTH_1R, all of them very relevant at vascular level [35,36,39–41]. Since most GPCRs contain at least one N-glycan chain in their extracellular domain, further investigation will be required to address their patho-physiological functions. As part of the GPCR network, heterotrimeric G proteins, such as Gαi and Gαq/11, seem to be the critical element in the orchestration of this mechanosensitive response [40,41]. A more detailed information is listed in Table 1 where GPCRs mainly coupled to Gq proteins are summarized. Adding to the complexity, mechanical forces can also be sensed by intracellular organelles [42]. Mechanosensitive organelles such as the nucleus, mitochondria or even lysosomes are also the residence for GPCRs and its downstream signaling, opening interesting new areas in GPCR field.

Table 1. Gq-coupled GPCRs and their stimuli.

GPCRs (Coupled to Gq Protein)	Chemical Class of Natural Ligand	Mechanical Stimulation	References
5HT2A 5HT2B 5HT2C	Serotonin	Mechanical stretch	[43]
ADRA1A	Adrenaline/Noradrenaline	Shear stress	[44]
BB1	Bombesin	No reported	[45]
BLT1	Leukotrienes	No reported	[46]
CCK1	Cholecystokinin gastrin	No reported	[47]
CysLT1	Leukotrienes	Hypotonicity/ Increased intravascular pressure	[48,49]
EP1	ProstaglandinE2	No reported	[50,51]
ET1AR	Endothelin	Stretch	[34,52]
PAR1	Thrombin	Laminar flow	[53,54]
Gal2	Galanin	No reported	[55]
GHSR1a	Ghrelin	No reported	[56]
GnRH1	Gonadotropin	Insensitive	[57,58]
GRP39	Obestatin/Zinc	No reported	[59]
GPR68 GPR4	Protons	Shear stress	[41,60]
H1R	Histamine	Hypotonicity, direct membrane stretches, shear stress, intravascular flow	[34,58,61]
M5R	Acetylcholine	Hypotoniticy and membrane stretch	[62,63]
M1R M3R	Acetylcholine	No reported	
AT1R	Angiotensin	Hypotonicity, direct membrane stretch, pressure overload, increased intravascular pressure Pressure overload	[64–66]
MCHR	Melanin	No reported	[67]
B2R	Bradykinin	Shear stress, hipotonicity, Increase in plasma membrane fluidity	[68]
GPER	Estrogen	Mechanical stress	[69]
FFAR1	Fatty acids	No reported	[70]
PTH1R	PTH	Fluid shear stress	[71,72]
V1AR	Oxytocin Vasopressin	Stretch, shear stress	[34,73]
ADGRG2	No identified	Luminal fluid	[74]
P2YR	nucleotides	Fluid shear stress/Mechanical stress	[33,75,76]

3. New Avenues in Gαq/11 Signaling Complexes

Despite the high diversity of GPCRs, there are relatively small number of G proteins involved in the initiation of different intracellular signaling cascades. As we have mention, G protein α-subunits are defined by their ability to bind and hydrolyze GTP [77,78], which is a central event in their functionality. On the basis of sequence similarity Gα subunits have been divided into five different families (Gαs, Gαi, Gαq, Gα12 and Gv) [79]. Recent structures of GPCRs in complex with G proteins have revealed novel insights into G-protein coupling, including sequence determinants, and the flexibility of critical contact points (e.g., transmembrane helix 6, TM6) regulating G-protein access [80,81]. In this regard, the characterization of the structure of fully active GPCRs complexed with Gα proteins are being solved by advances cryo-electron microscopy techniques [82–84]. Several studies identify the N-terminus of Gα proteins as a key determinant of selectivity in GPCR binding and subsequent activation, providing new insights into the molecular basis of G protein-coupling selectivity beyond the Gα carboxy terminus [77].While some models propose a specific Gα binding to a particular GPCR, more recently it has been established the possibility that GPCRs can activate several Gα subtypes, displaying certain selectivity for specific isoforms [77,78,85].

The Gαq family of G proteins comprises four family members. The ubiquitously expressed Gαq and Gα11, Gα14 mainly expressed in liver, lung and kidney, and Gα15/16 (orthologues in mouse/human), specifically expressed in hematopoietic cells [86–88]. In this review we will focus on Gαq, the most widely studied member.

Classically, Gαq activity has been linked to the binding and activation of the β-isoform of phospholipase C, but in the last years a complex and important Gαq interactome has unraveled the possibility of activating different signaling pathways in distinct cellular scenarios [89]. Indeed, as intracellular GPCRs, it has become more evident that G proteins can dynamically be modulated to localize at diverse subcellular compartments. These new localizations provide novel mechanisms for signaling by G proteins [90,91]. A great number of cellular components have been reported to interact with Gαq, resulting in either propagation or deactivation of Gαq signaling. Gαq is known also to interact with components of the cytoskeleton, with important organizers of membrane microdomains, an also to reside in different organelles (see [79] for more details). This includes proteins involved in the regulation of GTPase activity such as GAPs (GTPase-activating proteins) and GEFs (Guanine-nucleotide-exchange factors) which led to the modulation of G protein cycle [92]. The regulator of G protein signaling (RGS) proteins act as GAP for G proteins, accelerating endogenous GTPase activity of Gα subunits. More than 20 members of RGS have been described with different members regulating Gαq activity [93,94]. Although RGS2 showed selectivity for Gαq/11 over Gi/o in vitro and in intact cell assay, recent data reveals new rules governing RGS-Gα recognition and the structural basis of this selectivity [95]. In a yeast two-hybrid screening using Gαq as bait, Ric8 has been also reported as a novel regulator and Gαq effector. siRNA-gene silencing of Ric8 shows a reduction in Gαq-coupled receptor-mediated ERK activation and intercellular calcium mobilization [96,97]. Ric-8 proteins were also shown to positively influence both plasma membrane localization and abundance of G proteins [98]. In this sense, an additional GEF-independent function for Ric-8 has been described during the protein synthesis process where it serves as a molecular chaperone that aids Gα subunit biosynthesis and mediates the initial association of G protein α subunits with endomembranes [99].

Interestingly, multitude of physiological processes regulated by GPCRs signaling regulators are involved in the rearrangements of the cytoskeleton with Rho GTPases as key. The signaling from the stimulation of GPCR to the RhoA activation is another important pathway which is mediated by Dbl-family GEFs [100,101]. Both G12/13 and Gαq/11 family members are upstream activators of RhoA. Recently, p63RhoGEF has been identified as a novel effector of Gαq involved in the stimulation of SRF-dependent gene expression. Biochemical and biophysical approaches have shown that p63RhoGEF directly and specifically associates with activated Gαq to enhance the guanine nucleotide exchange of RhoA, RhoB and RhoC [102]. In addition, Trio and Duet members, also act as Gαq effectors involved in the activation of RhoA [103]. Gαq interacts also with filamentous actin (F-actin) and moreover, the stimulation of Gq-coupled GPCR recruits tubulin to the membrane, both fostering PLCβ activation [104,105]. Thus, there is a cooperative relationship between cytoskeletal components, GPCRs and G proteins to confine signaling molecules in specific domains.

The general view of G protein signaling usually centers on its association with the cytoplasmic surface of the plasma membrane and the mechanisms underlying G protein cycling to carry out their cellular signaling functions. Plasma membrane (PM) is an extremely complex cellular entity, characterized by a two-dimensional asymmetric distribution formed by glycerophospholipids, sphingolipids, proteins, cholesterol, and carbohydrates. Its composition confers PM a specific fluidity, which enables the control of lateral diffusion and mobility of embedded molecules due to liquid-ordered and liquid-disorder plasma membrane microdomains [106]. Despite the controversies about the nomenclature, organization and dynamic of these microdomains considered as lipid rafts, what it is common to all these membrane nanodomains is its enrichment in cholesterol, sphingolipids, and specific anchored proteins [107]. Caveolae, represent a subset of membrane lipid rafts characterized by an enrichment in the membrane organizers caveolins and cavins. These two principal components have emerged as critical elements in the control of PM topography, rendering PM invaginations of 20–100 nm size which can undergo fusion endocytic and exocytic events through a variety of pathways ensuring protein recycling and chemical communication with the outside microenvironment [108,109]. Interestingly, plasma membrane nanoplatforms and its lateral organization have been proposed as critical driver modulators in the maintenance of membrane tension and in cellular mechanical responses [110].

Caveolins can act as scaffold proteins in multiprotein complexes, and they have been described as regulators of GPCR-Gαq system [111]. It has been reported the enrichment of GPCR signaling components in lipid rafts or caveolae, restricting their mobility and increasing their concentration, thus promoting the interaction and the initiation of different signaling pathways [112]. Although caveolae-lipid rafts seem to be a determinant of receptor-effector coupling, not all GPCRs (or G proteins) are found in this liquid ordered domains. Different types of G proteins appear to segregate differently with Gαq protein preferentially localizing into caveolae, while Gs and Gi isoforms are mainly localized into lipid rafts [113]. Consistent with this idea, it has been reported that Gαq but not Gs can immunoprecipitate with Caveolin-1 (Cav1). Interestingly, while Cav1 depletion does not alter Gαq subcellular localization, Gαq-mediated GPCR signaling is impaired.

The Cav1 contribution in the mechanosensing and adaptation in response to various mechanical stimuli, such as membrane stretching, hypoosmotic shock, shear stress or detachment [114,115], raises the possibility that Cav1-GPCR-Gαq could be a novel integrated module in the regulation of mechanotransduction. Indeed, it has been described that when cells are subjected to osmotic pressure, the enhancement of Ca^{2+} signals due to Gαq-Cav1 interactions is ablated [116] and more recently, an interesting work shows that the activation of Gq-calcium dependent signaling by mechanical stretch is mediated by the type of stretch and the amount of caveolae [117]. Previous studies have demonstrated that mechanical stressors can result in conformational changes in Cav1 to cause the release of key signaling molecules such as eNOS and Gαq [113,118].

In addition to Cav1, Gαq has been reported to interact with flotillins, lipid rafts resident proteins, in a mechanism which is nucleotide-binding independent. It has been demonstrated the implication of flotillins in Gαq-medated p38 MAPK activation, through Src family tyrosine kinase [119]. The interesting finding that Gαq can interact with both caveolin and flotillins opens the possibility of a differential regulation in each specific type of microdomains, both at plasma membrane and internal compartments. Furthermore, caveolin and flotillins can act as scaffolding proteins in signal transduction mechanisms directly connected with multiple cellular processes including the control of autophagy.

Interestingly while many other Gαq interactors have been reported, our group has recently described an unanticipated role of Gαq/11 as a key regulator of autophagy via modulation and interaction with components of the mTORC1 signaling hub. We have described that Gαq is part of the autophagic and lysosomal compartments participating as a general modulator of autophagy in response to serum, amino acids or glucose via interaction with the multifunctional p62/SQSTM1 protein [21]. In the next sections we will focus on autophagy emphasizing the potential and novel contribution of GPCR-G protein signaling in this process.

4. Autophagy between Nutrient, Mechanical and Oxidative Stress: An Emerging Role of Gαq

Autophagy is a highly conserved mechanism for cellular degradation in which cytosolic waste, protein aggregates and organelles are sequestered into a double-membrane vesicle (autophagosome) and delivered into lysosomes for breakdown [120]. Autophagy is orchestrated by sequential steps tightly control machinery in which ATG and associated proteins regulate the formation and maturation of autophagosomes into autolysosomes [121]. Under pathophysiological conditions, damaged or excessive accumulation of organelles such as endoplasmic reticulum, ribosomes, mitochondria, lipid droplets and peroxisomes can be degraded through mechanisms mediated by a collection of specific autophagy-related proteins [13,120].

Typically, autophagy stimulation depends on the mTOR system modulation [122]. It is initiated by the ULK1 complex (Unc-51-like-autophagy-activating kinase) which can receive input from cellular energy balance, and the availability of nutrients from mTORC1 (Rapamycin Complex1) and AMPK signaling networks (AMP-activated protein kinase). Canonical initiation of autophagy entails those metabolic stresses (chemical stimuli) such as nutrient deprivation. This causes mTORC1 dissociation from ULK1 which becomes active and binds to ATG13 and FIP200 triggering autophagosome formation [123,124]. In addition, mechanical stresses are also involved in autophagic flux control. Although it is unclear whether mechanical stresses may play a direct role in ULK1 activation, it has been reported that the mechanosensitive mTORC2 complex can indirectly induce ULK1 activation via inactivation of mTORC1 repressor function, through a FAK (focal adhesion kinase)-dependent mechanisms [125–127].

The extracellular matrix (ECM) constitutes a dynamic and plastic network of biophysical and biochemical factors that maintains tissue homeostasis. Changes in ECM composition, elasticity, and structure have also been reported to impact on autophagic flux raising the potential of matrix biology modulation as a critical controller of this process [9]. Apart from the interaction with physicochemical environmental imposed by the ECM, cells are also subject, and they have to respond to a great variety of mechanical forces due to other external forces, such shear stresses of fluid pressure (e.g., blood vessels), lateral stretches and compression (such in the case of muscles). Overall, this plethora of short- and large- scale forces elicit an adaptive cellular response in which autophagy seems to be a critical player.

In this adaptive response, cells can sense extracellular mechanical cues in different ways. This includes cellular adhesion complexes with ECM and/or cells, mechanosensors such as proteoglycans localized at the cell surface or mechanically activated ion channels (e.g., Piezo) [128,129] even plasma membrane-associated structures such cilium, caveolae, and clathrin-coated pits [115,130,131]. Moreover, multiple intracellular organelles, including autophagosomes, can also sense mechanical forces [132–134].

Mechanical cues imposed by forces or microenvironmental cues may affect the autophagic process through specific crosstalk with autophagy regulatory proteins (such as mTORC system, or AMPK pathway) or via mechanical regulation of cytoskeletal elements or phospholipid membranes, which can be crucial in the autophagic process [135–137]. Interestingly, a direct link between cell attachment to ECM and autophagy has also been reported [138]. Loss of cell attachment with the ECM usually results in programmed cell death via anoikis. In some cases, ECM-cell detachment can rapidly activate autophagy allowing for survival and re-attachment to the substrate [139]. However, it remains elusive the mechanism controlling this process with integrin-mediated adhesions emerging as a critical element [140]. Furthermore, ECM and integrin-mediated adhesion may trigger autophagy via FAK and ILK (integrin linked kinase) being relevant in different processes such as for immunosurveillance [141].

Additionally, matrix constituents have been shown to regulate autophagy in both directions, promoting or inhibiting this process, depending on matrix stiffness but also on its specific composition [142]. Indeed, recent studies have demonstrated an active and dynamic signaling role of specific extracellular matrix components on autophagic regulation which can act in both positive (activators) or negative (inhibitors) ways. Among them, decorin, collagen VI, kringle 5, endorepellin and endostatin function as activators and pro-autophagic matrix constituents that engage a diverse array of cell surface receptor for autophagic initiation [143–145]. In contrast, laminin $\alpha 2$ acts as an inhibitor. Thus, absence of laminin $\alpha 2$ permits excessive autophagy [146]. Interestingly, the action of all these components seems to be independent of the predominant nutrient concentrations. This unique class of matrix molecules can function as an alternative mechanism to the classical nutrient deprivation mechanism to safeguard cellular homeostatic balance through autophagic control and providing a new mechanism through which GPCRs could also be participating in the regulation of autophagy.

As we mention in a previous section, GPCRs are directly linked with mechanotransduction mechanisms (see Table 1). Supporting this idea, it has been demonstrated that mechanical perturbations can modulate GPCR conformational transitions [36]. Moreover, the response to shear stress can be directly modulated through the Gαq-coupled GPCR, GPR68 [35,41]. Interestingly, PKCζ, a protein that we have described as new interactor of Gαq [147], can be regulated by shear stress and activated by disturbed flow in atheroprone areas [148]. Further evidence on this mechanical stimulation comes from studies on adhesion GPCRs which display a long extracellular N-terminus with adhesive properties to the extracellular matrix or N-glycans modifications. These glycan chains have been reported to be able to be activated in the context of mechanical traction forces [38,149]. Further investigations are required to really address how these forces can structurally activate GPCRs in different contexts.

As a sequential process involving membrane remodeling events, autophagy is mechanically linked to cytoskeletal dynamics that lead to mechanical deformation and transport. Actin filaments and fibers and microtubule network can act as critical modulators in the control of organelle dynamics and autophagy control [150].

Although the most classical autophagy process relies on the delivery of cytoplasmic material to lysosomes via the double-membraned autophagosome, another form of autophagy, known as chaperone-mediated autophagy (CMA), occurs in the lysosome directly [151]. Thus, lysosomes can be considered critical hubs in the modulation of autophagic control. Lysosomes constitute a highly dynamic organelle which display important changes, including acidification and enhanced enzymatic activity. Furthermore, this organelle can move from the perinuclear localization to the cell periphery with important implications in cell metabolic control [152].

An emerging aspect to be considered is how lysosomal position can directly modulate its function. Lysosomes are transported bidirectionally through the microtubule network by dynein and kinesin motors, with microtubule motors such as dynein modulating the movement of lysosomes from the periphery towards a perinuclear location, while kinesins promote the scattering of lysosomes through the cytoplasm [153,154]. Recent evidence suggest that the distribution of lysosomes can be controlled by stimulation with different inputs. Under cell starvation, autophagosomes and lysosomes move toward the center of the cell facilitating the fusion of both compartments and the degradative capacity [155].

Several protein complexes have been implicated in the regulation of lysosomal positioning. One important regulator is the Rab7, a small GTPase involved in the coordination of appropriate coordination and homeostatic control of late endosomes and lysosomes [156,157]. Recently, WDR91 a Rab7 effector has been reported to be essential for lysosomal function [158]. In addition, the transcription factors TFEB and TFE3 are essential to promote the expression of multiple lysosomal genes. They play critical roles in the modulation of lysosomal biogenesis and distribution through the control of the lysosomal transmembrane protein TMEM55B (transmembrane protein 55B) expression [159]. TFEB acts as a link between autophagic process and lysosome biology [160], interacting with mTORC1 complex but not with mTORC2 controlling the lysosomal localization and function of mTORC1 [161].

Our recent studies strongly suggest that Gαq is a critical autophagy regulator raising the potential to control the shift between mTORC1-mTORC2 switch through lysosome control. We have recently reported a higher number of autolysosomes in cell lacking Gαq/11 compared to the wild type of counterpart which is directly linked with the involvement of this protein in the modulation of autophagy [21]. Immunofluorescence with LAMP1, a lysosomal-endolysosomal compartment marker, revealed that Gαq KO cells showed a predominant perinuclear lysosomal distribution in basal conditions, a phenotype that was mimicked by upon starving conditions. These results are consistent with a not-yet described role of Gαq in the modulation of lysosome dynamic regulation.

Furthermore, although many components of the autophagic machinery and autophagy receptors which are involved in the regulation of the process are being subjected to lysosomal degradation, in the case of Gαq, its presence in lysosomes does not alter its protein expression levels neither in basal nor in nutrient stress conditions which reinforce a critical role of this protein in the modulation of the autophagic control process.

Recent studies have demonstrated by using a biosensor the presence of GTP-loaded Gαq/11 at endosomes [162]. This provides a powerful tool to be applied to other cellular organelles such lysosomal compartment to fully address the specific function of GPCR-Gαq signaling at these organelles. Furthermore, this raises the possibility that GPCR-Gαq signaling may act as a modulator of autophagy by acting as a switch between chemical and mechanical cellular responses (see Figure 1).

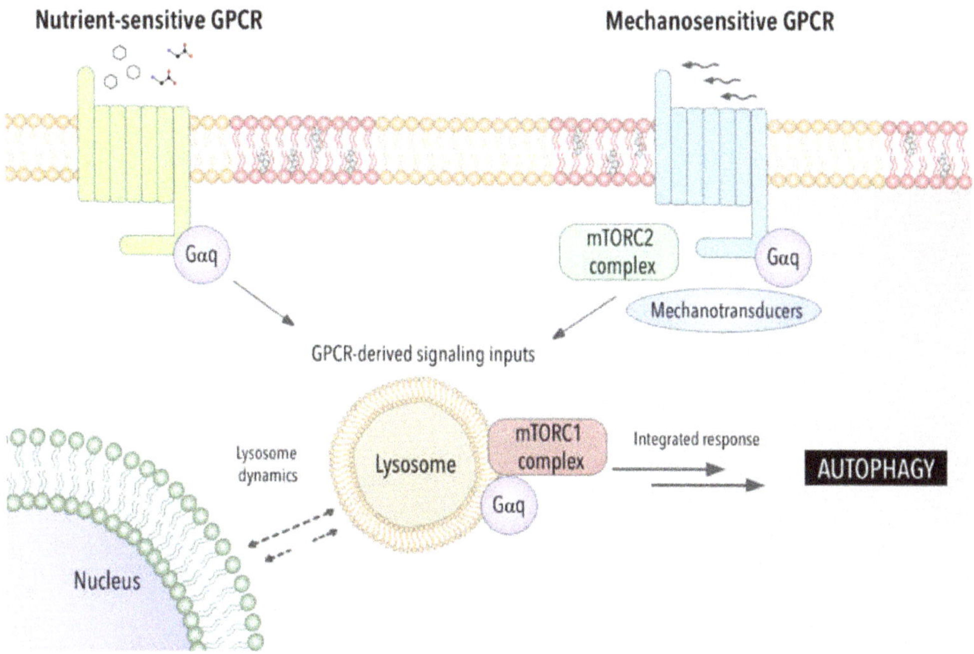

Figure 1. Gαq as a potential integrator of chemical and mechanical signals modulating autophagic process. Involvement of Gαq interactome-autophagy control in pathophysiological settings.

Importantly, in the last years there have been many reports that suggest that oxidative stress is also an important inducer of autophagy. Autophagy eliminates the toxic effects of reactive oxygen species (ROS) production to enable cell survival [163]. Reactive oxygen species (ROS) are produced in many cellular stress conditions such as hypoxia, nutrient deprivation or viral infection, among others. Under normal conditions, ROS levels participates in physiological processes regulating signaling pathways, to maintain cellular homeostasis. Excessive ROS production irreversibly oxidize organelles, proteins, lipids and DNA that can be partially counteracted by antioxidant enzymes, but high ROS eventually results in cellular damage by oxidative stress which is associated to the pathogenesis of diseases [164,165]. Autophagy serves the cell to clear off the damaged biomolecules and DNA produced by oxidative stress and there is a clear interplay between oxidative stress and induction of autophagy [166,167].

The main source of ROS in the cell (approximately 90%) is the respiratory chain in the inner membrane of the mitochondria. During oxidative phosphorylation the leaking of electrons from the electron transport chain (ETC) produce superoxide anion (O^{2-}), hydroxylperoxide (H_2O_2), then OH^- under the catalysis of superoxide dismutase (SOD) and glutathione peroxidase (GSH-Px) [167]. Several works have established complex I and III as the major sites to generate ROS [168–170]. Under normal physiological conditions, ROS production is low, and the antioxidant machinery is able to scavenge ROS, while under high ROS production, oxidative stress occurs.

There are several studies that describe the crosstalk between oxidative stress and the autophagic machinery [165]. A direct link was shown through the inactivation of ATG4 by H_2O_2-oxidation to ensure autophagosome elongation [170]. On the other hand, ROS has been proven to induce autophagy through mTORC1 inactivation or AMPK activation [171–173] and through transcriptional regulation (HIF-1a, NRF2, p53 and FOXO3) [165,174]. These transcription factors mediate the induction of autophagy genes, including Beclin-1, LC3, p62, the mitophagy associated BNIP3 and NIX. On the other hand, ROS can regulate autophagy through the oxidation and inactivation of the ATG7, ATG10 and ATG3 proteins involved in the process of autosomal maturation and fusion to lysosomes and inactivate autophagy modulators like TFEB and PTEN [166,173,175]. Not only does ROS regulate autophagy, but autophagy can modulate ROS through the Keap-Nrf2 system. p62 can interact with Keap1 and release Nrf2 that translocates to the nuclei and activates antioxidant genes [176]. In this way, ROS can induce autophagy that in turn activates antioxidant genes to control ROS.

Extensive ROS can turn on the selective removal of damaged mitochondria by autophagy (mitophagy) [177]. Thus, autophagy limits the production of ROS and protects the cell from oxidative damage by selectively removing mitochondria [178]. There are several mechanisms of mitochondrial removal by mitophagy that have been described [178]. The most studied is the PINK/Parkin axis. Thus, mitochondrial damage triggers the translocation and regulation of the PTEN-induced putative kinase 1 (PINK1) and the ubiquitin E3-ligase Parkin. Parkin ubiquitinates several outer mitochondrial proteins that in turn recruit autophagy cargo receptors [179]. Among these receptors it has been described NDP52, optineurin and p62. The Rab proteins Rab5 and Rab7 located at the mitochondria surface and the RabGAP protein TBC1D15, that contains LC3 interacting domain, also help in this membrane recruitment process [180]. Several lines of work show that ROS stress leads to Parkin translocation to mitochondria to initiate the removal of mitochondria by mitophagy [15,181,182]. ROS accumulation can lead to disruption of mitochondrial membrane potential that stabilizes PINK. PINK1/Parkin mediated mitophagy has also been proven to play a protective role against oxidative stress in human nucleus pulposus cells [183]. ROS stress can also induce the translocation of adaptor proteins like DJ-1 to mitochondria [184]. The other mitophagy mechanisms described to induce mitophagy utilizes the mitochondrial adaptor proteins FUNDC1, BNIP3 and NIX [185]. These proteins localized at the outer membrane and can interact with LC3 to promote membrane contact and mitophagy [15]. The expression of several of these cargo receptors are induced by oxidative stress.

As mentioned before, ROS and mitophagy have a clear interplay and, although oxidative stress induces mitophagy, ROS can also lead to decrease mitophagy. The interaction between these processes requires a tight regulation and as we have seen both the mTORC1-p62 axis and mitophagy are crucial elements. Interestingly, recently it has been shown that mTORC1 signaling regulates mitophagy through the PINK1/PARK2 pathway [186]. In that context, the Gαq-dependent pathways may provide a mechanism of cross-regulation [79]. As stated Gαq is found in autophagic compartments and lysosomes and is part of the mTORC1 multimolecular complexes, contributing to inhibition of autophagy under GPCR activation and physiological conditions [21]. Gαq is also present at the outer and inner mitochondrial membrane [61] and contributes to cristae integrity and respiratory chain function. Interestingly, the absence of Gαq/11 alters mitochondrial crests and super respiratory complexes containing complex I and III which in turn may induce ROS production [187] As we have mention, Gαq also interacts with p62 and it is present in the autophagosomes [21]. Gαq is a key component for how nutrients and activated receptors sense and control autophagy, and through p62, can be an important component of the autophagy and oxidative stress control.

On the other hand, the role of Gαq in mitochondrial ROS generation is well sustained. As such, Gαq-induced cardiac decompensation has been associated with mitochondrial dysfunction and increased of ROS, with either enhanced expression of Gαq or activation of Gαq-linked GPCRs [188–191]. Noteworthy, the implication of Gαq-intracellular pathways via mitochondria and through mTORC1 and p62 seems to be a crucial asset to control the balance between oxidative stress and autophagy, but further work needs to be done to link these processes.

5. Autophagy in Disease for Good and for Bad: Gαq Involvement

Organisms have to constantly adapt to external stimuli and changes in their intracellular environment. Organs, tissues and cells have to face both chemical (e.g., Ca^{2+}, amino acids, cytokines, chemokines and hormones) as well as physical challenges. Among the cytoplasmic responses to mechanical forces, recent studies have uncovered the role of autophagy in the translation of mechanical forces into biological responses [126,127,192,193]. Given the importance of autophagy regulation and dynamics of lysosomal system to ensure cellular fitness, it is not surprising that autophagy disruption can contribute to the development of several diseases such as metabolic disorders, cardiovascular or cancer diseases. Although the involvement of autophagy in these major diseases has been well studied (reviewed in [194], over the years autophagy regulation has grown in complexity and their consequences are less predictable. The importance of Gαq and Gαq-coupled GPCRs in all these contexts, together with its recently described importance in autophagy, strongly suggest that alteration in Gαq modulation signaling pathways can contribute to all these pathological situations. In this part of the review, we will focus on how autophagy may be involved in different pathologies, emphasizing as far as possible the influence of mechanical inputs.

Various metabolic disorders have shown functional defects in autophagy [195,196]. Since the lysosomal disposal of intracellular macromolecules leads to their breakdown into important metabolic intermediates, including amino acids, glucose, nucleotides, and free fatty acids (FAs), autophagy plays an important role in the response to energetic stresses, at both the tissue-specific and systemic levels [197]. Many studies have emphasized the importance of autophagy in conditions such as obesity, insulin resistance and diabetes that are characterized by metabolic alterations and intracellular stresses that have in common the accumulation of damaged cellular components. Silencing of ATG system promotes obesity and induces metabolic alterations [198]. Interestingly autophagy genes are differentially expressed and activated in a tissue and stage-specific manner. In general, nutrient limitation and different stress situations favor autophagy as a mechanism of cytoprotecting, reducing cellular death and limiting inflammatory response. Upon autophagy inhibition alteration of adipocyte differentiation, lipid metabolism and storing of lipids is drastically altered [199–201] For example, in obesity, autophagy is suppressed due to an increased in mTOR activity. Moreover, in patients with diabetes, changes in oxidative stress and autophagy have been reported [202]. Therefore, the enhancement of autophagy activity has been suggested as a novel therapeutic approach against organ failure associated to metabolic disorders.

In general, GPCRs regulate virtually all metabolic processes including glucose and energy homeostasis, particularly diabetes and obesity-related diseases. Several endogenous ligands such as free fatty acids and their receptors have been extensively studied in insulin secretion regulation, and glucose metabolism [203]. A growing number of GPCR are being identified as sensors of circulating of local concentrators of energy substrates or metabolic intermediates. Examples of these receptors include the amino-acid responsive receptors GPRC6A taste receptors type 1 members 1 and 3 (T1R1/T1R3), the calcium sensing receptor (CaSR) long chain fatty acid receptors GPR120 and GPR40, short fatty acid receptors GPR41 and GPR43 or hydroxy carboxylic acid receptors [204–206]. These GPCR nutrient receptors act via different G proteins including Gαq/11 and might be able to modulate the canonical metabolic regulators AMPK and mTORC1 [19]. In this sense, Gαq-coupled T1R1/T1R3 act as a direct sensor of the fed state and amino acids availability, leading to the activation of mTORC1 [207]. Nutrient and homeostasis fluctuations may also indicate the release of classical hormones and neurotransmitters that activated GPCRs, along a systemic regulation of autophagy. In this sense, β-adrenergic receptors activation has been related with autophagic flux favoring lipolysis [208], and hyperglycemia induces autophagy in pancreatic β cells through P_2Y purinergic receptors [209]. In addition, drugs that target metabolic tissues have emerged as attractive diabetes therapeutic targets as well. The p62-mTORC1-autophagy axis has been described to regulate adipogenesis and energy control in a complex manner [210]. The potential and reported connections of Gαq signaling that we have described with this axis [21] may provide new insights in the mechanisms underlying these metabolic alterations. Recent studies have further confirmed the relevance of Gαq signaling for driving metabolic reprogramming in uveal melanoma [211] and in the regulation of glucose and lipid homeostasis [212] reinforcing a critical role of GPCR-Gαq system in metabolic diseases. Further investigation will be required to define the mechanisms involved.

Interestingly, cell metabolism is sensitive also to the physical cell microenvironment [213]. Although cell metabolism has recently emerged as one of the processes regulated by mechanical cues, the link between cell mechanics and metabolism is still poorly understood when compared with other pathologies such as cardiovascular diseases or cancer (Figure 2). Thus, in addition to metabolic intermediates, autophagy can influence metabolic reprogramming in epithelial cells through the involvement of mechanical forces such as shear stress [214,215] and, mechanical stretching/tension exerted by exercise has been shown to induce also autophagy in peripheral tissues (liver, pancreas and adipose tissue) [216].

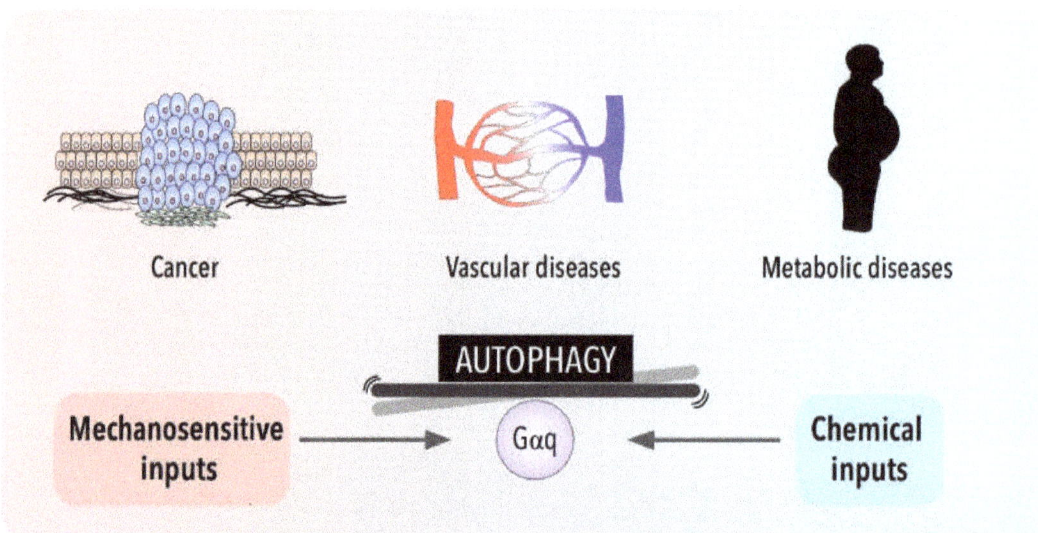

Figure 2. Gαq signaling and autophagy at the crossroads of a balance between mechanical and chemical cues and their impact in cancer, metabolic and cardiovascular pathologies.

The involvement of shear stress and mechanical forces in endothelial function has been well established and although, there is an increasing interest in the role of autophagic flux in vessel wall biology, the mechanosensors upstream of autophagy induction in endothelial cells are not well known [217]. Emerging evidence links alterations in autophagic flux with disease processes that include atherosclerosis, pulmonary hypertension and cardiovascular diseases [218]. Interestingly, a very recent study proposes a protective role of CMA (Chaperon Mediated Autophagy) against atherosclerosis [219]. Loss of autophagy may be a central mechanism through which risk factors elicit endothelial dysfunction. The role of autophagy in vascular development and in sprouting has been associated with a defective autophagy in mice lacking endothelial specific-TFEB factor [220]. The autophagic state of endothelial cells is also critical for vascular permeability [221]. Additionally, endothelial cells require autophagy to regulate tight junction proteins and maintain endothelial barrier integrity during inflammation [222]. Moreover, it has been reported that autophagy may be involved in the regulation of nitric oxide bioavailability, a crucial molecule that maintains vascular homeostasis in endothelial cells. Concomitant with a reduction in NO, loss of autophagy promotes and increase in endothelial ROS and inflammatory cytokine production [223,224]. Shear-stress-dependent autophagy is also important for NO production [224,225]. Indeed, in zones of low shear stress that are prone to develop atherosclerotic plaques, the impairment of autophagic flux induces endothelial NO synthase (eNOS) uncoupling, resulting in the production of superoxide instead of NO. Restoration of the autophagic flux favors the production of NO by endothelial NO synthase [226]. Interestingly, Gαq has been described as an important sensor of shear stress in endothelium [33,227]. Recent studies have demonstrated that changes in the type of flow can activate the same initial mechanosensing pathway involving Piezo1- and Gαq/11-mediated signaling with different atheroprotective response depending on the activation of α5 integrin, which is activated only by disturbed flow, but not by sustained laminar flow [228].

Regarding cardiovascular context, autophagy preserves cardiac structure and function under baseline conditions and is activated during stress, contributing to limit damage and preserve cardiac functionality during ischemia [229]. Cardiac cells are also subjected to tension. Several pathophysiological conditions, lead to an increase in cardiac workload and mechanical forces that are usually associated with pathological cardiac hypertrophy [230]. Mechanical forces can induce autophagy in cardiac cells being protective or detrimental depending on the context [217]. Indeed, during ischemia, autophagy has a protective effect on cardiomyocytes [231–233], while inhibition of autophagy improves cardiac function after reperfusion in an ischemia–reperfusion mouse model [232]. Much evidence places Gαq at the center of hypertrophic pathways in the heart (see [234], for more details). Indeed, Gαq signaling is both necessary and sufficient for the development of cardiac hypertrophy. The development of a cardiac hypertrophy phenotype has been correlated with a higher risk of heart failure [186]. Interestingly, it has become increasingly clear that both events are tied to the activation/presence of Gαq that can promote cardiomyocyte apoptosis and heart failure [235] by affecting vascular permeability and hypertension. Indeed, Gαq inhibition using specific drugs have been proposed to have anti-hypertensive role [236]. We have also described, in previous studies, a role of Gαq/PKCζ signaling axis in the development of cardiac hypertrophy in response to angiotensin II through a novel binding region on Gαq. Given the involvement of Gαq in cardiovascular function and in the process of autophagy, the potential participation of novel mechanisms downstream Gαq directly linked with autophagic flux in cardiovascular system is a relevant open question.

Autophagy is also recognized as a critical player in a context-dependent manner in cancer. Although it is well accepted that autophagy is important in many diseases, as described above, practically all clinical studies that involve autophagy manipulation are focused on cancer therapy. Autophagy networks are related to multiple aspects of cancer and may play a dual role with tumor-suppressive and tumor promoting functions depending on tumor cell type and stage [237]. Both inhibition of autophagy and its overstimulation are strategies tested in cancer, with the use of different drugs such as hydroxychloroquine, 3-methyl-adenina and everolimus as currently new strategies to be employed in clinics in combination with other chemotherapeutic treatments [238]. However, the high toxicity and adverse effects of these treatments urge a further understanding of the specific mechanisms by which autophagy modulates the different tumor progression steps.

Deficiency of autophagic genes has been found in various cancers. Impaired autophagy can promote tumorigenic environment through ROS dysregulation and inflammation processes [163,239,240]. On the other hand, at advanced cancer stages, increased autophagy can sustain tumor cell growth in nutrient-deficient, hypoxic tumor microenvironment and resistance to anoikis [241]. Upregulation confers chemoresistance and promotes the maintenance and survival of stem cell cancer status. Furthermore, autophagy inhibition can favor tumor cell invasiveness through the induction of de-differentiation mechanism. Thus, it seems that in premalignant lesions, enhanced autophagy might be beneficial preventing cancer, but in advance cancers most therapeutic strategies are focused on inhibiting autophagy [242]. Adding another layer of complexity, the novel Gαq role in modulating autophagy suggests that the balance between these processes characteristic of tumor growth might be altered in different cancer settings.

Furthermore, evidence identifies tumor microenvironment as a central driver of tumorigenesis in cancer [243]. Interestingly, cancer cells can also experience shear stress that can induce autophagy in different tumor cell lines [244–247]. Interstitial flow can promote the distribution of tumor-derived cells in primary tumor, while circulating tumor cells are also subjected to the shear stress from body fluids (blood, lymph and interstitial fluid) during metastasis [248,249]. It has been suggested that shear stress-induced autophagy can play an important role in controlling important cell responses from the regulation of cell size and metabolism to inflammation and cell death [217]. Moreover, the activation of tumor stromal fibroblasts to a state commonly known as cancer associated fibroblasts (CAFs) is critical. CAFs impact tumor progression by the modulation of multiple secretion functions of different factors (growth factors and inflammatory signals), by remodeling the extracellular matrix and even reprogramming their metabolism to provide nutrients and survival factors [250]. Moreover, autophagy can play a key role in CAFs activation [251]. Recent studies demonstrate that normal fibroblasts can differentiate into CAFs as protective responses to stresses under tumor microenvironment via the p62-Nrf2-pathway [252]. Furthermore, a molecular mechanism for CAFs activation has shown that tumor secreted lactate downregulates p62 in the stroma blocking AP-1-mediated p62 transcription [253]. Interestingly, we have described changes in the expression of Gαq affecting some partners such as p62 promoting its downregulation and favoring autophagic flux [21].

Altered GPCR pathways have increasingly been reported in cancer context and activating mutations in Gαq has been identified in approximately 80% of uveal melanomas. Considering Gαq as a component of the nutrient-sensing machinery, able to link nutrient availability with the activation of mTORC1 through its interaction with p62 [21] strongly reinforce its potential contribution in the modulation of both the tumor and its microenvironment, during tumor progression.

6. Conclusions

The highly conserved autophagy mechanisms are critical in cellular homeostasis by allowing degradation of cellular components in a lysosomal-dependent manner both in basal conditions or in response to internal or external fluctuations. In consequence, autophagy represents a central adaptation system.

Different studies have demonstrated that molecular mechanisms of autophagy are not only regulated by chemical stresses and metabolic challenges, such as starvation, but can also be modulated by mechanical stresses stemming from the environment. How these stimuli are integrated remains a challenge. In this review we propose a putative role of GPCR-Gαq signaling as a central integrator in this crosstalk suggesting its potential contribution as a balancer shaft (Figure 2). In addition to the canonical roles of Gαq and other heterotrimeric G proteins derived from their presence at the cytoplasmic surface of the plasma membrane, emerging evidence demonstrates that Gα subunit proteins can also localize in other cellular organelles such as endosomes, Golgi, ER, nucleus, mitochondria and most importantly in lysosomes. Determining whether such intracellular pools of heterotrimeric Gα protein subunits are dynamically generated via trafficking from the plasma membrane or represent resident stable subpopulations, as well as the identification of their location-specific interactors and functional roles, are active areas of research.

Furthermore, our recent study pointing out the critical involvement of Gαq as an autophagy regulator and the existence of GPCRs that can be directly modulated by physical forces providing new frontiers for deeper analysis. To understand how integration of chemical-mechanical-autophagic process occurs and the specific role of Gαq-GPCR in the interplay between metabolic cellular modulation and environmental cues will help us to understand the molecular basis of multiple diseases, in which autophagy represents a central element. Thus, a better comprehension of Gαq-mediated signaling pathways considering the context-specific modulation of autophagy will open new avenues for treating autophagy-related diseases based not only upon chemical but also mechanical inputs.

Funding: Our work was supported by: Instituto de Salud Carlos III (PI18/01662 to CR, co-funded with European FEDER contribution), CIBERCV-Instituto de Salud Carlos III, Spain (grant CB16/11/00278, co-funded with European FEDER contribution), Fundación Ramón Areces (to CR). Ministerio de Economía; Industria y Competitividad (MINECO) of Spain (grant BFU2017-83379-R to AMA), COVTRAVI-19-CM (Comunidad de Madrid, co-funded with European FEDER contribution). We also acknowledge institutional support to the CBMSO from Fundación Ramón Areces.

Conflicts of Interest: The authors declare no conflict of interest.

References

1. Gudipaty, S.A.; Conner, C.M.; Rosenblatt, J.; Montell, D.J. Unconventional Ways to Live and Die: Cell Death and Survival in Development, Homeostasis, and Disease. *Annu. Rev. Cell Dev. Biol.* **2018**, *34*, 311–332. [CrossRef] [PubMed]
2. Heldin, C.-H.; Lu, B.; Evans, R.; Gutkind, J.S. Signals and Receptors. *Cold Spring Harb. Perspect. Biol.* **2016**, *8*, a005900. [CrossRef] [PubMed]
3. Hilger, D.; Masureel, M.; Kobilka, B.K. Structure and Dynamics of GPCR Signaling Complexes. *Nat. Struct. Mol. Biol.* **2018**, *25*, 4–12. [CrossRef] [PubMed]
4. Marinissen, M.J.; Gutkind, J.S. G-Protein-Coupled Receptors and Signaling Networks: Emerging Paradigms. *Trends Pharmacol. Sci.* **2001**, *22*, 368–376. [CrossRef]
5. Humphrey, J.D.; Dufresne, E.R.; Schwartz, M.A. Mechanotransduction and Extracellular Matrix Homeostasis. *Nat. Rev. Mol. Cell Biol.* **2014**, *15*, 802–812. [CrossRef] [PubMed]
6. Cox, T.R.; Erler, J.T. Remodeling and Homeostasis of the Extracellular Matrix: Implications for Fibrotic Diseases and Cancer. *Dis. Model. Mech.* **2011**, *4*, 165–178. [CrossRef]
7. Mahoney, J.P.; Sunahara, R.K. Mechanistic Insights into GPCR–G Protein Interactions. *Curr. Opin. Struct. Biol.* **2016**, *41*, 247–254. [CrossRef]
8. Pierce, K.L.; Premont, R.T.; Lefkowitz, R.J. Seven-Transmembrane Receptors. *Nat. Rev. Mol. Cell Biol.* **2002**, *3*, 639–650. [CrossRef]
9. Chen, C.G.; Iozzo, R.V. Extracellular Matrix Guidance of Autophagy: A Mechanism Regulating Cancer Growth. *Open Biol.* **2022**, *12*, 210304. [CrossRef]
10. Hupfer, A.; Brichkina, A.; Koeniger, A.; Keber, C.; Denkert, C.; Pfefferle, P.; Helmprobst, F.; Pagenstecher, A.; Visekruna, A.; Lauth, M. Matrix Stiffness Drives Stromal Autophagy and Promotes Formation of a Protumorigenic Niche. *Proc. Natl. Acad. Sci. USA* **2021**, *118*, e2105367118. [CrossRef]
11. Ge, H.; Tian, M.; Pei, Q.; Tan, F.; Pei, H. Extracellular Matrix Stiffness: New Areas Affecting Cell Metabolism. *Front. Oncol.* **2021**, *11*, 631991. [CrossRef]
12. Parzych, K.R.; Klionsky, D.J. An Overview of Autophagy: Morphology, Mechanism, and Regulation. *Antioxid. Redox Signal.* **2014**, *20*, 460–473. [CrossRef]
13. Mizushima, N.; Levine, B.; Cuervo, A.M.; Klionsky, D.J. Autophagy Fights Disease through Cellular Self-Digestion. *Nature* **2008**, *451*, 1069–1075. [CrossRef] [PubMed]
14. Shintani, T.; Klionsky, D.J. Autophagy in Health and Disease: A Double-Edged Sword. *Science* **2004**, *306*, 990–995. [CrossRef]
15. Bellot, G.; Garcia-Medina, R.; Gounon, P.; Chiche, J.; Roux, D.; Pouysségur, J.; Mazure, N.M. Hypoxia-Induced Autophagy Is Mediated through Hypoxia-Inducible Factor Induction of BNIP3 and BNIP3L via Their BH3 Domains. *Mol. Cell. Biol.* **2009**, *29*, 2570–2581. [CrossRef]
16. Seyedabadi, M.; Gharghabi, M.; Gurevich, E.V.; Gurevich, V.V. Structural Basis of GPCR Coupling to Distinct Signal Transducers: Implications for Biased Signaling. *Trends Biochem. Sci.* **2022**, *47*, 570–581. [CrossRef]

17. Vizurraga, A.; Adhikari, R.; Yeung, J.; Yu, M.; Tall, G.G. Mechanisms of Adhesion G Protein–Coupled Receptor Activation. *J. Biol. Chem.* **2020**, *295*, 14065–14083. [CrossRef] [PubMed]
18. Wauson, E.M.; Dbouk, H.A.; Ghosh, A.B.; Cobb, M.H. G Protein-Coupled Receptors and the Regulation of Autophagy. *Trends Endocrinol. Metab.* **2014**, *25*, 274–282. [CrossRef]
19. Wauson, E.M.; Zaganjor, E.; Lee, A.-Y.; Guerra, M.L.; Ghosh, A.B.; Bookout, A.L.; Chambers, C.P.; Jivan, A.; McGlynn, K.; Hutchison, M.R.; et al. The G Protein-Coupled Taste Receptor T1R1/T1R3 Regulates MTORC1 and Autophagy. *Mol. Cell* **2012**, *47*, 851–862. [CrossRef] [PubMed]
20. Woo, J.-A.; Yan, Y.; Kee, T.R.; Cazzaro, S.; McGill Percy, K.C.; Wang, X.; Liu, T.; Liggett, S.B.; Kang, D.E. β-Arrestin1 Promotes Tauopathy by Transducing GPCR Signaling, Disrupting Microtubules and Autophagy. *Life Sci. Alliance* **2022**, *5*, e202101183. [CrossRef]
21. Cabezudo, S.; Sanz-Flores, M.; Caballero, A.; Tasset, I.; Rebollo, E.; Diaz, A.; Aragay, A.M.; Cuervo, A.M.; Mayor, F.; Ribas, C. Gαq Activation Modulates Autophagy by Promoting MTORC1 Signaling. *Nat. Commun.* **2021**, *12*, 4540. [CrossRef] [PubMed]
22. Attwood, T.K.; Findlay, J.B.C. Fingerprinting G-Protein-Coupled Receptors. *Protein Eng. Des. Sel.* **1994**, *7*, 195–203. [CrossRef] [PubMed]
23. Fredriksson, R.; Lagerström, M.C.; Lundin, L.-G.; Schiöth, H.B. The G-Protein-Coupled Receptors in the Human Genome Form Five Main Families. Phylogenetic Analysis, Paralogon Groups, and Fingerprints. *Mol. Pharmacol.* **2003**, *63*, 1256–1272. [CrossRef] [PubMed]
24. Latorraca, N.R.; Venkatakrishnan, A.J.; Dror, R.O. GPCR Dynamics: Structures in Motion. *Chem. Rev.* **2017**, *117*, 139–155. [CrossRef]
25. Culhane, K.J.; Gupte, T.M.; Madhugiri, I.; Gadgil, C.J.; Sivaramakrishnan, S. Kinetic Model of GPCR-G Protein Interactions Reveals Allokairic Modulation of Signaling. *Nat. Commun.* **2022**, *13*, 1202. [CrossRef]
26. Shenoy, S.K.; Lefkowitz, R.J. β-Arrestin-Mediated Receptor Trafficking and Signal Transduction. *Trends Pharmacol. Sci.* **2011**, *32*, 521–533. [CrossRef] [PubMed]
27. Ranjan, R.; Dwivedi, H.; Baidya, M.; Kumar, M.; Shukla, A.K. Novel Structural Insights into GPCR–β-Arrestin Interaction and Signaling. *Trends Cell Biol.* **2017**, *27*, 851–862. [CrossRef] [PubMed]
28. Gurevich, V.V.; Gurevich, E.V. GPCR Signaling Regulation: The Role of GRKs and Arrestins. *Front. Pharmacol.* **2019**, *10*, 125. [CrossRef] [PubMed]
29. Tanowitz, M.; von Zastrow, M. A Novel Endocytic Recycling Signal That Distinguishes the Membrane Trafficking of Naturally Occurring Opioid Receptors. *J. Biol. Chem.* **2003**, *278*, 45978–45986. [CrossRef]
30. Pavlos, N.J.; Friedman, P.A. GPCR Signaling and Trafficking: The Long and Short of It. *Trends Endocrinol. Metab.* **2017**, *28*, 213–226. [CrossRef]
31. Sadoshima, J.; Xu, Y.; Slayter, H.S.; Izumo, S. Autocrine Release of Angiotensin II Mediates Stretch-Induced Hypertrophy of Cardiac Myocytes in Vitro. *Cell* **1993**, *75*, 977–984. [CrossRef]
32. Kauffenstein, G.; Laher, I.; Matrougui, K.; Guerineau, N.C.; Henrion, D. Emerging Role of G Protein-Coupled Receptors in Microvascular Myogenic Tone. *Cardiovasc. Res.* **2012**, *95*, 223–232. [CrossRef] [PubMed]
33. Wang, S.; Iring, A.; Strilic, B.; Albarrán Juárez, J.; Kaur, H.; Troidl, K.; Tonack, S.; Burbiel, J.C.; Müller, C.E.; Fleming, I.; et al. P2Y2 and Gq/G11 Control Blood Pressure by Mediating Endothelial Mechanotransduction. *J. Clin. Investig.* **2015**, *125*, 3077–3086. [CrossRef]
34. Schnitzler, M.M.Y.; Storch, U.; Meibers, S.; Nurwakagari, P.; Breit, A.; Essin, K.; Gollasch, M.; Gudermann, T. Gq-Coupled Receptors as Mechanosensors Mediating Myogenic Vasoconstriction. *EMBO J.* **2008**, *27*, 3092–3103. [CrossRef] [PubMed]
35. Iliff, A.J.; Xu, X.Z.S. A Mechanosensitive GPCR That Detects the Bloody Force. *Cell* **2018**, *173*, 542–544. [CrossRef]
36. Marullo, S.; Doly, S.; Saha, K.; Enslen, H.; Scott, M.G.H.; Coureuil, M. Mechanical GPCR Activation by Traction Forces Exerted on Receptor *N*-Glycans. *ACS Pharmacol. Transl. Sci.* **2020**, *3*, 171–178. [CrossRef] [PubMed]
37. Langenhan, T.; Piao, X.; Monk, K.R. Adhesion G Protein-Coupled Receptors in Nervous System Development and Disease. *Nat. Rev. Neurosci.* **2016**, *17*, 550–561. [CrossRef]
38. Folts, C.J.; Giera, S.; Li, T.; Piao, X. Adhesion G Protein-Coupled Receptors as Drug Targets for Neurological Diseases. *Trends Pharmacol. Sci.* **2019**, *40*, 278–293. [CrossRef]
39. Makino, A.; Prossnitz, E.R.; Bünemann, M.; Wang, J.M.; Yao, W.; Schmid-Schönbein, G.W. G Protein-Coupled Receptors Serve as Mechanosensors for Fluid Shear Stress in Neutrophils. *Am. J. Physiol.-Cell Physiol.* **2006**, *290*, C1633–C1639. [CrossRef]
40. White, J.P.; Wrann, C.D.; Rao, R.R.; Nair, S.K.; Jedrychowski, M.P.; You, J.-S.; Martínez-Redondo, V.; Gygi, S.P.; Ruas, J.L.; Hornberger, T.A.; et al. G Protein-Coupled Receptor 56 Regulates Mechanical Overload-Induced Muscle Hypertrophy. *Proc. Natl. Acad. Sci. USA* **2014**, *111*, 15756–15761. [CrossRef]
41. Ozkan, A.D.; Gettas, T.; Sogata, A.; Phaychanpheng, W.; Zhou, M.; Lacroix, J.J. Mechanical and Chemical Activation of GPR68 Probed with a Genetically Encoded Fluorescent Reporter. *J. Cell Sci.* **2021**, *134*, jcs255455. [CrossRef] [PubMed]
42. Feng, Q.; Kornmann, B. Mechanical Forces on Cellular Organelles. *J. Cell Sci.* **2018**, *131*, jcs218479. [CrossRef] [PubMed]
43. Liang, Y.; Lai, L.; Wang, B.; Juang, S.; Chang, C.; Leu, J.; Shyu, K. Mechanical Stress Enhances Serotonin 2B Receptor Modulating Brain Natriuretic Peptide through Nuclear Factor-KB in Cardiomyocytes. *Cardiovasc. Res.* **2006**, *72*, 303–312. [CrossRef]

44. Myagmar, B.-E.; Ismaili, T.; Swigart, P.M.; Raghunathan, A.; Baker, A.J.; Sahdeo, S.; Blevitt, J.M.; Milla, M.E.; Simpson, P.C. Coupling to Gq Signaling Is Required for Cardioprotection by an Alpha-1A-Adrenergic Receptor Agonist. *Circ. Res.* **2019**, *125*, 699–706. [CrossRef] [PubMed]
45. Jian, X.; Sainz, E.; Clark, W.A.; Jensen, R.T.; Battey, J.F.; Northup, J.K. The Bombesin Receptor Subtypes Have Distinct G Protein Specificities. *J. Biol. Chem.* **1999**, *274*, 11573–11581. [CrossRef]
46. Wang, N.; He, X.; Zhao, J.; Jiang, H.; Cheng, X.; Xia, Y.; Eric Xu, H.; He, Y. Structural Basis of Leukotriene B4 Receptor 1 Activation. *Nat. Commun.* **2022**, *13*, 1156. [CrossRef]
47. Mobbs, J.I.; Belousoff, M.J.; Harikumar, K.G.; Piper, S.J.; Xu, X.; Furness, S.G.B.; Venugopal, H.; Christopoulos, A.; Danev, R.; Wootten, D.; et al. Structures of the Human Cholecystokinin 1 (CCK1) Receptor Bound to Gs and Gq Mimetic Proteins Provide Insight into Mechanisms of G Protein Selectivity. *PLoS Biol.* **2021**, *19*, e3001295. [CrossRef] [PubMed]
48. Capra, V. Molecular and Functional Aspects of Human Cysteinyl Leukotriene Receptors. *Pharmacol. Res.* **2004**, *50*, 1–11. [CrossRef] [PubMed]
49. Storch, U.; Blodow, S.; Gudermann, T.; Mederos y Schnitzler, M. Cysteinyl Leukotriene 1 Receptors as Novel Mechanosensors Mediating Myogenic Tone Together with Angiotensin II Type 1 Receptors—Brief Report. *Arterioscler. Thromb. Vasc. Biol.* **2015**, *35*, 121–126. [CrossRef]
50. Legler, D.F.; Bruckner, M.; Uetz-von Allmen, E.; Krause, P. Prostaglandin E2 at New Glance: Novel Insights in Functional Diversity Offer Therapeutic Chances. *Int. J. Biochem. Cell Biol.* **2010**, *42*, 198–201. [CrossRef] [PubMed]
51. Akhter, M.P.; Cullen, D.M.; Gong, G.; Recker, R.R. Bone Biomechanical Properties in Prostaglandin EP1 and EP2 Knockout Mice. *Bone* **2001**, *29*, 121–125. [CrossRef]
52. Eguchi, S.; Hirata, Y.; Imai, T.; Marumo, F. Endothelin Receptor Subtypes Are Coupled to Adenylate Cyclase via Different Guanyl Nucleotide-Binding Proteins in Vasculature. *Endocrinology* **1993**, *132*, 524–529. [CrossRef] [PubMed]
53. McCoy, K.L.; Gyoneva, S.; Vellano, C.P.; Smrcka, A.V.; Traynelis, S.F.; Hepler, J.R. Protease-Activated Receptor 1 (PAR1) Coupling to Gq/11 but Not to Gi/o or G12/13 Is Mediated by Discrete Amino Acids within the Receptor Second Intracellular Loop. *Cell. Signal.* **2012**, *24*, 1351–1360. [CrossRef] [PubMed]
54. Kim, S.; Han, J.-H.; Nam, D.-H.; Kim, G.-Y.; Lim, J.H.; Kim, J.-R.; Woo, C.-H. PAR-1 Is a Novel Mechano-Sensor Transducing Laminar Flow-Mediated Endothelial Signaling. *Sci. Rep.* **2018**, *8*, 15172. [CrossRef]
55. Liu, H.-X.; Hökfelt, T. The Participation of Galanin in Pain Processing at the Spinal Level. *Trends Pharmacol. Sci.* **2002**, *23*, 468–474. [CrossRef]
56. Abizaid, A.; Hougland, J.L. Ghrelin Signaling: GOAT and GHS-R1a Take a LEAP in Complexity. *Trends Endocrinol. Metab.* **2020**, *31*, 107–117. [CrossRef]
57. Liu, F.; Usui, I.; Evans, L.G.; Austin, D.A.; Mellon, P.L.; Olefsky, J.M.; Webster, N.J.G. Involvement of Both Gq/11 and Gs Proteins in Gonadotropin-Releasing Hormone Receptor-Mediated Signaling in LβT2 Cells. *J. Biol. Chem.* **2002**, *277*, 32099–32108. [CrossRef] [PubMed]
58. Erdogmus, S.; Storch, U.; Danner, L.; Becker, J.; Winter, M.; Ziegler, N.; Wirth, A.; Offermanns, S.; Hoffmann, C.; Gudermann, T.; et al. Helix 8 Is the Essential Structural Motif of Mechanosensitive GPCRs. *Nat. Commun.* **2019**, *10*, 5784. [CrossRef] [PubMed]
59. Jiang, Y.; Li, T.; Wu, Y.; Xu, H.; Xie, C.; Dong, Y.; Zhong, L.; Wang, Z.; Zhao, H.; Zhou, Y.; et al. GPR39 Overexpression in OSCC Promotes YAP-Sustained Malignant Progression. *J. Dent. Res.* **2020**, *99*, 949–958. [CrossRef]
60. Xu, J.; Mathur, J.; Vessières, E.; Hammack, S.; Nonomura, K.; Favre, J.; Grimaud, L.; Petrus, M.; Francisco, A.; Li, J.; et al. GPR68 Senses Flow and Is Essential for Vascular Physiology. *Cell* **2018**, *173*, 762–775.e16. [CrossRef]
61. Adjobo-Hermans, M.J.; Goedhart, J.; van Weeren, L.; Nijmeijer, S.; Manders, E.M.; Offermanns, S.; Gadella, T.W. Real-Time Visualization of Heterotrimeric G Protein Gq Activation in Living Cells. *BMC Biol.* **2011**, *9*, 32. [CrossRef] [PubMed]
62. Maeda, S.; Qu, Q.; Robertson, M.J.; Skiniotis, G.; Kobilka, B.K. Structures of the M1 and M2 Muscarinic Acetylcholine Receptor/G-Protein Complexes. *Science* **2019**, *364*, 552–557. [CrossRef] [PubMed]
63. Cui, Y.; Kassmann, M.; Nickel, S.; Zhang, C.; Alenina, N.; Anistan, Y.M.; Schleifenbaum, J.; Bader, M.; Welsh, D.G.; Huang, Y.; et al. Myogenic Vasoconstriction Requires Canonical $G_{q/11}$ Signaling of the Angiotensin II Type 1 Receptor. *J. Am. Heart Assoc.* **2022**, *11*, e022070. [CrossRef] [PubMed]
64. Hunyady, L.; Catt, K.J. Pleiotropic AT1 Receptor Signaling Pathways Mediating Physiological and Pathogenic Actions of Angiotensin II. *Mol. Endocrinol.* **2006**, *20*, 953–970. [CrossRef] [PubMed]
65. Zhang, H.; Han, G.W.; Batyuk, A.; Ishchenko, A.; White, K.L.; Patel, N.; Sadybekov, A.; Zamlynny, B.; Rudd, M.T.; Hollenstein, K.; et al. Structural Basis for Selectivity and Diversity in Angiotensin II Receptors. *Nature* **2017**, *544*, 327–332. [CrossRef]
66. Zou, Y.; Akazawa, H.; Qin, Y.; Sano, M.; Takano, H.; Minamino, T.; Makita, N.; Iwanaga, K.; Zhu, W.; Kudoh, S.; et al. Mechanical Stress Activates Angiotensin II Type 1 Receptor without the Involvement of Angiotensin II. *Nat. Cell Biol.* **2004**, *6*, 499–506. [CrossRef]
67. Hawes, B.E.; Kil, E.; Green, B.; O'Neill, K.; Fried, S.; Graziano, M.P. The Melanin-Concentrating Hormone Receptor Couples to Multiple G Proteins to Activate Diverse Intracellular Signaling Pathways. *Endocrinology* **2000**, *141*, 4524–4532. [CrossRef]
68. Chachisvilis, M.; Zhang, Y.-L.; Frangos, J.A. G Protein-Coupled Receptors Sense Fluid Shear Stress in Endothelial Cells. *Proc. Natl. Acad. Sci. USA* **2006**, *103*, 15463–15468. [CrossRef]
69. Sun, Y.; Leng, P.; Guo, P.; Gao, H.; Liu, Y.; Li, C.; Li, Z.; Zhang, H. G Protein Coupled Estrogen Receptor Attenuates Mechanical Stress-Mediated Apoptosis of Chondrocyte in Osteoarthritis via Suppression of Piezo1. *Mol. Med.* **2021**, *27*, 96. [CrossRef]

70. Gao, H.; Tian, K.; Feng, X.; Yan, M.; Gao, C.; Jiang, Y.; Zhu, C.; Zhu, H.; Liu, X.; Peng, Y. Free Fatty Acid Receptor 2 Promotes Cardiomyocyte Hypertrophy by Activating STAT3 and GATA4. *Food Sci. Hum. Wellness* **2022**, *11*, 405–417. [CrossRef]
71. Mahon, M.J. The Parathyroid Hormone Receptorsome and the Potential for Therapeutic Intervention. *Curr. Drug Targets* **2012**, *13*, 116–128. [CrossRef]
72. Ascolani, G.; Skerry, T.M.; Lacroix, D.; Dall'Ara, E.; Shuaib, A. Analysis of Mechanotransduction Dynamics during Combined Mechanical Stimulation and Modulation of the Extracellular-Regulated Kinase Cascade Uncovers Hidden Information within the Signalling Noise. *Interface Focus* **2021**, *11*, 20190136. [CrossRef]
73. Terrillon, S.; Barberis, C.; Bouvier, M. Heterodimerization of V1a and V2 Vasopressin Receptors Determines the Interaction with β-Arrestin and Their Trafficking Patterns. *Proc. Natl. Acad. Sci. USA* **2004**, *101*, 1548–1553. [CrossRef] [PubMed]
74. Zhang, D.-L.; Sun, Y.-J.; Ma, M.-L.; Wang, Y.; Lin, H.; Li, R.-R.; Liang, Z.-L.; Gao, Y.; Yang, Z.; He, D.-F.; et al. Gq Activity- and β-Arrestin-1 Scaffolding-Mediated ADGRG2/CFTR Coupling Are Required for Male Fertility. *eLife* **2018**, *7*, e33432. [CrossRef] [PubMed]
75. Erb, L.; Weisman, G.A. Coupling of P_2Y Receptors to G Proteins and Other Signaling Pathways. *Wiley Interdiscip. Rev. Membr. Transp. Signal.* **2012**, *1*, 789–803. [CrossRef]
76. Sathanoori, R.; Bryl-Gorecka, P.; Müller, C.E.; Erb, L.; Weisman, G.A.; Olde, B.; Erlinge, D. P2Y2 Receptor Modulates Shear Stress-Induced Cell Alignment and Actin Stress Fibers in Human Umbilical Vein Endothelial Cells. *Cell. Mol. Life Sci.* **2017**, *74*, 731–746. [CrossRef]
77. Jelinek, V.; Mösslein, N.; Bünemann, M. Structures in G Proteins Important for Subtype Selective Receptor Binding and Subsequent Activation. *Commun. Biol.* **2021**, *4*, 635. [CrossRef]
78. Campbell, A.P.; Smrcka, A.V. Targeting G Protein-Coupled Receptor Signalling by Blocking G Proteins. *Nat. Rev. Drug Discov.* **2018**, *17*, 789–803. [CrossRef] [PubMed]
79. Sánchez-Fernández, G.; Cabezudo, S.; García-Hoz, C.; Benincá, C.; Aragay, A.M.; Mayor, F.; Ribas, C. Gαq Signalling: The New and the Old. *Cell. Signal.* **2014**, *26*, 833–848. [CrossRef]
80. Kang, Y.; Kuybeda, O.; de Waal, P.W.; Mukherjee, S.; van Eps, N.; Dutka, P.; Zhou, X.E.; Bartesaghi, A.; Erramilli, S.; Morizumi, T.; et al. Cryo-EM Structure of Human Rhodopsin Bound to an Inhibitory G Protein. *Nature* **2018**, *558*, 553–558. [CrossRef] [PubMed]
81. Koehl, A.; Hu, H.; Maeda, S.; Zhang, Y.; Qu, Q.; Paggi, J.M.; Latorraca, N.R.; Hilger, D.; Dawson, R.; Matile, H.; et al. Structure of the M-Opioid Receptor–Gi Protein Complex. *Nature* **2018**, *558*, 547–552. [CrossRef]
82. Glukhova, A.; Draper-Joyce, C.J.; Sunahara, R.K.; Christopoulos, A.; Wootten, D.; Sexton, P.M. Rules of Engagement: GPCRs and G Proteins. *ACS Pharmacol. Transl. Sci.* **2018**, *1*, 73–83. [CrossRef]
83. Xia, R.; Wang, N.; Xu, Z.; Lu, Y.; Song, J.; Zhang, A.; Guo, C.; He, Y. Cryo-EM Structure of the Human Histamine H1 Receptor/Gq Complex. *Nat. Commun.* **2021**, *12*, 2086. [CrossRef]
84. Shen, J.; Zhang, D.; Fu, Y.; Chen, A.; Yang, X.; Zhang, H. Cryo-EM Structures of Human Bradykinin Receptor-Gq Proteins Complexes. *Nat. Commun.* **2022**, *13*, 714. [CrossRef]
85. Okashah, N.; Wan, Q.; Ghosh, S.; Sandhu, M.; Inoue, A.; Vaidehi, N.; Lambert, N.A. Variable G Protein Determinants of GPCR Coupling Selectivity. *Proc. Natl. Acad. Sci. USA* **2019**, *116*, 12054–12059. [CrossRef]
86. Hubbard, K.B.; Hepler, J.R. Cell Signalling Diversity of the Gqα Family of Heterotrimeric G Proteins. *Cell. Signal.* **2006**, *18*, 135–150. [CrossRef]
87. Davignon, I.; Barnard, M.; Gavrilova, O.; Sweet, K.; Wilkie, T.M. Gene Structure of MurineGna11andGna15:Tandemly Duplicated Gq Class G Protein α Subunit Genes. *Genomics* **1996**, *31*, 359–366. [CrossRef] [PubMed]
88. Wilkie, T.M.; Scherle, P.A.; Strathmann, M.P.; Slepak, V.Z.; Simon, M.I. Characterization of G-Protein Alpha Subunits in the Gq Class: Expression in Murine Tissues and in Stromal and Hematopoietic Cell Lines. *Proc. Natl. Acad. Sci. USA* **1991**, *88*, 10049–10053. [CrossRef]
89. Sánchez-Fernández, G.; Cabezudo, S.; Caballero, Á.; García-Hoz, C.; Tall, G.G.; Klett, J.; Michnick, S.W.; Mayor, F.; Ribas, C. Protein Kinase C ζ Interacts with a Novel Binding Region of Gαq to Act as a Functional Effector. *J. Biol. Chem.* **2016**, *291*, 9513–9525. [CrossRef] [PubMed]
90. Simon, M.I.; Strathmann, M.P.; Gautam, N. Diversity of G Proteins in Signal Transduction. *Science* **1991**, *252*, 802–808. [CrossRef] [PubMed]
91. Wedegaertner, P.B. G Protein Trafficking. In *GPCR Signalling Complexes—Synthesis, Assembly, Trafficking and Specificity*; Subcellular Biochemistry; Dupré, D., Hébert, T., Jockers, R., Eds.; Springer: Dordrecht, The Netherlands, 2012; Volume 63, pp. 193–223. [CrossRef]
92. Siderovski, D.P.; Willard, F.S. The GAPs, GEFs, and GDIs of Heterotrimeric G-Protein Alpha Subunits. *Int. J. Biol. Sci.* **2005**, *1*, 51–66. [CrossRef] [PubMed]
93. Hollinger, S. Cellular Regulation of RGS Proteins: Modulators and Integrators of G Protein Signaling. *Pharmacol. Rev.* **2002**, *54*, 527–559. [CrossRef] [PubMed]
94. Xu, X.; Zeng, W.; Popov, S.; Berman, D.M.; Davignon, I.; Yu, K.; Yowe, D.; Offermanns, S.; Muallem, S.; Wilkie, T.M. RGS Proteins Determine Signaling Specificity of Gq-Coupled Receptors. *J. Biol. Chem.* **1999**, *274*, 3549–3556. [CrossRef] [PubMed]
95. Masuho, I.; Balaji, S.; Muntean, B.S.; Skamangas, N.K.; Chavali, S.; Tesmer, J.J.G.; Babu, M.M.; Martemyanov, K.A. A Global Map of G Protein Signaling Regulation by RGS Proteins. *Cell* **2020**, *183*, 503–521.e19. [CrossRef] [PubMed]

96. Nishimura, A.; Okamoto, M.; Sugawara, Y.; Mizuno, N.; Yamauchi, J.; Itoh, H. Ric-8A Potentiates Gq-Mediated Signal Transduction by Acting Downstream of G Protein-Coupled Receptor in Intact Cells. *Genes Cells* **2006**, *11*, 487–498. [CrossRef] [PubMed]
97. Hampoelz, B.; Hoeller, O.; Bowman, S.K.; Dunican, D.; Knoblich, J.A. Drosophila Ric-8 Is Essential for Plasma-Membrane Localization of Heterotrimeric G Proteins. *Nat. Cell Biol.* **2005**, *7*, 1099–1105. [CrossRef]
98. Tall, G.G. Ric-8 Regulation of Heterotrimeric G Proteins. *J. Recept. Signal Transduct.* **2013**, *33*, 139–143. [CrossRef]
99. Gabay, M.; Pinter, M.E.; Wright, F.A.; Chan, P.; Murphy, A.J.; Valenzuela, D.M.; Yancopoulos, G.D.; Tall, G.G. Ric-8 Proteins Are Molecular Chaperones That Direct Nascent G Protein α Subunit Membrane Association. *Sci. Signal.* **2011**, *4*, ra79. [CrossRef] [PubMed]
100. Rojas, R.J.; Yohe, M.E.; Gershburg, S.; Kawano, T.; Kozasa, T.; Sondek, J. Gαq Directly Activates P63RhoGEF and Trio via a Conserved Extension of the Dbl Homology-Associated Pleckstrin Homology Domain. *J. Biol. Chem.* **2007**, *282*, 29201–29210. [CrossRef]
101. Fukuhara, S.; Murga, C.; Zohar, M.; Igishi, T.; Gutkind, J.S. A Novel PDZ Domain Containing Guanine Nucleotide Exchange Factor Links Heterotrimeric G Proteins to Rho. *J. Biol. Chem.* **1999**, *274*, 5868–5879. [CrossRef]
102. Lutz, S.; Freichel-Blomquist, A.; Yang, Y.; Rümenapp, U.; Jakobs, K.H.; Schmidt, M.; Wieland, T. The Guanine Nucleotide Exchange Factor P63RhoGEF, a Specific Link between Gq/11-Coupled Receptor Signaling and RhoA. *J. Biol. Chem.* **2005**, *280*, 11134–11139. [CrossRef] [PubMed]
103. Vaqué, J.P.; Dorsam, R.T.; Feng, X.; Iglesias-Bartolome, R.; Forsthoefel, D.J.; Chen, Q.; Debant, A.; Seeger, M.A.; Ksander, B.R.; Teramoto, H.; et al. A Genome-Wide RNAi Screen Reveals a Trio-Regulated Rho GTPase Circuitry Transducing Mitogenic Signals Initiated by G Protein-Coupled Receptors. *Mol. Cell* **2013**, *49*, 94–108. [CrossRef]
104. Popova, J.S.; Rasenick, M.M. Gβγ Mediates the Interplay between Tubulin Dimers and Microtubules in the Modulation of Gq Signaling. *J. Biol. Chem.* **2003**, *278*, 34299–34308. [CrossRef] [PubMed]
105. Ciruela, F.; McIlhinney, R.A.J. Metabotropic Glutamate Receptor Type 1α and Tubulin Assemble into Dynamic Interacting Complexes. *J. Neurochem.* **2008**, *76*, 750–757. [CrossRef] [PubMed]
106. Maxfield, F.R. Plasma Membrane Microdomains. *Curr. Opin. Cell Biol.* **2002**, *14*, 483–487. [CrossRef]
107. Simons, K.; Ikonen, E. Functional Rafts in Cell Membranes. *Nature* **1997**, *387*, 569–572. [CrossRef] [PubMed]
108. Kurzchalia, T.V.; Partan, R.G. Membrane Microdomains and Caveolae. *Curr. Opin. Cell Biol.* **1999**, *11*, 424–431. [CrossRef]
109. Parton, R.G.; del Pozo, M.A. Caveolae as Plasma Membrane Sensors, Protectors and Organizers. *Nat. Rev. Mol. Cell Biol.* **2013**, *14*, 98–112. [CrossRef]
110. Le Roux, A.-L.; Quiroga, X.; Walani, N.; Arroyo, M.; Roca-Cusachs, P. The Plasma Membrane as a Mechanochemical Transducer. *Philos. Trans. R. Soc. B Biol. Sci.* **2019**, *374*, 20180221. [CrossRef]
111. Ostrom, R.S.; Insel, P.A. The Evolving Role of Lipid Rafts and Caveolae in G Protein-Coupled Receptor Signaling: Implications for Molecular Pharmacology. *Br. J. Pharmacol.* **2004**, *143*, 235–245. [CrossRef]
112. Mizuno, N.; Itoh, H. Functions and Regulatory Mechanisms of Gq-Signaling Pathways. *Neurosignals* **2009**, *17*, 42–54. [CrossRef]
113. Oh, P.; Schnitzer, J.E. Segregation of Heterotrimeric G Proteins in Cell Surface Microdomains. *Mol. Biol. Cell* **2001**, *12*, 685–698. [CrossRef] [PubMed]
114. Moreno-Vicente, R.; Pavón, D.M.; Martín-Padura, I.; Català-Montoro, M.; Díez-Sánchez, A.; Quílez-Álvarez, A.; López, J.A.; Sánchez-Álvarez, M.; Vázquez, J.; Strippoli, R.; et al. Caveolin-1 Modulates Mechanotransduction Responses to Substrate Stiffness through Actin-Dependent Control of YAP. *Cell Rep.* **2018**, *25*, 1622–1635.e6. [CrossRef] [PubMed]
115. Sinha, B.; Köster, D.; Ruez, R.; Gonnord, P.; Bastiani, M.; Abankwa, D.; Stan, R.V.; Butler-Browne, G.; Vedie, B.; Johannes, L.; et al. Cells Respond to Mechanical Stress by Rapid Disassembly of Caveolae. *Cell* **2011**, *144*, 402–413. [CrossRef]
116. Calizo, R.C.; Scarlata, S. A Role for G-Proteins in Directing G-Protein-Coupled Receptor–Caveolae Localization. *Biochemistry* **2012**, *51*, 9513–9523. [CrossRef]
117. Qifti, A.; Garwain, O.; Scarlata, S. Mechanical Stretch Redefines Membrane Gαq–Calcium Signaling Complexes. *J. Membr. Biol.* **2019**, *252*, 307–315. [CrossRef]
118. Rizzo, V.; McIntosh, D.P.; Oh, P.; Schnitzer, J.E. In Situ Flow Activates Endothelial Nitric Oxide Synthase in Luminal Caveolae of Endothelium with Rapid Caveolin Dissociation and Calmodulin Association. *J. Biol. Chem.* **1998**, *273*, 34724–34729. [CrossRef]
119. Sugawara, Y.; Nishii, H.; Takahashi, T.; Yamauchi, J.; Mizuno, N.; Tago, K.; Itoh, H. The Lipid Raft Proteins Flotillins/Reggies Interact with Gαq and Are Involved in Gq-Mediated P38 Mitogen-Activated Protein Kinase Activation through Tyrosine Kinase. *Cell. Signal.* **2007**, *19*, 1301–1308. [CrossRef] [PubMed]
120. Galluzzi, L.; Baehrecke, E.H.; Ballabio, A.; Boya, P.; Bravo-San Pedro, J.M.; Cecconi, F.; Choi, A.M.; Chu, C.T.; Codogno, P.; Colombo, M.I.; et al. Molecular Definitions of Autophagy and Related Processes. *EMBO J.* **2017**, *36*, 1811–1836. [CrossRef] [PubMed]
121. Mizushima, N.; Yoshimori, T.; Ohsumi, Y. The Role of Atg Proteins in Autophagosome Formation. *Annu. Rev. Cell Dev. Biol.* **2011**, *27*, 107–132. [CrossRef] [PubMed]
122. Dunlop, E.A.; Tee, A.R. MTOR and Autophagy: A Dynamic Relationship Governed by Nutrients and Energy. *Semin. Cell Dev. Biol.* **2014**, *36*, 121–129. [CrossRef] [PubMed]
123. Park, J.-M.; Jung, C.H.; Seo, M.; Otto, N.M.; Grunwald, D.; Kim, K.H.; Moriarity, B.; Kim, Y.-M.; Starker, C.; Nho, R.S.; et al. The ULK1 Complex Mediates MTORC1 Signaling to the Autophagy Initiation Machinery via Binding and Phosphorylating ATG14. *Autophagy* **2016**, *12*, 547–564. [CrossRef] [PubMed]

124. Hosokawa, N.; Hara, T.; Kaizuka, T.; Kishi, C.; Takamura, A.; Miura, Y.; Iemura, S.; Natsume, T.; Takehana, K.; Yamada, N.; et al. Nutrient-Dependent MTORC1 Association with the ULK1–Atg13–FIP200 Complex Required for Autophagy. *Mol. Biol. Cell* **2009**, *20*, 1981–1991. [CrossRef]
125. Case, N.; Thomas, J.; Sen, B.; Styner, M.; Xie, Z.; Galior, K.; Rubin, J. Mechanical Regulation of Glycogen Synthase Kinase 3β (GSK3β) in Mesenchymal Stem Cells Is Dependent on Akt Protein Serine 473 Phosphorylation via MTORC2 Protein. *J. Biol. Chem.* **2011**, *286*, 39450–39456. [CrossRef]
126. Hernández-Cáceres, M.P.; Munoz, L.; Pradenas, J.M.; Pena, F.; Lagos, P.; Aceiton, P.; Owen, G.I.; Morselli, E.; Criollo, A.; Ravasio, A.; et al. Mechanobiology of Autophagy: The Unexplored Side of Cancer. *Front. Oncol.* **2021**, *11*, 632956. [CrossRef]
127. Dupont, N.; Codogno, P. Autophagy Transduces Physical Constraints into Biological Responses. *Int. J. Biochem. Cell Biol.* **2016**, *79*, 419–426. [CrossRef] [PubMed]
128. Martinac, B.; Poole, K. Mechanically Activated Ion Channels. *Int. J. Biochem. Cell Biol.* **2018**, *97*, 104–107. [CrossRef] [PubMed]
129. Saotome, K.; Murthy, S.E.; Kefauver, J.M.; Whitwam, T.; Patapoutian, A.; Ward, A.B. Structure of the Mechanically Activated Ion Channel Piezo1. *Nature* **2018**, *554*, 481–486. [CrossRef] [PubMed]
130. Praetorius, H.A. The Primary Cilium as Sensor of Fluid Flow: New Building Blocks to the Model. A Review in the Theme: Cell Signaling: Proteins, Pathways and Mechanisms. *Am. J. Physiol. -Cell Physiol.* **2015**, *308*, C198–C208. [CrossRef] [PubMed]
131. Baschieri, F.; Dayot, S.; Elkhatib, N.; Ly, N.; Capmany, A.; Schauer, K.; Betz, T.; Vignjevic, D.M.; Poincloux, R.; Montagnac, G. Frustrated Endocytosis Controls Contractility-Independent Mechanotransduction at Clathrin-Coated Structures. *Nat. Commun.* **2018**, *9*, 3825. [CrossRef]
132. Kirby, T.J.; Lammerding, J. Emerging Views of the Nucleus as a Cellular Mechanosensor. *Nat. Cell Biol.* **2018**, *20*, 373–381. [CrossRef] [PubMed]
133. Helle, S.C.J.; Feng, Q.; Aebersold, M.J.; Hirt, L.; Grüter, R.R.; Vahid, A.; Sirianni, A.; Mostowy, S.; Snedeker, J.G.; Šarić, A.; et al. Mechanical Force Induces Mitochondrial Fission. *eLife* **2017**, *6*, e30292. [CrossRef]
134. Guet, D.; Mandal, K.; Pinot, M.; Hoffmann, J.; Abidine, Y.; Sigaut, W.; Bardin, S.; Schauer, K.; Goud, B.; Manneville, J.-B. Mechanical Role of Actin Dynamics in the Rheology of the Golgi Complex and in Golgi-Associated Trafficking Events. *Curr. Biol.* **2014**, *24*, 1700–1711. [CrossRef]
135. Aguilera, M.O.; Berón, W.; Colombo, M.I. The Actin Cytoskeleton Participates in the Early Events of Autophagosome Formation upon Starvation Induced Autophagy. *Autophagy* **2012**, *8*, 1590–1603. [CrossRef]
136. Kast, D.J.; Dominguez, R. The Cytoskeleton–Autophagy Connection. *Curr. Biol.* **2017**, *27*, R318–R326. [CrossRef]
137. Hsu, P.; Shi, Y. Regulation of Autophagy by Mitochondrial Phospholipids in Health and Diseases. *Biochim. Et Biophys. Acta (BBA)—Mol. Cell Biol. Lipids* **2017**, *1862*, 114–129. [CrossRef]
138. Avivar-Valderas, A.; Bobrovnikova-Marjon, E.; Alan Diehl, J.; Bardeesy, N.; Debnath, J.; Aguirre-Ghiso, J.A. Regulation of Autophagy during ECM Detachment Is Linked to a Selective Inhibition of MTORC1 by PERK. *Oncogene* **2013**, *32*, 4932–4940. [CrossRef] [PubMed]
139. Cai, Z.; Zhao, D.; Sun, Y.; Gao, D.; Li, X.; Yang, J.; Ma, Z. Detachment-Based Equilibrium of Anoikic Cell Death and Autophagic Cell Survival through Adaptor Protein P66^{Shc}. *Anat. Rec.* **2016**, *299*, 325–333. [CrossRef] [PubMed]
140. Vlahakis, A.; Debnath, J. The Interconnections between Autophagy and Integrin-Mediated Cell Adhesion. *J. Mol. Biol.* **2017**, *429*, 515–530. [CrossRef]
141. Anlaş, A.A.; Nelson, C.M. Soft Microenvironments Induce Chemoresistance by Increasing Autophagy Downstream of Integrin-Linked Kinase. *Cancer Res.* **2020**, *80*, 4103–4113. [CrossRef]
142. Neill, T.; Schaefer, L.; Iozzo, R.V. Instructive Roles of Extracellular Matrix on Autophagy. *Am. J. Pathol.* **2014**, *184*, 2146–2153. [CrossRef]
143. Buraschi, S.; Neill, T.; Goyal, A.; Poluzzi, C.; Smythies, J.; Owens, R.T.; Schaefer, L.; Torres, A.; Iozzo, R.V. Decorin Causes Autophagy in Endothelial Cells via Peg3. *Proc. Natl. Acad. Sci. USA* **2013**, *110*, E2582–E2591. [CrossRef]
144. Nguyen, T.M.B.; Subramanian, I.V.; Xiao, X.; Ghosh, G.; Nguyen, P.; Kelekar, A.; Ramakrishnan, S. Endostatin Induces Autophagy in Endothelial Cells by Modulating Beclin 1 and β-Catenin Levels. *J. Cell. Mol. Med.* **2009**, *13*, 3687–3698. [CrossRef] [PubMed]
145. Castagnaro, S.; Gambarotto, L.; Cescon, M.; Bonaldo, P. Autophagy in the Mesh of Collagen VI. *Matrix Biol.* **2021**, *100–101*, 162–172. [CrossRef] [PubMed]
146. Carmignac, V.; Svensson, M.; Körner, Z.; Elowsson, L.; Matsumura, C.; Gawlik, K.I.; Allamand, V.; Durbeej, M. Autophagy Is Increased in Laminin A2 Chain-Deficient Muscle and Its Inhibition Improves Muscle Morphology in a Mouse Model of MDC1A. *Hum. Mol. Genet.* **2011**, *20*, 4891–4902. [CrossRef]
147. García-Hoz, C.; Sánchez-Fernández, G.; Díaz-Meco, M.T.; Moscat, J.; Mayor, F.; Ribas, C. Gαq Acts as an Adaptor Protein in Protein Kinase Cζ (PKCζ)-Mediated ERK5 Activation by G Protein-Coupled Receptors (GPCR). *J. Biol. Chem.* **2010**, *285*, 13480–13489. [CrossRef] [PubMed]
148. Nigro, P.; Abe, J.; Berk, B.C. Flow Shear Stress and Atherosclerosis: A Matter of Site Specificity. *Antioxid. Redox Signal.* **2011**, *15*, 1405–1414. [CrossRef] [PubMed]
149. Scholz, N. Cancer Cell Mechanics: Adhesion G Protein-Coupled Receptors in Action? *Front. Oncol.* **2018**, *8*, 59. [CrossRef] [PubMed]
150. Aplin, A.; Jasionowski, T.; Tuttle, D.L.; Lenk, S.E.; Dunn, W.A. Cytoskeletal Elements Are Required for the Formation and Maturation of Autophagic Vacuoles. *J. Cell. Physiol.* **1992**, *152*, 458–466. [CrossRef]

151. Kaushik, S.; Cuervo, A.M. The Coming of Age of Chaperone-Mediated Autophagy. *Nat. Rev. Mol. Cell Biol.* **2018**, *19*, 365–381. [CrossRef]
152. Lim, C.-Y.; Zoncu, R. The Lysosome as a Command-and-Control Center for Cellular Metabolism. *J. Cell Biol.* **2016**, *214*, 653–664. [CrossRef] [PubMed]
153. Cabukusta, B.; Neefjes, J. Mechanisms of Lysosomal Positioning and Movement. *Traffic* **2018**, *19*, 761–769. [CrossRef]
154. Kesisova, I.A.; Robinson, B.P.; Spiliotis, E.T. A Septin GTPase Scaffold of Dynein–Dynactin Motors Triggers Retrograde Lysosome Transport. *J. Cell Biol.* **2021**, *220*, e202005219. [CrossRef] [PubMed]
155. Korolchuk, V.I.; Saiki, S.; Lichtenberg, M.; Siddiqi, F.H.; Roberts, E.A.; Imarisio, S.; Jahreiss, L.; Sarkar, S.; Futter, M.; Menzies, F.M.; et al. Lysosomal Positioning Coordinates Cellular Nutrient Responses. *Nat. Cell Biol.* **2011**, *13*, 453–460. [CrossRef]
156. Jordens, I.; Fernandez-Borja, M.; Marsman, M.; Dusseljee, S.; Janssen, L.; Calafat, J.; Janssen, H.; Wubbolts, R.; Neefjes, J. The Rab7 Effector Protein RILP Controls Lysosomal Transport by Inducing the Recruitment of Dynein-Dynactin Motors. *Curr. Biol.* **2001**, *11*, 1680–1685. [CrossRef]
157. Bucci, C.; Thomsen, P.; Nicoziani, P.; McCarthy, J.; van Deurs, B. Rab7: A Key to Lysosome Biogenesis. *Mol. Biol. Cell* **2000**, *11*, 467–480. [CrossRef] [PubMed]
158. Xing, R.; Zhou, H.; Jian, Y.; Li, L.; Wang, M.; Liu, N.; Yin, Q.; Liang, Z.; Guo, W.; Yang, C. The Rab7 Effector WDR91 Promotes Autophagy-Lysosome Degradation in Neurons by Regulating Lysosome Fusion. *J. Cell Biol.* **2021**, *220*, e202007061. [CrossRef] [PubMed]
159. Willett, R.; Martina, J.A.; Zewe, J.P.; Wills, R.; Hammond, G.R.V.; Puertollano, R. TFEB Regulates Lysosomal Positioning by Modulating TMEM55B Expression and JIP4 Recruitment to Lysosomes. *Nat. Commun.* **2017**, *8*, 1580. [CrossRef]
160. Settembre, C.; di Malta, C.; Polito, V.A.; Arencibia, M.G.; Vetrini, F.; Erdin, S.; Erdin, S.U.; Huynh, T.; Medina, D.; Colella, P.; et al. TFEB Links Autophagy to Lysosomal Biogenesis. *Science* **2011**, *332*, 1429–1433. [CrossRef]
161. Settembre, C.; Zoncu, R.; Medina, D.L.; Vetrini, F.; Erdin, S.; Erdin, S.; Huynh, T.; Ferron, M.; Karsenty, G.; Vellard, M.C.; et al. A Lysosome-to-Nucleus Signalling Mechanism Senses and Regulates the Lysosome via MTOR and TFEB. *EMBO J.* **2012**, *31*, 1095–1108. [CrossRef]
162. Wright, S.C.; Lukasheva, V.; le Gouill, C.; Kobayashi, H.; Breton, B.; Mailhot-Larouche, S.; Blondel-Tepaz, É.; Antunes Vieira, N.; Costa-Neto, C.; Héroux, M.; et al. BRET-Based Effector Membrane Translocation Assay Monitors GPCR-Promoted and Endocytosis-Mediated G_q Activation at Early Endosomes. *Proc. Natl. Acad. Sci. USA* **2021**, *118*, e2025846118. [CrossRef] [PubMed]
163. Filomeni, G.; de Zio, D.; Cecconi, F. Oxidative Stress and Autophagy: The Clash between Damage and Metabolic Needs. *Cell Death Differ.* **2015**, *22*, 377–388. [CrossRef] [PubMed]
164. Pizzino, G.; Irrera, N.; Cucinotta, M.; Pallio, G.; Mannino, F.; Arcoraci, V.; Squadrito, F.; Altavilla, D.; Bitto, A. Oxidative Stress: Harms and Benefits for Human Health. *Oxidative Med. Cell. Longev.* **2017**, *2017*, 8416763. [CrossRef] [PubMed]
165. Chang, K.-C.; Liu, P.-F.; Chang, C.-H.; Lin, Y.-C.; Chen, Y.-J.; Shu, C.-W. The Interplay of Autophagy and Oxidative Stress in the Pathogenesis and Therapy of Retinal Degenerative Diseases. *Cell Biosci.* **2022**, *12*, 1. [CrossRef] [PubMed]
166. Filomeni, G.; Desideri, E.; Cardaci, S.; Rotilio, G.; Ciriolo, M.R. Under the ROS: Thiol Network Is the Principal Suspect for Autophagy Commitment. *Autophagy* **2010**, *6*, 999–1005. [CrossRef] [PubMed]
167. Fang, C.; Gu, L.; Smerin, D.; Mao, S.; Xiong, X. The Interrelation between Reactive Oxygen Species and Autophagy in Neurological Disorders. *Oxidative Med. Cell. Longev.* **2017**, *2017*, 8495160. [CrossRef] [PubMed]
168. Murphy, M.P. How Mitochondria Produce Reactive Oxygen Species. *Biochem. J.* **2009**, *417*, 1–13. [CrossRef] [PubMed]
169. Chen, Q.; Vazquez, E.J.; Moghaddas, S.; Hoppel, C.L.; Lesnefsky, E.J. Production of Reactive Oxygen Species by Mitochondria. *J. Biol. Chem.* **2003**, *278*, 36027–36031. [CrossRef] [PubMed]
170. Scherz-Shouval, R.; Shvets, E.; Fass, E.; Shorer, H.; Gil, L.; Elazar, Z. Reactive Oxygen Species Are Essential for Autophagy and Specifically Regulate the Activity of Atg4. *EMBO J.* **2007**, *26*, 1749–1760. [CrossRef] [PubMed]
171. Inoki, K.; Zhu, T.; Guan, K.-L. TSC2 Mediates Cellular Energy Response to Control Cell Growth and Survival. *Cell* **2003**, *115*, 577–590. [CrossRef]
172. Gwinn, D.M.; Shackelford, D.B.; Egan, D.F.; Mihaylova, M.M.; Mery, A.; Vasquez, D.S.; Turk, B.E.; Shaw, R.J. AMPK Phosphorylation of Raptor Mediates a Metabolic Checkpoint. *Mol. Cell* **2008**, *30*, 214–226. [CrossRef]
173. Cao, J.; Schulte, J.; Knight, A.; Leslie, N.R.; Zagozdzon, A.; Bronson, R.; Manevich, Y.; Beeson, C.; Neumann, C.A. Prdx1 Inhibits Tumorigenesis via Regulating PTEN/AKT Activity. *EMBO J.* **2009**, *28*, 1505–1517. [CrossRef]
174. Mahalingaiah, P.K.S.; Singh, K.P. Chronic Oxidative Stress Increases Growth and Tumorigenic Potential of MCF-7 Breast Cancer Cells. *PLoS ONE* **2014**, *9*, e87371. [CrossRef] [PubMed]
175. Su, Q.; Zheng, B.; Wang, C.; Yang, Y.; Luo, W.; Ma, S.; Zhang, X.; Ma, D.; Sun, Y.; Yang, Z.; et al. Oxidative Stress Induces Neuronal Apoptosis through Suppressing Transcription Factor EB Phosphorylation at Ser467. *Cell. Physiol. Biochem.* **2018**, *46*, 1536–1554. [CrossRef]
176. Komatsu, M.; Kurokawa, H.; Waguri, S.; Taguchi, K.; Kobayashi, A.; Ichimura, Y.; Sou, Y.-S.; Ueno, I.; Sakamoto, A.; Tong, K.I.; et al. The Selective Autophagy Substrate P62 Activates the Stress Responsive Transcription Factor Nrf2 through Inactivation of Keap1. *Nat. Cell Biol.* **2010**, *12*, 213–223. [CrossRef] [PubMed]
177. Pickles, S.; Vigié, P.; Youle, R.J. Mitophagy and Quality Control Mechanisms in Mitochondrial Maintenance. *Curr. Biol.* **2018**, *28*, R170–R185. [CrossRef] [PubMed]

178. Schofield, J.H.; Schafer, Z.T. Mitochondrial Reactive Oxygen Species and Mitophagy: A Complex and Nuanced Relationship. *Antioxid. Redox Signal.* **2021**, *34*, 517–530. [CrossRef] [PubMed]
179. Harper, J.W.; Ordureau, A.; Heo, J.-M. Building and Decoding Ubiquitin Chains for Mitophagy. *Nat. Rev. Mol. Cell Biol.* **2018**, *19*, 93–108. [CrossRef]
180. Yamano, K.; Fogel, A.I.; Wang, C.; van der Bliek, A.M.; Youle, R.J. Mitochondrial Rab GAPs Govern Autophagosome Biogenesis during Mitophagy. *eLife* **2014**, *3*, e01612. [CrossRef] [PubMed]
181. Xiao, B.; Goh, J.-Y.; Xiao, L.; Xian, H.; Lim, K.-L.; Liou, Y.-C. Reactive Oxygen Species Trigger Parkin/PINK1 Pathway–Dependent Mitophagy by Inducing Mitochondrial Recruitment of Parkin. *J. Biol. Chem.* **2017**, *292*, 16697–16708. [CrossRef]
182. Xiao, B.; Deng, X.; Lim, G.G.Y.; Xie, S.; Zhou, Z.D.; Lim, K.-L.; Tan, E.-K. Superoxide Drives Progression of Parkin/PINK1-Dependent Mitophagy Following Translocation of Parkin to Mitochondria. *Cell Death Dis.* **2017**, *8*, e3097. [CrossRef]
183. Wang, Y.; Shen, J.; Chen, Y.; Liu, H.; Zhou, H.; Bai, Z.; Hu, Z.; Guo, X. PINK1 Protects against Oxidative Stress Induced Senescence of Human Nucleus Pulposus Cells via Regulating Mitophagy. *Biochem. Biophys. Res. Commun.* **2018**, *504*, 406–414. [CrossRef]
184. Joselin, A.P.; Hewitt, S.J.; Callaghan, S.M.; Kim, R.H.; Chung, Y.-H.; Mak, T.W.; Shen, J.; Slack, R.S.; Park, D.S. ROS-Dependent Regulation of Parkin and DJ-1 Localization during Oxidative Stress in Neurons. *Hum. Mol. Genet.* **2012**, *21*, 4888–4903. [CrossRef] [PubMed]
185. Liu, L.; Feng, D.; Chen, G.; Chen, M.; Zheng, Q.; Song, P.; Ma, Q.; Zhu, C.; Wang, R.; Qi, W.; et al. Mitochondrial Outer-Membrane Protein FUNDC1 Mediates Hypoxia-Induced Mitophagy in Mammalian Cells. *Nat. Cell Biol.* **2012**, *14*, 177–185. [CrossRef]
186. Bartolomé, A.; García-Aguilar, A.; Asahara, S.-I.; Kido, Y.; Guillén, C.; Pajvani, U.B.; Benito, M. MTORC1 Regulates Both General Autophagy and Mitophagy Induction after Oxidative Phosphorylation Uncoupling. *Mol. Cell. Biol.* **2017**, *37*, e00441-17. [CrossRef] [PubMed]
187. Benincá, C.; Planagumà, J.; de Freitas Shuck, A.; Acín-Perez, R.; Muñoz, J.P.; de Almeida, M.M.; Brown, J.H.; Murphy, A.N.; Zorzano, A.; Enríquez, J.A.; et al. A New Non-Canonical Pathway of Gαq Protein Regulating Mitochondrial Dynamics and Bioenergetics. *Cell. Signal.* **2014**, *26*, 1135–1146. [CrossRef] [PubMed]
188. Lancel, S.; Qin, F.; Lennon, S.L.; Zhang, J.; Tong, X.; Mazzini, M.J.; Kang, Y.J.; Siwik, D.A.; Cohen, R.A.; Colucci, W.S. Short Communication: Oxidative Posttranslational Modifications Mediate Decreased SERCA Activity and Myocyte Dysfunction in Gαq-Overexpressing Mice. *Circ. Res.* **2010**, *107*, 228–232. [CrossRef] [PubMed]
189. Westenbrink, B.D.; Ling, H.; Divakaruni, A.S.; Gray, C.B.B.; Zambon, A.C.; Dalton, N.D.; Peterson, K.L.; Gu, Y.; Matkovich, S.J.; Murphy, A.N.; et al. Mitochondrial Reprogramming Induced by CaMKIIδ Mediates Hypertrophy Decompensation. *Circ. Res.* **2015**, *116*, e28–e39. [CrossRef]
190. Jhun, B.S.; O-Uchi, J.; Adaniya, S.M.; Mancini, T.J.; Cao, J.L.; King, M.E.; Landi, A.K.; Ma, H.; Shin, M.; Yang, D.; et al. Protein Kinase D Activation Induces Mitochondrial Fragmentation and Dysfunction in Cardiomyocytes. *J. Physiol.* **2018**, *596*, 827–855. [CrossRef] [PubMed]
191. Lin, C.-C.; Yang, C.-C.; Cho, R.-L.; Wang, C.-Y.; Hsiao, L.-D.; Yang, C.-M. Sphingosine 1-Phosphate-Induced ICAM-1 Expression via NADPH Oxidase/ROS-Dependent NF-KB Cascade on Human Pulmonary Alveolar Epithelial Cells. *Front. Pharmacol.* **2016**, *7*, 80. [CrossRef]
192. Kenific, C.M.; Wittmann, T.; Debnath, J. Autophagy in Adhesion and Migration. *J. Cell Sci.* **2016**, *129*, 3685–3693. [CrossRef] [PubMed]
193. King, J.S. Mechanical Stress Meets Autophagy: Potential Implications for Physiology and Pathology. *Trends Mol. Med.* **2012**, *18*, 583–588. [CrossRef]
194. Klionsky, D.J.; Petroni, G.; Amaravadi, R.K.; Baehrecke, E.H.; Ballabio, A.; Boya, P.; Bravo-San Pedro, J.M.; Cadwell, K.; Cecconi, F.; Choi, A.M.K.; et al. Autophagy in Major Human Diseases. *EMBO J.* **2021**, *40*, e108863. [CrossRef] [PubMed]
195. Ichimura, Y.; Komatsu, M. Pathophysiological Role of Autophagy: Lesson from Autophagy-Deficient Mouse Models. *Exp. Anim.* **2011**, *60*, 329–345. [CrossRef] [PubMed]
196. Kitada, M.; Koya, D. Autophagy in Metabolic Disease and Ageing. *Nat. Rev. Endocrinol.* **2021**, *17*, 647–661. [CrossRef]
197. Rabinowitz, J.D.; White, E. Autophagy and Metabolism. *Science* **2010**, *330*, 1344–1348. [CrossRef]
198. Van Beek, N.; Klionsky, D.J.; Reggiori, F. Genetic Aberrations in Macroautophagy Genes Leading to Diseases. *Biochim. Et Biophys. Acta (BBA)—Mol. Cell Res.* **2018**, *1865*, 803–816. [CrossRef] [PubMed]
199. Zhang, X.; Wu, D.; Wang, C.; Luo, Y.; Ding, X.; Yang, X.; Silva, F.; Arenas, S.; Weaver, J.M.; Mandell, M.; et al. Sustained Activation of Autophagy Suppresses Adipocyte Maturation via a Lipolysis-Dependent Mechanism. *Autophagy* **2020**, *16*, 1668–1682. [CrossRef]
200. Codogno, P.; Meijer, A.J. Autophagy: A Potential Link between Obesity and Insulin Resistance. *Cell Metab.* **2010**, *11*, 449–451. [CrossRef]
201. Madrigal-Matute, J.; Cuervo, A.M. Regulation of Liver Metabolism by Autophagy. *Gastroenterology* **2016**, *150*, 328–339. [CrossRef]
202. Park, E.-Y.; Park, J.-B. High Glucose-Induced Oxidative Stress Promotes Autophagy through Mitochondrial Damage in Rat Notochordal Cells. *Int. Orthop.* **2013**, *37*, 2507–2514. [CrossRef]
203. Oliveira de Souza, C.; Sun, X.; Oh, D. Metabolic Functions of G Protein-Coupled Receptors and β-Arrestin-Mediated Signaling Pathways in the Pathophysiology of Type 2 Diabetes and Obesity. *Front. Endocrinol.* **2021**, *12*, 1005. [CrossRef] [PubMed]

204. Blad, C.C.; Tang, C.; Offermanns, S. G Protein-Coupled Receptors for Energy Metabolites as New Therapeutic Targets. *Nat. Rev. Drug Discov.* **2012**, *11*, 603–619. [CrossRef]
205. Offermanns, S. Free Fatty Acid (FFA) and Hydroxy Carboxylic Acid (HCA) Receptors. *Annu. Rev. Pharmacol. Toxicol.* **2014**, *54*, 407–434. [CrossRef] [PubMed]
206. Wauson, E.M.; Zaganjor, E.; Cobb, M.H. Amino Acid Regulation of Autophagy through the GPCR TAS1R1-TAS1R3. *Autophagy* **2013**, *9*, 418–419. [CrossRef] [PubMed]
207. Wauson, E.M.; Lorente-Rodríguez, A.; Cobb, M.H. Minireview: Nutrient Sensing by G Protein-Coupled Receptors. *Mol. Endocrinol.* **2013**, *27*, 1188–1197. [CrossRef]
208. Lizaso, A.; Tan, K.-T.; Lee, Y.-H. β-Adrenergic Receptor-Stimulated Lipolysis Requires the RAB7-Mediated Autolysosomal Lipid Degradation. *Autophagy* **2013**, *9*, 1228–1243. [CrossRef] [PubMed]
209. Jung, H.S.; Chung, K.W.; Won Kim, J.; Kim, J.; Komatsu, M.; Tanaka, K.; Nguyen, Y.H.; Kang, T.M.; Yoon, K.-H.; Kim, J.-W.; et al. Loss of Autophagy Diminishes Pancreatic β Cell Mass and Function with Resultant Hyperglycemia. *Cell Metab.* **2008**, *8*, 318–324. [CrossRef]
210. Moscat, J.; Diaz-Meco, M.T. Feedback on Fat: P62-MTORC1-Autophagy Connections. *Cell* **2011**, *147*, 724–727. [CrossRef] [PubMed]
211. Onken, M.D.; Noda, S.E.; Kaltenbronn, K.M.; Frankfater, C.; Makepeace, C.M.; Fettig, N.; Piggott, K.D.; Custer, P.L.; Ippolito, J.E.; Blumer, K.J. Oncogenic Gq/11 Signaling Acutely Drives and Chronically Sustains Metabolic Reprogramming in Uveal Melanoma. *J. Biol. Chem.* **2022**, *298*, 101495. [CrossRef]
212. Kimura, T.; Pydi, S.P.; Wang, L.; Haspula, D.; Cui, Y.; Lu, H.; König, G.M.; Kostenis, E.; Steinberg, G.R.; Gavrilova, O.; et al. Adipocyte Gq Signaling Is a Regulator of Glucose and Lipid Homeostasis in Mice. *Nat. Commun.* **2022**, *13*, 1652. [CrossRef] [PubMed]
213. Romani, P.; Valcarcel-Jimenez, L.; Frezza, C.; Dupont, S. Crosstalk between Mechanotransduction and Metabolism. *Nat. Rev. Mol. Cell Biol.* **2021**, *22*, 22–38. [CrossRef]
214. Kim, S.W.; Ehrman, J.; Ahn, M.-R.; Kondo, J.; Lopez, A.A.M.; Oh, Y.S.; Kim, X.H.; Crawley, S.W.; Goldenring, J.R.; Tyska, M.J.; et al. Shear Stress Induces Noncanonical Autophagy in Intestinal Epithelial Monolayers. *Mol. Biol. Cell* **2017**, *28*, 3043–3056. [CrossRef] [PubMed]
215. Miceli, C.; Roccio, F.; Penalva-Mousset, L.; Burtin, M.; Leroy, C.; Nemazanyy, I.; Kuperwasser, N.; Pontoglio, M.; Friedlander, G.; Morel, E.; et al. The Primary Cilium and Lipophagy Translate Mechanical Forces to Direct Metabolic Adaptation of Kidney Epithelial Cells. *Nat. Cell Biol.* **2020**, *22*, 1091–1102. [CrossRef]
216. He, C.; Sumpter, R., Jr.; Levine, B. Exercise Induces Autophagy in Peripheral Tissues and in the Brain. *Autophagy* **2012**, *8*, 1548–1551. [CrossRef] [PubMed]
217. Claude-Taupin, A.; Codogno, P.; Dupont, N. Links between Autophagy and Tissue Mechanics. *J. Cell Sci.* **2021**, *134*, jcs258589. [CrossRef]
218. Mameli, E.; Martello, A.; Caporali, A. Autophagy at the Interface of Endothelial Cell Homeostasis and Vascular Disease. *FEBS J.* **2021**, *289*, 2976–2991. [CrossRef]
219. Madrigal-Matute, J.; de Bruijn, J.; van Kuijk, K.; Riascos-Bernal, D.F.; Diaz, A.; Tasset, I.; Martín-Segura, A.; Gijbels, M.J.J.; Sander, B.; Kaushik, S.; et al. Protective Role of Chaperone-Mediated Autophagy against Atherosclerosis. *Proc. Natl. Acad. Sci. USA* **2022**, *119*, e2121133119. [CrossRef]
220. Doronzo, G.; Astanina, E.; Corà, D.; Chiabotto, G.; Comunanza, V.; Noghero, A.; Neri, F.; Puliafito, A.; Primo, L.; Spampanato, C.; et al. TFEB Controls Vascular Development by Regulating the Proliferation of Endothelial Cells. *EMBO J.* **2019**, *38*, e98250. [CrossRef]
221. Shadab, M.; Millar, M.W.; Slavin, S.A.; Leonard, A.; Fazal, F.; Rahman, A. Autophagy Protein ATG7 Is a Critical Regulator of Endothelial Cell Inflammation and Permeability. *Sci. Rep.* **2020**, *10*, 13708. [CrossRef]
222. Reglero-Real, N.; Pérez-Gutiérrez, L.; Yoshimura, A.; Rolas, L.; Garrido-Mesa, J.; Barkaway, A.; Pickworth, C.; Saleeb, R.S.; Gonzalez-Nuñez, M.; Austin-Williams, S.N.; et al. Autophagy Modulates Endothelial Junctions to Restrain Neutrophil Diapedesis during Inflammation. *Immunity* **2021**, *54*, 1989–2004.e9. [CrossRef] [PubMed]
223. Deretic, V.; Saitoh, T.; Akira, S. Autophagy in Infection, Inflammation and Immunity. *Nat. Rev. Immunol.* **2013**, *13*, 722–737. [CrossRef] [PubMed]
224. Bharath, L.P.; Cho, J.M.; Park, S.-K.; Ruan, T.; Li, Y.; Mueller, R.; Bean, T.; Reese, V.; Richardson, R.S.; Cai, J.; et al. Endothelial Cell Autophagy Maintains Shear Stress–Induced Nitric Oxide Generation via Glycolysis-Dependent Purinergic Signaling to Endothelial Nitric Oxide Synthase. *Arterioscler. Thromb. Vasc. Biol.* **2017**, *37*, 1646–1656. [CrossRef] [PubMed]
225. Bharath, L.P.; Mueller, R.; Li, Y.; Ruan, T.; Kunz, D.; Goodrich, R.; Mills, T.; Deeter, L.; Sargsyan, A.; Anandh Babu, P.V.; et al. Impairment of Autophagy in Endothelial Cells Prevents Shear-Stress-Induced Increases in Nitric Oxide Bioavailability. *Can. J. Physiol. Pharmacol.* **2014**, *92*, 605–612. [CrossRef] [PubMed]
226. Zhang, J.-X.; Qu, X.-L.; Chu, P.; Xie, D.-J.; Zhu, L.-L.; Chao, Y.-L.; Li, L.; Zhang, J.-J.; Chen, S.-L. Low Shear Stress Induces Vascular ENOS Uncoupling via Autophagy-Mediated ENOS Phosphorylation. *Biochim. Et Biophys. Acta (BBA)—Mol. Cell Res.* **2018**, *1865*, 709–720. [CrossRef] [PubMed]

227. Dela Paz, N.G.; Melchior, B.; Frangos, J.A. Shear Stress Induces G$\alpha_{q/11}$ Activation Independently of G Protein-Coupled Receptor Activation in Endothelial Cells. *Am. J. Physiol. -Cell Physiol.* **2017**, *312*, C428–C437. [CrossRef] [PubMed]
228. Albarrán-Juárez, J.; Iring, A.; Wang, S.; Joseph, S.; Grimm, M.; Strilic, B.; Wettschureck, N.; Althoff, T.F.; Offermanns, S. Piezo1 and Gq/G11 Promote Endothelial Inflammation Depending on Flow Pattern and Integrin Activation. *J. Exp. Med.* **2018**, *215*, 2655–2672. [CrossRef]
229. Sciarretta, S.; Maejima, Y.; Zablocki, D.; Sadoshima, J. The Role of Autophagy in the Heart. *Annu. Rev. Physiol.* **2018**, *80*, 1–26. [CrossRef] [PubMed]
230. Nakamura, M.; Sadoshima, J. Mechanisms of Physiological and Pathological Cardiac Hypertrophy. *Nat. Rev. Cardiol.* **2018**, *15*, 387–407. [CrossRef]
231. Kanamori, H.; Takemura, G.; Goto, K.; Maruyama, R.; Tsujimoto, A.; Ogino, A.; Takeyama, T.; Kawaguchi, T.; Watanabe, T.; Fujiwara, T.; et al. The Role of Autophagy Emerging in Postinfarction Cardiac Remodelling. *Cardiovasc. Res.* **2011**, *91*, 330–339. [CrossRef]
232. Matsui, Y.; Takagi, H.; Qu, X.; Abdellatif, M.; Sakoda, H.; Asano, T.; Levine, B.; Sadoshima, J. Distinct Roles of Autophagy in the Heart During Ischemia and Reperfusion. *Circ. Res.* **2007**, *100*, 914–922. [CrossRef]
233. Yan, L.; Vatner, D.E.; Kim, S.-J.; Ge, H.; Masurekar, M.; Massover, W.H.; Yang, G.; Matsui, Y.; Sadoshima, J.; Vatner, S.F. Autophagy in Chronically Ischemic Myocardium. *Proc. Natl. Acad. Sci. USA* **2005**, *102*, 13807–13812. [CrossRef]
234. García-Hoz, C.; Sánchez-Fernández, G.; García-Escudero, R.; Fernández-Velasco, M.; Palacios-García, J.; Ruiz-Meana, M.; Díaz-Meco, M.T.; Leitges, M.; Moscat, J.; García-Dorado, D.; et al. Protein Kinase C (PKC) ζ-Mediated Gαq Stimulation of ERK5 Protein Pathway in Cardiomyocytes and Cardiac Fibroblasts. *J. Biol. Chem.* **2012**, *287*, 7792–7802. [CrossRef]
235. Adams, J.W.; Pagel, A.L.; Means, C.K.; Oksenberg, D.; Armstrong, R.C.; Brown, J.H. Cardiomyocyte Apoptosis Induced by Gαq Signaling Is Mediated by Permeability Transition Pore Formation and Activation of the Mitochondrial Death Pathway. *Circ. Res.* **2000**, *87*, 1180–1187. [CrossRef] [PubMed]
236. Meleka, M.M.; Edwards, A.J.; Xia, J.; Dahlen, S.A.; Mohanty, I.; Medcalf, M.; Aggarwal, S.; Moeller, K.D.; Mortensen, O.V.; Osei-Owusu, P. Anti-Hypertensive Mechanisms of Cyclic Depsipeptide Inhibitor Ligands for Gq/11 Class G Proteins. *Pharmacol. Res.* **2019**, *141*, 264–275. [CrossRef]
237. Yun, C.W.; Jeon, J.; Go, G.; Lee, J.H.; Lee, S.H. The Dual Role of Autophagy in Cancer Development and a Therapeutic Strategy for Cancer by Targeting Autophagy. *Int. J. Mol. Sci.* **2020**, *22*, 179. [CrossRef]
238. Chude, C.; Amaravadi, R. Targeting Autophagy in Cancer: Update on Clinical Trials and Novel Inhibitors. *Int. J. Mol. Sci.* **2017**, *18*, 1279. [CrossRef] [PubMed]
239. Qu, X.; Yu, J.; Bhagat, G.; Furuya, N.; Hibshoosh, H.; Troxel, A.; Rosen, J.; Eskelinen, E.-L.; Mizushima, N.; Ohsumi, Y.; et al. Promotion of Tumorigenesis by Heterozygous Disruption of the Beclin 1 Autophagy Gene. *J. Clin. Investig.* **2003**, *112*, 1809–1820. [CrossRef] [PubMed]
240. Shen, Y.; Li, D.-D.; Wang, L.-L.; Deng, R.; Zhu, X.-F. Decreased Expression of Autophagy-Related Proteins in Malignant Epithelial Ovarian Cancer. *Autophagy* **2008**, *4*, 1067–1068. [CrossRef] [PubMed]
241. Peng, Y.-F.; Shi, Y.-H.; Ding, Z.-B.; Ke, A.-W.; Gu, C.-Y.; Hui, B.; Zhou, J.; Qiu, S.-J.; Dai, Z.; Fan, J. Autophagy Inhibition Suppresses Pulmonary Metastasis of HCC in Mice via Impairing Anoikis Resistance and Colonization of HCC Cells. *Autophagy* **2013**, *9*, 2056–2068. [CrossRef] [PubMed]
242. Chmurska, A.; Matczak, K.; Marczak, A. Two Faces of Autophagy in the Struggle against Cancer. *Int. J. Mol. Sci.* **2021**, *22*, 2981. [CrossRef] [PubMed]
243. Tsai, M.-J.; Chang, W.-A.; Huang, M.-S.; Kuo, P.-L. Tumor Microenvironment: A New Treatment Target for Cancer. *ISRN Biochem.* **2014**, *2014*, 351959. [CrossRef]
244. Das, J.; Maji, S.; Agarwal, T.; Chakraborty, S.; Maiti, T.K. Hemodynamic Shear Stress Induces Protective Autophagy in HeLa Cells through Lipid Raft-Mediated Mechanotransduction. *Clin. Exp. Metastasis* **2018**, *35*, 135–148. [CrossRef]
245. Lien, S.-C.; Chang, S.-F.; Lee, P.-L.; Wei, S.-Y.; Chang, M.D.-T.; Chang, J.-Y.; Chiu, J.-J. Mechanical Regulation of Cancer Cell Apoptosis and Autophagy: Roles of Bone Morphogenetic Protein Receptor, Smad1/5, and P38 MAPK. *Biochim. Et Biophys. Acta (BBA)—Mol. Cell Res.* **2013**, *1833*, 3124–3133. [CrossRef] [PubMed]
246. Wang, K.; Wei, Y.; Liu, W.; Liu, L.; Guo, Z.; Fan, C.; Wang, L.; Hu, J.; Li, B. Mechanical Stress-Dependent Autophagy Component Release via Extracellular Nanovesicles in Tumor Cells. *ACS Nano* **2019**, *13*, 4589–4602. [CrossRef]
247. Yan, Z.; Su, G.; Gao, W.; He, J.; Shen, Y.; Zeng, Y.; Liu, X. Fluid Shear Stress Induces Cell Migration and Invasion via Activating Autophagy in HepG2 Cells. *Cell Adhes. Migr.* **2019**, *13*, 152–163. [CrossRef] [PubMed]
248. Follain, G.; Herrmann, D.; Harlepp, S.; Hyenne, V.; Osmani, N.; Warren, S.C.; Timpson, P.; Goetz, J.G. Fluids and Their Mechanics in Tumour Transit: Shaping Metastasis. *Nat. Rev. Cancer* **2020**, *20*, 107–124. [CrossRef] [PubMed]
249. Swartz, M.A.; Lund, A.W. Lymphatic and Interstitial Flow in the Tumour Microenvironment: Linking Mechanobiology with Immunity. *Nat. Rev. Cancer* **2012**, *12*, 210–219. [CrossRef] [PubMed]
250. Kalluri, R.; Zeisberg, M. Fibroblasts in Cancer. *Nat. Rev. Cancer* **2006**, *6*, 392–401. [CrossRef] [PubMed]
251. Goruppi, S.; Clocchiatti, A.; Dotto, G.P. A Role for Stromal Autophagy in Cancer-Associated Fibroblast Activation. *Autophagy* **2019**, *15*, 738–739. [CrossRef] [PubMed]

252. Kang, J.I.; Kim, D.H.; Sung, K.W.; Shim, S.M.; Cha-Molstad, H.; Soung, N.K.; Lee, K.H.; Hwang, J.; Lee, H.G.; Kwon, Y.T.; et al. P62-Induced Cancer-Associated Fibroblast Activation via the Nrf2-ATF6 Pathway Promotes Lung Tumorigenesis. *Cancers* **2021**, *13*, 864. [CrossRef] [PubMed]
253. Linares, J.F.; Cid-Diaz, T.; Duran, A.; Osrodek, M.; Martinez-Ordoñez, A.; Reina-Campos, M.; Kuo, H.-H.; Elemento, O.; Martin, M.L.; Cordes, T.; et al. The Lactate-NAD+ Axis Activates Cancer-Associated Fibroblasts by Downregulating P62. *Cell Rep.* **2022**, *39*, 110792. [CrossRef] [PubMed]

antioxidants

Review

Mitophagy and Oxidative Stress: The Role of Aging

Anna De Gaetano [1], Lara Gibellini [2], Giada Zanini [1], Milena Nasi [3], Andrea Cossarizza [2] and Marcello Pinti [1,*]

1 Department of Life Sciences, University of Modena and Reggio Emilia, 41125 Modena, Italy; anna.degaetano@unimore.it (A.D.G.); giada.zanini@unimore.it (G.Z.)
2 Department of Medical and Surgical Sciences of Children and Adults, University of Modena and Reggio Emilia, 41125 Modena, Italy; lara.gibellini@unimore.it (L.G.); andrea.cossarizza@unimore.it (A.C.)
3 Department of Surgery, Medicine, Dentistry and Morphological Sciences, University of Modena and Reggio Emilia, 41125 Modena, Italy; milena.nasi@unimore.it
* Correspondence: marcello.pinti@unimore.it; Tel.: +39-059-205-5386; Fax: +39-059-205-5426

Abstract: Mitochondrial dysfunction is a hallmark of aging. Dysfunctional mitochondria are recognized and degraded by a selective type of macroautophagy, named mitophagy. One of the main factors contributing to aging is oxidative stress, and one of the early responses to excessive reactive oxygen species (ROS) production is the induction of mitophagy to remove damaged mitochondria. However, mitochondrial damage caused at least in part by chronic oxidative stress can accumulate, and autophagic and mitophagic pathways can become overwhelmed. The imbalance of the delicate equilibrium among mitophagy, ROS production and mitochondrial damage can start, drive, or accelerate the aging process, either in physiological aging, or in pathological age-related conditions, such as Alzheimer's and Parkinson's diseases. It remains to be determined which is the prime mover of this imbalance, i.e., whether it is the mitochondrial damage caused by ROS that initiates the dysregulation of mitophagy, thus activating a vicious circle that leads to the reduced ability to remove damaged mitochondria, or an alteration in the regulation of mitophagy leading to the excessive production of ROS by damaged mitochondria.

Keywords: mitophagy; aging; Reactive Oxygen Species; PINK1; mitochondria; Alzheimer; Parkinson

Citation: De Gaetano, A.; Gibellini, L.; Zanini, G.; Nasi, M.; Cossarizza, A.; Pinti, M. Mitophagy and Oxidative Stress: The Role of Aging. *Antioxidants* **2021**, *10*, 794. https://doi.org/10.3390/antiox10050794

Academic Editors: Cecilia Gotor and Angeles Aroca

Received: 14 April 2021
Accepted: 15 May 2021
Published: 17 May 2021

1. Introduction

Mitochondrial (mt) dysfunction is considered a hallmark of aging [1]. Dysfunctional mitochondria are recognized and degraded either by non-selective autophagy or by a selective type of macroautophagy, named mitophagy [2]. This catabolic process allows the degradation of dysfunctional and damaged mitochondria [3,4], with the aim of recycling mitochondrial contents and macromolecules, such as amino acids and preserving ATP production [3,5–8]. Mitophagy is evolutionarily conserved and has been observed from yeast to mammals. Mitophagy process starts when dysfunctional mitochondria are targeted with specific receptors or adaptors and are engulfed in a double-membrane vacuole named autophagosome. Then, this vesicle fuses with a lysosome, forming an autolysosome in which specific enzymes degrade the organelle. Based on the ability of receptor to recruit ubiquitin, the mitophagy regulatory pathways could be classified as ubiquitin-dependent or ubiquitin-independent (receptor-dependent) [3,9]. This process is triggered by multiple stimuli and can be activated on the basis of cell requirement. Depending on the physiological condition of the cell, mitophagy can be classified in steady-state or basal mitophagy, programmed mitophagy, and stress-induced mitophagy [5,10]. The functions of basal mitophagy are not well understood and the process level differs between cells and tissues. However, it is likely that, in physiological conditions, mitophagy is required for mitochondrial turnover, cellular homeostasis, and metabolic demand [3,10–13]. Programmed mitophagy is necessary for development and differentiation processes, such as maturation of erythrocytes and cardiomyocytes and for allophagy [5,10,14–16]. Finally, stress-induced mitophagy is activated by stimuli, such as oxidative stress, starvation, hypoxia, and loss of the mitochondrial membrane potential (MMP), with the aim of reducing mitochondrial amount and, in turn, decreasing the production of reactive oxygen species (ROS) and oxygen consumption by damaged mitochondria [17].

Several lines of evidence indicate a close correlation between the increase in ROS observed with age and the modulation of age-dependent mitophagy. On the one hand, the study of mitophagy on different model species, from *S. cerevisiae* to *C. elegans*, clearly indicates that the aging process is related to an impairment of the regulation of mitophagy, and that targeting of genes that regulate mitophagy can modulate lifespan. On the other hand, several studies on cellular and in vivo models have shown a close correlation between ROS production, mitochondrial stress, and activation of mitophagy. Since oxidative stress is one of the main drivers of aging, and many diseases characterized by premature aging are characterized by an excess of ROS production or a defect in the scavenging processes of free radicals, it is likely that age-related increase of ROS can play a role in the impairment of mitophagy observed with aging.

In this review, we discuss the main findings linking mitophagy, oxidative stress, and aging, both in physiological aging, and in age-related diseases such as Alzheimer's disease (AD) and Parkinson's disease (PD).

2. Mitophagy and Its Regulation

Mitophagy is an extremely complex and finely regulated process; the detailed description of the molecular mechanisms underpinning mitophagy goes far beyond the purpose of this review For this reason, we will summarize only the pathways that are—or could be—relevant for the aging process (Figure 1).

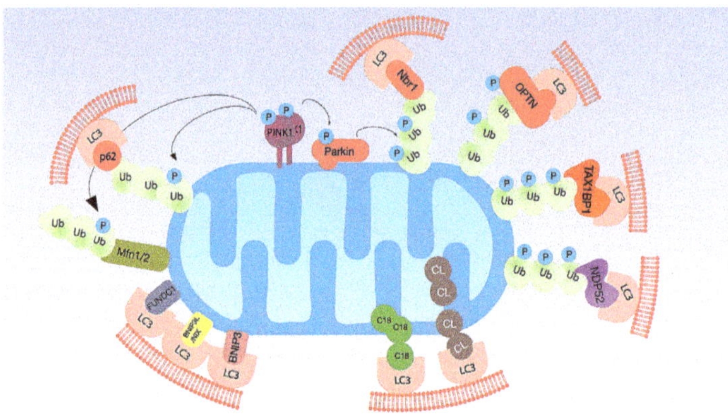

Figure 1. Main molecular mechanisms of mitophagy. Mitophagy is regulated by the interaction of mitochondrial proteins with LC3 through different mechanisms. In the PINK1/Parkin pathway, decreased MMP leads to the accumulation of PINK1 to the OMM. PINK1 phosphorylates both ubiquitin and Parkin. Activated Parkin polyubiquitinates specific proteins on the OMM, making available ubiquitins for PINK1 phosphorylation. The ubiquitinated proteins on the OMM allow the interaction of mitochondria with LC3 through specific adaptors, such as p62, Nbr1, OPTN, TAX1BP1, and NDP52. PINK1 can also phosphorylate Mfn2 and promote its ubiquitination by Parkin and rapid degradation, to prevent fusion of damaged mitochondria with healthy organelles. Besides PINK1/Parkin pathway, mitophagy is triggered by the mitochondrial receptors BNIP3, BNIP3L/NIX, FUNDC1, which can bind directly to LC3. Finally, mitochondrial lipids cardiolipin (CL) and C18-ceramide (C18) can move from the mitochondrial cristae to the OMM, where they interact with LC3.

2.1. Ubiquitin-Dependent Mitophagy

The phosphatase and tensin homologue (PTEN)-induced putative kinase 1 (PINK1)–Parkin pathway represents one of the most studied ubiquitin-dependent mechanism of mitophagy. In functional mitochondria, the serine/threonine kinase PINK1 is continually imported to mitochondria through the translocases of the outer and the inner membrane (TOM and TIM complexes) taking advantage of the mitochondrial targeting sequence (MTS). During the import, PINK1 is clipped first by matrix processing peptidases (MPP) and subsequently by a protease of the mitochondrial inner membrane (IMM), the presenilin-associated rhomboid like (PARL). Cleaved PINK1 moves to the cytosol, where is rapidly degraded [2,18,19]. In damaged or aged mitochondria, the decrease of the MMP locks the import of PINK1 in the mitochondrial matrix and its degradation, stabilizing it on the mitochondrial outer membrane (OMM) in a complex with the translocase TOM [2,7,19]. Stabilized PINK1 is auto-phosphorylated and: (i) phosphorylates ubiquitin and poly-ubiquitin, connected in a basal manner to proteins on the OMM, at Ser65, (ii) recruits and phosphorylates the cytosolic E3-ubiquitin ligase Parkin at Ser65. Phospho-ubiquitin, in turn, can recruit Parkin [3,10,20–24]. The active conformation of Parkin polyubiquitinates specific proteins on the OMM, making available ubiquitins for PINK1 phosphorylation, and triggering a feedback loop which leads to the recruitment of other molecules of Parkin on the mitochondrial surface finally activating mitophagy in a feedforward mechanism [24]. The ubiquitinated proteins of the OMM recruit five LC3 interacting region (LIR)-containing autophagy adapters: sequestosome-1 (p62/SQSTM1), Optineurin (OPTN), neighbor of BRCA1 gene 1 (NBR1), nuclear domain 10 protein 52 (NDP52), and TAX1 binding protein 1 (TAX1BP1). Through the LIR region, these ubiquitin-binding receptors recognize and bind LC3 driving mitochondria to mitophagy [2,25]. The PINK1/Parkin pathway is strictly interconnected with molecules and pathway regulating mitochondrial dynamics, such as Mitofusin (MFN) 1 and 2, GTPase of OMM involved in mitochondrial fusion MFN1 and MFN2 are highly susceptible to Parkin ubiquitination. MFN2 mediates the recruitment of Parkin to damaged mitochondria; PINK1 phosphorylates Mfn2 and promotes its ubiquitination by Parkin [26]. Thus, PINK1/Parkin activation causes their rapid ubiquitination and degradation, which prevents fusion of damaged mitochondria with healthy organelles [26].

In addition to Parkin, several other ubiquitin E3 ligases are able to ubiquitinate proteins on the mitochondrial surface: SMAD ubiquitination regulatory factor 1 (SMURF1), Glycoprotein 78 (Gp78), mitochondrial E3 ubiquitin protein ligase 1 (MUL1), HECT, UBA, and WWE domain-containing protein 1 (HUWE1), E3 ubiquitin-protein ligase SIAH1 (SIAH1), and Ariadne RBR E3 ubiquitin ligase homolog 1 (ARIH1) [5,9,10,27–36]. The ubiquitin chains generated recruit autophagy adaptors, such as p62, NDP52, and OPTN. Finally, other molecules such as unc-51-like autophagy activating kinase 1 (ULK1) mediates the biogenesis of the phagophore [9,25].

2.2. Ubiquitin-Independent Mitophagy (Receptor-Dependent)

An alternative mechanism to ubiquitin-dependent mitophagy takes advantage of protein receptors encompassing LIR motif and able to interact directly with LC3 or other autophagy-related proteins (ATGs) such as GABARAP in a ubiquitin-independent way.

In OMM, FUN14 domain containing 1 (FUNDC1), BCL2 interacting protein 3 (BNIP3), and NIP3-like protein X (NIX, also known as BNIP3-like, NIX/BNIP3L) are involved in the induction of mitophagy under stress condition, such as hypoxia [37–39]. The interaction of this group of protein with LC3/GABARAP is generally mediated by their phosphorylated/dephosphorylated status.

FUNDC1 contributes to mitophagy during hypoxia in mammalian cells by directly binding to LC3 [2,27]. In normoxic conditions, FUNDC1 is phosphorylated at Ser13 by casein kinase II (CK2) and at Tyr18 by SRC kinase, and this prevents its interaction with LC3 and, in turn, the activation of mitophagy pathways [27]. In hypoxic conditions, phosphoglycerate mutase 5 (PGAM5) dephosphorylates FUNDC1 at Ser13, while serine/threonine-protein kinase ULK1 phosphorylates it at Ser17. Dephosphorylation of

FUNDC1 mediated by PGAM5 improves its interaction with LC3 [28]. FUNDC1 is also involved in the intercross communication between mitochondrial fission and fusion and mitophagy. Specifically, during physiological conditions, FUNDC1 can bind the Dynamin-like 120 kDa protein (OPA1, GTPase involved in fusion and fission processes) to the inner surface of OMM taking advantage of Lys70. Conversely, under stress conditions, this interaction is reduced and FUNDC1 can also recruit Dynamin-1-like protein (DNM1L, GTPase involved in fission) from cytosol [29].

NIX and BNIP3 are proteins with homology to BCL2 in the BH3 domain. NIX (also called BNIP3L) can trigger mitophagy under both physiological and hypoxic conditions. During development, NIX plays an important role in the maturation process of erythrocytes, eliminating mitochondria [16,19]. BNIP3 and NIX mRNA and BNIP3 protein are induced by hypoxic environment in a wide range of human epithelial, endothelial, and macrophage cell lines but not in fibrosarcoma or lymphocyte cell lines [30]. Moreover, NIX and BNIP3 are induced at the transcriptional level also in CHO-K1 cells [31]. It is possible that BNIP3 is a direct target of hypoxia-inducible factor 1-alpha (HIF1α) [30,31]. In addition to its involvement in ubiquitin-independent mitophagy, NIX acts as a substrate of Parkin, recruiting NBR1 and finally triggering mitophagy [32].

Other mitophagy protein receptors of OMM are autophagy and beclin 1 regulator 1 (AMBRA1), FKBP prolyl isomerase 8 (FKBP8), and BCL2 like 13 (BCL2L13).

2.3. Mitophagy Triggered by Lipid Receptors

In addition to protein receptors, mitophagy can also be promoted by lipids such as Cardiolipin (CL), C18-ceramide, and SREBF1 [17]. CL triggers mitophagy moving from mitochondrial cristae (where is located in normal conditions) to OMM where interacts with LC3 taking advantages from LIR motif [17,33,34]. Interestingly, CL is particularly prone to peroxidative attack by ROS, and CL peroxidation has been shown to play a critical role in several physiopathological situations [35], including neurodegenerative diseases [36].

3. Effects of Oxidative Stress on Mitophagy

Mitochondria are the main source for cellular ROS [37]. NADH: ubiquinone oxidoreductase, or complex I (CI) is the main producer of superoxide anion ($O_2^{\bullet -}$) in the mitochondrial (mt) matrix [38]. The coenzyme Q:cytochrome c—oxidoreductase or complex III (CIII) is the main source of $O_2^{\bullet -}$ in the intermembrane space [39]. Mitochondria detoxifies the excess of $O_2^{\bullet -}$ by means of manganese superoxide dismutase (MnSOD or SOD2), which is located in the mt matrix, and of the copper-zinc SOD (CuZnSOD or SOD1), which is located in the intermembrane space. The product of SOD-mediated reactions is hydrogen peroxide (H_2O_2), which is less reactive and can diffuse across mitochondrial membranes. Because of its low reactivity, its relative specificity for cysteine residues, and its capability to diffuse across membranes, H_2O_2 can also act as a second messenger [40]. However, uncontrolled levels of H_2O_2 lead to hydroxyl radical ($OH^{\bullet}$) formation. For this reason, H_2O_2 levels are tightly regulated by robust detoxification systems. In the cytosol, it is converted to water primarily by catalase, while in the mitochondria it is detoxified by peroxiredoxin -3 (Prx3) and -5 (Prx5), by the Glutathione peroxidase 1 (GPx1) and Glutaredoxin 2 (Grx2) [41].

Oxidative stress occurs when the steady-state levels of ROS surpass their catabolism or detoxification, and can regulate mitophagy at multiple levels. Stress conditions stimulating ROS production, such as hypoxia, nutrient starvation, and ischemia/reperfusion (IR) can cause oxidative damage to mitochondria. One of the early responses to excessive ROS is to induce mitophagy, which can reduce oxidative damage and ROS production [42] and multiple lines of evidence have proved that ROS interacts with both ubiquitin- dependent and receptor-dependent mitophagy pathways. Generation of ROS within mitochondria using the mitochondrial-targeted photosensitizer mt KillerRed (mtKR) causes increased ROS levels in the mt matrix, the loss of MMP, and the activation of Parkin-dependent mitophagy. The overexpression of the mt antioxidant proteins, like mtSOD2, abolishes this effect [43]. ROS determines the recruitment of Peroxiredoxin-6 (PRDX6) to damaged mitochondria, where it controls ROS homeostasis in the initial step of PINK1-dependent mitophagy [44]. Although mtROS resulted not required for mitophagy [45] and not involved in mitochondrial translocation of Parkin, $O_2^{\bullet-}$ has been shown to be drive the progression of Parkin/PINK1-dependent mitophagy, once Parkin has translocated to mitochondria [46]. Mild oxidative stress selectively triggers mitophagy in the absence of autophagy, which is dependent on Dynamin-1-like protein (Drp1) [47].

Changes in redox balance of the cell can also affect mitophagy by modifying mitochondrial dynamics. When reduced glutathione (GSH) is oxidized to oxidized glutathione (GSSG), Mfn forms oligomers via disulfide bond formation and enhances membrane fusion [48]. Conversely, S-nitrosylation of Drp1 determines mitochondrial fragmentation [49]. The effect of oxidative stress and redox imbalance on PINK1/Parkin is less obvious and clear. The oxidation of Parkin has been reported to inhibit [50,51] or stimulate [52,53] the activity of E3-Ub ligase, depending on the model taken into account. It is likely that these opposite effects are due to different cysteine residues modified.

Another emerging link between oxidative stress, mitophagy, and aging is provided by the regulation of the ataxia-telangiectasia mutated (ATM)/Denitrosylase S-nitrosoglutathione reductase (GSNOR) axis. The downregulation of GSNOR during senescence and human aging promotes mitochondrial nitrosative stress, nitrosylation of Drp1 and Parkin, and impairment of mitochondrial dynamics and mitophagy [54]. GSNOR is also induced at the translational level in response to H_2O_2 and mt ROS [55].

Methionine is a sulfur-containing amino acid susceptible to reversible oxidation. It has been recently demonstrated that the mitochondrial matrix protein methionine sulfoxide reductase B2 (MsrB2) is a Parkin substrate, needed for mitophagy induction [56]. In condition of high oxidative stress, Parkin is oxidized at Met192, leading to protein inactivation and inhibition of mitophagy. MsrB2 released from damaged mitochondria reduces oxidized Met192, restoring Parkin function. Interestingly, Met192 is mutated in familial, early onset forms of PD [57], a typical age-related disease (see below), and levels of MsrB2 declines with age [58]. Thus, an imbalance in this axis due to the impaired capability to restore reduced Met192 could contribute to the age-related mitophagy dysregulation.

ROS can also modulate mitophagy at transcriptional level [59]. A crucial transcription factor for response to oxidative stress is the nuclear factor (erythroid-derived 2)-like 2 transcription factor (Nrf2). In unstressed conditions, Nrf2 is sequestered in the cytoplasm by the kelch-like ECH-associated protein 1 (Keap1)-Cul3 complex and targeted for ubiquitin-mediated degradation. Under oxidative stress conditions, Keap1 cysteines are oxidized, and Keap1 cannot ubiquitinate Nrf2, which is free to translocate to the nucleus and activate antioxidant response [60]. A tight interplay exists between Nrf2 and mitophagy upon oxidative stress. Nrf2 regulates PINK1 expression under oxidative stress conditions [61], and pharmacological inhibition of Keap1 triggers mitophagy [62]. Furthermore, autophagic degradation of Keap1 mediated by p62 activates Nrf2, which in turn increases the transcription of p62 gene [63,64]. In a model of liver damage, p62 can also promote mitochondrial ubiquitination in Parkin-independent mitophagy, by recruiting Keap1 and another cullin, E3 ubiquitin-protein ligase RBX1 (Rbx1), to mitochondria [65]. In glioblastoma, NRF2 activates NIX in conditions of hypoxia and oxidative stress, and silencing NIX promotes the production of superoxide under hypoxia, likely mediated by dysfunctional mitochondria [66]. Since the FUNDC1 phosphatase PGAM5 is a substrate for Keap1 [67], it is also possible that FUNDC1-dependent mitophagy is regulated by Keap1/Nrf2 axis, even if this hypothesis is still to be proved.

Another crucial transcriptional factor that modulates mitophagy in oxidative stress conditions is HIF-1α [30]. Low O_2 tension and subsequent increase of ROS inhibit prolyl hydroxylase (PHD) responsible for HIF-1α degradation [68]. High ROS levels can also stabilize Sentrin/SUMO specific proteases (SENPs) that enhance HIF-1α transcriptional activity [69]. Thus, HIF-1α induces the transcription of mitophagy receptors BNIP3 and NIX, producing a metabolic adaptation to a hypoxic environment [30]. Notably, the absence of BNIP3 in mammary cancer causes accumulation of dysfunctional mitochondria and elevated mtROS that upregulates HIF-1α and HIF-1α target genes, including those involved in cancer aggressiveness [70]. This observation further underlines that efficient turnover of mitochondria mediated by mitophagy is crucial in preventing ROS-mediated damage.

Mitophagy is also regulated by the action of Histone acetyltransferases (HATs) and histone deacetylases (HDACs). HATs acetylate conserved Lys on target proteins, while HDAC deacetylates Ac-Lys. Two classes of HAT and three classes of HDAC exist [71]. Of particular interest in the regulation of mitophagy is the class III HDAC, a family of NAD+ dependent deacetylases, known as sirtuins [72]. Seven sirtuins have been identified in humans (SIRT1-SIRT7) with different subcellular distribution; SIRT3, SIRT4, and SIRT5 are located in mitochondria [73]. SIRT 3, the best characterized mitochondrial sirtuin, is involved in the regulation of mitophagy in different ways [74]. By activating LKB1, SIRT 3 promotes the activation of AMPK–mTOR pathway, which in turn leads to autophagy. Furthermore, SIRT3 triggers mitophagy via deacetylation of FOX O3 under oxidative stress conditions. Once deacetylated, FOXO3 translocates to the nucleus, where it promotes the transcription of NIX, Bnip3, and LC3 [75]. Finally, SIRT 3 can mediate an antioxidant response by deacetylating superoxide dismutase 2 (SOD2), a crucial mitochondrial antioxidant enzyme, in two critical residues. A higher activity of SOD2 reduces mtROS, and inhibits Beclin-1. The inhibition of Beclin-1 reduces mitophagy. The activity of sirtuins declines with aging, and this decline can contribute to age-dependent impaired mitophagy [76,77]. The decline of Sirtuin activity could be due to the parallel decline of NAD+ levels with age, as the upregulation of NAD metabolism counteracts age-related diseases, and increasing intracellular NAD^+ improves mitochondrial quality via mitophagy and reverse cognitive deficits in models of AD [78,79].

Finally, a possible role of ROS in regulating mitophagy is played in immune response, and in particular in natural killer (NK) cells. After viral infection, the majority of effector NK cells undergo apoptosis; ROS triggers BNIP3- and NIX-dependent mitophagy, which in turn promotes the survival of virus-specific NK cells and seeding of memory, by removing dysfunctional mitochondria [80].

4. Mitophagy and Oxidative Stress in Physiological Aging

Several lines of evidence strongly suggest that mitophagy plays a role in aging, and recent studies have demonstrated that mitophagy is crucial in delaying physiological aging and age-related disorders, such as neurodegenerative and cardiovascular diseases. Mitophagy decline with aging has been observed in different tissues either from humans or mice, including (but not limited to) myocardium [81] skeletal muscle [82,83] skeletal muscle satellite cells [84], dentate gyrus [85], cultured fibroblasts [86].

A crucial contribution in understanding the crosstalk between aging, ROS, and mitophagy has been given by studies performed in model organisms, such as *Saccharomyces cerevisiae*, *Caenorhabditis elegans*, and *Drosophila melanogaster.* In *S. cerevisiae*, deficiency in non-selective autophagy (atg1Δ) and ubiquitin-independent mitophagy (atg32Δ or atg11Δ) causes accumulation of ROS upon starvation [87,88]. In mitophagy-deficient cells, excessive quantity of mitochondria is not degraded, produce ROS in excess, and spontaneously age [88] suggesting a link between mitophagy, ROS and aging. In *C. elegans*, mitophagy mediated by dct-1, the homolog of BNIP3 and NIX/BNIP3L, plays an important role during aging. DCT-1 is a key mediator of mitophagy and contribute to longevity in stress conditions. Deficiency of dct-1 causes accumulation of mitochondria in young adults, in a way similar to what observed in aged animals. Impairment of mitophagy triggers mitochondrial retrograde signaling, which coordinates the biogenesis and turnover of mitochondria and antagonizes the aging process [89].

The importance of mitophagy in aging of *C. elegans* is also demonstrated by the effects of mitophagy modulation on daf-2 mutants, which are characterized by extended lifespan. Induction of mitophagy in these mutants determines a lifespan shortening. Similarly, altered mitophagy by inactivating dct-1, PINK-1, and pdr-1 (the *C. elegans* Parkin homologs) significantly reduces their lifespan. Although none of these studies has analyzed in depth the consequence of ROS production and oxidative stress, it is interesting to note that dct-1 is transcriptionally regulated by skn-1 and daf-16, the counterparts of mammalian NRF2 and FOXO3, which are crucial regulators of oxidative stress response. In agreement with these observations, it has been reported that the glycoalkaloid tomatidine enhances lifespan in *C. elegans* through ROS-dependent induction of skn-1, which in turn induces mitophagy [90].

Furthermore, mitophagy is activated in reaction to mitochondrial stress in pdr-1, PINK1, and dct-1 dependent manner, as a response to iron starvation upon frataxin depletion. This response is similar to that one to hypoxia, and is involved in the extension of animal lifespan [91]. However, it must be noted that in human cells, loss of iron does not cause depolarization of mitochondria or extensive production of ROS, if compared to electron transport chain inhibitors [92]. Thus, it is possible that it is activated in response to free-iron deficiency stored in the organelle, rather than in response to ROS-induced damage or to metabolic reprogramming induced by HIF-1α.

In *D. melanogaster*, *Pink1*, and *Parkin* mutants are characterized by male sterility, loss of normal mitochondrial morphology and muscle degeneration [93]. In intestinal stem cells, depletion of *Pink1* or *Parkin* alters mitochondrial morphology and density, and results in higher levels of ROS in the intestinal progenitor cells; these changes are associated with an up-regulation of senescence-associated markers [94]. Conversely, the overexpression of *Pink1* and *Parkin* in indirect flight muscles leads to lifespan extension [95].

The importance of excessive ROS production in age-related impairment of mitophagy has been also observed in different mice models. Sarcopenia is one of the most evident phenomena that characterize the aging process, and is strictly associated with mitochondrial dysfunction and oxidative stress. In skeletal muscle, a dramatic impairment of PINK1/Parkin pathway has been observed with aging, and a crucial role of this pathway in counteracting the age-related mitochondrial dysfunction has been demonstrated. Knock out of Parkin in mice leads to an aging-like phenotype of skeletal muscle in adult animals [96], while Parkin overexpression increases mitochondrial enzyme activity, mitochondrial content in skeletal muscle of aged mice, and attenuates age-related oxidative stress [97]. Accordingly, a reduction in Parkin levels has been observed in atrophied muscle of elderly men [98] as well as a reduced expression of PARK2 gene in elderly, inactive women [82]. In the same tissue, an age-dependent decrease of MFN2 has been observed that impacts on mitophagy. Low levels of MFN2 impair mitochondrial fusion/fission regulation and quality control, and favor the accumulation of damaged mitochondria [99]. The subsequent higher ROS production initially activates a feedback loop that promotes mitochondrial turnover through the axis ROS/HIF1a/BNIP3/mitophagy, and minimizes mitochondrial damage. When MFN2 is absent or its levels are too low, mitophagy impairment occurs and age-related mitochondrial dysfunction are exacerbated [99]. The age-related reduction of MFN2 and the impairment of mitophagy impacts also other tissues. MFN2 promotes mitophagy and prevents mitochondrial dysfunction caused by ischemia/reperfusion in murine liver; the age-related reduction of MFN2, along with SIRT1, makes hepatocytes more susceptible to ischemia/reperfusion injury. A similar role has been shown in murine cultured neurons [100]. In chondrocytes, an age dependent reduction of MFN2 causes a reduction in mitochondrial fission, accompanied by dysfunctional mitochondria and oxidative stress [101]. Interestingly, in this model Parkin negatively regulates the levels of MFN2, and an age-related decrease of Parkin causes a post transcriptional increase of MFN2 and hyperfused mitochondria. These observations suggest that a limited production of ROS can activate mitophagy to prevent or reduce mitochondrial damage. When the production of ROS is excessive, activation of mitophagy is not sufficient to limit mitochondrial damage, leading to accumulation of dysfunctional mitochondria, producing increasing levels of ROS.

One of the most interesting phenomena that indirectly suggests a crucial role of oxidative stress in regulating mitophagy during aging is caloric restriction (CR). Caloric restriction is known to extend lifespan. Almost every study performed so far indicates that mtROS production is lower in liver, brain, heart, and other tissues of long-lived than of short-lived species [102], and that long-term CR reduces the rate of mtROS generation, and extends lifespan in different animal models, including *C. elegans* and mice. Not only CR prolongs lifespan in mice, but it also attenuates the effects of aging on different tissues, including skeletal muscle and myocardium [103]. In these animal models, CR causes the activation of AMPK-ULK1 pathway, which determines the removal of damaged mitochondria via mitophagy [97–99]. Aged mice kept for 20 weeks on a CR diet showed normal, not damaged mitochondria, low levels of oxidative stress and low levels of PINK1 in kidney, suggesting that CR can mitigate the mitochondrial damage observed with age, and making activation of the PINK1 pathway no longer necessary [104].

In some models, caloric restriction cannot be maintained for a long time. However, the use of molecules that mimics some of their effects can help in understanding the impact of CR on mitophagy [105]. An example of these molecules is spermidine, a dietary compound that has been shown to extend lifespan through induction of autophagy in *S. cerevisiae, C. elegans* and *D. melanogaster* [106]. Increased levels of Parkin-positive mitochondria have been found in aged hearts along with lower levels of Nrf1, a crucial factor for mitochondrial biogenesis, reduced Drp1-mediated mitochondrial fission, and formation of enlarged mitochondria [107]; spermidine feeding promotes protective autophagy and mitophagy in cardiomyocytes, counteracting mitochondrial respiration decline observed with aging [108]. Although it has not been shown that this effect on mitophagy is mediated by reduction or mtROS generation, the strict correlation between CR and mtROS suggests that this could be the case, at least in part. Notably, induction of autophagy by resveratrol, another molecule mimicking CR, is dependent on the nicotinamide adenine dinucleotide-dependent deacetylase sirtuin 1 (SIRT1) [109], a factor that is upregulated in response to oxidative stress in the heart [110]. Accordingly, moderate expression of Sirt1 induces resistance to oxidative stress and counteracts aging in mice [110]. Interestingly, in centenarians—exceptional humans who reach the age of 100 years or more—a decreased mitophagy has been observed, even in the presence of oxidative stress. This phenomenon is accompanied by a sort of "mitochondrial hypertrophy" within the cell that help to keep a full bioenergetic competence, even in the presence of OXPHOS defects [86].

5. Mitophagy and Oxidative Stress: Insights from Age-Related Diseases

The efficacy of the mitochondrial respiratory chain tends to diminish with aging, with a reduction in ATP synthesis and increase in the production of ROS [111]. Thus, cellular damage caused at least in part by chronic oxidative stress can accumulate, and autophagic and mitophagic pathways can become overwhelmed, particularly in non-dividing, high energy demanding cells such as neurons. As a consequence, cortical degeneration is commonly observed in aging. Thus, it is not surprising that many age-related diseases, such as AD or PD, show the simultaneous occurrence of chronic oxidative stress, mitochondrial dysfunctions and impaired mitophagy. The study of mitophagy in these pathologies, as well as in other diseases characterized by accelerated aging, provided crucial information about the mechanistic connections between mitophagy and oxidative stress during aging.

5.1. Mitophagy Defects and Oxidative Stress in Premature Aging Diseases

The observations concerning mitophagy defects in monogenic diseases characterized by premature aging are of particular interest to understand the interconnections between mitophagy and ROS. Loss of mitophagy was first described in Cockayne syndrome (CS), a progeroid syndrome characterized by progressive neurodegeneration that resembles that observed in mitochondrial disorders [112]. Mutations in CS complementation group B (CSB) gene cause the 80% of CS cases. It has been shown that CSB deficient cells are characterized by increased mitochondrial content, higher MMP and sustained production of ROS [113], accompanied by higher spare respiratory capacity and increased oxygen consumption rate. These changes did not appear to be related to increased mitochondrial biogenesis, but rather to an impairment of mitophagy that reduces the turnover of damaged mitochondria. This impairment is likely due to a reduced activity of PGC-1α, which is needed to the transcription of genes encoding uncoupling proteins (UCP). A lower UCP expression increases MMP and impairs PINK1-mediated mitophagy. In agreement with this hypothesis, the overexpression of UCP2 rescues mitophagic defects in CS [114]. A similar phenotype was reported in ATM deficient cells [115]: an increased oxygen consumption rate was associated with higher mitochondrial mass, higher ROS levels and decreased mitophagy. As ATM is present in mitochondria, it is possible that CSB and ATM contribute to a mtDNA damage response pathway by enhancing mitophagy. Notably, the same phenotype (higher mitochondrial metabolism, MMP and ROS formation, along with impaired mitophagy) has been found in related DNA repair disorder xeroderma pigmentosum group A [116].

5.2. Mitophagy and Oxidative Stress in Alzheimer's Disease

AD represents a paradigmatic example of age-related, multifactorial neurodegenerative disease. AD occurs in two forms: a familial early-onset and a sporadic, late-onset form, and is the most common cause of dementia, accounting for 50–75 % of cases [117]. Early onset AD, which represents around 5% of AD total cases, is caused by highly penetrant mutations of few genes, *PSEN1*, *PSEN2*, and *APP* [118] whereas age-related factors are responsible for disease process and clinical symptoms. AD is characterized by progressive accumulation of extracellular aggregates of the amyloid-β peptide (Aβ), which are generated from cleavage of the membrane-bound amyloid precursor protein (APP), and aggregates of tau proteins, which form neurofibrillary tangles in the cytoplasm. Both soluble Aβ and abnormally phosphorylated tau can directly impair mitochondrial functions.

Multiple lines of evidence indicate that mitophagy is involved in neurodegeneration observed in AD. ROS are among the players that drive mitophagy impairment, and markers of oxidative stress (such as protein carbonylation, lipid oxidation, and the oxidation of the mtDNA) that are increased with age, appears to be particularly evident in AD [119]. Consistently, enhanced oxidative stress was observed in animal models of AD [120,121] and mtROS are sufficient to trigger Aβ production in vitro and in vivo [122]. Abnormal mitophagy in AD patient brains have been evidenced by autophagic accumulation of mitochondria in vulnerable AD neurons [110] and then confirmed by different groups [123]. The overexpression of mutant APP (mAPP) in mouse primary hippocampal cells results in higher expression of mitochondrial fission genes, DRP1, and FISs1 and decreased levels of fusion genes (MFN1, MFN2, and OPA1) as well as of autophagy (ATG5 and LC3BI, LC3BII) and mitophagy (PINK1, TERT, BCL2, and BNIP3L) genes at the mRNA and protein level [124], suggesting that the initiation and cargo recognition component of mitophagy is inhibited by Aβ.

The involvement of PINK1/Parkin-dependent mitophagy in AD pathogenesis have been intensively studied in the last years. Progressive Aβ accumulation and subsequent mitochondrial damage strongly induce PINK1/Parkin pathway in animal models of AD, and its upregulation has been observed in AD patient brains [125,126]. Furthermore, cytosolic Parkin is depleted in AD brains during disease progression, resulting in mitophagic impairment and augmented mitochondrial defects. In neurons bearing a mutant hAPP, an increased recruitment of cytosolic Parkin to depolarized mitochondria has been observed in the absence of MMP dissipation reagents [127]. Moreover, Parkin translocation has been observed mainly in the somatodendritic regions of the cells. This imbalanced recruitment leads to a decreased anterograde and increased retrograde mitochondrial axonal transport. Along with the observation that mitophagy is enhanced in AD brains, accompanied by depletion of cytosolic Parkin over disease progression, these data suggest that impaired mitophagy significantly contributes to the accumulation of dysfunctional mitochondria in AD-affected neurons [127]. In agreement with these observations, skin fibroblasts and brain biopsies from AD patients showed high levels of oxidized proteins, which suggests the presence of mitochondrial damage caused by oxidative stress, low Parkin levels and accumulation of PINK1 [128]. The overexpression of Parkin in cultured patients' fibroblasts restored mitophagy, as evidenced by decreased PINK1 and accumulation of defective mitochondria, and recovery of MMP.

As far as the effects of abnormal Tau on mitophagy are concerned, first studies showed that the destabilization of microtubule networks and interruption of organelle migration determines accumulation of damaged organelles in the soma of neurons. In brains from AD patients with increased levels of Tau, an increase in the levels of different mitophagy markers has been observed, suggesting a mitophagy deficit within cells. The overexpression of hTau in a cellular model determines an increase of the MMP, associated with a decrease in the localization of Parkin to the mitochondria [129]. Tau has been shown to interact with Parkin and inhibit its translocation to defective mitochondria by sequestering it in the cytosol in neuroblastoma cells [130]. In a transgenic mouse model of AD, excessive levels of Tau can induce mitophagy by increasing MMP and Parkin levels [129]. Tau can also interact with DRP1, suggesting that it can contribute to increased mitochondrial fragmentation observed in AD [131]. In old transgenic tau mice bearing P301L mutation, increased levels of the fission proteins DRP1 and FIS1 and decreased levels of mitochondrial fusion proteins, MFNn1, MFN2, and OPA1 has been observed in the hippocampus. This change was associated with higher levels of mtROS and lipid peroxidation mice [132]. Although no data concerning mitophagy has been provided in this study, it is conceivable that the alterations of mitochondrial dynamics impact on mitophagy, and contribute to impaired capability to remove damaged mitochondria, which resulted more prone to produce ROS.

Finally, changes in the cardiolipin profile of synaptic mitochondria have been observed in a mouse model of AD, which were associated with mitochondrial dysfunction, in a way similar to what observed during aging in rat hearts [35,133,134]. Even if direct evidence has not been provided, this observation could also suggest that cardiolipin-related mitophagy could also be impaired in AD [135].

Defects in the proteolytic activities of lysosomes can impair mitophagy. Lysosomal defects have been repeatedly observed in brains specimens from AD patients. Studies in mouse models highlighted the importance of lysosome functionality in AD pathogenesis, as suppression of lysosomal proteolysis mimics AD neuropathology, while restoring normal lysosomal proteolysis and autophagy efficiency in AD mouse models improves neuronal function and cognitive performance, [136,137]. Accordingly, mutations of PSEN1—one of the genes causing early onset AD—in combination with ApoE4, a key genetic risk factor of AD, disrupt lysosomal function [138]. Other factors, including Aβ peptides, ROS, and oxidized lipids and lipoproteins, can also impair lysosomal proteolysis. Lysosomal deficits in AD have been also attributed to defects in protease targeting to lysosomes [139]. As a whole, these lysosomal defects reduce proteolytic removal of defective mitochondria, along with other autophagic cargoes, in neurons of AD patients [139]. Therefore, increased Parkin association with mitochondria and abnormal mitochondrial retention within lysosomes observed in AD neurons of patients, as well as in cells overexpressing mutant APP, could also be due to lysosomal deficiency [139].

Overall, these studies demonstrate that mitophagy impairment is clearly involved in AD pathogenesis, and contributes to the progressive loss of mitochondrial functionality observed in AD progression. To what extent the functional alteration of mitochondria associated with age is crucial for the development of AD, or to what extent it represents a concurrent but secondary phenomenon is still under discussion. According to the so-called "mitochondrial cascade hypothesis", mitochondria could represent the primary generator of AD [140]: since mitochondrial function declines during aging, it connects AD and aging by explaining why advanced age is the greatest AD risk factor. In physiological aging, an equilibrium exists between mtROS, mitochondrial damage and removal of dysfunctional mitochondria via mitophagy. In AD progression, the decrease below a critical threshold of mitochondrial function associated with age starts the events leading to the accumulation of Aβ. Mitophagy impairment can contribute to overcome this threshold, and in turn impaired mitochondrial function and associated bioenergetic changes alter Aβ homeostasis and lead to an accumulation of Aβ.

5.3. Mitophagy and Oxidative Stress in Parkinson's Disease

PD is the second most common progressive disorder of the CNS, that affects predominantly the dopaminergic neurons od the *substantia nigra* (SN) [141]. A typical sign of PD is the presence of the Lewy bodies, eosinophilic cytoplasmic inclusions in the SN [142]. PD is mainly sporadic and associated with aging, even if 5–15% is hereditary with an autosomal transmission, always with early onset [143]. In a way similar to AD, the early-onset form of PD is caused by gene mutations, whereas aging is the single, most important risk factor for the sporadic form. Indeed, the prevalence of PD is at 5% in people aged 80 years, at 2% in aged 65 years, and rare in aged 50 years or less [144,145]. The mechanisms at the basis of neuronal degeneration in PD have not been fully elucidated, but several lines of evidence suggest that deficiencies in mitochondrial homeostasis play a crucial role in neuronal degeneration characterizing PD pathogenesis [146]. The high request of energy of these cells is probably at the basis of their susceptibility to mitochondrial dysfunction [141] and both mitophagy impairment and oxidative stress have been proved to be involved in this process.

A possible involvement of mitophagy in the pathogenesis of PD was first suggested by a study showing that mitochondria accumulated abnormally in autophagosomes in neurons of patients with PD and Lewy Body Dementia (LBD) [147]. Numerous studies have described defects in mitophagy and an overall mitochondrial impairment with consequent increased ROS production in neurons of PD patients and/or in models [146,148–156]. Although oxidative stress is described as a key regulator of the neurodegenerative process in all forms of PD [146,157], it is not clear yet if the increase of ROS is a causative factor or a consequence of cells degeneration. Nevertheless, investigation in early-stage PD patients showed that oxidative stress is arising during the initial stages of the disease before the neuron loss, supporting the idea that ROS could be the cause of neuronal degeneration [158]. As oxidative stress is observed during physiological aging, molecular alterations occurring during PD could aggravate the imbalance between ROS production and scavenging observed with age, leading to nigral neurodegeneration [145].

The observations made on early onset, recessive familial PD are of particular interest to understand how impairment of mitophagy affects the delicate balance between damage caused by oxidative stress and mitochondrial turnover maintaining the organelle homeostasis. Early onset, recessive familial PD can be caused by mutations in the genes *Park2* (*Parkin*), *Park6* (*Pink1*), or *Park7* (*DJ-1*), among others [125,159–161]. All three proteins are crucial in resistance to oxidative stress and to maintain mitochondrial functions [125]. While the role of PINK1 and Parkin in mitophagy is clearly established, the possible involvement and the precise function of DJ-1 in this process is still a matter of discussion. DJ-1 is involved in anti-oxidative, anti-inflammatory, and anti-apoptotic pathways, and can protect *substantia nigra* from oxidative stress during PD onset. Missense mutations in this gene cause a very rare autosomal recessive PD, often with early onset [162], and fibroblasts and lymphoblasts from PD patients with mutated DJ-1 showed fragmentated mitochondria, revealing a role of DJ-1 in mitochondrial dynamics [163–165]. Interestingly, overexpression of PINK1 and Parkin rescues the aberrant mitochondrial phenotype observed in DJ-1 deficient cells, suggesting that DJ-1 functions in mitophagy are partially redundant [164].

Alpha-Synuclein, encoded by the *SNCA* gene, plays a role in compartmentalization of neurotransmitters and synaptic vesicle recycling and it is mainly located in neurons, in presynaptic terminals [146,166–170]. Multiplications and mutations in *SNCA* gene are related to autosomal dominant PD. Alpha-Synuclein is a natively unfolded protein, without a stable structure in aqueous solutions, and its aggregation represents a hallmark of PD [146,171]. Indeed, alpha-Synuclein fibrils are the main component of Lewy bodies. Mutated or high amount of protein can constitute aggregations and amyloid fibrils [172–175]. Numerous in vivo and in vitro studies connect oxidative stress with the formation of α-Synuclein aggregations and fibrils, supporting the idea that an unbalance redox state in brain may contribute to neurodegeneration [176,177]. Neurotoxins such as rotenone or MPP^+ (metabolite of 1-methyl-4-phenyl-1,2,3,6-tetrahydropyridine, MPTP) capable of inhibit mitochondrial complex I producing ROS have been shown to increase α-Synuclein in vitro and in vivo; a study revealed that this effect is due to the derepression of microRNAs (miRNAs) capable of inhibit α-Synuclein mRNA expression, leading to a *de novo* translation [172]. Moreover, an increase of α-Synuclein aggregation has been observed in a transgenic mouse overexpressing α-Synuclein in the presence of oxidative stress due to haploinsufficiency of SOD2 [178]. Furthermore, α-Synuclein misfolding is responsible of increasing in ROS production, triggering a vicious cycle that in turn leads to neurodegeneration [179,180].

LRKK2 is a kinase enzyme codified by the LRRK2 gene and mutations in LRRK2 are responsible for 1–2% of total PD cases and about 5% of total familial cases, even if LRKK2 represent also a risk locus for sporadic PD [181–183]. It has been showed that LRKK2 can interact with Miro, a protein of the OMM responsible for the link between microtubules and the mitochondrial surface and involved in mitochondria mobility [184–186]. Interaction of LRKK2 with Miro targets it for degradation and triggers mitophagy. In a study performed on fibroblasts from sporadic and familial LRRK2-mutated PD patients treated with CCCP (a mitochondrial uncoupler), Miro degradation, and the subsequent mitochondrial clearance were compromised. Delaying and impairment of mitochondrial clearance were also observed in induced pluripotent stem cell (iPSC)-derived neurons from PD patients with LRKK2G2019S mutations treated with antimycin A, an inhibitor of complex III able to start mitophagy. This demonstrated that mutations in LRRK2 could delay Miro degradation and the onset of mitophagy, leading to the increase of ROS, followed by cell death [187]. In another study, the effects of mutated PINK1 have been analyzed in dopaminergic neurons derived from iPS cells from skin fibroblasts of PD patients. Although iPSC were treated with valinomycin, a potassium ionophore capable of dissipate the transmembrane electrochemical gradients, they showed impaired Parkin translocation to mitochondria with an increase of mitochondrial copy number [188].

Multiple lines of evidence suggest that the imbalance of acetylases and deacetylases activity also impacts mitophagy regulation in PD pathogenesis, either in idiopathic or familiar forms of the disease. Hyperacetylation of SOD1 has been observed in post-mortem midbrains from PD patients [189]. SIRT3 overexpression, or administration of Nicotinamide Riboside (NR), a NAD+ precursor, counteracts the degeneration of dopaminergic neurons in PD [190]. CR, which is known to induce SIRT3, reduces neurodegeneration in animal models of both PD and AD [191]. Decreased sirtuin deacetylase activity was observed in iPSC-derived dopaminergic neurons from patients bearing the G2019S LRRK2 mutation [192]. Fibroblasts from PD patients with the same mutation displays increased mitophagy, due to the activation of SIRT3, clearly suggesting that impaired SIRT-induced mitophagy plays a major role in the pathogenesis of this form of early-onset PD. Conversely, idiopathic PD exhibits a reduced capability to remove defective mitochondria, associated with higher levels or ROS production reactive oxygen species (ROS) [71].

The crucial role of oxidative stress and mitophagy in the pathogenesis of PD has been also proved by the observation that NIX and AMBRA1 can help in delaying cell death in PD by exerting an antioxidant action. When PINK1-mediated mitophagy is abrogated in PD dopaminergic neurons, NIX can stimulate the removal of damaged mitochondria, so preserving dopaminergic neurons. Phorbol 12-myristate 13-acetate (PMA) induces NIX expression, and leads to ROS production in the same cells, clearly suggesting that induction of mitophagy helps in reducing oxidative stress [193]. In a similar model in which PINK1-mediated mitophagy is not functional, because of PINK1 or PARK2 mutations, AMBRA1 can rescue mitophagy [194] and reduce cell death in vitro induced by rotenone or 6-OHDA treatments, by limiting selectively the source of oxidative stress [195]. The main alterations of factors modulating mitophagy observed in AD and PD are depicted in Figure 2, while Table 1 summarizes the main findings described above concerning changes of mitophagy observed during aging or in PD and AD, which are related to oxidative stress.

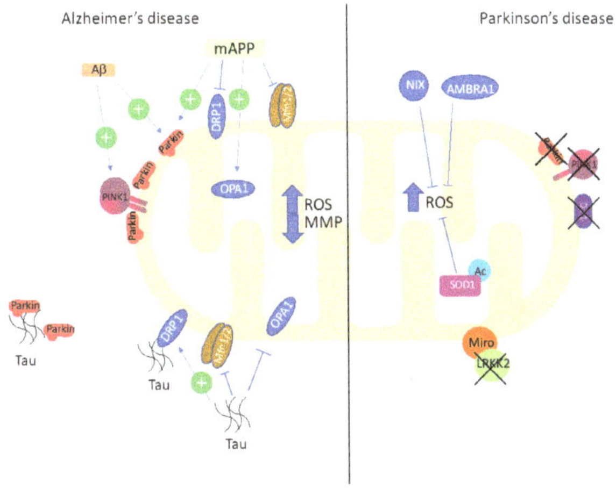

Figure 2. Main alterations of factors modulating mitophagy observed in AD and PD. In AD, Aβ accumulation upregulated PINK/Parkin pathway. Mutant human APP increases recruitment of Parkin to depolarized mitochondria. mAPP also causes upregulation of mitochondrial fission genes DRP1, and decrease of fusion genes MFN1/2 and OPA1. Tau interacts with Parkin and inhibits its translocation to defective mitochondria. Furthermore, it upregulates fission proteins and inhibits fusion proteins. In PD, defective mitophagy is determined by loss of function of Parkin, PINK1, and DJ- 1; their dysfunction is associated with high levels of ROS. Loss of LRKK2 blocks the degradation of the outer membrane protein Miro and triggers mitophagy. In this context, NIX and AMBRA, can limit the excessive production of ROS. See text for details.

Table 1. Changes of mitophagy observed with age or in PD and AD, which are related to oxidative stress.

Mitophagy Pathway or Protein	Age Related Changes	Changes in PD or AD
Pink/Parkin	• Depletion of Pink1 and Parkin leads to hallmarks of senescence in ISCs (Intestinal stem cells) of *D. melanogaster,* including high ROS levels [94]; • Overexpression of Pink1 or Parkin extends lifespan of *D. melanogaster* [95]; • Parkin overexpression attenuates molecular and biochemical markers of aging, extending lifespan in *D. melanogaster* [196] • mRNA levels of BNIP3, PINK1, Parkin and NIX, and the protein levels of BNIP3, PINK1 and Parkin decrease in the mouse auditory cortex with aging [197]; • Parkin levels are diminished in atrophied muscles of elderly men [98]; • Parkin overexpression attenuates the effects of advanced aging on myocardial function in transgenic mice [198] • Parkin overexpression attenuates aging-dependent loss of muscle strength and mass in transgenic mice [97].	• Upregulation of the PINK1/parkin pathway were showed in AD patients' brain [119,120]; • Depletion of Parkin during AD progression were found in AD patients' brain [121]; • Low levels of the Parkin protein were identified in skin fibroblasts and brain biopsies from AD patients [122]; • Increased levels of Parkin were revealed in a transgenic mouse model [123]; • Homozygous or compound heterozygous mutations and the consequent loss of function of Parkin and PINK1 genes are the main cause of recessive early-onset PD [119,159]; • Mutations in PINK1 gene are also a rare source of sporadic early-onset PD [160].

Table 1. *Cont.*

Mitophagy Pathway or Protein	Age Related Changes	Changes in PD or AD
Cardiolipin	• Cardiolipin levels in mitochondria decrease with aging [134]; • Changes to cardiolipin content and oxidative damage have been related to aging in hearts of rats; no direct evidence of cardiolipin-mitophagy impairment has been provided [35,133].	• Changes in the cardiolipin profile were described in mouse model of AD [126]; • Correlation between oxidative damage of CL by ROS and pathogenesis of PD, likely because of the impairment of mitophagy caused by damaged CL. No direct evidence is provided [199].
DJ-1	• DJ-1 mutants in *D. melanogaster* exhibit lifespan shortening and sensitivity to oxidative stress; • Repressed during aging in rat thymus tissues [200].	• Immunostaining revealed high levels of DJ-1 protein in hippocampal pyramidal neurons and astrocytes of AD brains [201]; • Mutations in DJ-1 gene are cause of autosomal recessive PD [161]; • Fibroblasts and lymphoblasts from PD patients with mutated DJ-1 showed fragmented mitochondria [151,153]. • Mutations in Dj-1 impaired protection against oxidative stress, a key regulator of the neurodegenerative process in PD and AD [202–210]
BNIP3	• mRNA levels of BNIP3, PINK1, Parkin and NIX, and the protein levels of BNIP3, PINK1 and Parkin in the mouse auditory cortex decrease with aging [197].	• No data available
MFN1	• Age-related increase of MFN1 and OPA1 in cultured fibroblasts.	• Increased expression in neurons from patients with AD.
MFN1/MFN2	• MFN2 expression is higher in rat and human chondrocytes during aging and OA (osteoarthritis) [101] • MFN2 decreases during aging in mouse skeletal muscle [99].	• Reduced levels of MFN1, MFN2 and OPA1 were found in aged tau mice [132]; • Changes of MFN1 and MFN2 were identified both in the PrP-hAPP/hPS1 AD mouse model brains and in an SH-SY5Y cell model of early-onset AD [211]. • Decreased levels of mitochondrial fusion proteins, MFN1, MFN2 and OPA1 were found in 12-month-old tau mice relative to age-matched WT mice [132].

6. Conclusions

The interplay between mitophagy, ROS production, and aging is complex and far from being completely elucidated. The central role of ROS production and consequent damage to mitochondria in the aging process has been clearly established in the last 50 years, despite some objections to this theory over the past 15 years [212], and mitophagy is a key mechanism for mitochondrial quality and quantity control, as it limits the production of ROS, the damage to mtDNA of transmembrane potential loss and the decrease in ATP production. The data and observations discussed in this review indicate that the imbalance of the delicate equilibrium among mitophagy, ROS production, and mitochondrial damage can start, drive, or accelerate the aging process, either in physiological or pathological conditions (Figure 3). It remains to be determined which is the prime mover of this imbalance, i.e., whether it is the mitochondrial damage caused by ROS that initiates the dysregulation of mitophagy, thus activating a vicious circle that leads to the reduced ability to remove damaged mitochondria, and further damage from ROS, or if, on the other hand, an alteration in the regulation of mitophagy constitutes one of the initial events leading to the main of the excessive production of ROS.

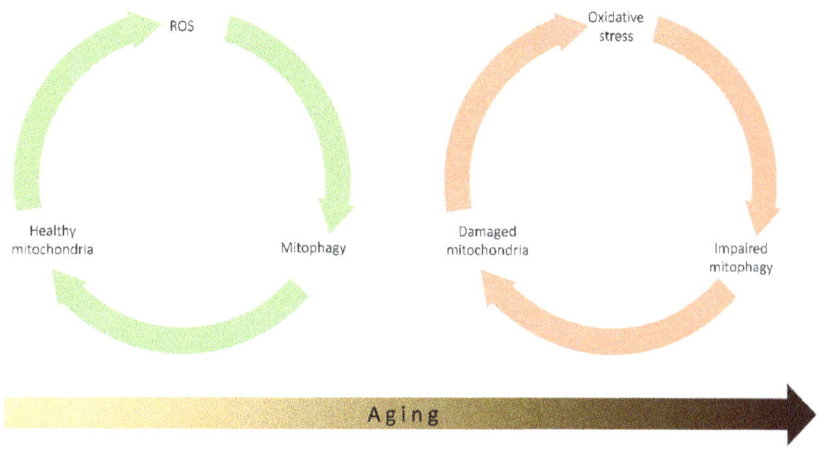

Figure 3. In normal conditions, ROS physiologically produced in the cell, and particularly by mitochondria, can induce mitophagy, which contributes to the normal homeostasis of the cells by removing damaged mitochondria, so maintaining the organelles healthy. The progressive increase in ROS production observed with age can lead to chronic oxidative stress, which in turn impairs mitophagy and reduces the capability to remove damaged mitochondria. Damaged organelles further produced ROS, so keeping a vicious cycle active.

Author Contributions: Conceptualization, A.D.G., M.P., A.C., and L.G.; writing—original draft preparation, A.D.G. and M.P.; writing—review and editing A.D.G., L.G., G.Z., M.N., A.C., and M.P., project administration, M.P.; funding acquisition, M.P. All authors have read and agreed to the published version of the manuscript.

Funding: This research was funded by a grant from Associazione Italiana per la Ricerca sul Cancro (AIRC), IG 19786 to MP.

Conflicts of Interest: The authors declare no conflict of interest.

Abbreviations

AD	Alzheimer's disease
AMBRA1	Autophagy and beclin 1 regulator 1
APP	Amyloid precursor protein
ARIH1	Ariadne RBR E3 ubiquitin ligase homolog 1
ATG	Autophagy-related protein
ATM	Ataxia-telangiectasia mutated
Aβ	Amyloid-β peptide
BCL2L13	BCL2 like 13
BNIP3	BCL2 interacting protein 3
CI	Complex I
CIII	Complex III
CK2	Casein kinase II
CL	Cardiolipin
CR	Caloric restriction
CS	Cockayne syndrome
CSB	CS complementation group B
CuZnSOD (SOD1)	Copper-zinc SOD
DNM1L	Dynamin-1-like protein
Drp1	Dynamin-1-like protein
FKBP8	FKBP prolyl isomerase 8
FUNDC1	FUN14 domain containing 1
Gp78	Glycoprotein 78
GPx1	Glutathione peroxidase 1
Grx2	Glutaredoxin 2
GSNOR	S-nitrosoglutathione reductase
HAT	Histone acetyltransferase
HDAC	Histone deacetylases
HIF1a	Hypoxia-inducible factor 1-alpha
HUWE1	HECT, UBA and WWE domain-containing protein 1
IMM	Mitochondrial inner membrane
iPSC	induced pluripotent stem cell
IR	Ischemia/reperfusion
Keap1	Kelch-like ECH-associated protein 1
LBD	Lewy Body Dementia
LIR	LC3 interacting region
mAPP	Mutant APP
MnSOD (SOD2)	Manganese superoxide dismutase
MPP	Matrix processing peptidase
MsrB2	Methionine sulfoxide reductase B2
mtKR	Mt KillerRed
MMP	Mitochondrial membrane potential
MTS	Mitochondrial targeting sequence
MUL1	Mitochondrial E3 ubiquitin protein ligase 1
NAD	Nicotinamide adenine dinucleotide
NBR1	Neighbor of BRCA1 gene 1
NDP52	Nuclear domain 10 protein 52
NIX (BNIP3-like, NIX/BNIP3L)	NIP3-like protein X
NK	Natural killer
NR	Nicotinamide riboflavide
Nrf2	Nuclear factor (erythroid-derived 2)-like 2 transcription factor
OMM	Mitochondrial outer membrane
OPA1	Dynamin-like 120 kDa protein
OPTN	Optineurin
p62/SQSTM1	Sequestosome-1
PARL	Presenilin-associated rhomboid like
PD	Parkinson's disease

PGAM5	Phosphoglycerate mutase 5
PHD	Prolyl hydroxylase
PINK1	Phosphatase and tensin homologue (PTEN)-induced putative kinase 1
PMA	Phorbol 12-myristate 13-acetate
PRDX6	Peroxiredoxin-6
Prx3	Peroxiredoxin-3
Prx5	Peroxiredoxin-5
Rbx1	E3 ubiquitin-protein ligase RBX1
ROS	Reactive oxygen species
SENPs	Sentrin/SUMO specific proteases
SIAH1	E3 ubiquitin-protein ligase SIAH1
SIRT1	Nicotinamide adenine dinucleotide–dependent deacetylase sirtuin 1
SMURF1	SMAD ubiquitination regulatory factor 1
SNc	Substantia nigra pars compacta
SNCA	Alpha-Synuclein
TAX1BP1	TAX1 binding protein 1
TIM	Translocase of the inner membrane
TOM	Translocase of the outer membrane
UCP	Uncoupling proteins
ULK1	Unc-51-like autophagy activating kinase 1

References

1. Lopez-Otin, C.; Blasco, M.A.; Partridge, L.; Serrano, M.; Kroemer, G. The hallmarks of aging. *Cell* **2013**, *153*, 1194–1217. [CrossRef] [PubMed]
2. Yoo, S.M.; Jung, Y.K. A Molecular Approach to Mitophagy and Mitochondrial Dynamics. *Mol. Cells* **2018**, *41*, 18–26. [CrossRef] [PubMed]
3. Vernucci, E.; Tomino, C.; Molinari, F.; Limongi, D.; Aventaggiato, M.; Sansone, L.; Tafani, M.; Russo, M.A. Mitophagy and Oxidative Stress in Cancer and Aging: Focus on Sirtuins and Nanomaterials. *Oxidative Med. Cell. Longev.* **2019**, *2019*, 6387357. [CrossRef] [PubMed]
4. Youle, R.J.; Narendra, D.P. Mechanisms of mitophagy. *Nat. Rev. Mol. Cell Biol.* **2011**, *12*, 9–14. [CrossRef]
5. Ke, P.Y. Mitophagy in the Pathogenesis of Liver Diseases. *Cells* **2020**, *9*, 831. [CrossRef]
6. Yu, X.; Long, Y.C. Autophagy modulates amino acid signaling network in myotubes: Differential effects on mTORC1 pathway and the integrated stress response. *Faseb J. Off. Publ. Fed. Am. Soc. Exp. Biol.* **2015**, *29*, 394–407. [CrossRef]
7. Shally, A.; McDonagh, B. The redox environment and mitochondrial dysfunction in age-related skeletal muscle atrophy. *Biogerontology* **2020**. [CrossRef]
8. Feng, Y.; He, D.; Yao, Z.; Klionsky, D.J. The machinery of macroautophagy. *Cell Res.* **2014**, *24*, 24–41. [CrossRef]
9. Palikaras, K.; Lionaki, E.; Tavernarakis, N. Mechanisms of mitophagy in cellular homeostasis, physiology and pathology. *Nat. Cell Biol.* **2018**, *20*, 1013–1022. [CrossRef]
10. Montava-Garriga, L.; Ganley, I.G. Outstanding Questions in Mitophagy: What We Do and Do Not Know. *J. Mol. Biol.* **2020**, *432*, 206–230. [CrossRef]
11. Tal, R.; Winter, G.; Ecker, N.; Klionsky, D.J.; Abeliovich, H. Aup1p, a yeast mitochondrial protein phosphatase homolog, is required for efficient stationary phase mitophagy and cell survival. *J. Biol. Chem.* **2007**, *282*, 5617–5624. [CrossRef]
12. McWilliams, T.G.; Prescott, A.R.; Allen, G.F.; Tamjar, J.; Munson, M.J.; Thomson, C.; Muqit, M.M.; Ganley, I.G. mito-QC illuminates mitophagy and mitochondrial architecture in vivo. *J. Cell Biol.* **2016**, *214*, 333–345. [CrossRef]
13. McWilliams, T.G.; Prescott, A.R.; Montava-Garriga, L.; Ball, G.; Singh, F.; Barini, E.; Muqit, M.M.K.; Brooks, S.P.; Ganley, I.G. Basal Mitophagy Occurs Independently of PINK1 in Mouse Tissues of High Metabolic Demand. *Cell Metab.* **2018**, *27*, 439–449.e435. [CrossRef]
14. Sato, K.; Sato, M. Multiple ways to prevent transmission of paternal mitochondrial DNA for maternal inheritance in animals. *J. Biochem.* **2017**, *162*, 247–253. [CrossRef]
15. Gottlieb, R.A.; Bernstein, D. METABOLISM. Mitochondria shape cardiac metabolism. *Science* **2015**, *350*, 1162–1163. [CrossRef]
16. Sandoval, H.; Thiagarajan, P.; Dasgupta, S.K.; Schumacher, A.; Prchal, J.T.; Chen, M.; Wang, J. Essential role for Nix in autophagic maturation of erythroid cells. *Nature* **2008**, *454*, 232–235. [CrossRef]
17. Panigrahi, D.P.; Praharaj, P.P.; Bhol, C.S.; Mahapatra, K.K.; Patra, S.; Behera, B.P.; Mishra, S.R.; Bhutia, S.K. The emerging, multifaceted role of mitophagy in cancer and cancer therapeutics. *Semin. Cancer Biol.* **2019**. [CrossRef]
18. Matsuda, N.; Sato, S.; Shiba, K.; Okatsu, K.; Saisho, K.; Gautier, C.A.; Sou, Y.S.; Saiki, S.; Kawajiri, S.; Sato, F.; et al. PINK1 stabilized by mitochondrial depolarization recruits Parkin to damaged mitochondria and activates latent Parkin for mitophagy. *J. Cell Biol.* **2010**, *189*, 211–221. [CrossRef]

19. Pickles, S.; Vigie, P.; Youle, R.J. Mitophagy and Quality Control Mechanisms in Mitochondrial Maintenance. *Curr. Biol. CB* **2018**, *28*, R170–R185. [CrossRef]
20. Kondapalli, C.; Kazlauskaite, A.; Zhang, N.; Woodroof, H.I.; Campbell, D.G.; Gourlay, R.; Burchell, L.; Walden, H.; Macartney, T.J.; Deak, M.; et al. PINK1 is activated by mitochondrial membrane potential depolarization and stimulates Parkin E3 ligase activity by phosphorylating Serine 65. *Open Biol.* **2012**, *2*, 120080. [CrossRef]
21. Kazlauskaite, A.; Kondapalli, C.; Gourlay, R.; Campbell, D.G.; Ritorto, M.S.; Hofmann, K.; Alessi, D.R.; Knebel, A.; Trost, M.; Muqit, M.M. Parkin is activated by PINK1-dependent phosphorylation of ubiquitin at Ser65. *Biochem. J.* **2014**, *460*, 127–139. [CrossRef]
22. Koyano, F.; Okatsu, K.; Kosako, H.; Tamura, Y.; Go, E.; Kimura, M.; Kimura, Y.; Tsuchiya, H.; Yoshihara, H.; Hirokawa, T.; et al. Ubiquitin is phosphorylated by PINK1 to activate parkin. *Nature* **2014**, *510*, 162–166. [CrossRef]
23. Kane, L.A.; Lazarou, M.; Fogel, A.I.; Li, Y.; Yamano, K.; Sarraf, S.A.; Banerjee, S.; Youle, R.J. PINK1 phosphorylates ubiquitin to activate Parkin E3 ubiquitin ligase activity. *J. Cell Biol.* **2014**, *205*, 143–153. [CrossRef]
24. Nguyen, T.N.; Padman, B.S.; Lazarou, M. Deciphering the Molecular Signals of PINK1/Parkin Mitophagy. *Trends Cell Biol.* **2016**, *26*, 733–744. [CrossRef]
25. Lazarou, M.; Sliter, D.A.; Kane, L.A.; Sarraf, S.A.; Wang, C.; Burman, J.L.; Sideris, D.P.; Fogel, A.I.; Youle, R.J. The ubiquitin kinase PINK1 recruits autophagy receptors to induce mitophagy. *Nature* **2015**, *524*, 309–314. [CrossRef]
26. Tanaka, A.; Cleland, M.M.; Xu, S.; Narendra, D.P.; Suen, D.-F.; Karbowski, M.; Youle, R.J. Proteasome and p97 mediate mitophagy and degradation of mitofusins induced by Parkin. *J. Cell Biol.* **2010**, *191*, 1367–1380. [CrossRef]
27. Liu, L.; Feng, D.; Chen, G.; Chen, M.; Zheng, Q.; Song, P.; Ma, Q.; Zhu, C.; Wang, R.; Qi, W.; et al. Mitochondrial outer-membrane protein FUNDC1 mediates hypoxia-induced mitophagy in mammalian cells. *Nat. Cell Biol.* **2012**, *14*, 177–185. [CrossRef]
28. Chen, G.; Han, Z.; Feng, D.; Chen, Y.; Chen, L.; Wu, H.; Huang, L.; Zhou, C.; Cai, X.; Fu, C.; et al. A regulatory signaling loop comprising the PGAM5 phosphatase and CK2 controls receptor-mediated mitophagy. *Mol. Cell* **2014**, *54*, 362–377. [CrossRef]
29. Chen, M.; Chen, Z.; Wang, Y.; Tan, Z.; Zhu, C.; Li, Y.; Han, Z.; Chen, L.; Gao, R.; Liu, L.; et al. Mitophagy receptor FUNDC1 regulates mitochondrial dynamics and mitophagy. *Autophagy* **2016**, *12*, 689–702. [CrossRef]
30. Sowter, H.M.; Ratcliffe, P.J.; Watson, P.; Greenberg, A.H.; Harris, A.L. HIF-1-dependent regulation of hypoxic induction of the cell death factors BNIP3 and NIX in human tumors. *Cancer Res.* **2001**, *61*, 6669–6673.
31. Bruick, R.K. Expression of the gene encoding the proapoptotic Nip3 protein is induced by hypoxia. *Proc. Natl. Acad. Sci. USA* **2000**, *97*, 9082–9087. [CrossRef] [PubMed]
32. Gao, F.; Chen, D.; Si, J.; Hu, Q.; Qin, Z.; Fang, M.; Wang, G. The mitochondrial protein BNIP3L is the substrate of PARK2 and mediates mitophagy in PINK1/PARK2 pathway. *Hum. Mol. Genet.* **2015**, *24*, 2528–2538. [CrossRef] [PubMed]
33. Praharaj, P.P.; Naik, P.P.; Panigrahi, D.P.; Bhol, C.S.; Mahapatra, K.K.; Patra, S.; Sethi, G.; Bhutia, S.K. Intricate role of mitochondrial lipid in mitophagy and mitochondrial apoptosis: Its implication in cancer therapeutics. *Cell Mol. Life Sci.* **2019**, *76*, 1641–1652. [CrossRef] [PubMed]
34. Anton, Z.; Landajuela, A.; Hervas, J.H.; Montes, L.R.; Hernandez-Tiedra, S.; Velasco, G.; Goni, F.M.; Alonso, A. Human Atg8-cardiolipin interactions in mitophagy: Specific properties of LC3B, GABARAPL2 and GABARAP. *Autophagy* **2016**, *12*, 2386–2403. [CrossRef] [PubMed]
35. Paradies, G.; Petrosillo, G.; Paradies, V.; Ruggiero, F.M. Oxidative stress, mitochondrial bioenergetics, and cardiolipin in aging. *Free Radic. Biol. Med.* **2010**, *48*, 1286–1295. [CrossRef]
36. Pope, S.; Land, J.M.; Heales, S.J. Oxidative stress and mitochondrial dysfunction in neurodegeneration; cardiolipin a critical target? *Biochim. Biophys. Acta* **2008**, *1777*, 794–799. [CrossRef]
37. Murphy, M.P. How mitochondria produce reactive oxygen species. *Biochem. J.* **2009**, *417*, 1–13. [CrossRef]
38. Grivennikova, V.G.; Vinogradov, A.D. Generation of superoxide by the mitochondrial Complex I. *Biochim. Biophys. Acta* **2006**, *1757*, 553–561. [CrossRef]
39. Muller, F.L.; Liu, Y.; Van Remmen, H. Complex III releases superoxide to both sides of the inner mitochondrial membrane. *J. Biol. Chem.* **2004**, *279*, 49064–49073. [CrossRef]
40. Reczek, C.R.; Chandel, N.S. ROS-dependent signal transduction. *Curr. Opin Cell Biol.* **2015**, *33*, 8–13. [CrossRef]
41. Ren, X.; Zou, L.; Zhang, X.; Branco, V.; Wang, J.; Carvalho, C.; Holmgren, A.; Lu, J. Redox Signaling Mediated by Thioredoxin and Glutathione Systems in the Central Nervous System. *Antioxid. Redox Signal.* **2017**, *27*, 989–1010. [CrossRef]
42. Shefa, U.; Jeong, N.Y.; Song, I.O.; Chung, H.J.; Kim, D.; Jung, J.; Huh, Y. Mitophagy links oxidative stress conditions and neurodegenerative diseases. *Neural. Regen. Res.* **2019**, *14*, 749–756. [CrossRef]
43. Wang, Y.; Nartiss, Y.; Steipe, B.; McQuibban, G.A.; Kim, P.K. ROS-induced mitochondrial depolarization initiates PARK2/PARKIN-dependent mitochondrial degradation by autophagy. *Autophagy* **2012**, *8*, 1462–1476. [CrossRef]
44. Ma, S.; Zhang, X.; Zheng, L.; Li, Z.; Zhao, X.; Lai, W.; Shen, H.; Lv, J.; Yang, G.; Wang, Q.; et al. Peroxiredoxin 6 Is a Crucial Factor in the Initial Step of Mitochondrial Clearance and Is Upstream of the PINK1-Parkin Pathway. *Antioxid. Redox Signal.* **2016**, *24*, 486–501. [CrossRef]
45. Jiang, J.; Maeda, A.; Ji, J.; Baty, C.J.; Watkins, S.C.; Greenberger, J.S.; Kagan, V.E. Are mitochondrial reactive oxygen species required for autophagy? *Biochem. Biophys. Res. Commun.* **2011**, *412*, 55–60. [CrossRef]
46. Xiao, B.; Deng, X.; Lim, G.G.; Xie, S.; Zhou, Z.D.; Lim, K.L.; Tan, E.K. Correction: Superoxide drives progression of Parkin/PINK1-dependent mitophagy following translocation of Parkin to mitochondria. *Cell Death Dis.* **2018**, *9*, 794. [CrossRef]

47. Frank, M.; Duvezin-Caubet, S.; Koob, S.; Occhipinti, A.; Jagasia, R.; Petcherski, A.; Ruonala, M.O.; Priault, M.; Salin, B.; Reichert, A.S. Mitophagy is triggered by mild oxidative stress in a mitochondrial fission dependent manner. *Biochim. Biophys. Acta* **2012**, *1823*, 2297–2310. [CrossRef]
48. Shutt, T.; Geoffrion, M.; Milne, R.; McBride, H.M. The intracellular redox state is a core determinant of mitochondrial fusion. *EMBO Rep.* **2012**, *13*, 909–915. [CrossRef]
49. Cho, D.H.; Nakamura, T.; Fang, J.; Cieplak, P.; Godzik, A.; Gu, Z.; Lipton, S.A. S-nitrosylation of Drp1 mediates beta-amyloid-related mitochondrial fission and neuronal injury. *Science* **2009**, *324*, 102–105. [CrossRef]
50. Chung, K.K.; Thomas, B.; Li, X.; Pletnikova, O.; Troncoso, J.C.; Marsh, L.; Dawson, V.L.; Dawson, T.M. S-nitrosylation of parkin regulates ubiquitination and compromises parkin's protective function. *Science* **2004**, *304*, 1328–1331. [CrossRef]
51. Meng, F.; Yao, D.; Shi, Y.; Kabakoff, J.; Wu, W.; Reicher, J.; Ma, Y.; Moosmann, B.; Masliah, E.; Lipton, S.A.; et al. Oxidation of the cysteine-rich regions of parkin perturbs its E3 ligase activity and contributes to protein aggregation. *Mol. Neurodegener.* **2011**, *6*, 34. [CrossRef] [PubMed]
52. Yao, D.; Gu, Z.; Nakamura, T.; Shi, Z.Q.; Ma, Y.; Gaston, B.; Palmer, L.A.; Rockenstein, E.M.; Zhang, Z.; Masliah, E.; et al. Nitrosative stress linked to spo*Radic.* Parkinson's disease: S-nitrosylation of parkin regulates its E3 ubiquitin ligase activity. *Proc. Natl. Acad. Sci. USA* **2004**, *101*, 10810–10814. [CrossRef] [PubMed]
53. Ozawa, K.; Komatsubara, A.T.; Nishimura, Y.; Sawada, T.; Kawafune, H.; Tsumoto, H.; Tsuji, Y.; Zhao, J.; Kyotani, Y.; Tanaka, T.; et al. S-nitrosylation regulates mitochondrial quality control via activation of parkin. *Sci. Rep.* **2013**, *3*, 2202. [CrossRef] [PubMed]
54. Rizza, S.; Cardaci, S.; Montagna, C.; Di Giacomo, G.; De Zio, D.; Bordi, M.; Maiani, E.; Campello, S.; Borreca, A.; Puca, A.A.; et al. S-nitrosylation drives cell senescence and aging in mammals by controlling mitochondrial dynamics and mitophagy. *Proc. Natl. Acad. Sci. USA* **2018**, *115*, E3388–E3397. [CrossRef]
55. Cirotti, C.; Rizza, S.; Giglio, P.; Poerio, N.; Allega, M.F.; Claps, G.; Pecorari, C.; Lee, J.H.; Benassi, B.; Barila, D.; et al. Redox activation of ATM enhances GSNOR translation to sustain mitophagy and tolerance to oxidative stress. *EMBO Rep.* **2021**, *22*, e50500. [CrossRef]
56. Lee, S.H.; Lee, S.; Du, J.; Jain, K.; Ding, M.; Kadado, A.J.; Atteya, G.; Jaji, Z.; Tyagi, T.; Kim, W.H.; et al. Mitochondrial MsrB2 serves as a switch and transducer for mitophagy. *EMBO Mol. Med.* **2019**, *11*, e10409. [CrossRef]
57. Wauer, T.; Komander, D. Structure of the human Parkin ligase domain in an autoinhibited state. *EMBO J.* **2013**, *32*, 2099–2112. [CrossRef]
58. Lourenco Dos Santos, S.; Petropoulos, I.; Friguet, B. The Oxidized Protein Repair Enzymes Methionine Sulfoxide Reductases and Their Roles in Protecting against Oxidative Stress, in Ageing and in Regulating Protein Function. *Antioxidants* **2018**, *7*, 191. [CrossRef]
59. Scherz-Shouval, R.; Elazar, Z. Regulation of autophagy by ROS: Physiology and pathology. *Trends Biochem. Sci.* **2011**, *36*, 30–38. [CrossRef]
60. Dinkova-Kostova, A.T.; Abramov, A.Y. The emerging role of Nrf2 in mitochondrial function. *Free Radic. Biol. Med.* **2015**, *88*, 179–188. [CrossRef]
61. Murata, H.; Takamatsu, H.; Liu, S.; Kataoka, K.; Huh, N.H.; Sakaguchi, M. NRF2 Regulates PINK1 Expression under Oxidative Stress Conditions. *PLoS ONE* **2015**, *10*, e0142438. [CrossRef]
62. Georgakopoulos, N.D.; Frison, M.; Alvarez, M.S.; Bertrand, H.; Wells, G.; Campanella, M. Reversible Keap1 inhibitors are preferential pharmacological tools to modulate cellular mitophagy. *Sci. Rep.* **2017**, *7*, 10303. [CrossRef]
63. Komatsu, M.; Kurokawa, H.; Waguri, S.; Taguchi, K.; Kobayashi, A.; Ichimura, Y.; Sou, Y.S.; Ueno, I.; Sakamoto, A.; Tong, K.I.; et al. The selective autophagy substrate p62 activates the stress responsive transcription factor Nrf2 through inactivation of Keap1. *Nat. Cell Biol.* **2010**, *12*, 213–223. [CrossRef]
64. Jain, A.; Lamark, T.; Sjottem, E.; Larsen, K.B.; Awuh, J.A.; Overvatn, A.; McMahon, M.; Hayes, J.D.; Johansen, T. p62/SQSTM1 is a target gene for transcription factor NRF2 and creates a positive feedback loop by inducing antioxidant response element-driven gene transcription. *J. Biol. Chem.* **2010**, *285*, 22576–22591. [CrossRef]
65. Yamada, T.; Murata, D.; Adachi, Y.; Itoh, K.; Kameoka, S.; Igarashi, A.; Kato, T.; Araki, Y.; Huganir, R.L.; Dawson, T.M.; et al. Mitochondrial Stasis Reveals p62-Mediated Ubiquitination in Parkin-Independent Mitophagy and Mitigates Nonalcoholic Fatty Liver Disease. *Cell Metab.* **2018**, *28*, 588–604.e585. [CrossRef]
66. Jung, J.; Zhang, Y.; Celiku, O.; Zhang, W.; Song, H.; Williams, B.J.; Giles, A.J.; Rich, J.N.; Abounader, R.; Gilbert, M.R.; et al. Mitochondrial NIX Promotes Tumor Survival in the Hypoxic Niche of Glioblastoma. *Cancer Res.* **2019**, *79*, 5218–5232. [CrossRef]
67. Lo, S.C.; Hannink, M. PGAM5, a Bcl-XL-interacting protein, is a novel substrate for the redox-regulated Keap1-dependent ubiquitin ligase complex. *J. Biol. Chem.* **2006**, *281*, 37893–37903. [CrossRef]
68. Lee, G.; Won, H.S.; Lee, Y.M.; Choi, J.W.; Oh, T.I.; Jang, J.H.; Choi, D.K.; Lim, B.O.; Kim, Y.J.; Park, J.W.; et al. Oxidative Dimerization of PHD2 is Responsible for its Inactivation and Contributes to Metabolic Reprogramming via HIF-1alpha Activation. *Sci. Rep.* **2016**, *6*, 18928. [CrossRef]
69. Huang, C.; Han, Y.; Wang, Y.; Sun, X.; Yan, S.; Yeh, E.T.; Chen, Y.; Cang, H.; Li, H.; Shi, G.; et al. SENP3 is responsible for HIF-1 transactivation under mild oxidative stress via p300 de-SUMOylation. *EMBO J.* **2009**, *28*, 2748–2762. [CrossRef]

70. Chourasia, A.H.; Tracy, K.; Frankenberger, C.; Boland, M.L.; Sharifi, M.N.; Drake, L.E.; Sachleben, J.R.; Asara, J.M.; Locasale, J.W.; Karczmar, G.S.; et al. Mitophagy defects arising from BNip3 loss promote mammary tumor progression to metastasis. *EMBO Rep.* **2015**, *16*, 1145–1163. [CrossRef]
71. Yakhine-Diop, S.M.S.; Niso-Santano, M.; Rodriguez-Arribas, M.; Gomez-Sanchez, R.; Martinez-Chacon, G.; Uribe-Carretero, E.; Navarro-Garcia, J.A.; Ruiz-Hurtado, G.; Aiastui, A.; Cooper, J.M.; et al. Impaired Mitophagy and Protein Acetylation Levels in Fibroblasts from Parkinson's Disease Patients. *Mol. Neurobiol.* **2019**, *56*, 2466–2481. [CrossRef] [PubMed]
72. Costantini, S.; Sharma, A.; Raucci, R.; Costantini, M.; Autiero, I.; Colonna, G. Genealogy of an ancient protein family: The Sirtuins, a family of disordered members. *BMC Evol. Biol.* **2013**, *13*, 60. [CrossRef] [PubMed]
73. Osborne, B.; Bentley, N.L.; Montgomery, M.K.; Turner, N. The role of mitochondrial sirtuins in health and disease. *Free Radic. Biol. Med.* **2016**, *100*, 164–174. [CrossRef] [PubMed]
74. Zheng, Y.; Shi, B.; Ma, M.; Wu, X.; Lin, X. The novel relationship between Sirt3 and autophagy in myocardial ischemia-reperfusion. *J. Cell Physiol.* **2019**, *234*, 5488–5495. [CrossRef] [PubMed]
75. Tseng, A.H.; Shieh, S.S.; Wang, D.L. SIRT3 deacetylates FOXO3 to protect mitochondria against oxidative damage. *Free Radic. Biol. Med.* **2013**, *63*, 222–234. [CrossRef] [PubMed]
76. Mouchiroud, L.; Houtkooper, R.H.; Moullan, N.; Katsyuba, E.; Ryu, D.; Canto, C.; Mottis, A.; Jo, Y.S.; Viswanathan, M.; Schoonjans, K.; et al. The NAD(+)/Sirtuin Pathway Modulates Longevity through Activation of Mitochondrial UPR and FOXO Signaling. *Cell* **2013**, *154*, 430–441. [CrossRef] [PubMed]
77. Kerr, J.S.; Adriaanse, B.A.; Greig, N.H.; Mattson, M.P.; Cader, M.Z.; Bohr, V.A.; Fang, E.F. Mitophagy and Alzheimer's Disease: Cellular and Molecular Mechanisms. *Trends Neurosci.* **2017**, *40*, 151–166. [CrossRef]
78. Fang, E.F.; Kassahun, H.; Croteau, D.L.; Scheibye-Knudsen, M.; Marosi, K.; Lu, H.; Shamanna, R.A.; Kalyanasundaram, S.; Bollineni, R.C.; Wilson, M.A.; et al. NAD(+) Replenishment Improves Lifespan and Healthspan in Ataxia Telangiectasia Models via Mitophagy and DNA Repair. *Cell Metab.* **2016**, *24*, 566–581. [CrossRef]
79. Fang, E.F.; Hou, Y.; Palikaras, K.; Adriaanse, B.A.; Kerr, J.S.; Yang, B.; Lautrup, S.; Hasan-Olive, M.M.; Caponio, D.; Dan, X.; et al. Mitophagy inhibits amyloid-beta and tau pathology and reverses cognitive deficits in models of Alzheimer's disease. *Nat. Neurosci.* **2019**, *22*, 401–412. [CrossRef]
80. O'Sullivan, T.E.; Johnson, L.R.; Kang, H.H.; Sun, J.C. BNIP3- and BNIP3L-Mediated Mitophagy Promotes the Generation of Natural Killer Cell Memory. *Immunity* **2015**, *43*, 331–342. [CrossRef]
81. Hoshino, A.; Mita, Y.; Okawa, Y.; Ariyoshi, M.; Iwai-Kanai, E.; Ueyama, T.; Ikeda, K.; Ogata, T.; Matoba, S. Cytosolic p53 inhibits Parkin-mediated mitophagy and promotes mitochondrial dysfunction in the mouse heart. *Nat. Commun.* **2013**, *4*, 2308. [CrossRef]
82. Drummond, M.J.; Addison, O.; Brunker, L.; Hopkins, P.N.; McClain, D.A.; LaStayo, P.C.; Marcus, R.L. Downregulation of E3 ubiquitin ligases and mitophagy-related genes in skeletal muscle of physically inactive, frail older women: A cross-sectional comparison. *J. Gerontol. A Biol. Sci. Med. Sci.* **2014**, *69*, 1040–1048. [CrossRef]
83. Drake, J.C.; Yan, Z. Mitophagy in maintaining skeletal muscle mitochondrial proteostasis and metabolic health with ageing. *J. Physiol.* **2017**, *595*, 6391–6399. [CrossRef]
84. Garcia-Prat, L.; Martinez-Vicente, M.; Perdiguero, E.; Ortet, L.; Rodriguez-Ubreva, J.; Rebollo, E.; Ruiz-Bonilla, V.; Gutarra, S.; Ballestar, E.; Serrano, A.L.; et al. Autophagy maintains stemness by preventing senescence. *Nature* **2016**, *529*, 37–42. [CrossRef]
85. Sun, N.; Yun, J.; Liu, J.; Malide, D.; Liu, C.; Rovira, I.I.; Holmstrom, K.M.; Fergusson, M.M.; Yoo, Y.H.; Combs, C.A.; et al. Measuring In Vivo Mitophagy. *Mol. Cell* **2015**, *60*, 685–696. [CrossRef]
86. Sgarbi, G.; Matarrese, P.; Pinti, M.; Lanzarini, C.; Ascione, B.; Gibellini, L.; Dika, E.; Patrizi, A.; Tommasino, C.; Capri, M.; et al. Mitochondria hyperfusion and elevated autophagic activity are key mechanisms for cellular bioenergetic preservation in centenarians. *Aging* **2014**, *6*, 296–310. [CrossRef]
87. Suzuki, S.W.; Onodera, J.; Ohsumi, Y. Starvation induced cell death in autophagy-defective yeast mutants is caused by mitochondria dysfunction. *PLoS ONE* **2011**, *6*, e17412. [CrossRef]
88. Kurihara, Y.; Kanki, T.; Aoki, Y.; Hirota, Y.; Saigusa, T.; Uchiumi, T.; Kang, D. Mitophagy plays an essential role in reducing mitochondrial production of reactive oxygen species and mutation of mitochondrial DNA by maintaining mitochondrial quantity and quality in yeast. *J. Biol. Chem.* **2012**, *287*, 3265–3272. [CrossRef]
89. Palikaras, K.; Lionaki, E.; Tavernarakis, N. Coordination of mitophagy and mitochondrial biogenesis during ageing in C. elegans. *Nature* **2015**, *521*, 525–528. [CrossRef]
90. Fang, E.F.; Waltz, T.B.; Kassahun, H.; Lu, Q.; Kerr, J.S.; Morevati, M.; Fivenson, E.M.; Wollman, B.N.; Marosi, K.; Wilson, M.A.; et al. Tomatidine enhances lifespan and healthspan in C. elegans through mitophagy induction via the SKN-1/Nrf2 pathway. *Sci. Rep.* **2017**, *7*, 46208. [CrossRef]
91. Schiavi, A.; Maglioni, S.; Palikaras, K.; Shaik, A.; Strappazzon, F.; Brinkmann, V.; Torgovnick, A.; Castelein, N.; De Henau, S.; Braeckman, B.P.; et al. Iron-Starvation-Induced Mitophagy Mediates Lifespan Extension upon Mitochondrial Stress in C. elegans. *Curr. Biol. CB* **2015**, *25*, 1810–1822. [CrossRef]
92. Allen, G.F.; Toth, R.; James, J.; Ganley, I.G. Loss of iron triggers PINK1/Parkin-independent mitophagy. *EMBO Rep.* **2013**, *14*, 1127–1135. [CrossRef]
93. Clark, I.E.; Dodson, M.W.; Jiang, C.; Cao, J.H.; Huh, J.R.; Seol, J.H.; Yoo, S.J.; Hay, B.A.; Guo, M. Drosophila pink1 is required for mitochondrial function and interacts genetically with parkin. *Nature* **2006**, *441*, 1162–1166. [CrossRef] [PubMed]

94. Koehler, C.L.; Perkins, G.A.; Ellisman, M.H.; Jones, D.L. Pink1 and Parkin regulate Drosophila intestinal stem cell proliferation during stress and aging. *J. Cell Biol.* **2017**, *216*, 2315–2327. [CrossRef] [PubMed]
95. Si, H.; Ma, P.; Liang, Q.; Yin, Y.; Wang, P.; Zhang, Q.; Wang, S.; Deng, H. Overexpression of pink1 or parkin in indirect flight muscles promotes mitochondrial proteostasis and extends lifespan in Drosophila melanogaster. *PLoS ONE* **2019**, *14*, e0225214. [CrossRef]
96. Gouspillou, G.; Godin, R.; Piquereau, J.; Picard, M.; Mofarrahi, M.; Mathew, J.; Purves-Smith, F.M.; Sgarioto, N.; Hepple, R.T.; Burelle, Y.; et al. Protective role of Parkin in skeletal muscle contractile and mitochondrial function. *J. Physiol.* **2018**, *596*, 2565–2579. [CrossRef]
97. Leduc-Gaudet, J.P.; Reynaud, O.; Hussain, S.N.; Gouspillou, G. Parkin overexpression protects from ageing-related loss of muscle mass and strength. *J. Physiol.* **2019**, *597*, 1975–1991. [CrossRef]
98. Gouspillou, G.; Sgarioto, N.; Kapchinsky, S.; Purves-Smith, F.; Norris, B.; Pion, C.H.; Barbat-Artigas, S.; Lemieux, F.; Taivassalo, T.; Morais, J.A.; et al. Increased sensitivity to mitochondrial permeability transition and myonuclear translocation of endonuclease G in atrophied muscle of physically active older humans. *Faseb J.: Off. Publ. Fed. Am. Soc. Exp. Biol.* **2014**, *28*, 1621–1633. [CrossRef]
99. Sebastian, D.; Sorianello, E.; Segales, J.; Irazoki, A.; Ruiz-Bonilla, V.; Sala, D.; Planet, E.; Berenguer-Llergo, A.; Munoz, J.P.; Sanchez-Feutrie, M.; et al. Mfn2 deficiency links age-related sarcopenia and impaired autophagy to activation of an adaptive mitophagy pathway. *EMBO J.* **2016**, *35*, 1677–1693. [CrossRef]
100. Peng, C.; Rao, W.; Zhang, L.; Gao, F.; Hui, H.; Wang, K.; Dai, S.; Yang, Y.; Luo, P.; Ma, Y.; et al. Mitofusin 2 Exerts a Protective Role in Ischemia Reperfusion Injury Through Increasing Autophagy. *Cell Physiol. Biochem.* **2018**, *46*, 2311–2324. [CrossRef]
101. Xu, L.; Wu, Z.; He, Y.; Chen, Z.; Xu, K.; Yu, W.; Fang, W.; Ma, C.; Moqbel, S.A.A.; Ran, J.; et al. MFN2 contributes to metabolic disorders and inflammation in the aging of rat chondrocytes and osteoarthritis. *Osteoarthr. Cartil.* **2020**, *28*, 1079–1091. [CrossRef] [PubMed]
102. Barja, G. Aging in vertebrates, and the effect of caloric restriction: A mitochondrial free radical production-DNA damage mechanism? *Biol. Rev. Camb. Philos. Soc.* **2004**, *79*, 235–251. [CrossRef] [PubMed]
103. Hepple, R.T.; Baker, D.J.; McConkey, M.; Murynka, T.; Norris, R. Caloric restriction protects mitochondrial function with aging in skeletal and cardiac muscles. *Rejuvenat. Res.* **2006**, *9*, 219–222. [CrossRef]
104. Cui, J.; Shi, S.; Sun, X.; Cai, G.; Cui, S.; Hong, Q.; Chen, X.; Bai, X.Y. Mitochondrial autophagy involving renal injury and aging is modulated by caloric intake in aged rat kidneys. *PLoS ONE* **2013**, *8*, e69720. [CrossRef]
105. Madeo, F.; Carmona-Gutierrez, D.; Hofer, S.J.; Kroemer, G. Caloric Restriction Mimetics against Age-Associated Disease: Targets, Mechanisms, and Therapeutic Potential. *Cell Metab.* **2019**, *29*, 592–610. [CrossRef]
106. Eisenberg, T.; Knauer, H.; Schauer, A.; Buttner, S.; Ruckenstuhl, C.; Carmona-Gutierrez, D.; Ring, J.; Schroeder, S.; Magnes, C.; Antonacci, L.; et al. Induction of autophagy by spermidine promotes longevity. *Nat. Cell Biol.* **2009**, *11*, 1305–1314. [CrossRef]
107. Liang, W.; Moyzis, A.G.; Lampert, M.A.; Diao, R.Y.; Najor, R.H.; Gustafsson, A.B. Aging is associated with a decline in Atg9b-mediated autophagosome formation and appearance of enlarged mitochondria in the heart. *Aging Cell* **2020**, e13187. [CrossRef]
108. Eisenberg, T.; Abdellatif, M.; Schroeder, S.; Primessnig, U.; Stekovic, S.; Pendl, T.; Harger, A.; Schipke, J.; Zimmermann, A.; Schmidt, A.; et al. Cardioprotection and lifespan extension by the natural polyamine spermidine. *Nat. Med.* **2016**, *22*, 1428–1438. [CrossRef]
109. Morselli, E.; Marino, G.; Bennetzen, M.V.; Eisenberg, T.; Megalou, E.; Schroeder, S.; Cabrera, S.; Benit, P.; Rustin, P.; Criollo, A.; et al. Spermidine and resveratrol induce autophagy by distinct pathways converging on the acetylproteome. *J. Cell Biol.* **2011**, *192*, 615–629. [CrossRef]
110. Alcendor, R.R.; Gao, S.; Zhai, P.; Zablocki, D.; Holle, E.; Yu, X.; Tian, B.; Wagner, T.; Vatner, S.F.; Sadoshima, J. Sirt1 regulates aging and resistance to oxidative stress in the heart. *Circ. Res.* **2007**, *100*, 1512–1521. [CrossRef]
111. LaFrance, R.; Brustovetsky, N.; Sherburne, C.; Delong, D.; Dubinsky, J.M. Age-related changes in regional brain mitochondria from Fischer 344 rats. *Aging Cell* **2005**, *4*, 139–145. [CrossRef]
112. Scheibye-Knudsen, M.; Scheibye-Alsing, K.; Canugovi, C.; Croteau, D.L.; Bohr, V.A. A novel diagnostic tool reveals mitochondrial pathology in human diseases and aging. *Aging* **2013**, *5*, 192–208. [CrossRef]
113. Scheibye-Knudsen, M.; Ramamoorthy, M.; Sykora, P.; Maynard, S.; Lin, P.C.; Minor, R.K.; Wilson, D.M., 3rd; Cooper, M.; Spencer, R.; de Cabo, R.; et al. Cockayne syndrome group B protein prevents the accumulation of damaged mitochondria by promoting mitochondrial autophagy. *J. Exp. Med.* **2012**, *209*, 855–869. [CrossRef]
114. Scheibye-Knudsen, M.; Fang, E.F.; Croteau, D.L.; Bohr, V.A. Contribution of defective mitophagy to the neurodegeneration in DNA repair-deficient disorders. *Autophagy* **2014**, *10*, 1468–1469. [CrossRef]
115. Valentin-Vega, Y.A.; Maclean, K.H.; Tait-Mulder, J.; Milasta, S.; Steeves, M.; Dorsey, F.C.; Cleveland, J.L.; Green, D.R.; Kastan, M.B. Mitochondrial dysfunction in ataxia-telangiectasia. *Blood* **2012**, *119*, 1490–1500. [CrossRef]
116. Fang, E.F.; Scheibye-Knudsen, M.; Brace, L.E.; Kassahun, H.; SenGupta, T.; Nilsen, H.; Mitchell, J.R.; Croteau, D.L.; Bohr, V.A. Defective mitophagy in XPA via PARP-1 hyperactivation and NAD(+)/SIRT1 reduction. *Cell* **2014**, *157*, 882–896. [CrossRef] [PubMed]
117. Lane, C.A.; Hardy, J.; Schott, J.M. Alzheimer's disease. *Eur. J. Neurol.* **2018**, *25*, 59–70. [CrossRef]
118. Cacace, R.; Sleegers, K.; Van Broeckhoven, C. Molecular genetics of early-onset Alzheimer's disease revisited. *Alzheimers Dement.* **2016**, *12*, 733–748. [CrossRef]

119. Navarro, A.; Boveris, A. Brain mitochondrial dysfunction in aging, neurodegeneration, and Parkinson's disease. *Front. Aging Neurosci.* **2010**, 2. [CrossRef]
120. Hauptmann, S.; Scherping, I.; Drose, S.; Brandt, U.; Schulz, K.L.; Jendrach, M.; Leuner, K.; Eckert, A.; Muller, W.E. Mitochondrial dysfunction: An early event in Alzheimer pathology accumulates with age in AD transgenic mice. *Neurobiol. Aging* **2009**, *30*, 1574–1586. [CrossRef]
121. Rhein, V.; Song, X.; Wiesner, A.; Ittner, L.M.; Baysang, G.; Meier, F.; Ozmen, L.; Bluethmann, H.; Drose, S.; Brandt, U.; et al. Amyloid-beta and tau synergistically impair the oxidative phosphorylation system in triple transgenic Alzheimer's disease mice. *Proc. Natl. Acad. Sci. USA* **2009**, *106*, 20057–20062. [CrossRef] [PubMed]
122. Leuner, K.; Schutt, T.; Kurz, C.; Eckert, S.H.; Schiller, C.; Occhipinti, A.; Mai, S.; Jendrach, M.; Eckert, G.P.; Kruse, S.E.; et al. Mitochondrion-derived reactive oxygen species lead to enhanced amyloid beta formation. *Antioxid. Redox Signal.* **2012**, *16*, 1421–1433. [CrossRef] [PubMed]
123. Moreira, P.I.; Siedlak, S.L.; Wang, X.; Santos, M.S.; Oliveira, C.R.; Tabaton, M.; Nunomura, A.; Szweda, L.I.; Aliev, G.; Smith, M.A.; et al. Increased autophagic degradation of mitochondria in Alzheimer disease. *Autophagy* **2007**, *3*, 614–615. [CrossRef] [PubMed]
124. Reddy, P.H.; Yin, X.; Manczak, M.; Kumar, S.; Pradeepkiran, J.A.; Vijayan, M.; Reddy, A.P. Mutant APP and amyloid beta-induced defective autophagy, mitophagy, mitochondrial structural and functional changes and synaptic damage in hippocampal neurons from Alzheimer's disease. *Hum. Mol. Genet.* **2018**, *27*, 2502–2516. [CrossRef]
125. Dodson, M.W.; Guo, M. Pink1, Parkin, DJ-1 and mitochondrial dysfunction in Parkinson's disease. *Curr. Opin. Neurobiol.* **2007**, *17*, 331–337. [CrossRef]
126. Cai, Q.; Jeong, Y.Y. Mitophagy in Alzheimer's Disease and Other Age-Related Neurodegenerative Diseases. *Cells* **2020**, *9*, 150. [CrossRef]
127. Ye, X.; Sun, X.; Starovoytov, V.; Cai, Q. Parkin-mediated mitophagy in mutant hAPP neurons and Alzheimer's disease patient brains. *Hum. Mol. Genet.* **2015**, *24*, 2938–2951. [CrossRef]
128. Martin-Maestro, P.; Gargini, R.; Perry, G.; Avila, J.; Garcia-Escudero, V. PARK2 enhancement is able to compensate mitophagy alterations found in spo*Radic.* Alzheimer's disease. *Hum. Mol. Genet.* **2016**, *25*, 792–806. [CrossRef]
129. Hu, Y.; Li, X.C.; Wang, Z.H.; Luo, Y.; Zhang, X.; Liu, X.P.; Feng, Q.; Wang, Q.; Yue, Z.; Chen, Z.; et al. Tau accumulation impairs mitophagy via increasing mitochondrial membrane potential and reducing mitochondrial Parkin. *Oncotarget* **2016**, *7*, 17356–17368. [CrossRef]
130. Cummins, N.; Tweedie, A.; Zuryn, S.; Bertran-Gonzalez, J.; Gotz, J. Disease-associated tau impairs mitophagy by inhibiting Parkin translocation to mitochondria. *EMBO J.* **2019**, 38. [CrossRef]
131. Manczak, M.; Reddy, P.H. Abnormal interaction between the mitochondrial fission protein Drp1 and hyperphosphorylated tau in Alzheimer's disease neurons: Implications for mitochondrial dysfunction and neuronal damage. *Hum. Mol. Genet.* **2012**, *21*, 2538–2547. [CrossRef]
132. Kandimalla, R.; Manczak, M.; Yin, X.; Wang, R.; Reddy, P.H. Hippocampal phosphorylated tau induced cognitive decline, dendritic spine loss and mitochondrial abnormalities in a mouse model of Alzheimer's disease. *Hum. Mol. Genet.* **2018**, *27*, 30–40. [CrossRef]
133. Paradies, G.; Ruggiero, F.M.; Dinoi, P.; Petrosillo, G.; Quagliariello, E. Decreased cytochrome oxidase activity and changes in phospholipids in heart mitochondria from hypothyroid rats. *Arch. Biochem. Biophys.* **1993**, *307*, 91–95. [CrossRef]
134. Schlame, M.; Hostetler, K.Y. Cardiolipin synthase from mammalian mitochondria. *Biochim. Biophys. Acta* **1997**, *1348*, 207–213. [CrossRef]
135. Monteiro-Cardoso, V.F.; Oliveira, M.M.; Melo, T.; Domingues, M.R.; Moreira, P.I.; Ferreiro, E.; Peixoto, F.; Videira, R.A. Cardiolipin profile changes are associated to the early synaptic mitochondrial dysfunction in Alzheimer's disease. *J. Alzheimers Dis.* **2015**, *43*, 1375–1392. [CrossRef]
136. Lee, S.; Sato, Y.; Nixon, R.A. Lysosomal proteolysis inhibition selectively disrupts axonal transport of degradative organelles and causes an Alzheimer's-like axonal dystrophy. *J. Neurosci.* **2011**, *31*, 7817–7830. [CrossRef]
137. Nixon, R.A.; Yang, D.S. Autophagy failure in Alzheimer's disease–locating the primary defect. *Neurobiol. Dis.* **2011**, *43*, 38–45. [CrossRef]
138. Nixon, R.A. The role of autophagy in neurodegenerative disease. *Nat. Med.* **2013**, *19*, 983–997. [CrossRef]
139. Tammineni, P.; Jeong, Y.Y.; Feng, T.; Aikal, D.; Cai, Q. Impaired axonal retrograde trafficking of the retromer complex augments lysosomal deficits in Alzheimer's disease neurons. *Hum. Mol. Genet.* **2017**, *26*, 4352–4366. [CrossRef]
140. Swerdlow, R.H.; Burns, J.M.; Khan, S.M. The Alzheimer's disease mitochondrial cascade hypothesis: Progress and perspectives. *Biochim. Biophys. Acta* **2014**, *1842*, 1219–1231. [CrossRef]
141. Clark, E.H.; Vazquez de la Torre, A.; Hoshikawa, T.; Briston, T. Targeting mitophagy in Parkinson's disease. *J. Biol. Chem.* **2020**. [CrossRef]
142. Nussbaum, R.L.; Polymeropoulos, M.H. Genetics of Parkinson's disease. *Hum. Mol. Genet.* **1997**, *6*, 1687–1691. [CrossRef] [PubMed]
143. Balestrino, R.; Schapira, A.H.V. Parkinson disease. *Eur. J. Neurol.* **2020**, *27*, 27–42. [CrossRef] [PubMed]
144. Tysnes, O.B.; Storstein, A. Epidemiology of Parkinson's disease. *J. Neural. Transm.* **2017**, *124*, 901–905. [CrossRef] [PubMed]
145. Trist, B.G.; Hare, D.J.; Double, K.L. Oxidative stress in the aging substantia nigra and the etiology of Parkinson's disease. *Aging Cell* **2019**, *18*, e13031. [CrossRef]

146. Dias, V.; Junn, E.; Mouradian, M.M. The role of oxidative stress in Parkinson's disease. *J. Parkinsons Dis.* **2013**, *3*, 461–491. [CrossRef]
147. Zhu, J.H.; Guo, F.; Shelburne, J.; Watkins, S.; Chu, C.T. Localization of phosphorylated ERK/MAP kinases to mitochondria and autophagosomes in Lewy body diseases. *Brain Pathol.* **2003**, *13*, 473–481. [CrossRef]
148. Elstner, M.; Morris, C.M.; Heim, K.; Bender, A.; Mehta, D.; Jaros, E.; Klopstock, T.; Meitinger, T.; Turnbull, D.M.; Prokisch, H. Expression analysis of dopaminergic neurons in Parkinson's disease and aging links transcriptional dysregulation of energy metabolism to cell death. *Acta Neuropathol.* **2011**, *122*, 75–86. [CrossRef]
149. Schapira, A.H.; Cooper, J.M.; Dexter, D.; Clark, J.B.; Jenner, P.; Marsden, C.D. Mitochondrial complex I deficiency in Parkinson's disease. *J. Neurochem.* **1990**, *54*, 823–827. [CrossRef]
150. Shults, C.W.; Haas, R.H.; Passov, D.; Beal, M.F. Coenzyme Q10 levels correlate with the activities of complexes I and II/III in mitochondria from parkinsonian and nonparkinsonian subjects. *Ann. Neurol.* **1997**, *42*, 261–264. [CrossRef]
151. Keeney, P.M.; Xie, J.; Capaldi, R.A.; Bennett, J.P., Jr. Parkinson's disease brain mitochondrial complex I has oxidatively damaged subunits and is functionally impaired and misassembled. *J. Neurosci.* **2006**, *26*, 5256–5264. [CrossRef]
152. Liu, J.; Liu, W.; Li, R.; Yang, H. Mitophagy in Parkinson's Disease: From Pathogenesis to Treatment. *Cells* **2019**, *8*, 712. [CrossRef]
153. Nakamura, K.; Nemani, V.M.; Azarbal, F.; Skibinski, G.; Levy, J.M.; Egami, K.; Munishkina, L.; Zhang, J.; Gardner, B.; Wakabayashi, J.; et al. Direct membrane association drives mitochondrial fission by the Parkinson disease-associated protein alpha-synuclein. *J. Biol. Chem.* **2011**, *286*, 20710–20726. [CrossRef]
154. Papkovskaia, T.D.; Chau, K.Y.; Inesta-Vaquera, F.; Papkovsky, D.B.; Healy, D.G.; Nishio, K.; Staddon, J.; Duchen, M.R.; Hardy, J.; Schapira, A.H.; et al. G2019S leucine-rich repeat kinase 2 causes uncoupling protein-mediated mitochondrial depolarization. *Hum. Mol. Genet.* **2012**, *21*, 4201–4213. [CrossRef]
155. Ramonet, D.; Podhajska, A.; Stafa, K.; Sonnay, S.; Trancikova, A.; Tsika, E.; Pletnikova, O.; Troncoso, J.C.; Glauser, L.; Moore, D.J. PARK9-associated ATP13A2 localizes to intracellular acidic vesicles and regulates cation homeostasis and neuronal integrity. *Hum. Mol. Genet.* **2012**, *21*, 1725–1743. [CrossRef]
156. Wang, W.; Wang, X.; Fujioka, H.; Hoppel, C.; Whone, A.L.; Caldwell, M.A.; Cullen, P.J.; Liu, J.; Zhu, X. Parkinson's disease-associated mutant VPS35 causes mitochondrial dysfunction by recycling DLP1 complexes. *Nat. Med.* **2016**, *22*, 54–63. [CrossRef]
157. Blesa, J.; Trigo-Damas, I.; Quiroga-Varela, A.; Jackson-Lewis, V.R. Oxidative stress and Parkinson's disease. *Front. Neuroanat.* **2015**, *9*, 91. [CrossRef]
158. Ferrer, I.; Martinez, A.; Blanco, R.; Dalfo, E.; Carmona, M. Neuropathology of spo*Radic.* Parkinson disease before the appearance of parkinsonism: Preclinical Parkinson disease. *J. Neural. Transm.* **2011**, *118*, 821–839. [CrossRef]
159. Lucking, C.B.; Durr, A.; Bonifati, V.; Vaughan, J.; De Michele, G.; Gasser, T.; Harhangi, B.S.; Meco, G.; Denefle, P.; Wood, N.W.; et al. Association between early-onset Parkinson's disease and mutations in the parkin gene. *N. Engl. J. Med.* **2000**, *342*, 1560–1567. [CrossRef]
160. Tan, E.K.; Yew, K.; Chua, E.; Puvan, K.; Shen, H.; Lee, E.; Puong, K.Y.; Zhao, Y.; Pavanni, R.; Wong, M.C.; et al. PINK1 mutations in spo*Radic.* early-onset Parkinson's disease. *Mov. Disord.* **2006**, *21*, 789–793. [CrossRef]
161. Bonifati, V.; Rizzu, P.; van Baren, M.J.; Schaap, O.; Breedveld, G.J.; Krieger, E.; Dekker, M.C.; Squitieri, F.; Ibanez, P.; Joosse, M.; et al. Mutations in the DJ-1 gene associated with autosomal recessive early-onset parkinsonism. *Science* **2003**, *299*, 256–259. [CrossRef]
162. Blauwendraat, C.; Nalls, M.A.; Singleton, A.B. The genetic architecture of Parkinson's disease. *Lancet Neurol.* **2020**, *19*, 170–178. [CrossRef]
163. Yan, M.H.; Wang, X.; Zhu, X. Mitochondrial defects and oxidative stress in Alzheimer disease and Parkinson disease. *Free Radic. Biol. Med.* **2013**, *62*, 90–101. [CrossRef] [PubMed]
164. Irrcher, I.; Aleyasin, H.; Seifert, E.L.; Hewitt, S.J.; Chhabra, S.; Phillips, M.; Lutz, A.K.; Rousseaux, M.W.; Bevilacqua, L.; Jahani-Asl, A.; et al. Loss of the Parkinson's disease-linked gene DJ-1 perturbs mitochondrial dynamics. *Hum. Mol. Genet.* **2010**, *19*, 3734–3746. [CrossRef] [PubMed]
165. Krebiehl, G.; Ruckerbauer, S.; Burbulla, L.F.; Kieper, N.; Maurer, B.; Waak, J.; Wolburg, H.; Gizatullina, Z.; Gellerich, F.N.; Woitalla, D.; et al. Reduced basal autophagy and impaired mitochondrial dynamics due to loss of Parkinson's disease-associated protein DJ-1. *PLoS ONE* **2010**, *5*, e9367. [CrossRef] [PubMed]
166. Cookson, M.R. The biochemistry of Parkinson's disease. *Annu. Rev. Biochem.* **2005**, *74*, 29–52. [CrossRef] [PubMed]
167. Goedert, M. Alpha-synuclein and neurodegenerative diseases. *Nat. Rev. Neurosci.* **2001**, *2*, 492–501. [CrossRef] [PubMed]
168. Yavich, L.; Jakala, P.; Tanila, H. Abnormal compartmentalization of norepinephrine in mouse dentate gyrus in alpha-synuclein knockout and A30P transgenic mice. *J. Neurochem.* **2006**, *99*, 724–732. [CrossRef] [PubMed]
169. Yavich, L.; Tanila, H.; Vepsalainen, S.; Jakala, P. Role of alpha-synuclein in presynaptic dopamine recruitment. *J. Neurosci.* **2004**, *24*, 11165–11170. [CrossRef] [PubMed]
170. Abeliovich, A.; Schmitz, Y.; Farinas, I.; Choi-Lundberg, D.; Ho, W.H.; Castillo, P.E.; Shinsky, N.; Verdugo, J.M.; Armanini, M.; Ryan, A.; et al. Mice lacking alpha-synuclein display functional deficits in the nigrostriatal dopamine system. *Neuron* **2000**, *25*, 239–252. [CrossRef]
171. Stefanis, L. alpha-Synuclein in Parkinson's disease. *Cold Spring Harb. Perspect. Med.* **2012**, *2*, a009399. [CrossRef]
172. Je, G.; Kim, Y.S. Mitochondrial ROS-mediated post-transcriptional regulation of alpha-synuclein through miR-7 and miR-153. *Neurosci. Lett.* **2017**, *661*, 132–136. [CrossRef]

173. Brettschneider, J.; Del Tredici, K.; Lee, V.M.; Trojanowski, J.Q. Spreading of pathology in neurodegenerative diseases: A focus on human studies. *Nat. Rev. Neurosci.* **2015**, *16*, 109–120. [CrossRef]
174. Shaltouki, A.; Hsieh, C.H.; Kim, M.J.; Wang, X. Alpha-synuclein delays mitophagy and targeting Miro rescues neuron loss in Parkinson's models. *Acta Neuropathol.* **2018**, *136*, 607–620. [CrossRef]
175. Jucker, M.; Walker, L.C. Self-propagation of pathogenic protein aggregates in neurodegenerative diseases. *Nature* **2013**, *501*, 45–51. [CrossRef]
176. Ischiropoulos, H.; Beckman, J.S. Oxidative stress and nitration in neurodegeneration: Cause, effect, or association? *J. Clin. Investig.* **2003**, *111*, 163–169. [CrossRef]
177. Paxinou, E.; Chen, Q.; Weisse, M.; Giasson, B.I.; Norris, E.H.; Rueter, S.M.; Trojanowski, J.Q.; Lee, V.M.; Ischiropoulos, H. Induction of alpha-synuclein aggregation by intracellular nitrative insult. *J. Neurosci.* **2001**, *21*, 8053–8061. [CrossRef]
178. Scudamore, O.; Ciossek, T. Increased Oxidative Stress Exacerbates alpha-Synuclein Aggregation In Vivo. *J. Neuropathol. Exp. Neurol.* **2018**, *77*, 443–453. [CrossRef]
179. Bose, A.; Beal, M.F. Mitochondrial dysfunction in Parkinson's disease. *J. Neurochem.* **2016**, *139* (Suppl. 1), 216–231. [CrossRef]
180. Junn, E.; Mouradian, M.M. Human alpha-synuclein over-expression increases intracellular reactive oxygen species levels and susceptibility to dopamine. *Neurosci. Lett.* **2002**, *320*, 146–150. [CrossRef]
181. Berg, D.; Schweitzer, K.J.; Leitner, P.; Zimprich, A.; Lichtner, P.; Belcredi, P.; Brussel, T.; Schulte, C.; Maass, S.; Nagele, T.; et al. Type and frequency of mutations in the LRRK2 gene in familial and spo*Radic.* Parkinson's disease*. *Brain* **2005**, *128*, 3000–3011. [CrossRef]
182. Bonifati, V. Deciphering Parkinson's disease–PARK8. *Lancet Neurol.* **2002**, *1*, 83. [CrossRef]
183. Cookson, M.R. The role of leucine-rich repeat kinase 2 (LRRK2) in Parkinson's disease. *Nat. Rev. Neurosci.* **2010**, *11*, 791–797. [CrossRef]
184. Glater, E.E.; Megeath, L.J.; Stowers, R.S.; Schwarz, T.L. Axonal transport of mitochondria requires milton to recruit kinesin heavy chain and is light chain independent. *J. Cell Biol.* **2006**, *173*, 545–557. [CrossRef]
185. Koutsopoulos, O.S.; Laine, D.; Osellame, L.; Chudakov, D.M.; Parton, R.G.; Frazier, A.E.; Ryan, M.T. Human Miltons associate with mitochondria and induce microtubule-dependent remodeling of mitochondrial networks. *Biochim. Biophys. Acta* **2010**, *1803*, 564–574. [CrossRef]
186. Wang, X.; Schwarz, T.L. The mechanism of Ca2+ -dependent regulation of kinesin-mediated mitochondrial motility. *Cell* **2009**, *136*, 163–174. [CrossRef]
187. Hsieh, C.H.; Shaltouki, A.; Gonzalez, A.E.; Bettencourt da Cruz, A.; Burbulla, L.F.; St Lawrence, E.; Schule, B.; Krainc, D.; Palmer, T.D.; Wang, X. Functional Impairment in Miro Degradation and Mitophagy Is a Shared Feature in Familial and Spo*Radic.* Parkinson's Disease. *Cell Stem. Cell* **2016**, *19*, 709–724. [CrossRef]
188. Seibler, P.; Graziotto, J.; Jeong, H.; Simunovic, F.; Klein, C.; Krainc, D. Mitochondrial Parkin recruitment is impaired in neurons derived from mutant PINK1 induced pluripotent stem cells. *J. Neurosci.* **2011**, *31*, 5970–5976. [CrossRef]
189. Song, W.; Song, Y.; Kincaid, B.; Bossy, B.; Bossy-Wetzel, E. Mutant SOD1G93A triggers mitochondrial fragmentation in spinal cord motor neurons: Neuroprotection by SIRT3 and PGC-1alpha. *Neurobiol. Dis.* **2013**, *51*, 72–81. [CrossRef]
190. Shi, H.; Deng, H.X.; Gius, D.; Schumacker, P.T.; Surmeier, D.J.; Ma, Y.C. Sirt3 protects dopaminergic neurons from mitochondrial oxidative stress. *Hum. Mol. Genet.* **2017**, *26*, 1915–1926. [CrossRef]
191. Mattson, M.P. Apoptosis in neurodegenerative disorders. *Nat. Rev. Mol. Cell Biol.* **2000**, *1*, 120–129. [CrossRef] [PubMed]
192. Schwab, A.J.; Sison, S.L.; Meade, M.R.; Broniowska, K.A.; Corbett, J.A.; Ebert, A.D. Decreased Sirtuin Deacetylase Activity in LRRK2 G2019S iPSC-Derived Dopaminergic Neurons. *Stem. Cell Rep.* **2017**, *9*, 1839–1852. [CrossRef] [PubMed]
193. Koentjoro, B.; Park, J.S.; Sue, C.M. Nix restores mitophagy and mitochondrial function to protect against PINK1/Parkin-related Parkinson's disease. *Sci. Rep.* **2017**, *7*, 44373. [CrossRef] [PubMed]
194. Strappazzon, F.; Nazio, F.; Corrado, M.; Cianfanelli, V.; Romagnoli, A.; Fimia, G.M.; Campello, S.; Nardacci, R.; Piacentini, M.; Campanella, M.; et al. AMBRA1 is able to induce mitophagy via LC3 binding, regardless of PARKIN and p62/SQSTM1. *Cell Death Differ.* **2015**, *22*, 419–432. [CrossRef] [PubMed]
195. Di Rita, A.; D'Acunzo, P.; Simula, L.; Campello, S.; Strappazzon, F.; Cecconi, F. AMBRA1-Mediated Mitophagy Counteracts Oxidative Stress and Apoptosis Induced by Neurotoxicity in Human Neuroblastoma SH-SY5Y Cells. *Front. Cell Neurosci.* **2018**, *12*, 92. [CrossRef] [PubMed]
196. Rana, A.; Rera, M.; Walker, D.W. Parkin overexpression during aging reduces proteotoxicity, alters mitochondrial dynamics, and extends lifespan. *Proc. Natl. Acad. Sci. USA* **2013**, *110*, 8638–8643. [CrossRef] [PubMed]
197. Youn, C.K.; Jun, Y.; Jo, E.R.; Cho, S.I. Age-Related Hearing Loss in C57BL/6J Mice Is Associated with Mitophagy Impairment in the Central Auditory System. *Int. J. Mol. Sci.* **2020**, *21*, 7202. [CrossRef]
198. Gao, B.; Yu, W.; Lv, P.; Liang, X.; Sun, S.; Zhang, Y. Parkin overexpression alleviates cardiac aging through facilitating K63-polyubiquitination of TBK1 to facilitate mitophagy. *Biochim. Biophys. Acta Mol. Basis Dis.* **2021**, *1867*, 165997. [CrossRef]
199. Song, C.; Zhang, J.; Qi, S.; Liu, Z.; Zhang, X.; Zheng, Y.; Andersen, J.P.; Zhang, W.; Strong, R.; Martinez, P.A.; et al. Cardiolipin remodeling by ALCAT1 links mitochondrial dysfunction to Parkinson's diseases. *Aging Cell* **2019**, *18*, e12941. [CrossRef]
200. Ismail, I.A.; El-Bakry, H.A.; Soliman, S.S. Melatonin and tumeric ameliorate aging-induced changes: Implication of immunoglobulins, cytokines, DJ-1/NRF2 and apoptosis regulation. *Int. J. Physiol. PathoPhysiol. Pharm.* **2018**, *10*, 70–82.

201. Baulac, S.; Lu, H.; Strahle, J.; Yang, T.; Goldberg, M.S.; Shen, J.; Schlossmacher, M.G.; Lemere, C.A.; Lu, Q.; Xia, W. Increased DJ-1 expression under oxidative stress and in Alzheimer's disease brains. *Mol. Neurodegener.* **2009**, *4*, 12. [CrossRef]
202. Hyun, D.H.; Lee, M.; Hattori, N.; Kubo, S.; Mizuno, Y.; Halliwell, B.; Jenner, P. Effect of wild-type or mutant Parkin on oxidative damage, nitric oxide, antioxidant defenses, and the proteasome. *J. Biol. Chem.* **2002**, *277*, 28572–28577. [CrossRef]
203. Jiang, H.; Ren, Y.; Zhao, J.; Feng, J. Parkin protects human dopaminergic neuroblastoma cells against dopamine-induced apoptosis. *Hum. Mol. Genet.* **2004**, *13*, 1745–1754. [CrossRef]
204. Jiang, H.; Jiang, Q.; Liu, W.; Feng, J. Parkin suppresses the expression of monoamine oxidases. *J. Biol. Chem.* **2006**, *281*, 8591–8599. [CrossRef]
205. Pridgeon, J.W.; Olzmann, J.A.; Chin, L.S.; Li, L. PINK1 protects against oxidative stress by phosphorylating mitochondrial chaperone TRAP1. *PLoS Biol.* **2007**, *5*, e172. [CrossRef]
206. Yokota, T.; Sugawara, K.; Ito, K.; Takahashi, R.; Ariga, H.; Mizusawa, H. Down regulation of DJ-1 enhances cell death by oxidative stress, ER stress, and proteasome inhibition. *Biochem. Biophys. Res. Commun.* **2003**, *312*, 1342–1348. [CrossRef]
207. Takahashi-Niki, K.; Niki, T.; Taira, T.; Iguchi-Ariga, S.M.; Ariga, H. Reduced anti-oxidative stress activities of DJ-1 mutants found in Parkinson's disease patients. *Biochem. Biophys. Res. Commun.* **2004**, *320*, 389–397. [CrossRef]
208. Meulener, M.; Whitworth, A.J.; Armstrong-Gold, C.E.; Rizzu, P.; Heutink, P.; Wes, P.D.; Pallanck, L.J.; Bonini, N.M. Drosophila DJ-1 mutants are selectively sensitive to environmental toxins associated with Parkinson's disease. *Curr. Biol. CB* **2005**, *15*, 1572–1577. [CrossRef]
209. Lavara-Culebras, E.; Paricio, N. Drosophila DJ-1 mutants are sensitive to oxidative stress and show reduced lifespan and motor deficits. *Gene* **2007**, *400*, 158–165. [CrossRef]
210. Kim, R.H.; Smith, P.D.; Aleyasin, H.; Hayley, S.; Mount, M.P.; Pownall, S.; Wakeham, A.; You-Ten, A.J.; Kalia, S.K.; Horne, P.; et al. Hypersensitivity of DJ-1-deficient mice to 1-methyl-4-phenyl-1,2,3,6-tetrahydropyrindine (MPTP) and oxidative stress. *Proc. Natl. Acad. Sci. USA* **2005**, *102*, 5215–5220. [CrossRef]
211. Wu, Z.; Zhu, Y.; Cao, X.; Sun, S.; Zhao, B. Mitochondrial toxic effects of Abeta through mitofusins in the early pathogenesis of Alzheimer's disease. *Mol. Neurobiol.* **2014**, *50*, 986–996. [CrossRef]
212. Perez, V.I.; Bokov, A.; Van Remmen, H.; Mele, J.; Ran, Q.; Ikeno, Y.; Richardson, A. Is the oxidative stress theory of aging dead? *Biochim. Biophys. Acta* **2009**, *1790*, 1005–1014. [CrossRef]

antioxidants

MDPI

Review

How Cells Deal with the Fluctuating Environment: Autophagy Regulation under Stress in Yeast and Mammalian Systems

Yuchen Lei [1,2], Yuxiang Huang [1,2,†], Xin Wen [1,2,†], Zhangyuan Yin [1,2,†], Zhihai Zhang [1,2,†] and Daniel J. Klionsky [1,2,*]

1 Life Sciences Institute, University of Michigan, Ann Arbor, MI 48109, USA; yclei@umich.edu (Y.L.); yxhuang@umich.edu (Y.H.); xinwen@umich.edu (X.W.); zyyin@umich.edu (Z.Y.); zhihaiz@umich.edu (Z.Z.)

2 Department of Molecular, Cellular and Developmental Biology, University of Michigan, Ann Arbor, MI 48109, USA

* Correspondence: klionsky@umich.edu

† These authors contributed equally to this work.

Abstract: Eukaryotic cells frequently experience fluctuations of the external and internal environments, such as changes in nutrient, energy and oxygen sources, and protein folding status, which, after reaching a particular threshold, become a type of stress. Cells develop several ways to deal with these various types of stress to maintain homeostasis and survival. Among the cellular survival mechanisms, autophagy is one of the most critical ways to mediate metabolic adaptation and clearance of damaged organelles. Autophagy is maintained at a basal level under normal growing conditions and gets stimulated by stress through different but connected mechanisms. In this review, we summarize the advances in understanding the autophagy regulation mechanisms under multiple types of stress including nutrient, energy, oxidative, and ER stress in both yeast and mammalian systems.

Keywords: autophagy; energy stress; ER stress; nutrient stress; oxidative stress; regulation

Citation: Lei, Y.; Huang, Y.; Wen, X.; Yin, Z.; Zhang, Z.; Klionsky, D.J. How Cells Deal with the Fluctuating Environment: Autophagy Regulation under Stress in Yeast and Mammalian Systems. *Antioxidants* **2022**, *11*, 304. https://doi.org/10.3390/antiox11020304

Academic Editors: Angeles Aroca and Cecilia Gotor

Received: 2 January 2022
Accepted: 31 January 2022
Published: 2 February 2022

1. Overview of Autophagy in Yeast and Mammals

Autophagy is a highly regulated cellular degradation and recycling process, conserved from yeast to more complex eukaryotes [1]. The proteasome is responsible for degrading most short-lived, individual proteins, therefore, autophagy can degrade and recycle long-lived proteins, large protein complexes, and organelles [2]. The key definition of autophagy is the delivery and, typically, the degradation of cytoplasmic cargo within the lysosome (or the vacuole in fungi and plants) [3]. Based on different types of cargo and various modes of cargo delivery, at least three types of autophagy have been characterized, including microautophagy, macroautophagy, and chaperone-mediated auto-phagy/CMA; the latter process occurs in birds, fish, and mammals, but is not present in fungi [4]. The most comprehensively studied of these processes is macroautophagy, and, hereafter, we will use the term autophagy to refer to macroautophagy. The morphological hallmark of autophagy is the formation of the autophagosome, a large cytoplasmic double-membrane vesicle, which originates through the generation of the phagophore and the latter's subsequent expansion and closure [5]. Once completed, the outer membrane of the autophagosome fuses with the lysosome/vacuole, while the inner membrane and cargo are exposed to the lumen of the degradative organelles for hydrolysis and the final efflux of the breakdown products into the cytosol [6].

The basic mechanism of autophagy has been well-documented, and the entire process of autophagy can be divided into the following stages: induction and nucleation of the phagophore, expansion and maturation of the phagophore into a completed autophagosome, docking and fusion with the lysosome/vacuole, and degradation and efflux of the breakdown products (Figure 1A) [7]. Initially identified in yeast, over 40 genes that have products primarily involved in the basic process of autophagy have been classified under the name autophagy-related (*ATG*) [8]. Many papers have provided details on this topic [2,4,9], therefore, here, we will briefly describe the autophagy process and the Atg proteins involved in both yeast and mammalian systems.

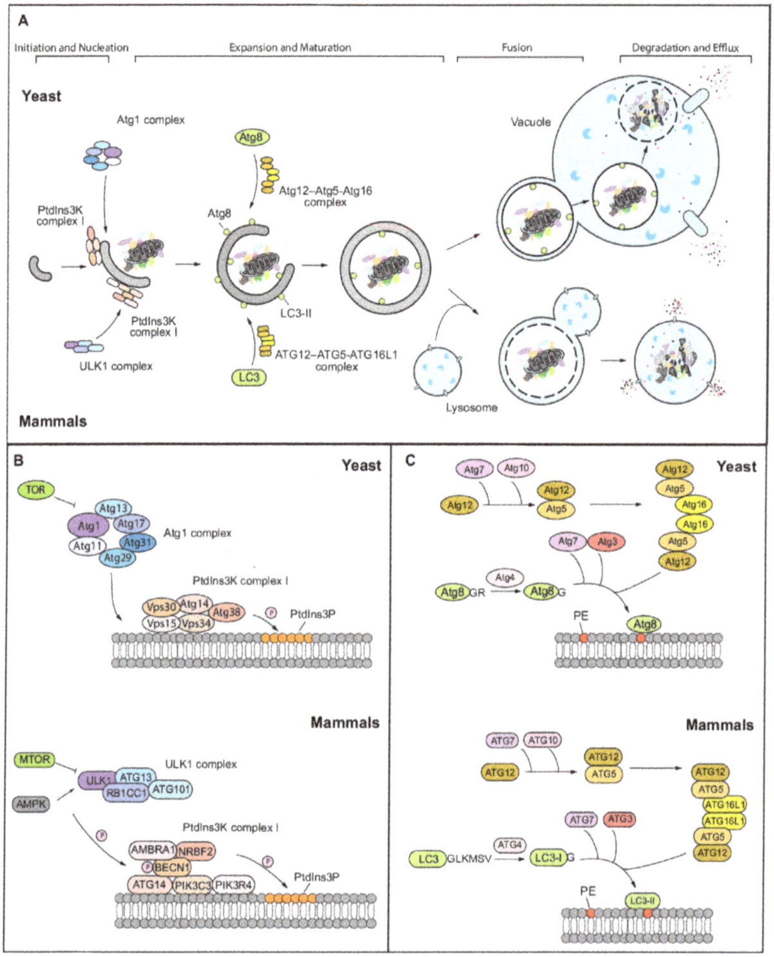

Figure 1. Autophagy in yeast and mammalian systems. (**A**) Four stages of autophagy. The upper and lower parts of each panel represent yeast and mammals, respectively. In both yeast and mammals, autophagy includes four stages, induction and nucleation of the phagophore, expansion and maturation of the phagophore, fusion with the vacuole (in yeast)/lysosome (in mammals), and degradation and efflux of the breakdown products. (**B**) Protein complexes involved in induction and nucleation of the phagophore. In yeast, Atg1 and PtdIns3K complex I will be recruited to the PAS and drive the formation of PtdIns3P on the phagophore. In the mammalian system, the ULK1 complex phosphorylates and activates PtdIns3K complex I, which contributes to the formation of the phagophore. (**C**) Two ubiquitin-like systems. In both yeast and mammals, the Atg12 complex (Atg12–Atg5-Atg16 in yeast and ATG12–ATG5-ATG16L1 in mammals) forms with the help of Atg7/ATG7 and Atg10/ATG10; this complex then functions as an E3 enzyme for the conjugation of Atg8 (in yeast) and Atg8-family proteins (LC3 and GABARAP subfamilies in mammals) with PE.

In yeast, the induction of autophagy begins at a single perivacuolar site, called the phagophore assembly site (PAS) which is proximal to the vacuole. This step is regulated by the Atg1 protein complex, including Atg1, Atg13, and the Atg17-Atg31-Atg29 ternary subcomplex [10,11]. In the nucleation stage, the Atg14-containing class III phosphatidylinositol (PtdIns) 3-kinase (PtdIns3K) complex I (consisting of Vps34, Vps30, Vps15, Atg14, and Atg38) is recruited to the PAS (Figure 1B) [12]. Next, the phagophore begins to expand and then seal to complete the formation of the autophagosome. Key components participating at this stage are two ubiquitin-like (Ubl) conjugation systems, which mediate the conjugation of Ubl proteins Atg12 and Atg8 [13,14] (Figure 1C). Through an enzymatic pathway involving Atg7 (an E1-like enzyme) for Atg12 activation and Atg10 (an E2-like enzyme) for Atg12–Atg5 conjugation, the C terminus of Atg12 is conjugated to an internal Lys of Atg5; Atg16 then noncovalently binds to Atg5 in the conjugate [15]. This system plays a role in membrane recruitment for the expanding phagophore. In contrast to Atg12, which is conjugated to another protein, Atg8 is conjugated to the lipid phosphatidylethanolamine (PE), allowing for its membrane association. Atg8 is initially synthesized with a C-terminal extension, which is removed by the Atg4 cysteine protease to expose a C-terminal Gly [16]. The modified Atg8 is activated with the help of Atg7, and then transferred to Atg3 (an E2 enzyme) that attaches the exposed C-terminal Gly to PE [13,15]. Atg8–PE is found on both sides of the phagophore and initially of the autophagosome; the portion on the autophagosome outer membrane will be deconjugated by a second Atg4-depedent cleavage when autophagosome formation is completed. The transmembrane protein Atg9 may cycle between the PAS and peripheral sites, thus carrying or directing the delivery of membrane for the expansion stage [17]. Upon maturation, the intact autophagosome fully surrounds the cargo, and ultimately delivers cargos to the vacuole by fusing with the vacuolar membrane. Finally, the cargo is degraded by various hydrolases in the vacuole, and breakdown products will be released back into the cytoplasm through permeases in the vacuole membrane.

There are some slight differences in components involved in the autophagy process in mammalian cells, whereas most components are homologs of Atg proteins in yeast. The activation of the ULK Ser/Thr kinase complex is required for autophagy induction, and the initiation begins with the ULK kinase complex (the catalytic subunits ULK1 or ULK2, the regulatory scaffold protein ATG13, RB1CC1, and the stabilizing protein ATG101) which can phosphorylate downstream factors for the induction of autophagy [9,18]. Next, the activated ULK1 complex phosphorylates and activates the PtdIns3K complex 1 (mainly composed of BECN1 [beclin 1], PIK3C3/VPS34, PIK3R4/VPS15, ATG14, NRBF2, and AMBRA1) [19]. The activated PIK3C3/VPS34 can phosphorylate PtdIns to produce phosphatidylinositol-3-phosphate (PtdIns3P), further contributing to formation of the phagophore [20]; ATG14 is directly associated with the ability of the PtdIns3K complex I to translocate to this site [21]. NRBF2 regulates PIK3C3 activity via promoting assembly of the complex, while AMBRA1 contributes to the interaction of BECN1 and PIK3C3 and the catalytic activity [22,23] (Figure 1B). Phagophores are nucleated on ER-emanating PtdIns3P-rich membrane domains called omegasomes [24]. In the expansion step, the ATG12 conjugation system (ATG12–ATG5-ATG16L1 complex) is like that in yeast (Figure 1C). In addition, the second Ubl system involves Atg8-family proteins including MAP1LC3/LC3 and GABARAP subfamilies, undergoing a similar process. With another protein, UVRAG, and core proteins from the PtdIns3K complex I, PtdIns3K complex II is formed, which is also important for autophagy. For example, UVRAG can bind to SH3GLB1 and promote autophagosome maturation [19].

Under normal conditions, autophagy keeps working constitutively at a basal state to maintain cellular homeostasis. When the cell is exposed to certain stress conditions, autophagy is massively induced and promotes the turnover of cytoplasmic materials required for cell survival or removing superfluous or damaged organelles. Too little or too much degradation from uncontrolled autophagy is harmful, and aberrant autophagy is associated with various diseases, such as cancer, aging, and neurodegeneration [25]. Autophagy can be either nonselective or selective: the nonselective mode degrades relatively random portions of the cytoplasm (although phase separation may be involved), whereas the selective mode is highly specific for certain components [26]. In particular, selective autophagy can degrade damaged and superfluous organelles or invasive microbes; these selective processes are given different names depending on the cargo, including mito-phagy (mitochondria), pexophagy (peroxisomes), aggrephagy (protein aggregates), lipophagy (lipid droplets), and xenophagy (intracellular pathogens) [26]. The selective mode of autophagy also plays a key role in cell physiology.

The frequently changing external environment causes cellular stress, and cells in a diseased state also experience stress from the unstable internal environment; therefore, stress biomarkers and their detection are important in the assessment of cell homeostasis (Table 1). To meditate metabolic adaptation and clear damaged organelles, autophagy is considered as one of the most important mechanisms to maintain cell survival under stress. Whereas the depletion of nutrients constitutes the main stimulus for massive autophagy induction, many other types of cellular stress are also involved in autophagy regulation. These different conditions can also regulate autophagy at different levels, including epigenetic, transcription, post-transcription, translation, and post-translation. In this review, we consider different stress stimuli and their relations with autophagy, with a goal of providing a more comprehensive understanding about this field.

Table 1. Stress biomarkers and their detection.

Stress Type	Organism	Biomarkers	Detection	Reference
Nutrient stress	Yeast	TORC1 inactivation	Sch9 dephosphorylation	[27]
			Mammalian EIF4EBP1 dephosphorylation in vitro	[28]
	Mammals	MTORC1 inactivation	RPS6KB1 dephosphorylation	[29]
			EIF4EBP1 dephosphorylation	[29]
Energy stress	Yeast and mammals	Lower ATP: ADP/AMP ratio	Liquid chromatography to detect ATP, ADP and AMP level	[30]
			ATP:ADP fluorescence reporter	[31,32]
			Bioluminescent detection	[33]
	Yeast	Snf1 activation	"SAMS" peptide phosphorylation	[34]
	Mammals	AMPK activation	AMPK phosphorylation	[35]
			Phosphorylation of downstream targets such as ACAC (acetyl-CoA carboxylase)	[35]
Oxidative stress	Yeast and Mammals	High level of ROS	Dichlorodihydrofluorescein fluorescence	[36]
			PG1 or PC1 fluorescence	[37]
			Calcein-acetoxymethylester (calcein-AM) fluorescence	[38]
			CellROX dye	[39]
		Increased GSSG:GSH ratio	High-performance liquid chromatography	[40]
			Capillary electrophoresis	[40]
			Bioluminescence	[41]
			Genetically-encoded fluorescent sensors	[42,43]
		Lipid peroxidation	Fluorescence shift of C11-BODIPY (581/591)	[44]
			TBA-MDA assay	[45,46]

Table 1. *Cont.*

Stress Type	Organism	Biomarkers	Detection	Reference
ER stress	Yeast	Misfolded protein accumulation	Kar2 sedimentation	[47]
		UPR pathway activation	Transcription reporter containing a UPR element promoter driving fluorescent proteins	[47]
			Ire1 clustering	[47]
	Mammals	Protein aggregates	Thioflavin T (ThT) fluorescence	[48]
		UPR pathway activation	Spliced *XBP1* mRNA detection using ER stress-activated indicator" (ERAI) construct	[49]
			Upregulated expression of UPR target genes, including *DDIT3* and *HSPA5/GRP78*	[50]
			ATF6 translocation	[51]

2. Autophagy Regulation under Nutrient Stress

2.1. Mechanisms of Autophagy Regulation by Nutrient Stress in Yeast

Nutrients, such as amino acids and other nitrogen sources, are crucial for yeast growth and the target of rapamycin (TOR) pathway is the central regulator [52]. Tor1 and Tor2 are conserved protein kinases that can be found in two protein complexes termed TOR complex 1 (TORC1) and TORC2. TORC1, which consists of Tor1 or Tor2, Kog1, Lst8, and Tco89, is particularly sensitive to rapamycin treatment and activated by nutrients [52]. Unlike TORC1, TORC2 can only utilize Tor2 as its catalytic component and is not sensitive to rapamycin [53]. Even though some studies indicate that TORC2 is involved in promoting autophagy [54], TORC1 is still considered as the master regulator of autophagy, especially under nutrient stress.

Nitrogen and amino acid signaling are transmitted to TORC1 via different mechanisms, which largely involve the conserved RAG GTPase, composed of Gtr1 and Gtr2 [55]. Gtr1 and Gtr2 form a heterodimer and are tethered on the vacuole membrane by the Ego1-Ego2-Ego3/EGO complex [52]. In the presence of sufficient nitrogen sources/amino acids, with the help of Vam6 as a guanine nucleotide exchange factor/GEF, Gtr1 and Gtr2 are in the GTP and GDP bound forms, respectively. Activated Gtr1 binds to Kog1 and Tco89 in the TORC1 complex, trapping and activating TORC1 on the vacuole [56]. When the amino acid level goes down, Seh1-associated subcomplex inhibiting TORC1/SEACIT, which is composed of Npr2, Npr3, and the catalytic subunit Iml1, functions as the GTPase activating protein (GAP) to induce GDP loading onto Gtr1; this form no longer activates TORC1 [57].

TORC1 regulation function with regard to autophagy can be summarized in three aspects (Figure 2A). First, TORC1 directly phosphorylates Atg13 under growing conditions, which prevents Atg13-Atg1 complex activity [58,59]. The inactivation of TORC1 by nutrient deprivation leads to the hypophosphorylation of Atg13, the induction of Atg1 kinase activity and autophagy stimulation [58].

Second, TORC1 regulates the transcription of *ATG* genes. TORC1 negatively regulates the expression of genes required for the adaptation to nutrient stress by regulating the expression and localization of several transcription factors (TFs), including Gcn4, and the GATA-binding proteins Gln3 and Gat1 [60,61]. When TORC1 is inactivated by the depletion of nutrients, the activated phosphatases Sit4 and PP2A (Pph21/22-Tpd3-Cdc55) mediate the translocation of Gln3 and Gat1 to the nucleus [60], where they are required for the successful induction of *ATG7*, *ATG8*, *ATG9*, *ATG29*, and *ATG32* expression [62]; Gln3 is necessary for promoting *ATG14* expression [63]. TORC1 inhibition also induces Gcn4 expression [61] and *GCN4* deletion leads to decreased *ATG1* mRNA level and autophagy activity during starvation [62]. Additionally, Gcn4 is responsible for the increased expression of *ATG41* during starvation, which is required for efficient autophagy [64]. In addition to directly regulating the localization of TFs, TORC1 and one of its downstream targets, Sch9, inhibit the translocation of Rim15 from the cytosol to the nucleus; Rim15 is a kinase, which controls the association between several TFs and *ATG* genes [27,65]. For instance, Rph1 is a transcriptional repressor of several *ATG* genes, including *ATG7, ATG8, ATG9, ATG14,* and *ATG29* under nutrient-rich conditions. Upon nutrient stress, Rph1 is phosphorylated by the nuclear-localized Rim15 and dissociates from the *ATG* genes to induce autophagy [66]. Similarly, Ume6 inhibits *ATG8* transcription when nutrients are abundant, and this inhibition is relieved by Rim15-dependent phosphorylation during starvation [67].

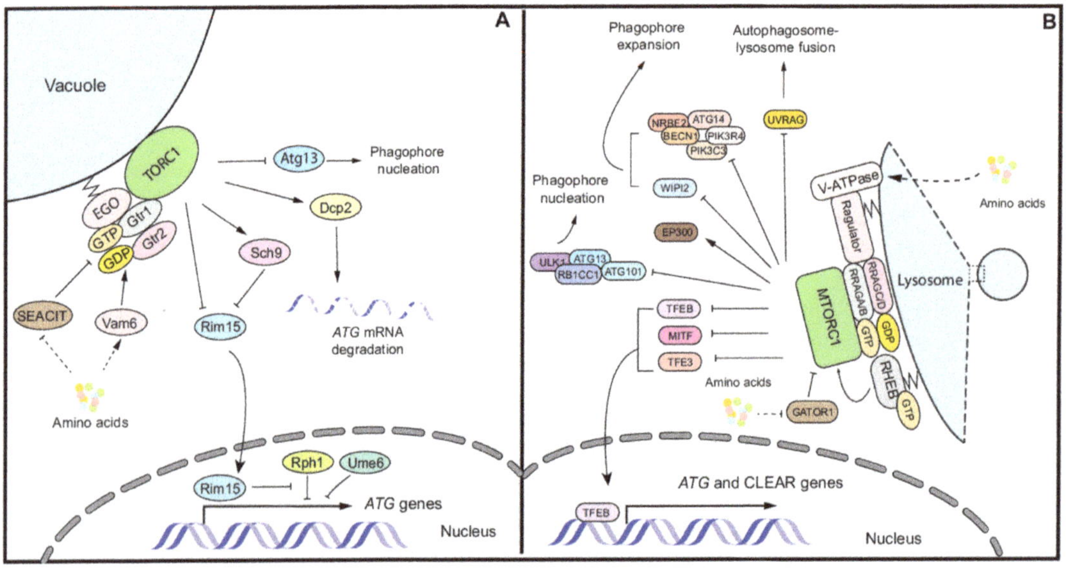

Figure 2. Autophagy regulation by TORC1 and MTORC1. (**A**) In yeast, nutrient sources activate TORC1 basically through activated Gtr1-Gtr2 heterodimer. The activated MTORC1 inhibits autophagy through phosphorylating Atg13, inducing degradation of *ATG* mRNA and inhibiting Rim15 translocation to nucleus, which releases some repressing transcription factors such as Rph1 and Ume6 and induces *ATG* genes transcription. (**B**) In mammalian cells, amino acids activate RRAG GTPase and recruit MTORC1 to lysosomes, where it is activated by RHEB. Activated MTORC1 phosphorylates several ATG proteins and inhibits their functions. Meanwhile, MTORC1 dependent phosphorylation in TFEB, MITF, and TFE3 blocks their translocation into nucleus and the induction of *ATG* and CLEAR genes transcription. Solid and dashed lines represent direct and indirect regulations respectively.

Third, TORC1 controls the posttranscriptional regulation of *ATG* genes. In a report by Hu et al., a temperature sensitive mutation in the decapping enzyme Dcp2 results in increasing mRNA level of multiple *ATG* genes under nutrient-replete conditions, including *ATG1, ATG8, ATG9,* and *ATG13,* and a higher autophagy activity [68]. In the same study, the researchers also found that in *C. neoformans*, TORC1 phosphorylates Dcp2 under nutrient-rich condition, thus promoting *ATG8* mRNA degradation [68].

Here, we only summarized autophagy regulation related to TORC1. Multiple layers of regulation happen in both growing and/or starvation conditions to regulate autophagy as listed in Table 2, but their relations to nutrient stress signals are not completely elucidated.

Table 2. Autophagy regulation in yeast under nutrient-rich and starvation conditions.

Type of Regulation		Regulatory Factors	Conditions	Effects (↑, Positive; ↓, Negative)	Target Genes or Proteins	Reference
Transcriptional regulation		Pho23	Nutrient-rich	↓	*ATG1, 7, 8, 9, 12, 14, 29*	[69]
		Spt10	Nutrient-rich	↓	*ATG1, 7, 9,14, 32*	[62]
		Fyv5	Nutrient-rich and starvation	↓	*ATG1, 7, 8, 9,14, 29, 32*	
		Sfl1	Nutrient-rich	↓	*ATG1, 7, 8, 9, 14, 29, 32*	
		Sko1	Nutrient-rich and starvation	↓	*ATG1, 7, 8, 32*	
		Zap1	Nutrient-rich	↓	*ATG1, 7, 8, 9, 14, 29, 32*	
		Swi5	Nutrient-rich	↑	*ATG7, 8, 9, 14, 29*	
		Rsc1	Starvation	↑	*ATG8*	[70]
		Spt4/5	Nutrient-rich	↓	*ATG8, 41*	[71]
			Starvation	↑	*ATG41*	
Post-transcriptional regulation		Xrn1	Nutrient-rich	↓	*ATG1, 4, 5, 7, 8, 12, 14, 16, 29, 31*	[72]
		Dhh1	Nutrient-rich	↓	*ATG3, 7, 8, 19, 20, 22, 24*	[68]
			Starvation	↑	*ATG1,13*	[73]
		Pat1	Starvation	↑	*ATG1, 2,7, 9*	[74]
		Psp2	Starvation	↑	*ATG1, 13*	[75]
		Ded1	Starvation	↑	*ATG1*	[76]
Post-translational regulation	Phosphorylation	Hrr25	Cvt pathway and pexophagy induction	↑	Atg19,36	[77]
	Ubiquitination	Met30	Nutrient-rich	↓	Atg9	[78]
	Acetylation	Esa1	Starvation	↑	Atg3	[79]
	Deacetylation	Rpd3	Starvation	↓	Atg3	[79]
Epigenetic regulation	Acetylation	Sas2	Nutrient-rich	↓	Histone H4 Lys16	[80]
	Methylation	Unclear	Nutrient-rich	↓	Histone H3 Lys4	[80]

Although additional factors have been identified that regulate autophagy in yeast during starvation, many questions still remain. First, some autophagy regulators play dual roles under growing and starvation conditions (Table 2). For example, Dhh1 contributes to the degradation of multiple *ATG* mRNAs under nutrient-rich conditions, but, on the contrary, promotes the translation of *ATG* genes during starvation [68,73,81]. However, how this transition happens and how it connects with a nutrient-sensing pathway is still unclear. Second, TORC1 has different localizations based on nutrient status. When nutrients are replete, TORC1 is activated and disperses along the vacuole membrane; several studies indicate that TORC1 forms punctate structure on the vacuole in response to starvation [82–84]. However, it is still unclear whether the change in TORC1 localization contributes to autophagy regulation. One model proposes that because the PAS is formed close to the vacuole, the dispersed TORC1 localization along the vacuole prevents Atg13 recruitment to the PAS. Conversely, TORC1 puncta formation during starvation limits its access to Atg13, providing more opportunities for hypophosphorylated Atg13 to be recruited to the PAS [82]. A recent study indicates that the EGO complex and TORC1 have two localizations, both on endosomes and the vacuole but only the TORC1 on endosomes controls autophagy through targeting Atg13 [85]. However, it is not clear why TORC1 has these two different pools and, considering that the distribution between them does not change during nitrogen starvation, why these two populations of TORC1 function differently. Third, epigenetic regulation is another critical way to control autophagy at an appropriate level. Although not much is known in yeast, some evidence indicates the relationship between TORC1, histone modifications, and autophagy regulation. A study from Füllgrabe et al. identified that autophagy occurs concomitant with the reduction in histone H4 Lys16 acetylation/H4K16ac, which may result from the autophagic degradation of the acetyltransferase Sas2 [80]. More recently, Set2, a histone methyltransferase, was shown to be necessary for the transcriptional response to nutrient stress; Set2 genetically interacts with Tor1 and Tor2, indicating a potential role in autophagy regulation [86], but further studies are needed to reveal the mechanism of this type of regulation.

2.2. *Autophagy Regulation in Mammalian Cells*

In mammalian cells, nutrient starvation is also a common stress that induces autophagy. Multiple important nutrient-response molecules have been reported to regulate autophagy, among which MTOR (mechanistic target of rapamycin kinase) complex 1 (MTORC1) is the best characterized. In this subsection, we will summarize autophagy regulation mechanisms mediated by MTORC1 and briefly introduce some other molecules that contribute to autophagy regulation under nutrient stress.

2.2.1. Autophagy Regulation by MTORC1

Like in yeast, there are two TOR complexes in mammalian cells, MTORC1 and MTORC2. MTORC1, which consists of MTOR, RPTOR/raptor, DEPTOR, LST8, and PRAS40, is the general responder to growth factors and nutrients [87]. Amino acids are essential for the activation of MTORC1 through RRAG GTPases [88,89]. Mammalian cells contain four RRAG GTPase members, RRAGA, RRAGB, RRAGC, and RRAGD and they functions in a heterodimer, in which one monomer of either RRAGA or RRAGB partners with either RRAGC or RRAGD [90]. Amino acids in the lysosomal lumen activate the Ragulator complex, possibly through the vacuolar-type H^+-translocating ATPase (V-ATPase) [91,92], and the Ragulator complex functions as a guanine nucleotide exchange factor that promotes the active conformation of RRAG GTPase; where RRAGA/B binds with GTP and RRAGC/D is loaded with GDP [91–93]. In addition, amino acids inhibit the GATOR1 complex (analogous to yeast SEACIT), the GAP for RRAGA/B, therefore facilitating the activation of the RRAG GTPase [94]. Once activated, RRAG heterodimer binds with RPTOR and brings MTORC1 into proximity with RHEB (Ras homolog, mTORC1 binding) GTPase on lysosomes [88]. MTORC1 activity is coupled with growth factors; the removal of TSC1-TSC2, the GAP of RHEB GTPase, in response to growth factors, allows the activation of MTORC1 by RHEB [95]. In contrast, the lack of nutrients results in the conversion of RRAG GTPase into its inactive form and the lysosomal localization of TSC2, which inhibits RHEB GTPase activity [96]. In addition, the absence of amino acids inhibits the polyubiquitination of RHEB, which is important for its binding with MTORC1 [97]. As a result, MTORC1 becomes inactivated and displays a cytosolic localization.

MTORC1 is considered as the master regulator of autophagy when cells are facing nutrient stress. Autophagy regulation by MTORC1 can be summarized in the following aspects (Figure 2B): First, MTORC1 regulates the posttranslational modification of autophagy-associated proteins. Several ATG proteins are the direct targets of MTORC1 and the best known are ULK1 and ATG13. MTORC1-dependent phosphorylation of ULK1 and ATG13 reduces ULK1 complex activity. During starvation, the inactivated MTORC1 disassociates from ULK1, relieving inhibition of the latter; subsequent triggering of ULK1 complex activity promotes autophagy [98–100]. In return, activated ULK1 inhibits MTORC1 activity via phosphorylating RPTOR and reducing its substrate-binding ability [101,102].

This feedback loop maintains the inactivation of MTORC1 and is important for the full activation of autophagy during nutrient deprivation.

Several components of the PtdIns3K complex are also targets of MTORC1. In complex I, MTORC1-dependent phosphorylation of ATG14 inhibits the kinase activity of the complex [103]. Another component, NRBF2, can be phosphorylated by MTORC1 at Ser113 and Ser120. MTORC1 inhibition suppresses NRBF2 phosphorylation and changes it binding preference from PIK3C3/VPS34 and PIK3R4/VPS15 to ATG14 and BECN1, supporting PtdIns3K complex I assembly and its association with the ULK1 complex [104]. In addition, AMBRA1 is phosphorylated by MTORC1 at Ser52 and becomes inactivated. Upon nutrient stress, activated AMBRA1 interacts with the E3 ligase TRAF6 and ubiquitinates ULK1, enhancing its activity [105]. In complex II, UVRAG is the direct target of MTORC1 [106]. MTORC1-dependent phosphorylation on Ser498 has a positive effect on the interaction between UVRAG and RUBCN, which negatively regulates PIK3C3/VPS34 kinase activity and the interaction between HOPS (a complex involved in tethering) and UVRAG, therefore inhibiting the fusion between autophagosomes and lysosomes [106]. Apart from the proteins in these two complexes, WIPI2 can be phosphorylated by MTORC1 at Ser395, which promotes its polyubiquitination by HUWE1 and subsequent degradation [107].

Besides directly phosphorylating ATG proteins, MTORC1-dependent phosphorylation of the acetyltransferase EP300 prevents its intra-molecular inhibition, thus activating its catalytic activity [108]. Several ATG proteins, including ATG5, ATG7, ATG8, and ATG12 are the targets of EP300, and the acetylation of these ATG proteins inhibits autophagy [109]. Therefore, under nutrient stress, MTORC1 inactivation reduces the acetylation of essential ATG proteins to fully activate autophagy.

Second, MTORC1 controls the transcription of *ATG* genes and lysosomal genes via regulating the localization of several TFs. TFEB is a member of the basic helix-loop-helix leucine-zipper family of TFs that promotes the transcription of genes in lysosomal biogenesis and autophagy [110]. When nutrients are repleted, TFEB will be recruited to the lysosome by active RRAG GTPase and phosphorylated by MTORC1 [111], whereas starvation leads to a rapid translocation of TFEB from the cytosol to the nucleus and the induction of transcription of autophagy-associated genes such as *UVRAG, WIPI, MAPLC3B, SQSTM1, VPS11, VPS18,* and *ATG9B* [112]. Several MTORC1-dependent phosphorylation sites are found on TFEB, including Ser122, Ser138, Ser142, and Ser211 [113–116], which regulate TFEB cellular localization through different but coordinated mechanisms. Phosphorylation on Ser211 by MTORC1 promotes TFEB association with YWHA/14-3-3 (tyrosine 3-monooxygenase/tryptophan 5-monooxygenase activation protein) proteins, which traps TFEB in the cytosol. Inactivation of MTORC1 leads to the transport of TFEB to the nucleus, thus stimulating the transcription of autophagy-associated genes [113,114]. Ser122 is another MTORC1-dependent phosphorylation site. The phosphorylation mimetic mutation Ser122Asp reduces nuclear TFEB when Ser211 is dephosphorylated but Ser122 dephosphorylation is not sufficient, in itself, to result in the nuclear localization of TFEB [115], indicating that Ser122 coordinates with Ser211 to control TFEB localization. Ser138 and Ser142 are localized in proximity to the nuclear export signal/NES on TFEB, and MTORC1-dependent phosphorylation at these two sites is critical for TFEB nuclear export [116]. Besides the direct regulation of TFEB, MTORC1 inhibits TFEB activity through activating KAT2B/GCN5, an acetyltransferase that acetylates TFEB and inhibits its DNA-binding activity [117]. Therefore, under nutrient stress, MTORC1 inhibition stimulates not only TFEB accumulation in the nucleus but its binding to target genes as well, thus promoting lysosomal biogenesis and autophagy flux.

TFE3 and MITF are additional TFs that drive the expression of genes involved in lysosome biogenesis and autophagy [118–120]. Similar to TFEB, TFE3 and MITF are recruited to lysosomes by activated RRAG GTPase, and MTORC1 inactivation during starvation is necessary for their release from YWHA/14-3-3 proteins and nuclear localization [111,118]. Ser321 on TFE3 is apparently an MTORC1-dependent phosphorylation site because both MTORC1 inactivation and Ser321Ala mutation abolish its interaction with YWHA/14-3-3 proteins and stimulate nuclear localization [118]. On MITF, Ser280, a residue that corresponds to TFEB Ser211 by homology analysis, is also nominated as a potential MTORC1 phosphorylation site [111]. Interestingly, a study found that TFEB and TFE3 positively regulate MTORC1 activity by promoting RRAGD expression and recruiting MTORC1 to the lysosome when nutrients are provided to the starved cells. Even though it remains as an open question as to how modulating RRAGD expression is sufficient for regulating MTORC1 activity, this mechanism may be important for cells to prepare for nutrient refeeding during starvation [121].

In addition to directly phosphorylating TFs and affecting their localization, MTORC1 also regulates the EIF2A (eukaryotic translation initiation factor 2A)-ATF4 pathway, which induces the expression of *ATG* genes [122,123]. During nutrient deprivation, MTORC1 inactivation induces PPP6C (protein phosphatase 6 catalytic subunit) phosphatase activity, which dephosphorylates and activates EIF2AK4/GCN2. EIF2AK4/GCN2 further phosphorylates and activates EIF2A, leading to the subsequent increase of ATF4 expression, *ATG* gene transcription induction and activated autophagy [122].

Third, MTORC1 is responsible for the posttranscriptional regulation of *ATG* genes. Recently, MTORC1 is reported to regulate autophagy via controlling mRNA N^6-methyladenosine (m^6A) methylation. In this study, the researchers found that MTORC1 activates the CCT (chaperonin containing TCP1) complex, which helps in the folding of proteins in the m^6A methyltransferase complex, resulting in more m^6A RNA methylation, the degradation of *ATG* transcripts and the suppression of autophagy [124].

Finally, in recent years epigenetics has been proposed to be an important regulatory aspect of autophagy [125]. Even though not much is known about the regulation of histone modifications by MTORC1 during nutrient stress, some studies provide initial clues. One example is that during starvation or rapamycin treatment, acetylation of histone H4 Lys16/H4K16ac and the corresponding acetyltransferase KAT8/hMOF are both downregulated, which is important for cell survival during starvation although the detailed mechanism is not known [80]. In addition, MTORC1 enhances the nuclear localization of FOXK1 and FOXK2, which recruit the SIN3A-HDAC complex to restrict the acetylation of histones and the expression of *ATG* genes [126]. These two examples suggest a close connection between MTORC1 and the epigenetic regulation of autophagy, but more studies are still needed to better understand this relationship.

2.2.2. Other Autophagy Regulation during Nutrient Stress

In addition to MTORC1, the main sensor of nutrients, several other stress-response kinases regulate autophagy during nutrient stress. For instance, the stress-activated signaling molecule MAPK8/JNK1 phosphorylates BCL2 during starvation, which prevents its interaction with BECN1, thus promoting autophagy [127]. MAPKAPK2 and MAPKAPK3, which belong to the stress-response kinase MAPK family, phosphorylates BECN1 Ser90 during starvation and this phosphorylation is important for BECN1 function [128]. Additionally, the IKK complex gets activated by starvation and induces the expression of several autophagy genes, therefore stimulating autophagy [129]. As mentioned above, multiple stress-response kinases contribute to the stimulation of autophagy during nutrient starvation, but whether, and how, these signaling pathways coordinate to regulate autophagy requires further attention.

3. Autophagy Regulation under Energy Stress

Recycling by autophagy is essential for yeast and mammals to survive starvation. The breakdown products and materials can be further used to provide building blocks for the synthesis of essential proteins and to produce ATP through catabolic pathways. Therefore, autophagy is essential for the maintenance of energy homeostasis and is finely regulated upon energy deprivation.

AMP-activated protein kinase (AMPK) is an evolutionarily conserved serine/threonine protein kinase [130], sensing low cellular ATP levels and controlling turnover of cellular materials and metabolism, thus being essential for cellular adaptation to energy limitation [131]. AMPK is a heterotrimeric complex composed of a catalytic subunit (PRKAA/α) and two regulatory subunits (PRKAB/β and PRKAG/γ). The PRKAA-subunit contains the kinase domain and the critical residue Thr172 whose phosphorylation by upstream kinases activates AMPK activity [132]. The PRKAG-subunit possesses four cystathionine β-synthase/CBS motifs that can bind to all forms of adenosine-containing ligands, enabling it to sense the changes in the ATP:AMP/ADP ratio [133,134], and the activity of AMPK is precisely regulated by these ratios in the cell [130]. When cells are in the fed state, AMPK is mostly bound by ATP and its activity is inhibited. Under energy-starvation conditions, the cellular concentration of ATP decreases whereas levels of ADP and AMP increase. AMP or ADP binding to the PRKAG-subunit activates the kinase through three distinct mechanisms: (1) it promotes the STK11/LKB1 (serine/threonine kinase 11)-mediated phosphorylation of the PRKAA subunit at Thr172, which can increase AMPK activity up to 100-fold in vitro [135,136]; (2) it protects phosphorylated Thr172 from dephosphorylation by phosphatases [137]; and (3) it causes allosteric activation of the AMPK complex [138]. Once activated, AMPK serves as a central metabolic regulator to restore energy homeostasis by inhibiting anabolic pathways and promoting catabolic pathways, including autophagy. AMPK promotes autophagy at various steps by phosphorylating autophagy-related proteins or autophagy regulators.

Activated AMPK can induce the autophagic process by inhibiting the activity of MTOR in two ways (Figure 3): (1) AMPK directly phosphorylates the MTORC1 component RPTOR on Ser722 and Ser792. This phosphorylation induces YWHA/14-3-3 binding to RPTOR, thus hindering the binding of RPTOR to MTOR and MTOR substrates, leading to suppression of MTORC1 activity [139]. (2) AMPK phosphorylates the MTOR upstream regulator TSC2 on Thr1227 and Ser1345, which promotes the GTPase-activating function of the TSC1-TSC2 complex, leading to the transformation of RHEB into an inactive RHEB-GDP state, which consequently reduces MTOR activity [140,141]. As mentioned above, reduced MTOR activity relieves the inhibition on ULK1 to activate autophagy.

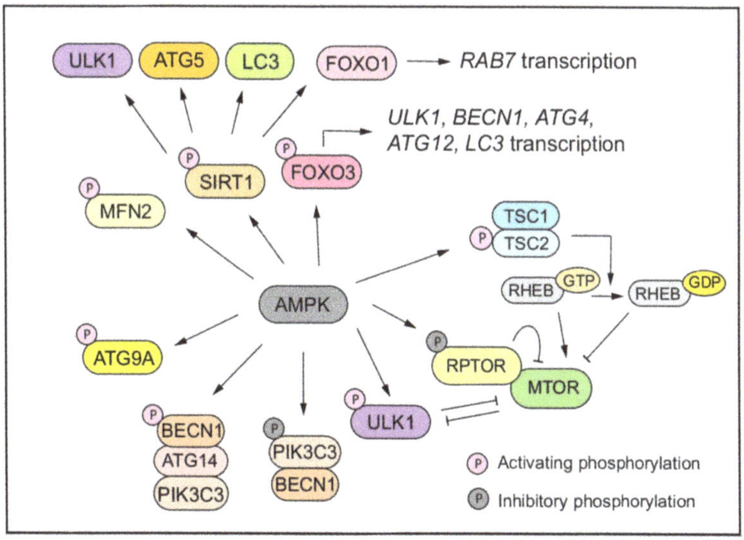

Figure 3. AMPK drives autophagy through three layers of regulation. (1) AMPK suppresses MTOR activity by phosphorylating TSC2 and RPTOR. (2) AMPK directly phosphorylates and activates proteins involved in autophagy including ULK1, BECN1, PIK3C3/VPS34, and ATG9A. (3) AMPK activates the positive regulators of autophagy, for example, phosphorylation of FOXO3 leads to increased transcription of autophagy-related genes.

AMPK can also stimulate autophagy through phosphorylating autophagy related proteins including ULK1, BECN1, and PIK3C3/VPS34 (Figure 3). Under energy-starvation conditions, AMPK directly activates ULK1 through phosphorylation of Ser317, Ser467, Ser555, Ser574, Ser637, and Ser777 [98]. This activation is prevented by MTOR activity during normal physiological conditions as MTORC1 phosphorylates ULK1 at Ser757, which is located in the AMPK-ULK1 binding region (amino acids 711-828), thereby inhibiting the interaction between AMPK and ULK1 [98]. AMPK also regulates the PIK3C3/VPS34 lipid kinase complex upon glucose withdrawal: AMPK activates the pro-autophagy PIK3C3/VPS34 complex by phosphorylating Ser91 and Ser94 in BECN1, which increases autophagosome formation. In the meantime, AMPK inhibits the PIK3C3/VPS34 complexes not involved in autophagy by phosphorylating Thr163 and Ser165 in PIK3C3/VPS34 to suppresses overall PtdIns3P production. The presence of ATG14 dictates the differential regulation by inhibiting PIK3C3/VPS34 phosphorylation and increasing BECN1 phosphorylation by AMPK during glucose starvation [142]. Furthermore, AMPK can phosphorylate other core components of the autophagy pathway. For example, activated AMPK can phosphorylate ATG9A at Ser761, which recruits ATG9A to LC3-positive autophagosomes and enhances autophagosome production [143].

Apart from direct phosphorylation of the core components of the autophagy machinery, AMPK can also promote autophagy through activating autophagy regulators (Figure 3). For example, human transcription factor FOXO3 is phosphorylated by AMPK at Thr179, Ser399, Ser413, Ser555, Ser588, and Ser626, which promotes the nuclear translocation of FOXO3 and its activity, thus upregulating the transcription of downstream auto-phagy-related genes such as *ATG4*, *ATG12*, *BECN1*, *LC3*, and *ULK1* [144,145]. The NAD-dependent deacetylase SIRT1 (sirtuin 1), an essential regulator of autophagy during energy deprivation, is also under the regulation of AMPK. Under glucose starvation conditions, cytoplasmic GAPDH is phosphorylated by activated AMPK and redistributes into the nucleus, where it interacts with SIRT1 and displaces SIRT1′s repressor CCAR2/DBC1 leading to the activation of SIRT1 [146]. SIRT1 can also be activated by the increased level of NAD^+ during starvation. The targets of SIRT1 include, but are not limited to, autophagy pathway components ULK1, ATG5, and LC3 and the transcription factor FOXO1, which induces the expression of the GTPase RAB7 that mediates the fusion of autophagosomes with lysosomes [80,147,148]. During energy stress, a considerable amount of AMPK is translocated from the cytosol to mitochondrial-associated ER membrane/MAM, where it interacts with and phosphorylates the mitochondrial fusion protein MFN2. This AMPK-MFN2 axis is required for mitochondrial-associated ER membrane dynamics and auto-phagy induction [149].

4. Autophagy Regulation under Oxidative/Nitrosative Stress

Reactive oxygen and nitrogen species (hereafter ROS and RNS) are highly reactive molecules that can cause oxidative damages on macromolecules and biological membranes [150,151]. Cells have developed very sophisticated mechanisms to regulate the homeostasis of ROS and RNS, including endogenous antioxidants, such as glutathione and TXN (thioredoxin), and detoxifying enzymes, such as GPX (glutathione peroxidase), CAT (catalase), and SOD (superoxide dismutase), to efficiently resolve the excessive oxidative stress [151,152]. In coordination with the ubiquitin–proteasome system, autophagy plays essential roles in sequestering oxidized proteins in the lysosome/vacuole for degradation to maintain homeostasis [153–155]. However, autophagy is also involved in oxidative stress-induced cell death [156]. For example, the free iron released by ferritinophagy could promote lipid ROS accumulation, thus triggering ferroptosis and the increased autophagic flux in $SOD1^{G93A}$ transgenic lead to muscle atrophy [157–159]. To date, there is abundant evidence showing that the autophagy activity is tightly regulated by the oxidative stress [154,160–162].

As for the ROS, the direct reaction between oxygen and extra electron gives rise to the superoxide ($O_2^{\bullet -}$), which is highly reactive and is rapidly converted into hydrogen peroxide by the endogenous SOD [163]. H_2O_2 is relatively stable and is considered as an important signaling molecule for the ROS responsive pathways [164,165]. In the presence of iron, H_2O_2 can generate the unstable hydroxyl radical ($HO^{\bullet}$) via a process called the Fenton reaction. Hydroxyl radicals can further react with polyunsaturated fatty acid to form various lipid peroxides [166,167]. Autophagy is activated by each of these different ROS species as well as numerous RNS species [156,168–176].

In this section, we cover the current understanding of the regulatory mechanisms of autophagy in yeast and mammalian cells. The relationship between mitophagy and oxidative stress has been reviewed by De Gaetano et al., in the same special issue [177].

4.1. Mechanisms of Autophagy Regulation by Oxidative Stress in Yeast

The regulatory role of oxidative stress on autophagy is evolutionarily conserved in the budding yeast *Saccharomyces cerevisiae*. For example, yeast mitophagy induced by nitrogen starvation and ethanol challenge both can be prevented by adding the antioxidant N-acetylcysteine/NAC [178,179].

Yap1 signaling is the most well-characterized oxidative stress responsive pathway in yeast [180,181]. Yap encompasses a transcription factor family of eight basic leucine zipper (bZIP) domain proteins [182,183]. Among them, Yap1 can directly translocate into the nucleus to activate the expression of various antioxidant genes such as *TRX2* by the stimulation of oxidative stress [184,185]. Under basal conditions, Yap1 is enriched in the cytosol as the nuclear exporter Crm1 efficiently pumps Yap1 out of the nucleus [185]. Upon H_2O_2 activation, however, Yap1 is oxidized, and several disulfide bonds are formed on its C-terminal cysteine-rich domain/c-CRD and amino-terminal cysteine-rich domain/n-CRD so that the Crm1-cognate nuclear export signal is masked. As a result, Yap1 is trapped in the nucleus where it activates the expression of stress-responsive genes [186,187]. The oxidation of Yap1 (especially the covalent bonds between Cys303 and Cys598) requires the participation of the thiol peroxidase Hyr1/Gpx3/Orp1, which acts as a direct receptor for H_2O_2 [188]. Yap1 recognizes a consensus DNA element in the promoter region called the Yap response element (YRE, which includes TGACTAA, TTAGTCA, TTACTAA, and T[T/G]ACAAA) [180]. Among all currently known *ATG* genes, only *ATG15* contains a potential binding site for Yap1. *ATG15* encodes a vacuolar phospholipase that can break down the inner autophagosome membrane in the vacuole lumen, and its direct activation by Yap1 has been experimentally verified (Figure 4A) [189,190].

Atg4 is a cysteine protease, and the mammalian homolog has been reported to be directly regulated by H_2O_2 (see below for further details) [191]. However, the Cys81 residue on human ATG4A and ATG4B that is proposed to play important roles in this process is not conserved in yeast. Yeast Atg4 is also redox regulated through a different mechanism: site-directed mutagenesis reveals that a single disulfide bond formed by Cys338 and Cys394 has a very low redox potential and is required for Atg4 redox regulation in yeast; the formation of this disulfide bond decreases the Atg4 protease activity and can be rapidly reduced by thioredoxin [192].

4.2. Mechanisms of Autophagy Regulation by Oxidative Stress in Mammalian Cells

In mammalian cells, ROS accumulation can be triggered by several different stimuli, such as hypoxia, nutrient stress or cytokines including TNF/TNFα [193–197]. Therefore, many upstream signaling pathways have been proposed to affect the activity of autophagy, including signaling by NFKB/NF-κB, AMPK, HIF1A/HIF-1, ATM, AKT-MTOR, and MAPK [194,195,197–204]. The regulation of autophagy by these upstream signaling pathways has been well summarized elsewhere [153,154,160,205].

The NFE2L2/Nrf2 (NFE2 like2 bZIP transcription factor 2)-KEAP1 (kelch ECH associated protein 1)-antioxidant signaling pathway can be directly activated by oxidative stress and can stimulate the expression of various stress-responsive genes including several detoxifying enzymes and autophagy proteins [206–208]. As a functional ortholog for Yap1, NFE2L2 is also a bZIP transcription factor; heterodimers of NFE2L2 and MAF proteins recognize a specific antioxidant response element/ARE in the promoter region of target genes [209,210]. Under basal conditions, the NFE2L2 is localized in the cytoplasm and maintained at a very low level. This dynamic regulation is achieved by its interacting partner KEAP1 which is a CUL3 (cullin 3) E3 ubiquitin ligase adaptor [211–213]. KEAP1 interacts with NFE2L2 via the carboxy-terminal Kelch domain with a 2:1 stoichiometry. At the same time, KEAP1 interacts with the CUL3 ligase via the amino-terminal bric-a-brac, tramtrack, and broad complex/BTB domain, thus promoting the efficient degradation of NFE2L2 [211,214–216]. KEAP1 acts as the redox sensor and the interaction between KEAP1 and NFE2L2 is directly regulated by environmental cues via a mechanism called the "hinge and latch model": in response to H_2O_2, KEAP1 Cys226, Cys613, and Cys622/624 residues form disulfide bonds that impair the interaction between KEAP1-CUL3 and NFE2L2, thus stabilizing NFE2L2 and releasing it into the nucleus where it is active [212,213,217–219].

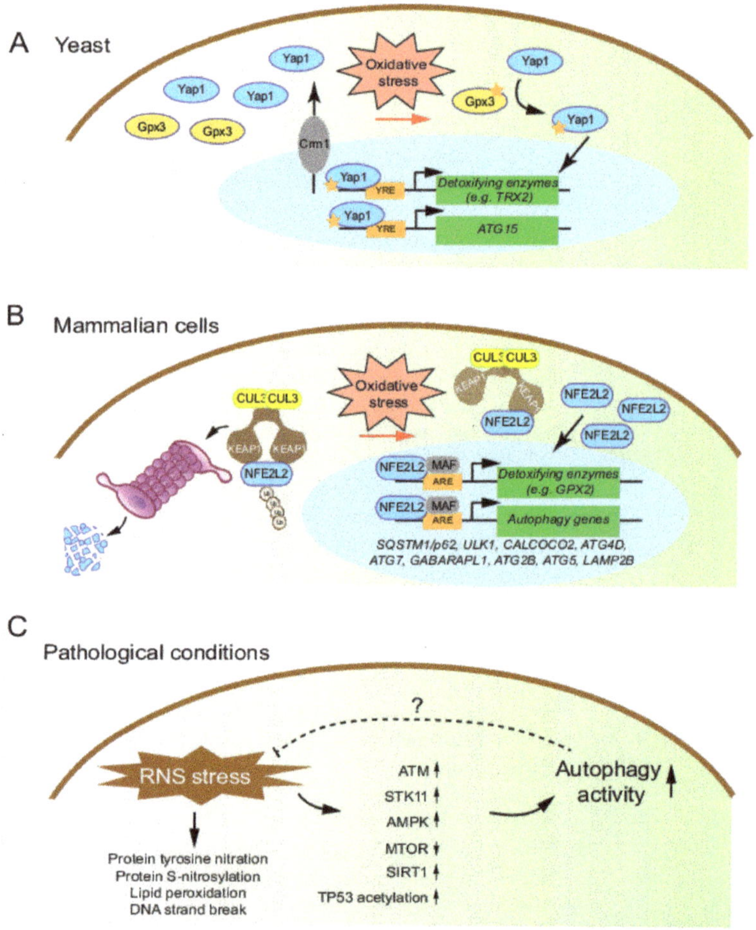

Figure 4. The transcriptional regulation of autophagy by oxidative stress. (**A**) In yeast, upon oxidative stress, oxidized Yap1 is accumulated in the nucleus and drives the expression of detoxifying enzymes and the autophagy-related gene *ATG15*. YRE: Yap response element. (**B**) In mammalian cells, upon oxidative stress, the oxidation of cysteine residues of KEAP1 prevent the ubiquitination of NFE2L2/NRF2, thus allowing NFE2L2 to enter the nucleus and activate several autophagy genes. ARE: antioxidant response element. (**C**) In pathological conditions, nitrosative stress activates autophagy activity via multiple signaling cascades.

The SQSTM1/p62 (sequestosome 1) protein contains an LC3-interacting region (LIR) domain thus allows it to act as an autophagy receptor to facilitate delivery of cargos into the phagophore for subsequent degradation [220,221]. The antioxidant response element has been identified in the promoter of *SQSTM1* that makes it a target for NFE2L2 activation [222]. Interestingly, KEAP1 is among the autophagic substrates of SQSTM1. Therefore, SQSTM1-mediated autophagy can degrade KEAP1 to further activate NFE2L2 signaling in a positive feedback loop [223–226]. Recently, more NFE2L2-targeted autophagy genes have been reported, including *ULK1*, *CALCOCO2*, *ATG4D*, *ATG7*, *GABARAPL1*, *ATG2B*, *ATG5*, and *LAMP2B*, suggesting that, in addition to TFEB and FOXO, NFE2L2 is an important autophagy regulator, perhaps in a more oxidative-stress relevant context (Figure 4B) [227,228].

In addition, ROS can regulate autophagy by directly oxidizing the cysteine residues on the core autophagy components. For example, the cysteine protease ATG4A and ATG4B can be inactivated by H_2O_2. The possible mechanism is that Cys81 is sensitive to oxidation, triggering a conformation change on Cys77 which is the catalytic residue, thus inhibiting the cysteine protease activity of ATG4 [191]. As a result, the transient blockade of ATG4 activity stabilizes the lipidated forms of its substrates LC3 and GABARAPL2/GATE-16 so that autophagosome biogenesis is promoted [191]. Of note, this reversible inhibition of ATG4 is spatially and temporally regulated. As the autophagosome is trafficked towards lysosomes where the local H_2O_2 is lower, ATG4 is reactivated to deconjugate and recycle the LC3 and GABARAPL2. Similarly, in the context of mito-phagy, the cysteine residues of the ubiquitin E3 ligase PRKN/parkin can be oxidized by sulfhydration that is required for full PRKN ligase activity and normal mitophagy flux [229,230].

4.3. Mechanisms of Autophagy Regulation by Nitrosative Stress

In both yeast and mammals, nitric oxide (NO) mediates critical physiological functions as a signaling molecule at low concentrations, but causes nitrosative stress at high concentrations [231–233]. Imbalance of reactive nitrogen species (RNS) results in accumulation of protein tyrosine nitration, protein S-nitrosylation on cysteine residues, and damage to lipids and DNA [234,235]. In mammalian systems, nitrosative stress is correlated with many pathological conditions, such as cancer, neurodegenerative diseases, and ischemia, and upregulated autophagy activity is observed in several nitrosative stress models [236–238]. For example, in a microsphere embolism rat model, the increased autophagy signaling (protein level of BECN1, LC3, LAMP2, and CTSB [cathepsin B]) is accompanied by nitrosative stress, which can be partially resolved by adding the peroxynitrite ($ONOO^-$) scavenger melatonin [239]. Furthermore, RNS can attenuate MTORC1 activity to promote autophagy via the ATM-AMPK-TSC2 and AKT signaling axis [240,241]. A recent study challenged MCF7 cells with the NO donor compound DETA-NONOate and observed an increased NAD^+:NADH ratio. The pharmacological and genetic inhibition of the NAD^+-dependent deacetylase SIRT1 reduces autophagy activity, and the acetylation of TP53, and promotes cell survival, suggesting the complex interplay among SIRT1, TP53, and autophagy upon nitrosative stress (Figure 4C) [242]. However, whether autophagy is required for RNS homeostasis will requires further loss-of-function studies of autophagy genes and their products.

5. Autophagy Regulation under ER Stress

5.1. Endoplasmic Reticulum Stress and Autophagy

The endoplasmic reticulum (ER) is a central membrane-bound organelle, and its membrane structure was first documented by Porter et al. in 1945 using electron microscopy [243]. The ER is an important organelle in eukaryotic cells with various functions, such as protein synthesis, modification and processing of proteins, secretion of correctly folded proteins, calcium homeostasis, and lipid and carbohydrate metabolism [244,245]. Therefore, the ER is essential for cell homeostasis.

Normally, the ER utilizes chaperones to properly fold newly synthesized proteins and identify misfolded proteins for destruction. However, the ER homeostasis is disrupted under numerous pathological conditions including nutrient deprivation, perturbation of cellular ATP level, calcium metabolic imbalance, redox imbalance, viral infection, and the presence of environmental toxins. In addition, the protein-folding capacity of the ER can also be compromised and eventually cause the accumulation of unfolded or misfolded proteins in the ER lumen, also known as ER stress [246].

The ER stress triggers an adaptive response referred to as the unfolded protein response (UPR). After sensing the ER stress, the UPR transduces the signal to the regulation of downstream transcription factors and then induces ER chaperone genes to upregulate the folding capacity. Additionally, the cells can also begin a process termed ER-associated degradation (ERAD) to mediate the transport of unfolded or misfolded proteins into the cytosol for degradation. ERAD mainly consists of two mechanisms: ubiquitin–proteasome-dependent ERAD/ERAD(I) and autophagy–lysosome dependent ERAD/ERAD(II) [247–249]. ER-to-lysosome-associated degradation/ERLAD is the name currently used for the autophagy-dependent mechanism that is employed to handle proteins that cannot be degraded by ERAD.

As mentioned in the previous section in this review, autophagy can be induced by different types of cellular stress, which includes the ER stress discussed here. The relationship between ER stress and autophagy was first described in 2006 in yeast [250,251]. Here, we name the autophagy activated by ER stress as "ER stress-mediated autophagy" because this is the term used in most studies.

5.2. The Mechanisms of ER Stress-Mediated Autophagy in Yeast

In yeast, the ER stress can be sensed by an ER-resident type 1 transmembrane protein called Ire1, which plays a critical role in the UPR induced by ER stress [252,253]. It is noteworthy that Ire1 was initially identified as an mRNA splicing factor in yeast [254,255]. In addition, Ire1 is also capable of sensing unfolded or misfolded proteins in the ER lumen because Ire1 has both an endoribonuclease domain and an ER lumenal stress-sensing domain.

Ire1 is localized in the ER membrane with its C terminus facing into the cytosol and the N terminus residing in the ER lumen. Under normal conditions, the N-terminal region of Ire1 is bound to Kar2 unless Ire1 senses the accumulation of unfolded proteins in the ER lumen. Ire1 is activated by autophosphorylation after dissociation from Kar2, leading to the expression of activated Hac1, a transcription factor.

Upon the UPR induced by ER stress, a non-classical intronic sequence near the 3′ end of the open reading frame of *HAC1* mRNA is excised by activated Ire1 and then the two ends of the mRNA are ligated by the tRNA ligase Trl1 [256]. The spliced *HAC1* encodes an activated form of the Hac1 protein containing 238 amino acids, which contains 18 amino acids more than the Hac1 protein encoded by un-spliced *HAC1* mRNA; these 18 amino acids play a key role for TF activation [257]. The difference in the properties of the two types of Hac1 protein is mainly caused by the C terminus. The N terminus of both types of Hac1 has a DNA-binding function while the C terminus of activated Hac1 has an active transcriptional activation domain due to the cleavage and splicing reaction [258]. Eventually, activated Hac1 is exported to the nucleus and binds to the unfolded protein response elements/UPREs to promote the transcription of UPR-related genes [259]. Furthermore, studies have reported that these unfolded protein response elements are commonly found in the promoters of some UPR-related genes including *FPR2/FKB2*, *KAR2*, and *PDI1* [260].

Studies carried out in 2006 show that ER stress can induce autophagy through this Ire1-Hac1 pathway in yeast [250,251]. Yorimitsu et al. used two types of drugs: dithiothreitol/DTT (an inhibitor of disulfide bond formation) and tunicamycin (an inhibitor of glycosylation) to induce ER stress. Then GFP-Atg8 processing and precursor Ape1 maturation assays were applied to monitor autophagic induction after the drug treatment. Both assays showed increased autophagic flux, indicating an induction of autophagy caused by ER stress. Additionally, the necessity of the Ire1-Hac1 signaling pathway for this ER stress-mediated autophagy was also explored in this study. The authors found that deletion of either *IRE1* or *HAC1* does not affect the capability for inducing autophagy caused by nutrient depletion; however, either knockout did block ER stress-mediated autophagy, suggesting that Ire1 and Hac1 are involved in the induction of this pathway probably through the UPR.

5.3. The Mechanisms of ER Stress-Mediated Autophagy in Mammalian Cells

The UPR is a highly conserved mechanism and mammalian cells also utilize it to alleviate ER stress by enhancing the protein-folding capacity of, and reducing the protein synthetic load on this organelle to restore ER homeostasis [259]. Unlike the UPR in yeast, that consists of the Ire1 signaling pathway, the UPR in mammalian cells is characterized by three major branches involving three ER membrane resident proteins: the serine/threonine-protein kinase/endoribonuclease ERN1/IRE1α (endoplasmic reticulum to nucleus signaling 1), EIF2AK3/PERK (eukaryotic translation initiation factor 2 alpha kinase 3), and the cyclic AMP-dependent transcription factor ATF6 (activating transcription factor 6).

Under normal physiological conditions, all three ER stress sensors are inactive due to the binding of an ER-resident chaperone protein, HSPA5/BIP/GRP78. Due to the high affinity of unfolded or misfolded proteins for HSPA5, ERN1, EIF2AK3, and ATF6 become active when there is an accumulation of unfolded or misfolded proteins in the ER lumen [261]. Moreover, the increased activity of these three ER sensors is also partially contributed to by unfolded or misfolded proteins acting as active ligands for their activation [262]. The activation of these three UPR signaling pathways alleviates ER stress by partially overlapping but distinct mechanisms, including autophagy.

5.3.1. ERN1

ERN1 is a bifunctional protein in mammalian cells consisting of three domains: an N-terminal lumenal domain, a cytosolic endoribonuclease domain, and a cytosolic serine/threonine kinase domain [263]. Similar to the Ire1 in yeast, active ERN1 can excise a 26-nucletide intron from *XBP1* mRNA. The spliced *XBP1* mRNA allows the expression of an active and stable form of XBP1 (X-box binding protein 1). The transcription factor XBP1 is then translocated to the nucleus and upregulates the expression of target genes in response to ER stress [264]. Among the target genes, *BECN1* plays a central role in autophagy, suggesting that the splicing of *XBP1* mRNA mediated by ERN1 under ER stress is important for autophagy induction [265]. Consistently, studies report that the un-spliced *XBP1* mRNA can interact with FOXO1 (forkhead box O1), resulting in a decreased level of this TF, finally leading to the downregulation of autophagy [266–268].

Additionally, ERN1 can interact with TRAF2 (TNF receptor associated factor 2) and form a complex, which can phosphorylate MAP3K5/ASK1 (mitogen-activated protein kinase kinase kinase 5). Next, the phosphorylated MAP3K5 actives MAPK8/JNK1 (mitogen-activated protein kinase 8) by phosphorylation. Subsequently, the phosphorylated MAPK8-mediated phosphorylation of BCL2 can increase the level of free BECN1 by disrupting the BECN1-BCL2 complex, or by elevating *BECN1* transcription, which leads to autophagosome formation [127,269]. Moreover, it is reported that the activation of AMPK mediated by ERN1 is involved in autophagy initiation [270].

5.3.2. EIF2AK3

Under ER stress conditions, the activation of the EIF2AK3 UPR signaling pathway upregulates many autophagy-related genes. The active EIF2AK3 can mediate the phosphorylation of EIF2A, which can elevate both *ATG12* mRNA and protein levels [271]. In addition, EIF2AK3-mediated EIF2A phosphorylation also enables the selective translation of *ATF4* mRNA, and the transcription factor ATF4 is then translocated to the nucleus where it upregulates the expression of multiple proteins, such as several autophagy-related proteins (ATG3, ATG12, ATG16L1, BECN1, and LC3) and DDIT3/CHOP (DNA damage inducible transcript 3) [272]. The expression of DDIT3 can also transcriptionally increase the expression of some proteins involved in autophagy (ATG5, ATG10, and GABARAP). In addition, DDIT3 can downregulate the expression of BCL2, a protein that binds to BECN1 and inhibits autophagosome formation [273,274]. Interestingly, the complex formed by ATF4 and DDIT3 can also induce the expression of some proteins involved in autophagy, including ATG7, NBR1 (NBR1 autophagy cargo receptor), and SQSTM1 [123]. Furthermore, the active EIF2AK3 pathway can initiate autophagy via the activation of AMPK and the inhibition of MTORC1 [275]. Consistent with this finding, the activation of ATF4-DDIT3 mediated by EIF2AK3 inhibits MTORC1 activity resulting in the induction of autophagy [276].

5.3.3. ATF6

Under ER stress conditions, ATF6 is translocated to the Golgi apparatus where it is cleaved by MBTPS1/S1P and MBTPS2/S2P. The N-terminal domain of ATF6 after the cleavage is translocated to the nucleus to induce the expression of UPR genes, including *DDIT3* and *XBP1* [277–279]. Therefore, ATF6 can indirectly regulate autophagy through the DDIT3 and XBP1 signaling pathway as mentioned above. In addition, ATF6 might regulate autophagy in the initiation step by the inhibition of AKT activity [280]. Additionally, ATF6 can interact with the transcription factor CEBPB (CCAAT enhancer binding protein beta) and then stimulate the expression of DAPK1 (death associated protein kinase 1) [281,282], which can phosphorylate BECN1 so that it will be released from the auto-phagy inhibitory BECN1-BCL2 complex, promoting the induction of autophagy.

5.3.4. Calcium

The ER is a multifunctional organelle which plays a pivotal role in maintaining intracellular calcium homeostasis. Under ER stress conditions, the calcium homeostasis is disrupted and the release of calcium from the ER to the cytosol is also elevated, which can induce autophagy. When calcium is released from the ER through ITPR (inositol 1,4,5-trisphosphate receptor), the CAMKK-AMPK-dependent signaling pathway is activated and the inhibitory effect of MTOR on the ULK1 complex is relieved [283,284]. Moreover, calcium release can activate DAPK1 [285], which is involved in the induction step of autophagy as noted above.

6. Conclusions

In this review, we summarized autophagy regulation under different types of stress, including that involving nutrients, energy, oxidation, and the ER (Figure 5). Besides the ones mentioned in this review, other types of stress, such as DNA damage and pathogen infection, are also able to induce autophagy [286,287]. The fact that autophagy is induced by multiple stresses highlights the importance of autophagy in allowing cells to maintain homeostasis in response to changes in the environment.

Of note, in this review, we mainly focus on how stress-response molecules and/or pathways regulate autophagy. There are a wide range of regulatory mechanisms affecting autophagy-associated genes under stress conditions, especially during nutrient deprivation, but the connections with stress-sensing pathways have not been established [288]. Additionally, apart from stress, the activity of stress-sensing molecules can also be regulated by other factors, and all these contribute to autophagy regulation under stress conditions. We did not discuss these factors in detail because they are beyond the scope of this review, but their roles in autophagy regulation cannot be ignored.

Even though we introduced different types of stress separately, it does not mean that the stress-responding molecules or pathways function alone. In fact, stress-responding pathways have very close connections, regulating each other or sharing the same downstream effector. Using the nutrient sensor MTORC1 and energy sensor AMPK as an example, it is well known that AMPK inhibits MTORC1 and both kinases target ULK1 to regulate its activity [53,98]. The regulatory network among AMPK, ULK1, and MTORC1 are important for the oscillation of autophagy [289]. Recently, AMPK was shown to be inhibited by MTORC1 [290], further indicating the complex interaction between these two important stress responders. Another example is seen with EIF4A, which is activated by multiple stresses and induces the expression level of ATF4, a transcription factor that promotes the transcription of multiple *ATG* genes [272]. Similarly, NFE2L2 is also in the center of the stress response, as the expression of this transcription factor is activated not only by oxidative stress, but also by other conditions such as ER stress [291]. The interaction between different stress-responding pathways and the existence of a common response pathway makes autophagy induction by stress a rapid and well-controlled process.

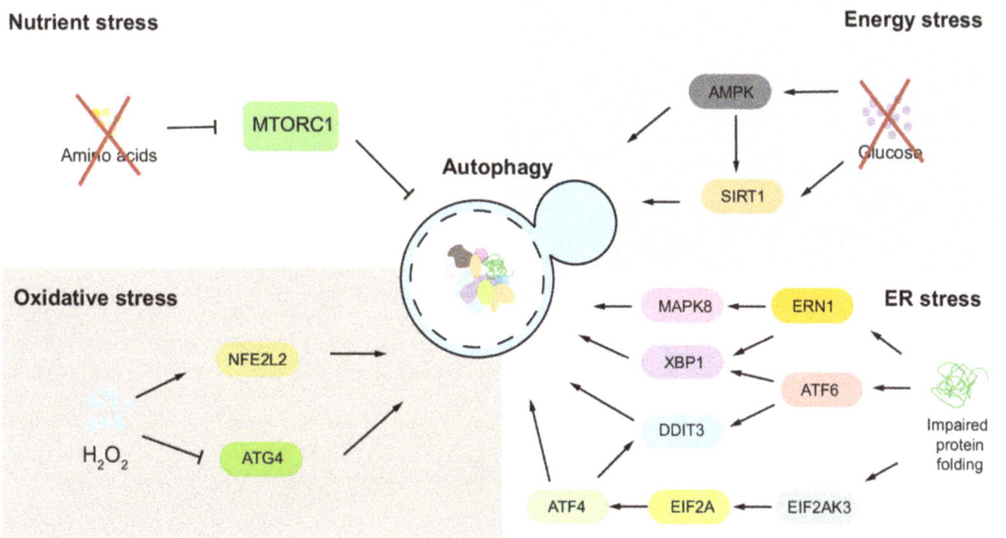

Figure 5. Summary of autophagy regulation under stress. Mammalian systems are presented here to demonstrate the major causes of stress, the main stress-sensing molecules and pathways and their regulation of autophagy.

Thanks to advanced studies in recent decades, we can now draw a clearer picture of the autophagy regulation network under stress conditions. However, more studies focusing on this field are still needed for a better understanding on how autophagy is controlled, because too much or too little autophagy can harm cells. More importantly, the insights on autophagy regulation under stress may shed light on understanding the relation between autophagy and disease because cells in a diseased state usually undergo stress. For example, cancer cells in the interior of a tumor usually experience nutrient and oxidative stress because of the lack of proximal blood vessels. ER stress and oxidative stress are also proposed to contribute to neurodegenerative diseases [292,293]. Autophagy has a close connection with these diseases [18], therefore, a deeper understanding of autophagy regulation under stress conditions may help us find more potential autophagy-targeting therapeutic approaches.

Author Contributions: Preparation, writing and editing of the manuscript, Y.L., Y.H., X.W., Z.Y., Z.Z., D.J.K. All authors have read and agreed to the published version of the manuscript.

Funding: This work was funded by the National Institutes of Health grant number GM131919.

Conflicts of Interest: The authors declare no conflict of interest.

Abbreviations

AMPK	AMP-activated protein kinase
ATG	autophagy related
ER	endoplasmic reticulum
ERAD	ER-associated degradation
GAP	GTPase activating protein
H_2O_2	hydrogen peroxide
m^6A	N^6-methyl-adenosine
MTORC1	mechanistic target of rapamycin kinase complex 1
NO	nitric oxide
PAS	phagophore assembly site
PE	phosphatidylethanolamine
PtdIns3K	class III phosphatidylinositol kinase
RNS	reactive nitrogen species
ROS	reactive oxygen species
TFs	transcription factors
Ubl	ubiquitin-like
UPR	unfolded protein response
V-ATPase	vacuolar-type H^+-translocating ATPase

References

1. Yorimitsu, T.; Klionsky, D.J. Autophagy: Molecular machinery for self-eating. *Cell Death Differ.* **2005**, *12* (Suppl. 2), 1542–1552. [CrossRef] [PubMed]
2. Wen, X.; Klionsky, D.J. An overview of macroautophagy in yeast. *J. Mol. Biol.* **2016**, *428*, 1681–1699. [CrossRef] [PubMed]
3. Yang, Z.; Klionsky, D.J. An overview of the molecular mechanism of autophagy. *Curr. Top. Microbiol. Immunol.* **2009**, *335*, 1–32. [CrossRef] [PubMed]
4. Parzych, K.R.; Klionsky, D.J. An overview of autophagy: Morphology, mechanism, and regulation. *Antioxid. Redox Signal.* **2014**, *20*, 460–473. [CrossRef]
5. Nakatogawa, H. Mechanisms governing autophagosome biogenesis. *Nat. Rev. Mol. Cell Biol.* **2020**, *21*, 439–458. [CrossRef]
6. Yim, W.W.; Mizushima, N. Lysosome biology in autophagy. *Cell Discov.* **2020**, *6*, 6. [CrossRef]
7. Ariosa, A.R.; Lahiri, V.; Lei, Y.; Yang, Y.; Yin, Z.; Zhang, Z.; Klionsky, D.J. A perspective on the role of autophagy in cancer. *Biochim. Biophys. Acta Mol. Basis Dis.* **2021**, *1867*, 166262. [CrossRef]
8. Ohsumi, Y. Historical landmarks of autophagy research. *Cell Res.* **2014**, *24*, 9–23. [CrossRef]
9. Gatica, D.; Chiong, M.; Lavandero, S.; Klionsky, D.J. The role of autophagy in cardiovascular pathology. *Cardiovasc. Res.* **2022**, *in press*. [CrossRef]
10. Suzuki, K.; Kirisako, T.; Kamada, Y.; Mizushima, N.; Noda, T.; Ohsumi, Y. The pre-autophagosomal structure organized by concerted functions of APG genes is essential for autophagosome formation. *EMBO J.* **2001**, *20*, 5971–5981. [CrossRef]
11. Mizushima, N. The role of the Atg1/ULK1 complex in autophagy regulation. *Curr. Opin. Cell Biol.* **2010**, *22*, 132–139. [CrossRef] [PubMed]
12. Burman, C.; Ktistakis, N.T. Regulation of autophagy by phosphatidylinositol 3-phosphate. *FEBS Lett.* **2010**, *584*, 1302–1312. [CrossRef] [PubMed]
13. Geng, J.; Klionsky, D.J. The Atg8 and Atg12 ubiquitin-like conjugation systems in macroautophagy. 'Protein modifications: Beyond the usual suspects' review series. *EMBO Rep.* **2008**, *9*, 859–864. [CrossRef] [PubMed]
14. Shpilka, T.; Mizushima, N.; Elazar, Z. Ubiquitin-like proteins and autophagy at a glance. *J. Cell Sci.* **2012**, *125*, 2343–2348. [CrossRef] [PubMed]
15. Yin, Z.; Popelka, H.; Lei, Y.; Yang, Y.; Klionsky, D.J. The Roles of Ubiquitin in Mediating Autophagy. *Cells* **2020**, *9*, 2025. [CrossRef]
16. Ichimura, Y.; Kirisako, T.; Takao, T.; Satomi, Y.; Shimonishi, Y.; Ishihara, N.; Mizushima, N.; Tanida, I.; Kominami, E.; Ohsumi, M.; et al. A ubiquitin-like system mediates protein lipidation. *Nature* **2000**, *408*, 488–492. [CrossRef]
17. Reggiori, F.; Tucker, K.A.; Stromhaug, P.E.; Klionsky, D.J. The Atg1-Atg13 complex regulates Atg9 and Atg23 retrieval transport from the pre-autophagosomal structure. *Dev. Cell* **2004**, *6*, 79–90. [CrossRef]
18. Lei, Y.; Klionsky, D.J. The Emerging Roles of Autophagy in Human Diseases. *Biomedicines* **2021**, *9*, 1651. [CrossRef]
19. Yu, X.; Long, Y.C.; Shen, H.M. Differential regulatory functions of three classes of phosphatidylinositol and phosphoinositide 3-kinases in autophagy. *Autophagy* **2015**, *11*, 1711–1728. [CrossRef]
20. Nascimbeni, A.C.; Codogno, P.; Morel, E. Phosphatidylinositol-3-phosphate in the regulation of autophagy membrane dynamics. *FEBS J.* **2017**, *284*, 1267–1278. [CrossRef]
21. Sun, Q.; Fan, W.; Chen, K.; Ding, X.; Chen, S.; Zhong, Q. Identification of Barkor as a mammalian autophagy-specific factor for Beclin 1 and class III phosphatidylinositol 3-kinase. *Proc. Natl. Acad. Sci. USA* **2008**, *105*, 19211–19216. [CrossRef] [PubMed]

22. Lu, J.; He, L.; Behrends, C.; Araki, M.; Araki, K.; Jun Wang, Q.; Catanzaro, J.M.; Friedman, S.L.; Zong, W.X.; Fiel, M.I.; et al. NRBF2 regulates autophagy and prevents liver injury by modulating Atg14L-linked phosphatidylinositol-3 kinase III activity. *Nat. Commun.* **2014**, *5*, 3920. [CrossRef] [PubMed]
23. Fimia, G.M.; Stoykova, A.; Romagnoli, A.; Giunta, L.; Di Bartolomeo, S.; Nardacci, R.; Corazzari, M.; Fuoco, C.; Ucar, A.; Schwartz, P.; et al. Ambra1 regulates autophagy and development of the nervous system. *Nature* **2007**, *447*, 1121–1125. [CrossRef] [PubMed]
24. Polson, H.E.; de Lartigue, J.; Rigden, D.J.; Reedijk, M.; Urbe, S.; Clague, M.J.; Tooze, S.A. Mammalian Atg18 (WIPI2) localizes to omegasome-anchored phagophores and positively regulates LC3 lipidation. *Autophagy* **2010**, *6*, 506–522. [CrossRef]
25. Klionsky, D.J.; Petroni, G.; Amaravadi, R.K.; Baehrecke, E.H.; Ballabio, A.; Boya, P.; Bravo-San Pedro, J.M.; Cadwell, K.; Cecconi, F.; Choi, A.M.K.; et al. Autophagy in major human diseases. *EMBO J.* **2021**, *40*, e108863. [CrossRef]
26. Gatica, D.; Lahiri, V.; Klionsky, D.J. Cargo recognition and degradation by selective autophagy. *Nat. Cell Biol.* **2018**, *20*, 233–242. [CrossRef]
27. Urban, J.; Soulard, A.; Huber, A.; Lippman, S.; Mukhopadhyay, D.; Deloche, O.; Wanke, V.; Anrather, D.; Ammerer, G.; Riezman, H.; et al. Sch9 is a major target of TORC1 in Saccharomyces cerevisiae. *Mol. Cell* **2007**, *26*, 663–674. [CrossRef]
28. Tanigawa, M.; Maeda, T. An In Vitro TORC1 Kinase Assay That Recapitulates the Gtr-Independent Glutamine-Responsive TORC1 Activation Mechanism on Yeast Vacuoles. *Mol. Cell. Biol.* **2017**, *37*, e00075-17. [CrossRef]
29. Ikenoue, T.; Hong, S.; Inoki, K. Monitoring mammalian target of rapamycin (mTOR) activity. *Methods Enzymol.* **2009**, *452*, 165–180. [CrossRef]
30. Zhang, C.S.; Hawley, S.A.; Zong, Y.; Li, M.; Wang, Z.; Gray, A.; Ma, T.; Cui, J.; Feng, J.W.; Zhu, M.; et al. Fructose-1,6-bisphosphate and aldolase mediate glucose sensing by AMPK. *Nature* **2017**, *548*, 112–116. [CrossRef]
31. Berg, J.; Hung, Y.P.; Yellen, G. A genetically encoded fluorescent reporter of ATP:ADP ratio. *Nat. Methods* **2009**, *6*, 161–166. [CrossRef] [PubMed]
32. Tantama, M.; Martinez-Francois, J.R.; Mongeon, R.; Yellen, G. Imaging energy status in live cells with a fluorescent biosensor of the intracellular ATP-to-ADP ratio. *Nat. Commun.* **2013**, *4*, 2550. [CrossRef] [PubMed]
33. Zegzouti, H.; Zdanovskaia, M.; Hsiao, K.; Goueli, S.A. ADP-Glo: A Bioluminescent and homogeneous ADP monitoring assay for kinases. *Assay Drug Dev. Technol.* **2009**, *7*, 560–572. [CrossRef] [PubMed]
34. Woods, A.; Munday, M.R.; Scott, J.; Yang, X.; Carlson, M.; Carling, D. Yeast SNF1 is functionally related to mammalian AMP-activated protein kinase and regulates acetyl-CoA carboxylase in vivo. *J. Biol. Chem.* **1994**, *269*, 19509–19515. [CrossRef]
35. Lim, C.T.; Lolli, F.; Thomas, J.D.; Kola, B.; Korbonits, M. Measurement of AMP-activated protein kinase activity and expression in response to ghrelin. *Methods Enzymol.* **2012**, *514*, 271–287. [CrossRef]
36. Hempel, S.L.; Buettner, G.R.; O'Malley, Y.Q.; Wessels, D.A.; Flaherty, D.M. Dihydrofluorescein diacetate is superior for detecting intracellular oxidants: Comparison with 2′,7′-dichlorodihydrofluorescein diacetate, 5(and 6)-carboxy-2′,7′-dichlorodihydrofluorescein diacetate, and dihydrorhodamine 123. *Free Radic. Biol. Med.* **1999**, *27*, 146–159. [CrossRef]
37. Miller, E.W.; Tulyathan, O.; Isacoff, E.Y.; Chang, C.J. Molecular imaging of hydrogen peroxide produced for cell signaling. *Nat. Chem. Biol.* **2007**, *3*, 263–267. [CrossRef]
38. Uggeri, J.; Gatti, R.; Belletti, S.; Scandroglio, R.; Corradini, R.; Rotoli, B.M.; Orlandini, G. Calcein-AM is a detector of intracellular oxidative activity. *Histochem. Cell Biol.* **2004**, *122*, 499–505. [CrossRef]
39. Tian, R.; Abarientos, A.; Hong, J.; Hashemi, S.H.; Yan, R.; Drager, N.; Leng, K.; Nalls, M.A.; Singleton, A.B.; Xu, K.; et al. Genome-wide CRISPRi/a screens in human neurons link lysosomal failure to ferroptosis. *Nat. Neurosci.* **2021**, *24*, 1020–1034. [CrossRef]
40. Camera, E.; Picardo, M. Analytical methods to investigate glutathione and related compounds in biological and pathological processes. *J. Chromatogr. B Anal. Technol. Biomed. Life Sci.* **2002**, *781*, 181–206. [CrossRef]
41. Owen, J.B.; Butterfield, D.A. Measurement of oxidized/reduced glutathione ratio. *Methods Mol. Biol.* **2010**, *648*, 269–277. [CrossRef] [PubMed]
42. Dooley, C.T.; Dore, T.M.; Hanson, G.T.; Jackson, W.C.; Remington, S.J.; Tsien, R.Y. Imaging dynamic redox changes in mammalian cells with green fluorescent protein indicators. *J. Biol. Chem.* **2004**, *279*, 22284–22293. [CrossRef] [PubMed]
43. Meyer, A.J.; Dick, T.P. Fluorescent protein-based redox probes. *Antioxid. Redox Signal.* **2010**, *13*, 621–650. [CrossRef] [PubMed]
44. Pap, E.H.; Drummen, G.P.; Winter, V.J.; Kooij, T.W.; Rijken, P.; Wirtz, K.W.; Op den Kamp, J.A.; Hage, W.J.; Post, J.A. Ratio-fluorescence microscopy of lipid oxidation in living cells using C11-BODIPY(581/591). *FEBS Lett.* **1999**, *453*, 278–282. [CrossRef]
45. Dahle, L.K.; Hill, E.G.; Holman, R.T. The thiobarbituric acid reaction and the autoxidations of polyunsaturated fatty acid methyl esters. *Arch. Biochem. Biophys.* **1962**, *98*, 253–261. [CrossRef]
46. Tang, Q.; Su, Y.W.; Xian, C.J. Determining Oxidative Damage by Lipid Peroxidation Assay in Rat Serum. *Bio Protoc.* **2019**, *9*, e3263. [CrossRef]
47. Lajoie, P.; Fazio, E.N.; Snapp, E.L. Approaches to imaging unfolded secretory protein stress in living cells. *Endoplasmic Reticulum Stress Dis.* **2014**, *1*, 27–39. [CrossRef]
48. Beriault, D.R.; Werstuck, G.H. Detection and quantification of endoplasmic reticulum stress in living cells using the fluorescent compound, Thioflavin T. *Biochim. Biophys. Acta* **2013**, *1833*, 2293–2301. [CrossRef]
49. Iwawaki, T.; Akai, R.; Kohno, K.; Miura, M. A transgenic mouse model for monitoring endoplasmic reticulum stress. *Nat. Med.* **2004**, *10*, 98–102. [CrossRef]

50. Samali, A.; Fitzgerald, U.; Deegan, S.; Gupta, S. Methods for monitoring endoplasmic reticulum stress and the unfolded protein response. *Int. J. Cell Biol.* **2010**, *2010*, 830307. [CrossRef]
51. Nadanaka, S.; Yoshida, H.; Kano, F.; Murata, M.; Mori, K. Activation of mammalian unfolded protein response is compatible with the quality control system operating in the endoplasmic reticulum. *Mol. Biol. Cell* **2004**, *15*, 2537–2548. [CrossRef] [PubMed]
52. Gonzalez, A.; Hall, M.N. Nutrient sensing and TOR signaling in yeast and mammals. *EMBO J.* **2017**, *36*, 397–408. [CrossRef] [PubMed]
53. Gonzalez, A.; Hall, M.N.; Lin, S.C.; Hardie, D.G. AMPK and TOR: The Yin and Yang of Cellular Nutrient Sensing and Growth Control. *Cell Metab.* **2020**, *31*, 472–492. [CrossRef]
54. Vlahakis, A.; Graef, M.; Nunnari, J.; Powers, T. TOR complex 2-Ypk1 signaling is an essential positive regulator of the general amino acid control response and autophagy. *Proc. Natl. Acad. Sci. USA* **2014**, *111*, 10586–10591. [CrossRef] [PubMed]
55. Stracka, D.; Jozefczuk, S.; Rudroff, F.; Sauer, U.; Hall, M.N. Nitrogen source activates TOR (target of rapamycin) complex 1 via glutamine and independently of Gtr/Rag proteins. *J. Biol. Chem.* **2014**, *289*, 25010–25020. [CrossRef] [PubMed]
56. Binda, M.; Peli-Gulli, M.P.; Bonfils, G.; Panchaud, N.; Urban, J.; Sturgill, T.W.; Loewith, R.; De Virgilio, C. The Vam6 GEF controls TORC1 by activating the EGO complex. *Mol. Cell* **2009**, *35*, 563–573. [CrossRef]
57. Panchaud, N.; Peli-Gulli, M.P.; De Virgilio, C. Amino acid deprivation inhibits TORC1 through a GTPase-activating protein complex for the Rag family GTPase Gtr1. *Sci. Signal.* **2013**, *6*, ra42. [CrossRef]
58. Kamada, Y.; Funakoshi, T.; Shintani, T.; Nagano, K.; Ohsumi, M.; Ohsumi, Y. Tor-mediated induction of autophagy via an Apg1 protein kinase complex. *J. Cell Biol.* **2000**, *150*, 1507–1513. [CrossRef]
59. Kamada, Y.; Yoshino, K.; Kondo, C.; Kawamata, T.; Oshiro, N.; Yonezawa, K.; Ohsumi, Y. Tor directly controls the Atg1 kinase complex to regulate autophagy. *Mol. Cell. Biol.* **2010**, *30*, 1049–1058. [CrossRef]
60. Tate, J.J.; Georis, I.; Dubois, E.; Cooper, T.G. Distinct phosphatase requirements and GATA factor responses to nitrogen catabolite repression and rapamycin treatment in Saccharomyces cerevisiae. *J. Biol. Chem.* **2010**, *285*, 17880–17895. [CrossRef]
61. Valenzuela, L.; Aranda, C.; Gonzalez, A. TOR modulates GCN4-dependent expression of genes turned on by nitrogen limitation. *J. Bacteriol.* **2001**, *183*, 2331–2334. [CrossRef] [PubMed]
62. Bernard, A.; Jin, M.; Xu, Z.; Klionsky, D.J. A large-scale analysis of autophagy-related gene expression identifies new regulators of autophagy. *Autophagy* **2015**, *11*, 2114–2122. [CrossRef] [PubMed]
63. Chan, T.F.; Bertram, P.G.; Ai, W.; Zheng, X.F. Regulation of APG14 expression by the GATA-type transcription factor Gln3p. *J. Biol. Chem.* **2001**, *276*, 6463–6467. [CrossRef] [PubMed]
64. Yao, Z.; Delorme-Axford, E.; Backues, S.K.; Klionsky, D.J. Atg41/Icy2 regulates autophagosome formation. *Autophagy* **2015**, *11*, 2288–2299. [CrossRef]
65. Swinnen, E.; Ghillebert, R.; Wilms, T.; Winderickx, J. Molecular mechanisms linking the evolutionary conserved TORC1-Sch9 nutrient signalling branch to lifespan regulation in Saccharomyces cerevisiae. *FEMS Yeast Res.* **2014**, *14*, 17–32. [CrossRef]
66. Bernard, A.; Jin, M.; Gonzalez-Rodriguez, P.; Fullgrabe, J.; Delorme-Axford, E.; Backues, S.K.; Joseph, B.; Klionsky, D.J. Rph1/KDM4 mediates nutrient-limitation signaling that leads to the transcriptional induction of autophagy. *Curr. Biol.* **2015**, *25*, 546–555. [CrossRef]
67. Bartholomew, C.R.; Suzuki, T.; Du, Z.; Backues, S.K.; Jin, M.; Lynch-Day, M.A.; Umekawa, M.; Kamath, A.; Zhao, M.; Xie, Z.; et al. Ume6 transcription factor is part of a signaling cascade that regulates autophagy. *Proc. Natl. Acad. Sci. USA* **2012**, *109*, 11206–11210. [CrossRef]
68. Hu, G.; McQuiston, T.; Bernard, A.; Park, Y.D.; Qiu, J.; Vural, A.; Zhang, N.; Waterman, S.R.; Blewett, N.H.; Myers, T.G.; et al. A conserved mechanism of TOR-dependent RCK-mediated mRNA degradation regulates autophagy. *Nat. Cell Biol.* **2015**, *17*, 930–942. [CrossRef]
69. Jin, M.; He, D.; Backues, S.K.; Freeberg, M.A.; Liu, X.; Kim, J.K.; Klionsky, D.J. Transcriptional regulation by Pho23 modulates the frequency of autophagosome formation. *Curr. Biol.* **2014**, *24*, 1314–1322. [CrossRef]
70. Yu, F.; Imamura, Y.; Ueno, M.; Suzuki, S.W.; Ohsumi, Y.; Yukawa, M.; Tsuchiya, E. The yeast chromatin remodeler Rsc1-RSC complex is required for transcriptional activation of autophagy-related genes and inhibition of the TORC1 pathway in response to nitrogen starvation. *Biochem. Biophys. Res. Commun.* **2015**, *464*, 1248–1253. [CrossRef]
71. Wen, X.; Gatica, D.; Yin, Z.; Hu, Z.; Dengjel, J.; Klionsky, D.J. The transcription factor Spt4-Spt5 complex regulates the expression of ATG8 and ATG41. *Autophagy* **2020**, *16*, 1172–1185. [CrossRef]
72. Delorme-Axford, E.; Abernathy, E.; Lennemann, N.J.; Bernard, A.; Ariosa, A.; Coyne, C.B.; Kirkegaard, K.; Klionsky, D.J. The exoribonuclease Xrn1 is a post-transcriptional negative regulator of autophagy. *Autophagy* **2018**, *14*, 898–912. [CrossRef] [PubMed]
73. Liu, X.; Yao, Z.; Jin, M.; Namkoong, S.; Yin, Z.; Lee, J.H.; Klionsky, D.J. Dhh1 promotes autophagy-related protein translation during nitrogen starvation. *PLoS Biol.* **2019**, *17*, e3000219. [CrossRef]
74. Gatica, D.; Hu, G.; Liu, X.; Zhang, N.; Williamson, P.R.; Klionsky, D.J. The Pat1-Lsm Complex Stabilizes ATG mRNA during Nitrogen Starvation-Induced Autophagy. *Mol. Cell* **2019**, *73*, 314–324.e4. [CrossRef] [PubMed]
75. Yin, Z.; Liu, X.; Ariosa, A.; Huang, H.; Jin, M.; Karbstein, K.; Klionsky, D.J. Psp2, a novel regulator of autophagy that promotes autophagy-related protein translation. *Cell Res.* **2019**, *29*, 994–1008. [CrossRef]
76. Lahiri, V.; Metur, S.P.; Hu, Z.; Song, X.; Mari, M.; Hawkins, W.D.; Bhattarai, J.; Delorme-Axford, E.; Reggiori, F.; Tang, D.; et al. Post-transcriptional regulation of ATG1 is a critical node that modulates autophagy during distinct nutrient stresses. *Autophagy* **2022**, *in press*. [CrossRef] [PubMed]

77. Tanaka, C.; Tan, L.J.; Mochida, K.; Kirisako, H.; Koizumi, M.; Asai, E.; Sakoh-Nakatogawa, M.; Ohsumi, Y.; Nakatogawa, H. Hrr25 triggers selective autophagy-related pathways by phosphorylating receptor proteins. *J. Cell Biol.* **2014**, *207*, 91–105. [CrossRef] [PubMed]
78. Feng, Y.; Ariosa, A.R.; Yang, Y.; Hu, Z.; Dengjel, J.; Klionsky, D.J. Downregulation of autophagy by Met30-mediated Atg9 ubiquitination. *Proc. Natl. Acad. Sci. USA* **2021**, *118*, e2005539118. [CrossRef]
79. Yi, C.; Ma, M.; Ran, L.; Zheng, J.; Tong, J.; Zhu, J.; Ma, C.; Sun, Y.; Zhang, S.; Feng, W.; et al. Function and molecular mechanism of acetylation in autophagy regulation. *Science* **2012**, *336*, 474–477. [CrossRef]
80. Fullgrabe, J.; Lynch-Day, M.A.; Heldring, N.; Li, W.; Struijk, R.B.; Ma, Q.; Hermanson, O.; Rosenfeld, M.G.; Klionsky, D.J.; Joseph, B. The histone H4 lysine 16 acetyltransferase hMOF regulates the outcome of autophagy. *Nature* **2013**, *500*, 468–471. [CrossRef]
81. Liu, X.; Jin, M.; Yao, Z.; Bernard, A.; Klionsky, D.J. Bidirectional roles of Dhh1 in regulating autophagy. *Autophagy* **2019**, *15*, 1838–1839. [CrossRef] [PubMed]
82. Noda, T. Regulation of Autophagy through TORC1 and mTORC1. *Biomolecules* **2017**, *7*, 52. [CrossRef] [PubMed]
83. Sullivan, A.; Wallace, R.L.; Wellington, R.; Luo, X.; Capaldi, A.P. Multilayered regulation of TORC1-body formation in budding yeast. *Mol. Biol. Cell* **2019**, *30*, 400–410. [CrossRef] [PubMed]
84. Ukai, H.; Araki, Y.; Kira, S.; Oikawa, Y.; May, A.I.; Noda, T. Gtr/Ego-independent TORC1 activation is achieved through a glutamine-sensitive interaction with Pib2 on the vacuolar membrane. *PLoS Genet.* **2018**, *14*, e1007334. [CrossRef]
85. Hatakeyama, R.; Peli-Gulli, M.P.; Hu, Z.; Jaquenoud, M.; Garcia Osuna, G.M.; Sardu, A.; Dengjel, J.; De Virgilio, C. Spatially Distinct Pools of TORC1 Balance Protein Homeostasis. *Mol. Cell* **2019**, *73*, 325–338.e8. [CrossRef]
86. McDaniel, S.L.; Hepperla, A.J.; Huang, J.; Dronamraju, R.; Adams, A.T.; Kulkarni, V.G.; Davis, I.J.; Strahl, B.D. H3K36 Methylation Regulates Nutrient Stress Response in Saccharomyces cerevisiae by Enforcing Transcriptional Fidelity. *Cell Rep.* **2017**, *19*, 2371–2382. [CrossRef]
87. Bar-Peled, L.; Sabatini, D.M. Regulation of mTORC1 by amino acids. *Trends Cell Biol.* **2014**, *24*, 400–406. [CrossRef]
88. Sancak, Y.; Peterson, T.R.; Shaul, Y.D.; Lindquist, R.A.; Thoreen, C.C.; Bar-Peled, L.; Sabatini, D.M. The Rag GTPases bind raptor and mediate amino acid signaling to mTORC1. *Science* **2008**, *320*, 1496–1501. [CrossRef]
89. Kim, E.; Goraksha-Hicks, P.; Li, L.; Neufeld, T.P.; Guan, K.L. Regulation of TORC1 by Rag GTPases in nutrient response. *Nat. Cell Biol.* **2008**, *10*, 935–945. [CrossRef]
90. Powis, K.; De Virgilio, C. Conserved regulators of Rag GTPases orchestrate amino acid-dependent TORC1 signaling. *Cell Discov.* **2016**, *2*, 15049. [CrossRef]
91. Zoncu, R.; Bar-Peled, L.; Efeyan, A.; Wang, S.; Sancak, Y.; Sabatini, D.M. mTORC1 senses lysosomal amino acids through an inside-out mechanism that requires the vacuolar H(+)-ATPase. *Science* **2011**, *334*, 678–683. [CrossRef] [PubMed]
92. Bar-Peled, L.; Schweitzer, L.D.; Zoncu, R.; Sabatini, D.M. Ragulator is a GEF for the rag GTPases that signal amino acid levels to mTORC1. *Cell* **2012**, *150*, 1196–1208. [CrossRef] [PubMed]
93. Shen, K.; Choe, A.; Sabatini, D.M. Intersubunit Crosstalk in the Rag GTPase Heterodimer Enables mTORC1 to Respond Rapidly to Amino Acid Availability. *Mol. Cell* **2017**, *68*, 552–565. [CrossRef] [PubMed]
94. Bar-Peled, L.; Chantranupong, L.; Cherniack, A.D.; Chen, W.W.; Ottina, K.A.; Grabiner, B.C.; Spear, E.D.; Carter, S.L.; Meyerson, M.; Sabatini, D.M. A Tumor suppressor complex with GAP activity for the Rag GTPases that signal amino acid sufficiency to mTORC1. *Science* **2013**, *340*, 1100–1106. [CrossRef]
95. Menon, S.; Dibble, C.C.; Talbott, G.; Hoxhaj, G.; Valvezan, A.J.; Takahashi, H.; Cantley, L.C.; Manning, B.D. Spatial control of the TSC complex integrates insulin and nutrient regulation of mTORC1 at the lysosome. *Cell* **2014**, *156*, 771–785. [CrossRef]
96. Demetriades, C.; Plescher, M.; Teleman, A.A. Lysosomal recruitment of TSC2 is a universal response to cellular stress. *Nat. Commun.* **2016**, *7*, 10662. [CrossRef]
97. Yao, Y.; Hong, S.; Ikeda, T.; Mori, H.; MacDougald, O.A.; Nada, S.; Okada, M.; Inoki, K. Amino Acids Enhance Polyubiquitination of Rheb and Its Binding to mTORC1 by Blocking Lysosomal ATXN3 Deubiquitinase Activity. *Mol. Cell* **2020**, *80*, 437–451.e6. [CrossRef]
98. Kim, J.; Kundu, M.; Viollet, B.; Guan, K.L. AMPK and mTOR regulate autophagy through direct phosphorylation of Ulk1. *Nat. Cell Biol.* **2011**, *13*, 132–141. [CrossRef]
99. Hosokawa, N.; Hara, T.; Kaizuka, T.; Kishi, C.; Takamura, A.; Miura, Y.; Iemura, S.; Natsume, T.; Takehana, K.; Yamada, N.; et al. Nutrient-dependent mTORC1 association with the ULK1-Atg13-FIP200 complex required for autophagy. *Mol. Biol. Cell* **2009**, *20*, 1981–1991. [CrossRef]
100. Chan, E.Y. mTORC1 phosphorylates the ULK1-mAtg13-FIP200 autophagy regulatory complex. *Sci. Signal.* **2009**, *2*, pe51. [CrossRef]
101. Dunlop, E.A.; Hunt, D.K.; Acosta-Jaquez, H.A.; Fingar, D.C.; Tee, A.R. ULK1 inhibits mTORC1 signaling, promotes multisite Raptor phosphorylation and hinders substrate binding. *Autophagy* **2011**, *7*, 737–747. [CrossRef] [PubMed]
102. Jung, C.H.; Seo, M.; Otto, N.M.; Kim, D.H. ULK1 inhibits the kinase activity of mTORC1 and cell proliferation. *Autophagy* **2011**, *7*, 1212–1221. [CrossRef] [PubMed]
103. Yuan, H.X.; Russell, R.C.; Guan, K.L. Regulation of PIK3C3/VPS34 complexes by MTOR in nutrient stress-induced autophagy. *Autophagy* **2013**, *9*, 1983–1995. [CrossRef] [PubMed]

104. Ma, X.; Zhang, S.; He, L.; Rong, Y.; Brier, L.W.; Sun, Q.; Liu, R.; Fan, W.; Chen, S.; Yue, Z.; et al. MTORC1-mediated NRBF2 phosphorylation functions as a switch for the class III PtdIns3K and autophagy. *Autophagy* **2017**, *13*, 592–607. [CrossRef]
105. Nazio, F.; Strappazzon, F.; Antonioli, M.; Bielli, P.; Cianfanelli, V.; Bordi, M.; Gretzmeier, C.; Dengjel, J.; Piacentini, M.; Fimia, G.M.; et al. mTOR inhibits autophagy by controlling ULK1 ubiquitylation, self-association and function through AMBRA1 and TRAF6. *Nat. Cell Biol.* **2013**, *15*, 406–416. [CrossRef] [PubMed]
106. Kim, Y.M.; Jung, C.H.; Seo, M.; Kim, E.K.; Park, J.M.; Bae, S.S.; Kim, D.H. mTORC1 phosphorylates UVRAG to negatively regulate autophagosome and endosome maturation. *Mol. Cell* **2015**, *57*, 207–218. [CrossRef]
107. Wan, W.; You, Z.; Zhou, L.; Xu, Y.; Peng, C.; Zhou, T.; Yi, C.; Shi, Y.; Liu, W. mTORC1-Regulated and HUWE1-Mediated WIPI2 Degradation Controls Autophagy Flux. *Mol. Cell* **2018**, *72*, 303–315.e6. [CrossRef]
108. Wan, W.; You, Z.; Xu, Y.; Zhou, L.; Guan, Z.; Peng, C.; Wong, C.C.L.; Su, H.; Zhou, T.; Xia, H.; et al. mTORC1 Phosphorylates Acetyltransferase p300 to Regulate Autophagy and Lipogenesis. *Mol. Cell* **2017**, *68*, 323–335. [CrossRef]
109. Lee, I.H.; Finkel, T. Regulation of autophagy by the p300 acetyltransferase. *J. Biol. Chem.* **2009**, *284*, 6322–6328. [CrossRef]
110. Napolitano, G.; Ballabio, A. TFEB at a glance. *J. Cell Sci.* **2016**, *129*, 2475–2481. [CrossRef]
111. Martina, J.A.; Puertollano, R. Rag GTPases mediate amino acid-dependent recruitment of TFEB and MITF to lysosomes. *J. Cell Biol.* **2013**, *200*, 475–491. [CrossRef] [PubMed]
112. Settembre, C.; Di Malta, C.; Polito, V.A.; Garcia Arencibia, M.; Vetrini, F.; Erdin, S.; Erdin, S.U.; Huynh, T.; Medina, D.; Colella, P.; et al. TFEB links autophagy to lysosomal biogenesis. *Science* **2011**, *332*, 1429–1433. [CrossRef]
113. Roczniak-Ferguson, A.; Petit, C.S.; Froehlich, F.; Qian, S.; Ky, J.; Angarola, B.; Walther, T.C.; Ferguson, S.M. The transcription factor TFEB links mTORC1 signaling to transcriptional control of lysosome homeostasis. *Sci. Signal.* **2012**, *5*, ra42. [CrossRef] [PubMed]
114. Martina, J.A.; Chen, Y.; Gucek, M.; Puertollano, R. MTORC1 functions as a transcriptional regulator of autophagy by preventing nuclear transport of TFEB. *Autophagy* **2012**, *8*, 903–914. [CrossRef] [PubMed]
115. Vega-Rubin-de-Celis, S.; Pena-Llopis, S.; Konda, M.; Brugarolas, J. Multistep regulation of TFEB by MTORC1. *Autophagy* **2017**, *13*, 464–472. [CrossRef]
116. Napolitano, G.; Esposito, A.; Choi, H.; Matarese, M.; Benedetti, V.; Di Malta, C.; Monfregola, J.; Medina, D.L.; Lippincott-Schwartz, J.; Ballabio, A. mTOR-dependent phosphorylation controls TFEB nuclear export. *Nat. Commun.* **2018**, *9*, 3312. [CrossRef]
117. Wang, Y.; Huang, Y.; Liu, J.; Zhang, J.; Xu, M.; You, Z.; Peng, C.; Gong, Z.; Liu, W. Acetyltransferase GCN5 regulates autophagy and lysosome biogenesis by targeting TFEB. *EMBO Rep.* **2020**, *21*, e48335. [CrossRef] [PubMed]
118. Martina, J.A.; Diab, H.I.; Lishu, L.; Jeong, A.L.; Patange, S.; Raben, N.; Puertollano, R. The nutrient-responsive transcription factor TFE3 promotes autophagy, lysosomal biogenesis, and clearance of cellular debris. *Sci. Signal.* **2014**, *7*, ra9. [CrossRef] [PubMed]
119. Ploper, D.; Taelman, V.F.; Robert, L.; Perez, B.S.; Titz, B.; Chen, H.W.; Graeber, T.G.; von Euw, E.; Ribas, A.; De Robertis, E.M. MITF drives endolysosomal biogenesis and potentiates Wnt signaling in melanoma cells. *Proc. Natl. Acad. Sci. USA* **2015**, *112*, E420–E429. [CrossRef] [PubMed]
120. Moller, K.; Sigurbjornsdottir, S.; Arnthorsson, A.O.; Pogenberg, V.; Dilshat, R.; Fock, V.; Brynjolfsdottir, S.H.; Bindesboll, C.; Bessadottir, M.; Ogmundsdottir, H.M.; et al. MITF has a central role in regulating starvation-induced autophagy in melanoma. *Sci. Rep.* **2019**, *9*, 1055. [CrossRef]
121. Di Malta, C.; Siciliano, D.; Calcagni, A.; Monfregola, J.; Punzi, S.; Pastore, N.; Eastes, A.N.; Davis, O.; De Cegli, R.; Zampelli, A.; et al. Transcriptional activation of RagD GTPase controls mTORC1 and promotes cancer growth. *Science* **2017**, *356*, 1188–1192. [CrossRef] [PubMed]
122. Wengrod, J.; Wang, D.; Weiss, S.; Zhong, H.; Osman, I.; Gardner, L.B. Phosphorylation of eIF2alpha triggered by mTORC1 inhibition and PP6C activation is required for autophagy and is aberrant in PP6C-mutated melanoma. *Sci. Signal.* **2015**, *8*, ra27. [CrossRef] [PubMed]
123. B'Chir, W.; Maurin, A.C.; Carraro, V.; Averous, J.; Jousse, C.; Muranishi, Y.; Parry, L.; Stepien, G.; Fafournoux, P.; Bruhat, A. The eIF2alpha/ATF4 pathway is essential for stress-induced autophagy gene expression. *Nucleic Acids Res.* **2013**, *41*, 7683–7699. [CrossRef] [PubMed]
124. Tang, H.W.; Weng, J.H.; Lee, W.X.; Hu, Y.; Gu, L.; Cho, S.; Lee, G.; Binari, R.; Li, C.; Cheng, M.E.; et al. mTORC1-chaperonin CCT signaling regulates m(6)A RNA methylation to suppress autophagy. *Proc. Natl. Acad. Sci. USA* **2021**, *118*, e2021945118. [CrossRef] [PubMed]
125. Baek, S.H.; Kim, K.I. Epigenetic Control of Autophagy: Nuclear Events Gain More Attention. *Mol. Cell* **2017**, *65*, 781–785. [CrossRef]
126. Bowman, C.J.; Ayer, D.E.; Dynlacht, B.D. Foxk proteins repress the initiation of starvation-induced atrophy and autophagy programs. *Nat. Cell Biol.* **2014**, *16*, 1202–1214. [CrossRef] [PubMed]
127. Wei, Y.; Pattingre, S.; Sinha, S.; Bassik, M.; Levine, B. JNK1-mediated phosphorylation of Bcl-2 regulates starvation-induced autophagy. *Mol. Cell* **2008**, *30*, 678–688. [CrossRef]
128. Wei, Y.; An, Z.; Zou, Z.; Sumpter, R.; Su, M.; Zang, X.; Sinha, S.; Gaestel, M.; Levine, B. The stress-responsive kinases MAPKAPK2/MAPKAPK3 activate starvation-induced autophagy through Beclin 1 phosphorylation. *Elife* **2015**, *4*, e05289. [CrossRef]
129. Comb, W.C.; Cogswell, P.; Sitcheran, R.; Baldwin, A.S. IKK-dependent, NF-kappaB-independent control of autophagic gene expression. *Oncogene* **2011**, *30*, 1727–1732. [CrossRef]
130. Hardie, D.G.; Carling, D.; Halford, N. Roles of the Snf1/Rkin1/AMP-activated protein kinase family in the response to environmental and nutritional stress. *Semin. Cell Biol.* **1994**, *5*, 409–416. [CrossRef]

131. Herzig, S.; Shaw, R.J. AMPK: Guardian of metabolism and mitochondrial homeostasis. *Nat. Rev. Mol. Cell Biol.* **2018**, *19*, 121–135. [CrossRef] [PubMed]
132. Hawley, S.A.; Davison, M.; Woods, A.; Davies, S.P.; Beri, R.K.; Carling, D.; Hardie, D.G. Characterization of the AMP-activated protein kinase kinase from rat liver and identification of threonine 172 as the major site at which it phosphorylates AMP-activated protein kinase. *J. Biol. Chem.* **1996**, *271*, 27879–27887. [CrossRef] [PubMed]
133. Xiao, B.; Heath, R.; Saiu, P.; Leiper, F.C.; Leone, P.; Jing, C.; Walker, P.A.; Haire, L.; Eccleston, J.F.; Davis, C.T.; et al. Structural basis for AMP binding to mammalian AMP-activated protein kinase. *Nature* **2007**, *449*, 496–500. [CrossRef] [PubMed]
134. Hardie, D.G.; Scott, J.W.; Pan, D.A.; Hudson, E.R. Management of cellular energy by the AMP-activated protein kinase system. *FEBS Lett.* **2003**, *546*, 113–120. [CrossRef]
135. Hawley, S.A.; Selbert, M.A.; Goldstein, E.G.; Edelman, A.M.; Carling, D.; Hardie, D.G. 5′-AMP activates the AMP-activated protein kinase cascade, and Ca^{2+}/calmodulin activates the calmodulin-dependent protein kinase I cascade, via three independent mechanisms. *J. Biol. Chem.* **1995**, *270*, 27186–27191. [CrossRef]
136. Shaw, R.J.; Kosmatka, M.; Bardeesy, N.; Hurley, R.L.; Witters, L.A.; DePinho, R.A.; Cantley, L.C. The tumor suppressor LKB1 kinase directly activates AMP-activated kinase and regulates apoptosis in response to energy stress. *Proc. Natl. Acad. Sci. USA* **2004**, *101*, 3329–3335. [CrossRef]
137. Davies, S.P.; Helps, N.R.; Cohen, P.T.; Hardie, D.G. 5′-AMP inhibits dephosphorylation, as well as promoting phosphorylation, of the AMP-activated protein kinase. Studies using bacterially expressed human protein phosphatase-2C alpha and native bovine protein phosphatase-2AC. *FEBS Lett.* **1995**, *377*, 421–425. [CrossRef]
138. Gowans, G.J.; Hawley, S.A.; Ross, F.A.; Hardie, D.G. AMP is a true physiological regulator of AMP-activated protein kinase by both allosteric activation and enhancing net phosphorylation. *Cell Metab.* **2013**, *18*, 556–566. [CrossRef]
139. Gwinn, D.M.; Shackelford, D.B.; Egan, D.F.; Mihaylova, M.M.; Mery, A.; Vasquez, D.S.; Turk, B.E.; Shaw, R.J. AMPK phosphorylation of raptor mediates a metabolic checkpoint. *Mol. Cell* **2008**, *30*, 214–226. [CrossRef]
140. Inoki, K.; Zhu, T.; Guan, K.L. TSC2 mediates cellular energy response to control cell growth and survival. *Cell* **2003**, *115*, 577–590. [CrossRef]
141. Inoki, K.; Li, Y.; Xu, T.; Guan, K.L. Rheb GTPase is a direct target of TSC2 GAP activity and regulates mTOR signaling. *Genes Dev.* **2003**, *17*, 1829–1834. [CrossRef] [PubMed]
142. Russell, R.C.; Tian, Y.; Yuan, H.; Park, H.W.; Chang, Y.Y.; Kim, J.; Kim, H.; Neufeld, T.P.; Dillin, A.; Guan, K.L. ULK1 induces autophagy by phosphorylating Beclin-1 and activating VPS34 lipid kinase. *Nat. Cell Biol.* **2013**, *15*, 741–750. [CrossRef] [PubMed]
143. Weerasekara, V.K.; Panek, D.J.; Broadbent, D.G.; Mortenson, J.B.; Mathis, A.D.; Logan, G.N.; Prince, J.T.; Thomson, D.M.; Thompson, J.W.; Andersen, J.L. Metabolic-stress-induced rearrangement of the 14-3-3zeta interactome promotes autophagy via a ULK1- and AMPK-regulated 14-3-3zeta interaction with phosphorylated Atg9. *Mol. Cell. Biol.* **2014**, *34*, 4379–4388. [CrossRef] [PubMed]
144. Greer, E.L.; Oskoui, P.R.; Banko, M.R.; Maniar, J.M.; Gygi, M.P.; Gygi, S.P.; Brunet, A. The energy sensor AMP-activated protein kinase directly regulates the mammalian FOXO3 transcription factor. *J. Biol. Chem.* **2007**, *282*, 30107–30119. [CrossRef] [PubMed]
145. Tamargo-Gomez, I.; Marino, G. AMPK: Regulation of Metabolic Dynamics in the Context of Autophagy. *Int. J. Mol. Sci.* **2018**, *19*, 3812. [CrossRef] [PubMed]
146. Chang, C.; Su, H.; Zhang, D.; Wang, Y.; Shen, Q.; Liu, B.; Huang, R.; Zhou, T.; Peng, C.; Wong, C.C.; et al. AMPK-Dependent Phosphorylation of GAPDH Triggers Sirt1 Activation and Is Necessary for Autophagy upon Glucose Starvation. *Mol. Cell* **2015**, *60*, 930–940. [CrossRef] [PubMed]
147. Lee, I.H.; Cao, L.; Mostoslavsky, R.; Lombard, D.B.; Liu, J.; Bruns, N.E.; Tsokos, M.; Alt, F.W.; Finkel, T. A role for the NAD-dependent deacetylase Sirt1 in the regulation of autophagy. *Proc. Natl. Acad. Sci. USA* **2008**, *105*, 3374–3379. [CrossRef]
148. Hariharan, N.; Maejima, Y.; Nakae, J.; Paik, J.; Depinho, R.A.; Sadoshima, J. Deacetylation of FoxO by Sirt1 Plays an Essential Role in Mediating Starvation-Induced Autophagy in Cardiac Myocytes. *Circ. Res.* **2010**, *107*, 1470–1482. [CrossRef]
149. Hu, Y.; Chen, H.; Zhang, L.; Lin, X.; Li, X.; Zhuang, H.; Fan, H.; Meng, T.; He, Z.; Huang, H.; et al. The AMPK-MFN2 axis regulates MAM dynamics and autophagy induced by energy stresses. *Autophagy* **2021**, *17*, 1142–1156. [CrossRef]
150. Finkel, T. Signal transduction by reactive oxygen species. *J. Cell Biol.* **2011**, *194*, 7–15. [CrossRef]
151. Sies, H.; Berndt, C.; Jones, D.P. Oxidative Stress. *Annu. Rev. Biochem.* **2017**, *86*, 715–748. [CrossRef] [PubMed]
152. Ray, P.D.; Huang, B.W.; Tsuji, Y. Reactive oxygen species (ROS) homeostasis and redox regulation in cellular signaling. *Cell. Signal.* **2012**, *24*, 981–990. [CrossRef] [PubMed]
153. Gao, Q. Oxidative Stress and Autophagy. *Adv. Exp. Med. Biol.* **2019**, *1206*, 179–198. [CrossRef] [PubMed]
154. Filomeni, G.; De Zio, D.; Cecconi, F. Oxidative stress and autophagy: The clash between damage and metabolic needs. *Cell Death Differ.* **2015**, *22*, 377–388. [CrossRef] [PubMed]
155. Kiffin, R.; Bandyopadhyay, U.; Cuervo, A.M. Oxidative stress and autophagy. *Antioxid. Redox Signal.* **2006**, *8*, 152–162. [CrossRef] [PubMed]
156. Chen, Y.; McMillan-Ward, E.; Kong, J.; Israels, S.J.; Gibson, S.B. Oxidative stress induces autophagic cell death independent of apoptosis in transformed and cancer cells. *Cell Death Differ.* **2008**, *15*, 171–182. [CrossRef]
157. Hou, W.; Xie, Y.; Song, X.; Sun, X.; Lotze, M.T.; Zeh, H.J., 3rd; Kang, R.; Tang, D. Autophagy promotes ferroptosis by degradation of ferritin. *Autophagy* **2016**, *12*, 1425–1428. [CrossRef]

158. Dobrowolny, G.; Aucello, M.; Rizzuto, E.; Beccafico, S.; Mammucari, C.; Boncompagni, S.; Belia, S.; Wannenes, F.; Nicoletti, C.; Del Prete, Z.; et al. Skeletal muscle is a primary target of SOD1G93A-mediated toxicity. *Cell Metab.* **2008**, *8*, 425–436. [CrossRef]
159. Park, E.; Chung, S.W. ROS-mediated autophagy increases intracellular iron levels and ferroptosis by ferritin and transferrin receptor regulation. *Cell Death Dis.* **2019**, *10*, 822. [CrossRef]
160. Scherz-Shouval, R.; Elazar, Z. Regulation of autophagy by ROS: Physiology and pathology. *Trends Biochem. Sci.* **2011**, *36*, 30–38. [CrossRef]
161. Scherz-Shouval, R.; Elazar, Z. ROS, mitochondria and the regulation of autophagy. *Trends Cell Biol.* **2007**, *17*, 422–427. [CrossRef] [PubMed]
162. Lee, J.; Giordano, S.; Zhang, J. Autophagy, mitochondria and oxidative stress: Cross-talk and redox signalling. *Biochem. J.* **2012**, *441*, 523–540. [CrossRef]
163. Hayyan, M.; Hashim, M.A.; AlNashef, I.M. Superoxide Ion: Generation and Chemical Implications. *Chem. Rev.* **2016**, *116*, 3029–3085. [CrossRef] [PubMed]
164. Groeger, G.; Quiney, C.; Cotter, T.G. Hydrogen peroxide as a cell-survival signaling molecule. *Antioxid. Redox Signal.* **2009**, *11*, 2655–2671. [CrossRef]
165. Marinho, H.S.; Real, C.; Cyrne, L.; Soares, H.; Antunes, F. Hydrogen peroxide sensing, signaling and regulation of transcription factors. *Redox Biol.* **2014**, *2*, 535–562. [CrossRef] [PubMed]
166. Lloyd, R.V.; Hanna, P.M.; Mason, R.P. The origin of the hydroxyl radical oxygen in the Fenton reaction. *Free Radic. Biol. Med.* **1997**, *22*, 885–888. [CrossRef]
167. Halliwell, B.; Chirico, S. Lipid peroxidation: Its mechanism, measurement, and significance. *Am. J. Clin. Nutr.* **1993**, *57*, 715S–724S, discussion 724S—725S. [CrossRef] [PubMed]
168. Zhang, H.; Kong, X.; Kang, J.; Su, J.; Li, Y.; Zhong, J.; Sun, L. Oxidative stress induces parallel autophagy and mitochondria dysfunction in human glioma U251 cells. *Toxicol. Sci.* **2009**, *110*, 376–388. [CrossRef]
169. Chen, Y.; Azad, M.B.; Gibson, S.B. Superoxide is the major reactive oxygen species regulating autophagy. *Cell Death Differ.* **2009**, *16*, 1040–1052. [CrossRef]
170. Kim, E.H.; Sohn, S.; Kwon, H.J.; Kim, S.U.; Kim, M.J.; Lee, S.J.; Choi, K.S. Sodium selenite induces superoxide-mediated mitochondrial damage and subsequent autophagic cell death in malignant glioma cells. *Cancer Res.* **2007**, *67*, 6314–6324. [CrossRef]
171. Chen, J.W.; Ni, B.B.; Li, B.; Yang, Y.H.; Jiang, S.D.; Jiang, L.S. The responses of autophagy and apoptosis to oxidative stress in nucleus pulposus cells: Implications for disc degeneration. *Cell. Physiol. Biochem.* **2014**, *34*, 1175–1189. [CrossRef] [PubMed]
172. Yu, Y.; Fan, S.M.; Song, J.K.; Tashiro, S.; Onodera, S.; Ikejima, T. Hydroxyl radical (·OH) played a pivotal role in oridonin-induced apoptosis and autophagy in human epidermoid carcinoma A431 cells. *Biol. Pharm. Bull.* **2012**, *35*, 2148–2159. [CrossRef] [PubMed]
173. Barsoum, M.J.; Yuan, H.; Gerencser, A.A.; Liot, G.; Kushnareva, Y.; Gräber, S.; Kovacs, I.; Lee, W.D.; Waggoner, J.; Cui, J.; et al. Nitric oxide-induced mitochondrial fission is regulated by dynamin-related GTPases in neurons. *EMBO J.* **2006**, *25*, 3900–3911. [CrossRef] [PubMed]
174. Liot, G.; Bossy, B.; Lubitz, S.; Kushnareva, Y.; Sejbuk, N.; Bossy-Wetzel, E. Complex II inhibition by 3-NP causes mitochondrial fragmentation and neuronal cell death via an NMDA- and ROS-dependent pathway. *Cell Death Differ.* **2009**, *16*, 899–909. [CrossRef] [PubMed]
175. Tyagi, N.; Vacek, J.C.; Givvimani, S.; Sen, U.; Tyagi, S.C. Cardiac specific deletion of N-methyl-d-aspartate receptor 1 ameliorates mtMMP-9 mediated autophagy/mitophagy in hyperhomocysteinemia. *J. Recept. Signal Transduct. Res.* **2010**, *30*, 78–87. [CrossRef] [PubMed]
176. Dodson, M.; Wani, W.Y.; Redmann, M.; Benavides, G.A.; Johnson, M.S.; Ouyang, X.; Cofield, S.S.; Mitra, K.; Darley-Usmar, V.; Zhang, J. Regulation of autophagy, mitochondrial dynamics, and cellular bioenergetics by 4-hydroxynonenal in primary neurons. *Autophagy* **2017**, *13*, 1828–1840. [CrossRef]
177. De Gaetano, A.; Gibellini, L.; Zanini, G.; Nasi, M.; Cossarizza, A.; Pinti, M. Mitophagy and Oxidative Stress: The Role of Aging. *Antioxidants* **2021**, *10*, 794. [CrossRef]
178. Deffieu, M.; Bhatia-Kissová, I.; Salin, B.; Galinier, A.; Manon, S.; Camougrand, N. Glutathione participates in the regulation of mitophagy in yeast. *J. Biol. Chem.* **2009**, *284*, 14828–14837. [CrossRef]
179. Jing, H.; Liu, H.; Zhang, L.; Gao, J.; Song, H.; Tan, X. Ethanol Induces Autophagy Regulated by Mitochondrial ROS in Saccharomyces cerevisiae. *J. Microbiol. Biotechnol.* **2018**, *28*, 1982–1991. [CrossRef]
180. Rodrigues-Pousada, C.; Devaux, F.; Caetano, S.M.; Pimentel, C.; da Silva, S.; Cordeiro, A.C.; Amaral, C. Yeast AP-1 like transcription factors (Yap) and stress response: A current overview. *Microb. Cell* **2019**, *6*, 267–285. [CrossRef]
181. Toledano, M.B.; Delaunay, A.; Monceau, L.; Tacnet, F. Microbial H_2O_2 sensors as archetypical redox signaling modules. *Trends Biochem. Sci.* **2004**, *29*, 351–357. [CrossRef] [PubMed]
182. Harshman, K.D.; Moye-Rowley, W.S.; Parker, C.S. Transcriptional activation by the SV40 AP-1 recognition element in yeast is mediated by a factor similar to AP-1 that is distinct from GCN4. *Cell* **1988**, *53*, 321–330. [CrossRef]
183. Fernandes, L.; Rodrigues-Pousada, C.; Struhl, K. Yap, a novel family of eight bZIP proteins in Saccharomyces cerevisiae with distinct biological functions. *Mol. Cell. Biol.* **1997**, *17*, 6982–6993. [CrossRef] [PubMed]

184. Kuge, S.; Jones, N. YAP1 dependent activation of TRX2 is essential for the response of Saccharomyces cerevisiae to oxidative stress by hydroperoxides. *EMBO J.* **1994**, *13*, 655–664. [CrossRef]
185. Kuge, S.; Jones, N.; Nomoto, A. Regulation of yAP-1 nuclear localization in response to oxidative stress. *EMBO J.* **1997**, *16*, 1710–1720. [CrossRef] [PubMed]
186. Delaunay, A.; Isnard, A.D.; Toledano, M.B. H_2O_2 sensing through oxidation of the Yap1 transcription factor. *EMBO J.* **2000**, *19*, 5157–5166. [CrossRef] [PubMed]
187. Okazaki, S.; Tachibana, T.; Naganuma, A.; Mano, N.; Kuge, S. Multistep disulfide bond formation in Yap1 is required for sensing and transduction of H2O2 stress signal. *Mol. Cell* **2007**, *27*, 675–688. [CrossRef]
188. Delaunay, A.; Pflieger, D.; Barrault, M.B.; Vinh, J.; Toledano, M.B. A thiol peroxidase is an H_2O_2 receptor and redox-transducer in gene activation. *Cell* **2002**, *111*, 471–481. [CrossRef]
189. Maeda, Y.; Oku, M.; Sakai, Y. A defect of the vacuolar putative lipase Atg15 accelerates degradation of lipid droplets through lipolysis. *Autophagy* **2015**, *11*, 1247–1258. [CrossRef]
190. Ramya, V.; Rajasekharan, R. ATG15 encodes a phospholipase and is transcriptionally regulated by YAP1 in Saccharomyces cerevisiae. *FEBS Lett.* **2016**, *590*, 3155–3167. [CrossRef]
191. Scherz-Shouval, R.; Shvets, E.; Fass, E.; Shorer, H.; Gil, L.; Elazar, Z. Reactive oxygen species are essential for autophagy and specifically regulate the activity of Atg4. *EMBO J.* **2007**, *26*, 1749–1760. [CrossRef] [PubMed]
192. Pérez-Pérez, M.E.; Zaffagnini, M.; Marchand, C.H.; Crespo, J.L.; Lemaire, S.D. The yeast autophagy protease Atg4 is regulated by thioredoxin. *Autophagy* **2014**, *10*, 1953–1964. [CrossRef] [PubMed]
193. Scherz-Shouval, R.; Elazar, Z. Monitoring starvation-induced reactive oxygen species formation. *Methods Enzymol.* **2009**, *452*, 119–130. [CrossRef]
194. Mungai, P.T.; Waypa, G.B.; Jairaman, A.; Prakriya, M.; Dokic, D.; Ball, M.K.; Schumacker, P.T. Hypoxia triggers AMPK activation through reactive oxygen species-mediated activation of calcium release-activated calcium channels. *Mol. Cell. Biol.* **2011**, *31*, 3531–3545. [CrossRef] [PubMed]
195. Djavaheri-Mergny, M.; Amelotti, M.; Mathieu, J.; Besançon, F.; Bauvy, C.; Souquère, S.; Pierron, G.; Codogno, P. NF-kappaB activation represses tumor necrosis factor-alpha-induced autophagy. *J. Biol. Chem.* **2006**, *281*, 30373–30382. [CrossRef]
196. Djavaheri-Mergny, M.; Javelaud, D.; Wietzerbin, J.; Besançon, F. NF-kappaB activation prevents apoptotic oxidative stress via an increase of both thioredoxin and MnSOD levels in TNFalpha-treated Ewing sarcoma cells. *FEBS Lett.* **2004**, *578*, 111–115. [CrossRef] [PubMed]
197. Zhang, H.; Bosch-Marce, M.; Shimoda, L.A.; Tan, Y.S.; Baek, J.H.; Wesley, J.B.; Gonzalez, F.J.; Semenza, G.L. Mitochondrial autophagy is an HIF-1-dependent adaptive metabolic response to hypoxia. *J. Biol. Chem.* **2008**, *283*, 10892–10903. [CrossRef]
198. Zmijewski, J.W.; Banerjee, S.; Bae, H.; Friggeri, A.; Lazarowski, E.R.; Abraham, E. Exposure to hydrogen peroxide induces oxidation and activation of AMP-activated protein kinase. *J. Biol. Chem.* **2010**, *285*, 33154–33164. [CrossRef]
199. Bellot, G.; Garcia-Medina, R.; Gounon, P.; Chiche, J.; Roux, D.; Pouysségur, J.; Mazure, N.M. Hypoxia-induced autophagy is mediated through hypoxia-inducible factor induction of BNIP3 and BNIP3L via their BH3 domains. *Mol. Cell. Biol.* **2009**, *29*, 2570–2581. [CrossRef]
200. Alexander, A.; Kim, J.; Walker, C.L. ATM engages the TSC2/mTORC1 signaling node to regulate autophagy. *Autophagy* **2010**, *6*, 672–673. [CrossRef]
201. Cirotti, C.; Rizza, S.; Giglio, P.; Poerio, N.; Allega, M.F.; Claps, G.; Pecorari, C.; Lee, J.H.; Benassi, B.; Barilà, D.; et al. Redox activation of ATM enhances GSNOR translation to sustain mitophagy and tolerance to oxidative stress. *EMBO Rep.* **2021**, *22*, e50500. [CrossRef] [PubMed]
202. Dames, S.A.; Mulet, J.M.; Rathgeb-Szabo, K.; Hall, M.N.; Grzesiek, S. The solution structure of the FATC domain of the protein kinase target of rapamycin suggests a role for redox-dependent structural and cellular stability. *J. Biol. Chem.* **2005**, *280*, 20558–20564. [CrossRef] [PubMed]
203. Bensaad, K.; Cheung, E.C.; Vousden, K.H. Modulation of intracellular ROS levels by TIGAR controls autophagy. *EMBO J.* **2009**, *28*, 3015–3026. [CrossRef] [PubMed]
204. McClung, J.M.; Judge, A.R.; Powers, S.K.; Yan, Z. p38 MAPK links oxidative stress to autophagy-related gene expression in cachectic muscle wasting. *Am. J. Physiol. Cell Physiol.* **2010**, *298*, C542–C549. [CrossRef] [PubMed]
205. Pajares, M.; Cuadrado, A.; Engedal, N.; Jirsova, Z.; Cahova, M. The Role of Free Radicals in Autophagy Regulation: Implications for Ageing. *Oxid. Med. Cell. Longev.* **2018**, *2018*, 2450748. [CrossRef] [PubMed]
206. Ma, Q. Role of nrf2 in oxidative stress and toxicity. *Annu. Rev. Pharmacol. Toxicol.* **2013**, *53*, 401–426. [CrossRef] [PubMed]
207. Madden, S.K.; Itzhaki, L.S. Structural and mechanistic insights into the Keap1-Nrf2 system as a route to drug discovery. *Biochim. Biophys. Acta Proteins Proteom.* **2020**, *1868*, 140405. [CrossRef]
208. Jiang, T.; Harder, B.; Rojo de la Vega, M.; Wong, P.K.; Chapman, E.; Zhang, D.D. p62 links autophagy and Nrf2 signaling. *Free Radic. Biol. Med.* **2015**, *88*, 199–204. [CrossRef]
209. Fuse, Y.; Kobayashi, M. Conservation of the Keap1-Nrf2 System: An Evolutionary Journey through Stressful Space and Time. *Molecules* **2017**, *22*, 436. [CrossRef]
210. Itoh, K.; Chiba, T.; Takahashi, S.; Ishii, T.; Igarashi, K.; Katoh, Y.; Oyake, T.; Hayashi, N.; Satoh, K.; Hatayama, I.; et al. An Nrf2/small Maf heterodimer mediates the induction of phase II detoxifying enzyme genes through antioxidant response elements. *Biochem. Biophys. Res. Commun.* **1997**, *236*, 313–322. [CrossRef]

211. Kobayashi, A.; Kang, M.I.; Okawa, H.; Ohtsuji, M.; Zenke, Y.; Chiba, T.; Igarashi, K.; Yamamoto, M. Oxidative stress sensor Keap1 functions as an adaptor for Cul3-based E3 ligase to regulate proteasomal degradation of Nrf2. *Mol. Cell. Biol.* **2004**, *24*, 7130–7139. [CrossRef]
212. Zhang, D.D.; Hannink, M. Distinct cysteine residues in Keap1 are required for Keap1-dependent ubiquitination of Nrf2 and for stabilization of Nrf2 by chemopreventive agents and oxidative stress. *Mol. Cell. Biol.* **2003**, *23*, 8137–8151. [CrossRef]
213. Zhang, D.D.; Lo, S.C.; Cross, J.V.; Templeton, D.J.; Hannink, M. Keap1 is a redox-regulated substrate adaptor protein for a Cul3-dependent ubiquitin ligase complex. *Mol. Cell. Biol.* **2004**, *24*, 10941–10953. [CrossRef] [PubMed]
214. Itoh, K.; Wakabayashi, N.; Katoh, Y.; Ishii, T.; Igarashi, K.; Engel, J.D.; Yamamoto, M. Keap1 represses nuclear activation of antioxidant responsive elements by Nrf2 through binding to the amino-terminal Neh2 domain. *Genes Dev.* **1999**, *13*, 76–86. [CrossRef] [PubMed]
215. Lo, S.C.; Li, X.; Henzl, M.T.; Beamer, L.J.; Hannink, M. Structure of the Keap1:Nrf2 interface provides mechanistic insight into Nrf2 signaling. *EMBO J.* **2006**, *25*, 3605–3617. [CrossRef] [PubMed]
216. Furukawa, M.; Xiong, Y. BTB protein Keap1 targets antioxidant transcription factor Nrf2 for ubiquitination by the Cullin 3-Roc1 ligase. *Mol. Cell. Biol.* **2005**, *25*, 162–171. [CrossRef]
217. Suzuki, T.; Muramatsu, A.; Saito, R.; Iso, T.; Shibata, T.; Kuwata, K.; Kawaguchi, S.I.; Iwawaki, T.; Adachi, S.; Suda, H.; et al. Molecular Mechanism of Cellular Oxidative Stress Sensing by Keap1. *Cell Rep.* **2019**, *28*, 746–758.e4. [CrossRef] [PubMed]
218. Hourihan, J.M.; Kenna, J.G.; Hayes, J.D. The gasotransmitter hydrogen sulfide induces nrf2-target genes by inactivating the keap1 ubiquitin ligase substrate adaptor through formation of a disulfide bond between cys-226 and cys-613. *Antioxid. Redox Signal.* **2013**, *19*, 465–481. [CrossRef]
219. Fourquet, S.; Guerois, R.; Biard, D.; Toledano, M.B. Activation of NRF2 by nitrosative agents and H2O2 involves KEAP1 disulfide formation. *J. Biol. Chem.* **2010**, *285*, 8463–8471. [CrossRef]
220. Komatsu, M.; Waguri, S.; Koike, M.; Sou, Y.S.; Ueno, T.; Hara, T.; Mizushima, N.; Iwata, J.; Ezaki, J.; Murata, S.; et al. Homeostatic levels of p62 control cytoplasmic inclusion body formation in autophagy-deficient mice. *Cell* **2007**, *131*, 1149–1163. [CrossRef]
221. Katsuragi, Y.; Ichimura, Y.; Komatsu, M. p62/SQSTM1 functions as a signaling hub and an autophagy adaptor. *FEBS J.* **2015**, *282*, 4672–4678. [CrossRef] [PubMed]
222. Jain, A.; Lamark, T.; Sjøttem, E.; Larsen, K.B.; Awuh, J.A.; Øvervatn, A.; McMahon, M.; Hayes, J.D.; Johansen, T. p62/SQSTM1 is a target gene for transcription factor NRF2 and creates a positive feedback loop by inducing antioxidant response element-driven gene transcription. *J. Biol. Chem.* **2010**, *285*, 22576–22591. [CrossRef] [PubMed]
223. Copple, I.M.; Lister, A.; Obeng, A.D.; Kitteringham, N.R.; Jenkins, R.E.; Layfield, R.; Foster, B.J.; Goldring, C.E.; Park, B.K. Physical and functional interaction of sequestosome 1 with Keap1 regulates the Keap1-Nrf2 cell defense pathway. *J. Biol. Chem.* **2010**, *285*, 16782–16788. [CrossRef]
224. Fan, W.; Tang, Z.; Chen, D.; Moughon, D.; Ding, X.; Chen, S.; Zhu, M.; Zhong, Q. Keap1 facilitates p62-mediated ubiquitin aggregate clearance via autophagy. *Autophagy* **2010**, *6*, 614–621. [CrossRef]
225. Komatsu, M.; Kurokawa, H.; Waguri, S.; Taguchi, K.; Kobayashi, A.; Ichimura, Y.; Sou, Y.S.; Ueno, I.; Sakamoto, A.; Tong, K.I.; et al. The selective autophagy substrate p62 activates the stress responsive transcription factor Nrf2 through inactivation of Keap1. *Nat. Cell Biol.* **2010**, *12*, 213–223. [CrossRef] [PubMed]
226. Lau, A.; Wang, X.J.; Zhao, F.; Villeneuve, N.F.; Wu, T.; Jiang, T.; Sun, Z.; White, E.; Zhang, D.D. A noncanonical mechanism of Nrf2 activation by autophagy deficiency: Direct interaction between Keap1 and p62. *Mol. Cell. Biol.* **2010**, *30*, 3275–3285. [CrossRef] [PubMed]
227. Pajares, M.; Jiménez-Moreno, N.; García-Yagüe, Á.J.; Escoll, M.; de Ceballos, M.L.; Van Leuven, F.; Rábano, A.; Yamamoto, M.; Rojo, A.I.; Cuadrado, A. Transcription factor NFE2L2/NRF2 is a regulator of macroautophagy genes. *Autophagy* **2016**, *12*, 1902–1916. [CrossRef]
228. Pajares, M.; Rojo, A.I.; Arias, E.; Díaz-Carretero, A.; Cuervo, A.M.; Cuadrado, A. Transcription factor NFE2L2/NRF2 modulates chaperone-mediated autophagy through the regulation of LAMP2A. *Autophagy* **2018**, *14*, 1310–1322. [CrossRef]
229. Meng, F.; Yao, D.; Shi, Y.; Kabakoff, J.; Wu, W.; Reicher, J.; Ma, Y.; Moosmann, B.; Masliah, E.; Lipton, S.A.; et al. Oxidation of the cysteine-rich regions of parkin perturbs its E3 ligase activity and contributes to protein aggregation. *Mol. Neurodegener.* **2011**, *6*, 34. [CrossRef]
230. Vandiver, M.S.; Paul, B.D.; Xu, R.; Karuppagounder, S.; Rao, F.; Snowman, A.M.; Ko, H.S.; Lee, Y.I.; Dawson, V.L.; Dawson, T.M.; et al. Sulfhydration mediates neuroprotective actions of parkin. *Nat. Commun.* **2013**, *4*, 1626. [CrossRef]
231. Astuti, R.I.; Nasuno, R.; Takagi, H. Nitric oxide signaling in yeast. *Appl. Microbiol. Biotechnol.* **2016**, *100*, 9483–9497. [CrossRef] [PubMed]
232. Tillmann, A.; Gow, N.A.; Brown, A.J. Nitric oxide and nitrosative stress tolerance in yeast. *Biochem. Soc. Trans.* **2011**, *39*, 219–223. [CrossRef]
233. Shahani, N.; Sawa, A. Nitric oxide signaling and nitrosative stress in neurons: Role for S-nitrosylation. *Antioxid. Redox Signal.* **2011**, *14*, 1493–1504. [CrossRef] [PubMed]
234. Ferrer-Sueta, G.; Campolo, N.; Trujillo, M.; Bartesaghi, S.; Carballal, S.; Romero, N.; Alvarez, B.; Radi, R. Biochemistry of Peroxynitrite and Protein Tyrosine Nitration. *Chem. Rev.* **2018**, *118*, 1338–1408. [CrossRef] [PubMed]
235. Radi, R. Protein tyrosine nitration: Biochemical mechanisms and structural basis of functional effects. *Acc. Chem. Res.* **2013**, *46*, 550–559. [CrossRef] [PubMed]

236. Wang, F.; Yuan, Q.; Chen, F.; Pang, J.; Pan, C.; Xu, F.; Chen, Y. Fundamental Mechanisms of the Cell Death Caused by Nitrosative Stress. *Front. Cell Dev. Biol.* **2021**, *9*, 742483. [CrossRef]
237. Nakamura, T.; Tu, S.; Akhtar, M.W.; Sunico, C.R.; Okamoto, S.; Lipton, S.A. Aberrant protein s-nitrosylation in neurodegenerative diseases. *Neuron* **2013**, *78*, 596–614. [CrossRef] [PubMed]
238. Seneviratne, U.; Nott, A.; Bhat, V.B.; Ravindra, K.C.; Wishnok, J.S.; Tsai, L.H.; Tannenbaum, S.R. S-nitrosation of proteins relevant to Alzheimer's disease during early stages of neurodegeneration. *Proc. Natl. Acad. Sci. USA* **2016**, *113*, 4152–4157. [CrossRef]
239. Han, F.; Chen, Y.X.; Lu, Y.M.; Huang, J.Y.; Zhang, G.S.; Tao, R.R.; Ji, Y.L.; Liao, M.H.; Fukunaga, K.; Qin, Z.H. Regulation of the ischemia-induced autophagy-lysosome processes by nitrosative stress in endothelial cells. *J. Pineal Res.* **2011**, *51*, 124–135. [CrossRef]
240. Tripathi, D.N.; Chowdhury, R.; Trudel, L.J.; Tee, A.R.; Slack, R.S.; Walker, C.L.; Wogan, G.N. Reactive nitrogen species regulate autophagy through ATM-AMPK-TSC2-mediated suppression of mTORC1. *Proc. Natl. Acad. Sci. USA* **2013**, *110*, E2950–E2957. [CrossRef]
241. Ma, L.L.; Ma, X.; Kong, F.J.; Guo, J.J.; Shi, H.T.; Zhu, J.B.; Zou, Y.Z.; Ge, J.B. Mammalian target of rapamycin inhibition attenuates myocardial ischaemia-reperfusion injury in hypertrophic heart. *J. Cell. Mol. Med.* **2018**, *22*, 1708–1719. [CrossRef] [PubMed]
242. Chakraborty, S.; Datta, S.; Ghosh, S. Induction of autophagy under nitrosative stress: A complex regulatory interplay between SIRT1 and AMPK in MCF7 cells. *Cell. Signal.* **2019**, *64*, 109411. [CrossRef] [PubMed]
243. Porter, K.R.; Claude, A.; Fullam, E.F. A Study of Tissue Culture Cells by Electron Microscopy: Methods and Preliminary Observations. *J. Exp. Med.* **1945**, *81*, 233–246. [CrossRef] [PubMed]
244. Wang, M.; Kaufman, R.J. Protein misfolding in the endoplasmic reticulum as a conduit to human disease. *Nature* **2016**, *529*, 326–335. [CrossRef]
245. Hebert, D.N.; Molinari, M. In and out of the ER: Protein folding, quality control, degradation, and related human diseases. *Physiol. Rev.* **2007**, *87*, 1377–1408. [CrossRef]
246. Merksamer, P.I.; Trusina, A.; Papa, F.R. Real-time redox measurements during endoplasmic reticulum stress reveal interlinked protein folding functions. *Cell* **2008**, *135*, 933–947. [CrossRef]
247. Fujita, E.; Kouroku, Y.; Isoai, A.; Kumagai, H.; Misutani, A.; Matsuda, C.; Hayashi, Y.K.; Momoi, T. Two endoplasmic reticulum-associated degradation (ERAD) systems for the novel variant of the mutant dysferlin: Ubiquitin/proteasome ERAD(I) and autophagy/lysosome ERAD(II). *Hum. Mol. Genet.* **2007**, *16*, 618–629. [CrossRef]
248. Schroder, M.; Kaufman, R.J. ER stress and the unfolded protein response. *Mutat. Res.* **2005**, *569*, 29–63. [CrossRef]
249. Werner, E.D.; Brodsky, J.L.; McCracken, A.A. Proteasome-dependent endoplasmic reticulum-associated protein degradation: An unconventional route to a familiar fate. *Proc. Natl. Acad. Sci. USA* **1996**, *93*, 13797–13801. [CrossRef]
250. Yorimitsu, T.; Nair, U.; Yang, Z.; Klionsky, D.J. Endoplasmic reticulum stress triggers autophagy. *J. Biol. Chem.* **2006**, *281*, 30299–30304. [CrossRef]
251. Bernales, S.; McDonald, K.L.; Walter, P. Autophagy counterbalances endoplasmic reticulum expansion during the unfolded protein response. *PLoS Biol.* **2006**, *4*, e423. [CrossRef] [PubMed]
252. Welihinda, A.A.; Kaufman, R.J. The unfolded protein response pathway in Saccharomyces cerevisiae: Oligomerization and trans-phosphorylation of Ire1p (Ern1p) are required for kinase activation. *J. Biol. Chem.* **1996**, *271*, 18181–18187. [PubMed]
253. Okamura, K.; Kimata, Y.; Higashio, H.; Tsuru, A.; Kohno, K. Dissociation of Kar2p/BiP from an ER sensory molecule, Ire1p, triggers the unfolded protein response in yeast. *Biochem. Biophys. Res. Commun.* **2000**, *279*, 445–450. [CrossRef] [PubMed]
254. Cox, J.S.; Walter, P. A novel mechanism for regulating activity of a transcription factor that controls the unfolded protein response. *Cell* **1996**, *87*, 391–404. [CrossRef]
255. Sidrauski, C.; Walter, P. The transmembrane kinase Ire1p is a site-specific endonuclease that initiates mRNA splicing in the unfolded protein response. *Cell* **1997**, *90*, 1031–1039. [PubMed]
256. Sidrauski, C.; Cox, J.S.; Walter, P. tRNA ligase is required for regulated mRNA splicing in the unfolded protein response. *Cell* **1996**, *87*, 405–413.
257. Mori, K. Signalling pathways in the unfolded protein response: Development from yeast to mammals. *J. Biochem.* **2009**, *146*, 743–750. [CrossRef]
258. Kawahara, T.; Yanagi, H.; Yura, T.; Mori, K. Endoplasmic reticulum stress-induced mRNA splicing permits synthesis of transcription factor Hac1p/Ern4p that activates the unfolded protein response. *Mol. Biol. Cell* **1997**, *8*, 1845–1862. [CrossRef]
259. Patil, C.; Walter, P. Intracellular signaling from the endoplasmic reticulum to the nucleus: The unfolded protein response in yeast and mammals. *Curr. Opin. Cell Biol.* **2001**, *13*, 349–355. [CrossRef]
260. Chakraborty, R.; Baek, J.H.; Bae, E.Y.; Kim, W.-Y.; Lee, S.Y.; Kim, M.G. Comparison and contrast of plant, yeast, and mammalian ER stress and UPR. *Appl. Biol. Chem.* **2016**, *59*, 337–347. [CrossRef]
261. Reddy, R.K.; Mao, C.; Baumeister, P.; Austin, R.C.; Kaufman, R.J.; Lee, A.S. Endoplasmic reticulum chaperone protein GRP78 protects cells from apoptosis induced by topoisomerase inhibitors: Role of ATP binding site in suppression of caspase-7 activation. *J. Biol. Chem.* **2003**, *278*, 20915–20924. [CrossRef] [PubMed]
262. Gardner, B.M.; Walter, P. Unfolded proteins are Ire1-activating ligands that directly induce the unfolded protein response. *Science* **2011**, *333*, 1891–1894. [CrossRef] [PubMed]
263. Calfon, M.; Zeng, H.; Urano, F.; Till, J.H.; Hubbard, S.R.; Harding, H.P.; Clark, S.G.; Ron, D. IRE1 couples endoplasmic reticulum load to secretory capacity by processing the XBP-1 mRNA. *Nature* **2002**, *415*, 92–96. [CrossRef] [PubMed]

264. Acosta-Alvear, D.; Zhou, Y.; Blais, A.; Tsikitis, M.; Lents, N.H.; Arias, C.; Lennon, C.J.; Kluger, Y.; Dynlacht, B.D. XBP1 controls diverse cell type-and condition-specific transcriptional regulatory networks. *Mol. Cell* **2007**, *27*, 53–66. [CrossRef] [PubMed]
265. Margariti, A.; Li, H.; Chen, T.; Martin, D.; Vizcay-Barrena, G.; Alam, S.; Karamariti, E.; Xiao, Q.; Zampetaki, A.; Zhang, Z. XBP1 mRNA splicing triggers an autophagic response in endothelial cells through BECLIN-1 transcriptional activation. *J. Biol. Chem.* **2013**, *288*, 859–872. [CrossRef] [PubMed]
266. Zhou, Y.; Lee, J.; Reno, C.M.; Sun, C.; Park, S.W.; Chung, J.; Lee, J.; Fisher, S.J.; White, M.F.; Biddinger, S.B. Regulation of glucose homeostasis through a XBP-1–FoxO1 interaction. *Nat. Med.* **2011**, *17*, 356–365. [CrossRef] [PubMed]
267. Vidal, R.L.; Figueroa, A.; Court, F.A.; Thielen, P.; Molina, C.; Wirth, C.; Caballero, B.; Kiffin, R.; Segura-Aguilar, J.; Cuervo, A.M. Targeting the UPR transcription factor XBP1 protects against Huntington's disease through the regulation of FoxO1 and autophagy. *Hum. Mol. Genet.* **2012**, *21*, 2245–2262. [CrossRef]
268. Zhao, Y.; Li, X.; Cai, M.-Y.; Ma, K.; Yang, J.; Zhou, J.; Fu, W.; Wei, F.-Z.; Wang, L.; Xie, D. XBP-1u suppresses autophagy by promoting the degradation of FoxO1 in cancer cells. *Cell Res.* **2013**, *23*, 491–507. [CrossRef]
269. Li, D.D.; Wang, L.L.; Deng, R.; Tang, J.; Shen, Y.; Guo, J.F.; Wang, Y.; Xia, L.P.; Feng, G.K.; Liu, Q.Q.; et al. The pivotal role of c-Jun NH2-terminal kinase-mediated Beclin 1 expression during anticancer agents-induced autophagy in cancer cells. *Oncogene* **2009**, *28*, 886–898. [CrossRef]
270. Meares, G.P.; Hughes, K.J.; Naatz, A.; Papa, F.R.; Urano, F.; Hansen, P.A.; Benveniste, E.N.; Corbett, J.A. IRE1-dependent activation of AMPK in response to nitric oxide. *Mol. Cell. Biol.* **2011**, *31*, 4286–4297. [CrossRef]
271. Kouroku, Y.; Fujita, E.; Tanida, I.; Ueno, T.; Isoai, A.; Kumagai, H.; Ogawa, S.; Kaufman, R.J.; Kominami, E.; Momoi, T. ER stress (PERK/eIF2α phosphorylation) mediates the polyglutamine-induced LC3 conversion, an essential step for autophagy formation. *Cell Death Differ.* **2007**, *14*, 230–239. [CrossRef]
272. Pakos-Zebrucka, K.; Koryga, I.; Mnich, K.; Ljujic, M.; Samali, A.; Gorman, A.M. The integrated stress response. *EMBO Rep.* **2016**, *17*, 1374–1395. [CrossRef] [PubMed]
273. Pattingre, S.; Tassa, A.; Qu, X.; Garuti, R.; Liang, X.H.; Mizushima, N.; Packer, M.; Schneider, M.D.; Levine, B. Bcl-2 antiapoptotic proteins inhibit Beclin 1-dependent autophagy. *Cell* **2005**, *122*, 927–939. [CrossRef] [PubMed]
274. Rzymski, T.; Milani, M.; Pike, L.; Buffa, F.; Mellor, H.; Winchester, L.; Pires, I.; Hammond, E.; Ragoussis, I.; Harris, A. Regulation of autophagy by ATF4 in response to severe hypoxia. *Oncogene* **2010**, *29*, 4424–4435. [CrossRef] [PubMed]
275. Avivar-Valderas, A.; Bobrovnikova-Marjon, E.; Diehl, J.A.; Bardeesy, N.; Debnath, J.; Aguirre-Ghiso, J. Regulation of autophagy during ECM detachment is linked to a selective inhibition of mTORC1 by PERK. *Oncogene* **2013**, *32*, 4932–4940. [CrossRef] [PubMed]
276. Brüning, A.; Rahmeh, M.; Friese, K. Nelfinavir and bortezomib inhibit mTOR activity via ATF4-mediated sestrin-2 regulation. *Mol. Oncol.* **2013**, *7*, 1012–1018. [CrossRef] [PubMed]
277. Adachi, Y.; Yamamoto, K.; Okada, T.; Yoshida, H.; Harada, A.; Mori, K. ATF6 is a transcription factor specializing in the regulation of quality control proteins in the endoplasmic reticulum. *Cell Struct. Funct.* **2008**, *33*, 75–89. [CrossRef]
278. Guo, F.-J.; Xiong, Z.; Lu, X.; Ye, M.; Han, X.; Jiang, R. ATF6 upregulates XBP1S and inhibits ER stress-mediated apoptosis in osteoarthritis cartilage. *Cell. Signal.* **2014**, *26*, 332–342. [CrossRef]
279. Hirsch, I.; Weiwad, M.; Prell, E.; Ferrari, D.M. ERp29 deficiency affects sensitivity to apoptosis via impairment of the ATF6–CHOP pathway of stress response. *Apoptosis* **2014**, *19*, 801–815. [CrossRef]
280. Yung, H.W.; Charnock-Jones, D.S.; Burton, G.J. Regulation of AKT phosphorylation at Ser473 and Thr308 by endoplasmic reticulum stress modulates substrate specificity in a severity dependent manner. *PLoS ONE* **2011**, *6*, e17894. [CrossRef]
281. Kalvakolanu, D.V.; Gade, P. IFNG and autophagy: A critical role for the ER-stress mediator ATF6 in controlling bacterial infections. *Autophagy* **2012**, *8*, 1673–1674. [CrossRef] [PubMed]
282. Gade, P.; Manjegowda, S.B.; Nallar, S.C.; Maachani, U.B.; Cross, A.S.; Kalvakolanu, D.V. Regulation of the death-associated protein kinase 1 expression and autophagy via ATF6 requires apoptosis signal-regulating kinase 1. *Mol. Cell. Biol.* **2014**, *34*, 4033–4048. [CrossRef] [PubMed]
283. Høyer-Hansen, M.; Bastholm, L.; Szyniarowski, P.; Campanella, M.; Szabadkai, G.; Farkas, T.; Bianchi, K.; Fehrenbacher, N.; Elling, F.; Rizzuto, R. Control of macroautophagy by calcium, calmodulin-dependent kinase kinase-β, and Bcl-2. *Mol. Cell* **2007**, *25*, 193–205. [CrossRef] [PubMed]
284. Borodkina, A.V.; Shatrova, A.N.; Deryabin, P.I.; Griukova, A.A.; Abushik, P.A.; Antonov, S.M.; Nikolsky, N.N.; Burova, E.B. Calcium alterations signal either to senescence or to autophagy induction in stem cells upon oxidative stress. *Aging* **2016**, *8*, 3400–3418. [CrossRef]
285. Simon, B.; Huart, A.-S.; Temmerman, K.; Vahokoski, J.; Mertens, H.D.; Komadina, D.; Hoffmann, J.-E.; Yumerefendi, H.; Svergun, D.I.; Kursula, P. Death-associated protein kinase activity is regulated by coupled calcium/calmodulin binding to two distinct sites. *Structure* **2016**, *24*, 851–861. [CrossRef]
286. Gomes, L.R.; Menck, C.F.M.; Leandro, G.S. Autophagy Roles in the Modulation of DNA Repair Pathways. *Int. J. Mol. Sci.* **2017**, *18*, 2351. [CrossRef]
287. Sharma, V.; Verma, S.; Seranova, E.; Sarkar, S.; Kumar, D. Selective Autophagy and Xenophagy in Infection and Disease. *Front. Cell Dev. Biol.* **2018**, *6*, 147. [CrossRef]
288. Feng, Y.; Yao, Z.; Klionsky, D.J. How to control self-digestion: Transcriptional, post-transcriptional, and post-translational regulation of autophagy. *Trends Cell Biol.* **2015**, *25*, 354–363. [CrossRef]

289. Holczer, M.; Hajdu, B.; Lorincz, T.; Szarka, A.; Banhegyi, G.; Kapuy, O. Fine-tuning of AMPK-ULK1-mTORC1 regulatory triangle is crucial for autophagy oscillation. *Sci. Rep.* **2020**, *10*, 17803. [CrossRef]
290. Ling, N.X.Y.; Kaczmarek, A.; Hoque, A.; Davie, E.; Ngoei, K.R.W.; Morrison, K.R.; Smiles, W.J.; Forte, G.M.; Wang, T.; Lie, S.; et al. mTORC1 directly inhibits AMPK to promote cell proliferation under nutrient stress. *Nat. Metab.* **2020**, *2*, 41–49. [CrossRef]
291. He, F.; Ru, X.; Wen, T. NRF2, a Transcription Factor for Stress Response and Beyond. *Int. J. Mol. Sci.* **2020**, *21*, 4777. [CrossRef] [PubMed]
292. Kim, G.H.; Kim, J.E.; Rhie, S.J.; Yoon, S. The Role of Oxidative Stress in Neurodegenerative Diseases. *Exp. Neurobiol.* **2015**, *24*, 325–340. [CrossRef] [PubMed]
293. Ghemrawi, R.; Khair, M. Endoplasmic Reticulum Stress and Unfolded Protein Response in Neurodegenerative Diseases. *Int. J. Mol. Sci.* **2020**, *21*, 6127. [CrossRef] [PubMed]

antioxidants

MDPI

Review

From Intestinal Epithelial Homeostasis to Colorectal Cancer: Autophagy Regulation in Cellular Stress

Qiuluo Liu [1,2,†], Yan Chen [2,†], Li Zhou [2], Haining Chen [1,2,*] and Zongguang Zhou [1,2,*]

1 Department of Gastrointestinal Surgery, West China Hospital, Sichuan University, No. 17, Block 3, Southern Renmin Road, Chengdu 610041, China; liuqiuluo@stu.scu.edu.cn
2 State Key Laboratory of Biotherapy and Cancer Center, West China Hospital, Sichuan University, and Collaborative Innovation Center for Biotherapy, Chengdu 610041, China; yanchen0524@scu.edu.cn (Y.C.); 2015224060079@stu.scu.edu.cn (L.Z.)
* Correspondence: hnchen@scu.edu.cn (H.C.); zhou767@scu.edu.cn (Z.Z.)
† These authors contributed equally to this work.

Abstract: The intestinal epithelium is continuously exposed to abundant stress stimuli, which relies on an evolutionarily conserved process, autophagy, to maintain its homeostasis by degrading and recycling unwanted and damaged intracellular substances. Otherwise, disruption of this balance will result in the development of a wide range of disorders, including colorectal cancer (CRC). Dysregulated autophagy is implicated in the regulation of cellular responses to stress during the development, progression, and treatment of CRC. However, experimental investigations addressing the impact of autophagy in different phases of CRC have generated conflicting results, showing that autophagy is context-dependently related to CRC. Thus, both inhibition and activation of autophagy have been proposed as therapeutic strategies against CRC. Here, we will discuss the multifaceted role of autophagy in intestinal homeostasis and CRC, which may provide insights for future research directions.

Keywords: autophagy; intestinal homeostasis; colorectal cancer; stress response

Citation: Liu, Q.; Chen, Y.; Zhou, L.; Chen, H.; Zhou, Z. From Intestinal Epithelial Homeostasis to Colorectal Cancer: Autophagy Regulation in Cellular Stress. *Antioxidants* **2022**, *11*, 1308. https://doi.org/10.3390/antiox11071308

Academic Editors: Alessandra Napolitano and Stanley Omaye

Received: 6 June 2022
Accepted: 27 June 2022
Published: 30 June 2022

1. Introduction

The mucosal surface of the gastrointestinal tract continuously encounters abundant stimuli originating from both endogenous and exogenous sources, including metabolic alterations, a variety of bacterial species, chemical irritants, and agents that produce oxidative stress. Autophagy, a stress-responsive process, is tightly linked to the maintenance of intestinal cellular homeostasis [1,2] (Figure 1). Under conditions of physiological stress, cells in the intestinal mucosa frequently accumulate unwanted and damaged intracellular substances. In this case, autophagy can be triggered to transport them to the lysosomes for degradation and recycling [3]. Intestinal epithelial cells (IECs) and intestinal stem cells rely on this mechanism to ensure their survival, as it helps maintain protein and organelle quality by selectively degrading and recycling aggregates of impaired or unnecessary proteins, mitochondria, peroxisomes, and endoplasmic reticulum (known as selective autophagy) [4–6]. Moreover, autophagic degradation of the intestinal tight junction proteins governs the intensity of the intestinal barrier. Apart from this, autophagy plays a central role in the host–microbiota interactions, where it eliminates potential pathogens and forms an integral component of anti-infectious immunity [2].

In contrast, defective autophagy predisposes normal IECs to undergo malignant transformation. Although the exact etiological mechanisms underlying CRC remain multifactorial and largely unknown, it is well established that both genetic predisposition and environmental factors contribute to its initiation and development. The genetic basis underpinning sporadic CRCs is well defined by theories such as the adenoma–carcinoma sequence model, suggesting that CRC is driven by sequential genetic and epigenetic mutations, arising from normal epithelial cells to dysplastic adenomas and, ultimately, carcinomas [7–10]. Various genetic events are required during the malignant transformation, involving mutations of *Adenomatous polyposis coli (APC)*, *KRAS*, and *P53* [7,8,10]. It is essential to perceive that across stages of CRC tumorigenesis, alteration of autophagy-related genes plays a significant role. A large genome-wide association study identified genetic variants of transcription factor EB (*TFEB*), a positive regulator of the autophagic pathway that promotes the expression of autophagy genes [11,12], as a novel risk factor associated with CRC susceptibility [13]. Mutation of another autophagy regulator, UV-radiation-resistance associated gene (*UVRAG*), which activates the Beclin1-PI3KC3 complex, also underpins the genetic basis of CRC tumorigenesis [14]. Similarly, genetic alterations involved in the endocytosis-autophagy network were frequently observed in *KRAS*-wild-type CRC [15].

Given that it generally takes years to decades for adenomas to transform into carcinomas, the mutated precursor cells constantly endure endogenous and exogenous stress [16,17]. A clear role has emerged for autophagy in CRC cells, where it exerts diverse effects on cellular adaptation to tumor microenvironmental cues and therapeutic stress, which ultimately results in cell survival, death, or growth inhibition [18,19]. First, highly proliferative CRC cells tend to have a limited supply of nutrients. In the context of nutrient deprivation, autophagy is triggered to provide energy sources and metabolites to sustain metabolism and tumor growth [20–22]. Moreover, insufficient and irregular neovascularization of rapidly proliferating CRC cells causes a hypoxic microenvironment, where autophagy is harnessed to eliminate protein aggregates and damaged endoplasmic reticulum (ER) and mitochondria [4,5]. This contributes to the prevention of the overproduction of reactive oxygen species (ROS) and reduction of oxidative and ER stress, thereby preserving genomic integrity [23]. In addition, intestinal microorganisms with oncogenic properties continuously cause an abnormal microenvironment that profoundly affects the initiation and progression of sporadic CRC [24]. The involvement of autophagy in the interaction of microbiota and CRC is complicated and differs in a temporal manner [1,2]. Finally, abnormal autophagy is activated in response to treatment and confers resistance to therapeutic challenges. In this context, autophagy protects CRC cells from drug-induced apoptosis and induces them into a slow-cycling, drug-tolerant state.

In this review, we focus on the regulatory roles of autophagy in the maintenance of intestinal homeostasis. Meanwhile, we discuss how dysregulation of this conserved process orchestrates different stress factors in a context-dependent manner in distinct stages of CRC development and progression and under therapeutic pressure, with the aim of providing a perspective for future research.

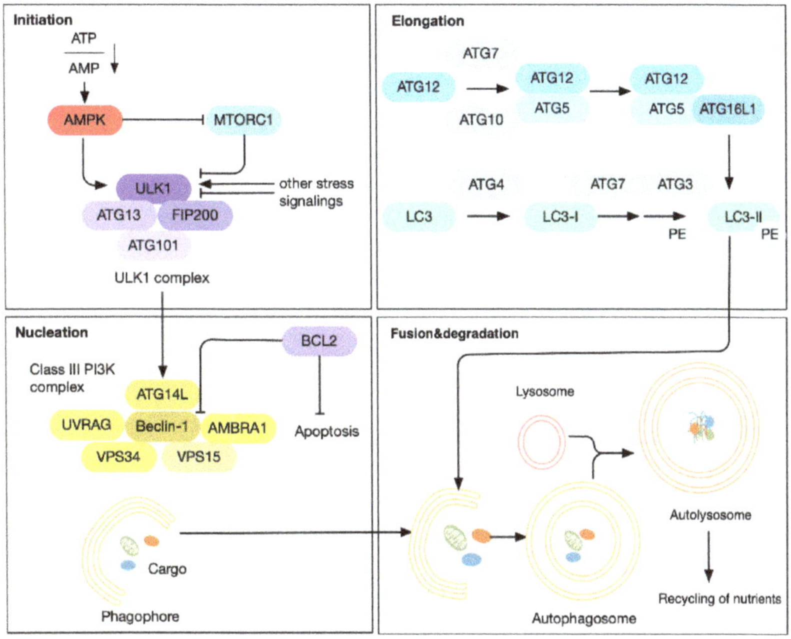

Figure 1. Functional mechanisms of autophagy. The most-studied form of autophagy is macroautophagy, a multistage and dynamic process. Autophagy is negatively regulated by the cell growth promoter rapamycin complex 1 (mTORC1), while the sensor of energy deprivation AMP-activated kinase (AMPK) activates autophagy. The canonical autophagy consists of four sequential stages: initiation, phagophore formation, phagophore elongation, and autophagosome–lysosome fusion [25]. During the initiation phase, the UNC-like autophagy-activating kinase 1 (ULK1) complex serves as a bridge between the upstream mTOR and AMPK and downstream autophagosome formation [26,27]. The complex is composed of ATG13, focal adhesion kinase-family-interacting protein of 200 kDa (FIP200), ATG101, and ULK1, in which ULK1 is the core protein with serine/threonine kinase activity [28]. After being stimulated by nutrient deficiency and stress-related pathways, phosphorylated ULK1 subsequently leads to membrane nucleation, which requires activation of class III phosphatidylinositol-3 kinase complex I (PI3KC3-CI). Formed by beclin-1, vacuolar protein sorting 34 (VPS34), autophagy-related protein 14- like protein (ATG14L), p150, and nuclear receptor-binding factor 2 (NRBF2), this multiprotein complex can be activated through ULK-dependent phosphorylation [29]. The nucleation of the isolation membrane, known as the phagophore, further expands with the support of PI3KC3-CI [30]. At the phagophore assembly site, the complex produces PI3P, favoring the recruitment of the effector proteins (such as WIPI/II), thus resulting in the phagophore elongation [31]. This phase is further promoted by the ubiquitin-like conjugation system, involving the E1 ligase, ATG7, the E2 ligase, ATG3, and the E3 ligase complex, ATG12/ATG5/ATG16L [32]. Through ATG4-dependent proteolytic cleavage, followed by the action of the conjugation system, microtubule-associated proteins 1A/1B light chain 3 (LC3) can be transformed to lipidated LC3 (LC3-II), which is instrumental for elongation and closure of the phagophore [33]. Meanwhile, LC3-II physically links to substrates that contain the LC3-interacting region (LIR) motif, thereby targeting them for degradation. Once phagophores are closed, the ensuing autophagosomes fuse with lysosomes to form autolysosomes; within them, the delivered contents are degraded and recycled [34].

2. Autophagy Maintains Intestinal Epithelial Homeostasis under Physiological Stress

The intestinal mucosa is constantly exposed to alimentary and bacterial antigens as well as mechanical stress, which relies on an intact intestinal barrier and healthy gut microbiota to maintain intestinal homeostasis that would otherwise cause infection, inflammation, and cellular damage [1,35–37]. Regulation of autophagy plays a key role in the ability of the gut epithelium to cope with cell stress, as elucidated by lines of evidence from experimental and clinical studies (Figure 2).

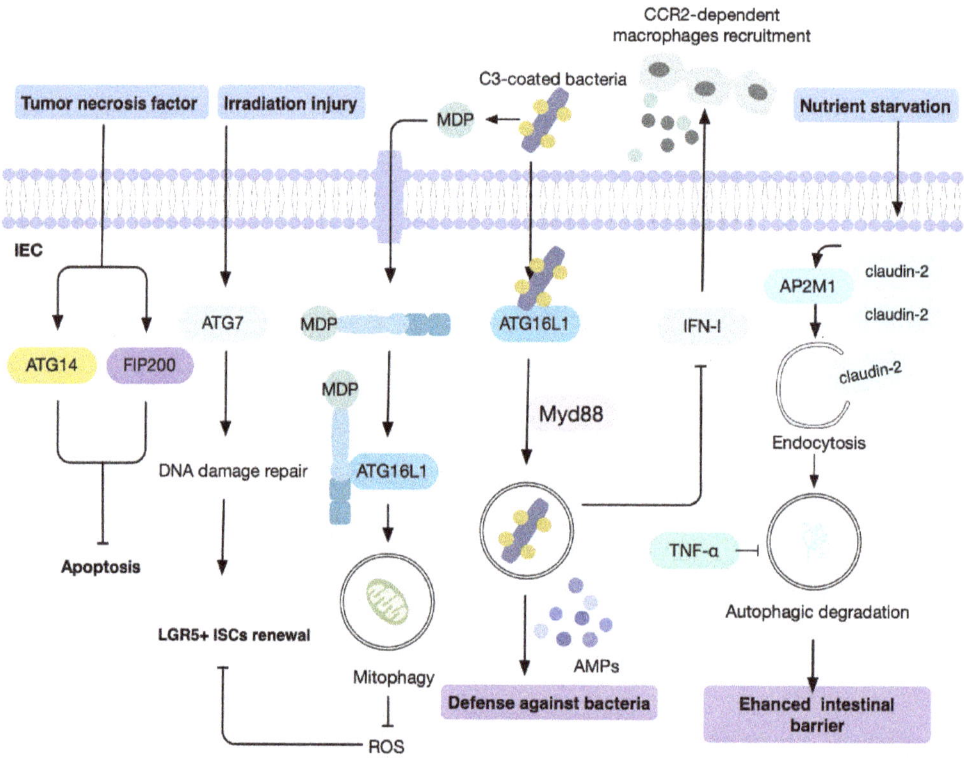

Figure 2. Role of autophagy in intestinal homeostasis maintenance Multiple roles of autophagy in intestinal homeostasis are shown, including regulating the survival of intestinal epithelial cells and intestinal stem cells, the host–microbiota interactions, and the intestinal tight junctions.

The IECs constitute the first line of defense, which includes the formation of the physical barrier as well as the integration of regulatory mechanisms. As the intestinal epithelium is one of the most vigorously regenerative tissues in adults, its turnover serves as a crucial mechanism for the protective effect provided by the mucosal barrier, which is achieved through a balance of cell apoptosis and proliferation in crypts [38]. Mouse models with ATG14 or Rb1cc1/Fip200 deleted in the intestinal epithelium exhibited extensive intestinal villous atrophy, suggesting that autophagy is protective against cell death during homeostasis in the intestinal epithelium [39]. Mechanistically, these autophagy-related proteins defended intestinal epithelial cells from TNF (tumor necrosis factor)-triggered apoptosis [39]. Moreover, intestinal homeostasis is maintained by leucine-rich repeat-containing G protein-coupled receptor 5-positive intestinal stem cells (LGR5$^+$ ISCs) for constant tissue regeneration. Notably, autophagy has been demonstrated to play a cytoprotective role in the LGR5$^+$ ISCs against toxic and infectious injuries. During irradiation damage, muramyl dipeptide (MDP), a microbiota-derived product, can be recognized by NOD2 in LGR5$^+$ ISCs, thereby promoting cell survival by mediating the clearance of ROS. This reduction of ROS was achieved via mitophagy induction coordinated by NOD2 and ATG16L1, which eliminate damaged mitochondria in ISCs and therefore enhance epithelial repair [40]. In addition, autophagy contributes to LGR5$^+$ ISCs maintenance under conditions of irradiation and chemotherapy. Following these stresses, ATG7-dependent DNA damage repair was stimulated, facilitating ISCs survival. Activation of autophagy on fasting showed a protective effect on LGR5$^+$ ISCs against oxaliplatin and doxorubicin-induced DNA damage and cell death [41].

The balance of host–microbiota interactions has profound impacts on the host's intestinal health. Notably, the crucial role of autophagy lies in maintaining intestinal microbiota homeostasis, and dysfunctional autophagy is known to cause gut microbial dysbiosis [42]. As previously demonstrated in the mouse model with conditional inactivation of Atg5 in IECs, blockade of the autophagic flux led to a remarkable alteration and reduced the diversity of gut microbiota [43]. The altered colonization pattern involved decreased abundances of anti-inflammatory microorganisms and enrichment of proinflammatory bacterial groups, many of which are believed to be associated with inflammatory bowel disease (IBD) and colorectal cancer (CRC) [43].

The proposed mechanisms by which autophagy modulates the balance of bacterial flora include direct degradation of harmful bacteria and regulation of the antibacterial immune response. Under physiological conditions, all intestinal bacteria are coated with complement protein C3 [44]. Following the invasion of potentially pathogenic microorganisms into the intestinal mucosa, C3 on the bacterial surfaces can be targeted by host cytosol ATG16L1, thereby activating the autophagy system [45]. Apart from this, it has been reported that MyD88, the canonical adaptor for inflammatory signaling pathways, was also required during the process of autophagy induction [46]. It is worth mentioning that autophagy in IECs can affect the expression and secretion of antimicrobial peptides (AMPs) to restrict bacterial dissemination [47]. Interestingly, several mouse models in which different autophagy genes are deleted in IECs (including *ATG16L1*, *ATG4B* and *LC3B)* showed an enhanced response to microbiota-induced type I interferon (IFN-I) signaling [48]. This spontaneous activation of IFN-I in IECs conferred protection against the pathogen *Citrobacter rodentium* and chemical injury via C-C motif chemokine receptor 2 (CCR2)-dependent monocyte recruitment, fortifying the intestinal barrier in response to both infectious and non-infectious stress [48]. Although autophagy was demonstrated to have an adverse function in antimicrobial activity and tissue repair, as evidenced by this study, the immunomodulatory properties of IFN-I signaling may be far more nuanced under different circumstances, such as autoimmune diseases and tumor immunity [48,49].

Epithelial cells in the intestinal tract attach via tight junctions (Tjs) including claudin, occludin, etc. TJ modulation is closely linked to intestinal permeability, and autophagy has been implicated in enhancing intestinal barrier function via TJ regulation [1]. By mediating the lysosomal-dependent degradation of claudin-2, a pore-forming protein, starvation-induced autophagy reduced intestinal permeability of ions and small molecules in IECs [50]. Further mechanistic investigation revealed that autophagy-triggered claudin-2 degradation was dependent on clathrin-mediated endocytosis, where claudin-2 directly binds to adaptor related protein complex 2 subunit mu 1 (AP2M1), and an increased claudin-2-AP2M1-LC3 association was observed [51]. In contrast, proinflammatory cytokine tumor necrosis factor alpha (TNF-α) weakened the intestinal barrier. This was mediated by the inhibitory effect of TNF-α on autophagy, which resulted in elevated claudin-2 expression and impaired epithelial tight junction [52]. Another TJ-associated protein, occludin, is also tightly regulated by autophagy in IECs. Notably, beclin 1 interacted with occludin on the cell membrane, leading to the endocytosis of occludin and, subsequently, defective TJ barrier function. While this process was autophagy independent, autophagy activation was shown to counteract the effect of Beclin 1 and restore the endothelial barrier [53].

Overall, these findings revealed that autophagy is required for the maintenance of intestinal homeostasis, but its beneficial or deleterious nature can vary depending on the setting.

3. Autophagy Coordinates Cellular Adaptation to Stress in the Progression of CRC

3.1. Autophagy Enables Adaptation to Metabolic Alteration-Induced Stress

After oncogenic transformation, the established tumor is highly proliferative and metabolically active, requiring large amounts of energy and metabolic precursors. Unlike other normal cells, cancer cells constitutively utilize glycolysis to sustain oncogenic metabolism [54]. In such a situation, autophagy serves as a mechanism of survival [22]. Given that a key feature of autophagy has been suggested to supply substrates to fuel metabolism, it is constantly active under nutrient-competent conditions to enhance tumor growth [55]. Cancer cells utilize it as an alternative source of nearly all aspects of metabolic fuel and reduce oxidative stress, creating "autophagy addiction" [55–57]. Otherwise, autophagy-deficient tumor cells suffer from metabolic vulnerabilities and energy crises in the stressed microenvironment. Indeed, in the transformed IECs, not the adjacent normal IECs, autophagy is indispensable for cell metabolism [42]. Deletion of *ATG7* in intestinal adenoma blocked its progression to malignancy via p53-induced growth arrest and AMPK-dependent downregulation of glycolytic genes [42], consistent with the theory that cancer cells exhibit a particular addiction to autophagy [58,59].

Starvation-induced reduction of nutrient inputs, such as glucose and amino acids, leads to decreased intermediate metabolites of various metabolic pathways; for example, the tricarboxylic acid (TCA) cycle [60]. This will ultimately lower the ratio of adenosine triphosphate (ATP) to adenosine diphosphate (ADP) and adenosine monophosphate (AMP), energy stress that can be sensed by AMPK [61]. Once activated, AMPK suppresses ATP-consuming pathways and upregulates energy-generating processes, including autophagy [62]. In contrast, another major autophagy modulator, mTOR, is sensitive to the abundance of amino acids and is activated by available nutrients. In addition, other regulators including ATF4, SIRT1, and TFEB govern the transcription of autophagy-related genes, in response to nutrient availability and reduction status [62]. Together, there is an intricate regulatory network that integrates autophagy with response to metabolic cues.

Interestingly, the metabolic reliance on autophagy of tumors may be dependent on their mutational status. Prior studies have revealed different degrees of autophagy addiction across multiple carcinoma types, most notably in Ras-driven tumors, including lung cancer, pancreatic cancer, prostate cancer, and CRC [58,63–65]. *KRAS* is mutated in approximately 40% of CRC patients and is associated with poor prognosis and therapy resistance [66]. For Ras-transformed CRC cells, enhanced glucose metabolism is required for their high rates of proliferation in starvation. Autophagy has been shown to facilitate glycolytic flux in *H-RAS*^{V12} cells [63], likely due to its potential to degrade macromolecules and provide metabolic substrates. Interestingly, autophagy is able to protect mitochondrial function in *H-RAS*^{V12} and *K-RAS*^{V12} cells. This was achieved through the supply of substrates for mitochondrial metabolism, presumably TCA-cycle metabolites, via conversion of pyruvate and fatty acids into acetyl-CoA [64]. Defective autophagy in *H-RAS* ^{V12} and *K-RAS*^{V12} models also impaired mitochondrial respiration, causing reduced energy levels and increased oxidative stress [64]. However, while lack of autophagy in a mouse model of *KRAS*-driven lung cancer resulted in impaired fatty acid oxidation, it was absent in the *BRAF*-driven mouse model [67]. This raises the question as to the influence of genotype on the metabolic role of autophagy, which is broadly unknown in CRC.

Apart from glucose metabolism, autophagy is closely associated with fatty acid metabolism. Fatty acid β-oxidation in mitochondria produces acetyl CoA, thereby fueling the TCA cycle [68]. Indeed, CRC patient-derived adipocytes were shown to favor the survival of CRC cells under the condition of nutrient deprivation. Mechanistically, the adipocytes secreted free fatty acids, which in turn are absorbed and utilized by colon cancer cells by inducing autophagy and mitochondrial fatty acid β-oxidation via AMPK activation [69]. In addition to cell-autonomous autophagy, host autophagy has a metabolic role in the antitumor immunity of CRC [70]. In activated T_{reg} cells, autophagy was functionally stimulated and negatively regulated mTORC1-dependent glycolytic metabolism, thus promoting their metabolic homeostasis and immunosuppressive function [71]. Together, these studies suggested the metabolic vulnerabilities mediated by autophagy and provided opportunities for therapeutic intervention in CRC (Figure 3).

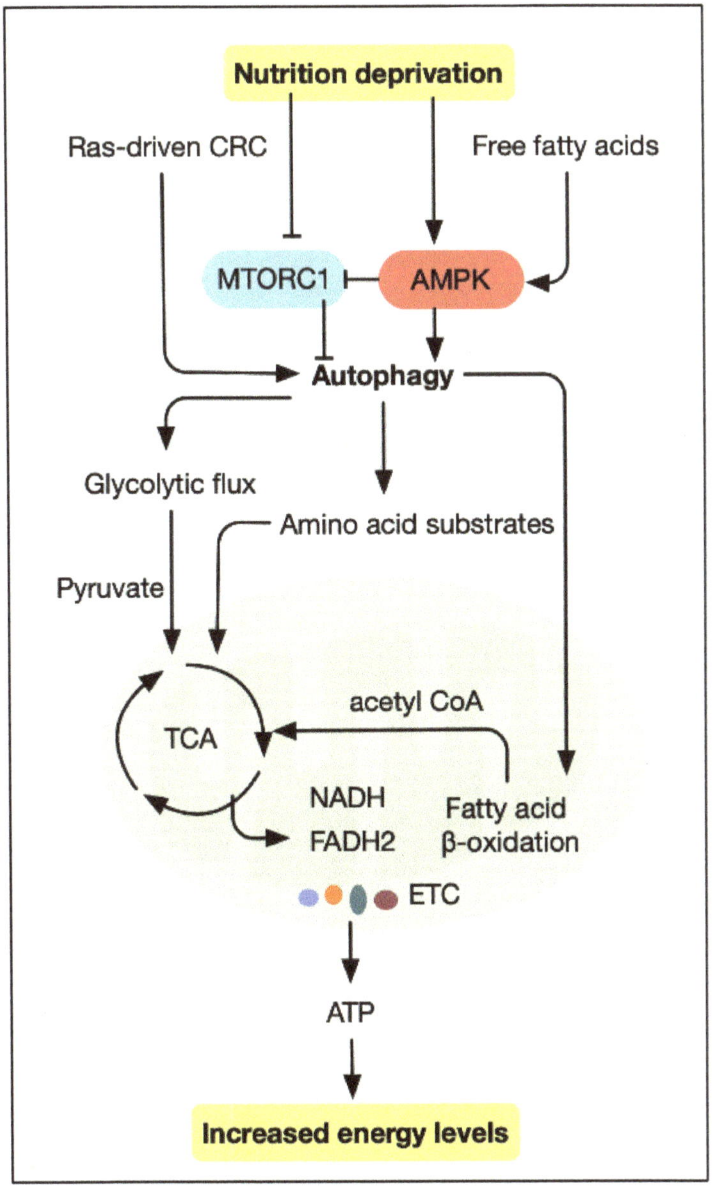

Figure 3. The crosstalk between autophagy and metabolic reprogramming CRC cells is highly proliferative and metabolically active, requiring large amounts of energy and metabolic precursors. In this context, autophagy serves as an alternative source of nearly all aspects of metabolic fuel. By coordinating glycolysis, fatty acid β-oxidation, and tricarboxylic acid cycle, autophagy is intimately connected to the metabolic reprogramming of CRC.

3.2. Autophagy Enables Adaptation to Hypoxia-Induced Stress

Due to rapid proliferation, excessive oxygen consumption, and abnormal microvasculature, the tumor mass of CRC is constantly exposed to reduced oxygen levels. Hypoxia is one of the major hallmarks of the CRC microenvironment, and autophagy is elicited to enable tumor cells to thrive in this situation. Hypoxia-inducible factor (HIF-1), the major transcription factor complex in response to hypoxic conditions, can induce autophagy through upregulation of autophagy-related genes, crosstalk with the mTOR signaling, and production of reactive oxygen species (ROS) [72–74]. Notably, under hypoxia stress, functional mitochondria play an indispensable role in ROS generation and subsequent HIF-1 stabilization.

The association of autophagy dysfunction with CRC initiation is evident in prior studies, in which essential autophagy genes, including Atg7 [42], Atg16l1 [75], and UVRAG [14], were edited in mice. In these cases, the arisen neoplasms would accumulate large amounts of autophagic cargo, most obviously damaged mitochondria [76]. Mitochondria is responsible for the adaptation of cells to a variety of stressors, and autophagy functions to eliminate defective mitochondria, a process known as "mitophagy" [77]. Indeed, during the onset of CRC, enhanced mitophagy in IECs was demonstrated to cause lysosomal membrane permeabilization via an iron(II)-dependent mechanism. In turn, the elevated lysosomal permeability led to release of proteases and subsequent antigen presentation, thereby activating $CD8^+$ T cells and antitumor immunity [78]. Excess ROS is another type of stress generated by abnormal cellular metabolism, hypoxia, and proteotoxic stress during intestinal tumorigenesis [6]. It was shown that autophagy in IECs was essential for counteracting ROS to enhance barrier integrity, and therefore attenuated the development of CRC [79].

Nevertheless, although autophagy prevents tumor formation at the early stage of intestinal carcinogenesis, it is not the case once the malignant transformation is established. Tumor-initiating cells (TICs), a cell subpopulation endowed with unlimited self-renewal and enhanced tumor-formation capacities, are known to greatly favor CRC initiation [80]. Under hypoxic conditions, autophagy promoted the self-renewal of TICs and their tumorigenic potential [81]. Conversely, autophagy suppressed the growth of the more differentiated counterpart cells [81]. As another example, Atg7-deficiency in IECs attenuated intestinal tumorigenesis in Apc(+/−) mice by the regulation of microbiome-mediated antitumor responses [42]. In samples taken from CRC patients, upregulation of Beclin 1 was related to HIF-1α overexpression, which further correlated with higher histological grade, disease stage, and poor prognosis [82]. Regarding the molecular mechanism, under normoxic conditions, Bcl-xL and Bcl-2 interacted with beclin-1, thereby inhibiting autophagy. In contrast, under hypoxic conditions, HIF-1α promoted the expression of proapoptotic genes BNIP3 and BNIP3L, which are associated with Bcl-xL and Bcl-2 to release beclin-1, thus triggering prosurvival autophagy in CRC cells [72]. In addition, HIF-1α upregulated miR-210, which suppressed Bcl-2 and induced autophagy to reduce the radiosensitivity of CRC [83]. Hypoxia also causes the accumulation of misfolded or unfolded proteins, leading to the unfolded protein response (UPR), and extended UPR signaling promotes cellular apoptosis. In CRC cells, hypoxia elicited UPR and the downstream key factor eukaryotic translation initiation factor 2 alpha kinase 3 (EIF2AK3). Subsequently, EIF2AK3 upregulated transcription factors ATF4 and CHOP to enhance the expression of LC3 and ATG5, thereby triggering cytoprotective autophagy [84]. Moreover, a recent study also demonstrated a sequential activation of AMPK, HIF-1α, HIF-2α, and JNK that accounted for the autophagy induction in CRC cells exposed to low oxygen levels [85]. Thus, these studies highlighted the distinct roles of autophagy in coordinating hypoxia stress response at different stages of colorectal development (Figure 4).

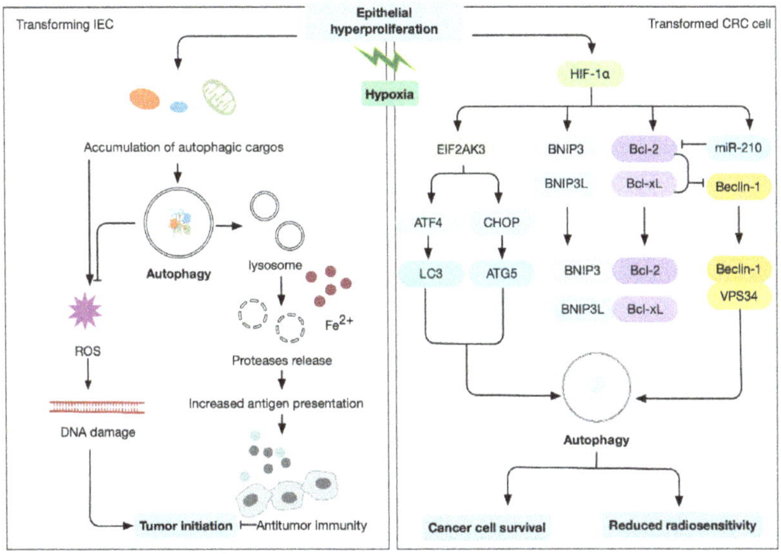

Figure 4. Autophagy coordinates cellular adaptation to hypoxia. Epithelial hyperproliferation results in a reduced level of oxygen, and dysregulated autophagy is involved in response to hypoxia. In transforming IECs, autophagy prevents cancer initiation via the elimination of hypoxia-induced accumulation of damaged cellular substances; while in transformed CRC cells, it promotes cancer cell survival by orchestrating multiple stress response pathways.

3.3. Autophagy Enables Adaptation to Oncogenic Microorganism-Induced Stress

As discussed above, accumulating evidence has supported the involvement of gut microorganisms in intestinal homeostasis and the etiology of sporadic CRC [2,86]. Bacterium pattern is different in CRC patients compared to healthy individuals. Through sequencing studies of the intestinal microbiota, the contribution of certain bacteria, including Fusobacterium nucleatum, Escherichia coli, and Bacteroides fragilis in CRC has been well established [87,88]. These infectious agents trigger DNA damage in host genetics by producing genotoxins, generating carcinogenic metabolites, regulating host cell signaling pathways, and shaping the cancer immune landscape in CRC [89–94].

A prime example of the role of microbiota-mediated autophagy in CRC is Fusobacterium nucleatum. Fusobacterium nucleatum, a Gram-negative anaerobe, is frequently present in the oral cavity and is commonly involved in dental plaques and periodontal disease [95]. Of note, Fusobacterium nucleatum was found in approximately 30% of CRC tissues in patients [96], and its abundance was positively associated with lymph node metastasis [97] and worse prognosis [98]. Interestingly, Fusobacterium nucleatum was enriched in CRC tissues from patients who relapsed after chemotherapy [24]. Although 5-fluorouracil (5-FU) in combination with platinum-based chemotherapy has been the first-line therapy for CRC patients [99], most patients develop chemoresistance during treatment and relapse after the initial response [100]. Mechanistic investigations revealed that infection with Fusobacterium nucleatum activated the innate immune response via TLR4 and MYD88-dependent signaling, which resulted in downregulation of miR-4802 and miR-18a*. Subsequently, reduction of these microRNAs attenuated their target on the 3′UTR regions of ULK1 and ATG7 genes, thus alleviating the silencing of autophagy. Eventually, activated autophagy gave rise to chemoresistance to oxaliplatin and 5-fu by protecting CRC cells from drug-induced apoptosis [24]. This has highlighted the prominent role of Fusobacterium nucleatum in coordinating a network of immune responses and autophagy to govern chemoresistance in CRC. Similarly, this network has also been implicated in CRC metastasis. Upon infection, Fusobacterium nucleatum induced the expression of

CARD3 in CRC cells, an essential kinase involved in innate and adaptive immune signaling. Upregulation of CARD3 then enhanced autophagic flux, thereby promoting the formation of liver and lung metastases in mouse models [101]. Nevertheless, the specific mechanism by which CARD3 regulates autophagy has been elusive and warrants further investigation.

Another mucosa-associated bacterium, Escherichia coli, is likely to exert an oncogenic phenotype in CRC through crosstalk with autophagy in a time-dependent manner. Activated autophagy can protect against CRC initiation in response to bacterial-induced stress. During the early stage of CRC initiation, increased epithelial autophagy eliminated the intracellular Colibactin-producing Escherichia coli to ameliorate malignant transformation in ApcMin/+ mice [75]. Colibactin-producing *E coli* (CoPEC), a colonic mucosa-associated E coli frequently detected in CRC patients, are able to promote CRC development by inducing genomic instability and inflammation [93,102,103]. There was evidence that autophagy-mediated elimination of Colibactin-producing Escherichia coli (CoPEC) limited the carcinogenesis process in ApcMin/+ mice by stimulating bacteria-induced DNA damage repair via RAD51 and reducing the secretion of inflammatory cytokines IL 6 and IL 8 [75]. Following this, at the mid phase of CRC development, the invasive *E. coli* that successfully colonized the colonic epithelium blocked autophagy to avoid clearance, achieving persistent infection. The repression of autophagy by *E. coli*, in turn, led to increased generation of ROS and epithelial hyperproliferation. However, as the dysplasia tissue progressed, autophagy was upregulated to eradicate the pathogen, suggesting an *E. coli* -independent tumor growth in the late stage of CRC development [104]. In line with this, the only time window when antibiotic intervention exhibited a tumor-suppressive effect was in the middle stages of tumor development [104]. Hence, the interplay between gut microbes and autophagy changes over time in the course of CRC development and therefore awaits future studies (Figure 5).

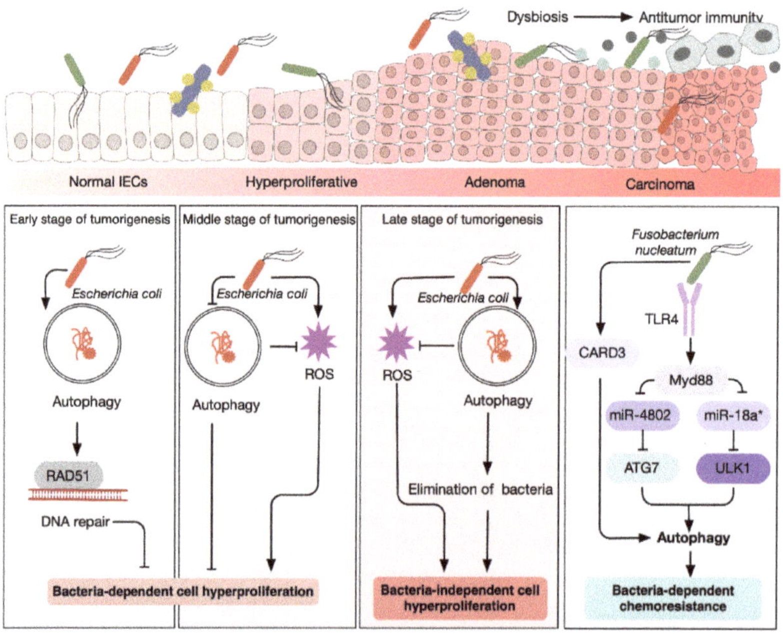

Figure 5. Distinct roles of autophagy in microbiota-induced stress. Sporadic CRC is driven by sequential genetic and epigenetic mutations, and environmental factors, arising from normal epithelial cells to dysplastic adenomas and, ultimately, carcinomas. During this process, a link between CRC tumorigenesis, infection with certain bacteria, and autophagy has been established. Time-dependent interactions between autophagy and intestinal bacteria are shown.

4. Autophagy Modulates Response to Therapeutic Stress in CRC

Although the above studies have highlighted the different roles of autophagy in coordinating environmental cues with CRC tumorigenesis and development, extensive laboratory evidence supported the stimulation of autophagy under therapeutic stress in CRC [105]. Given that numerous stress-sensing signaling pathways that elicit autophagy are utilized by CRC treatment approaches, many of these drugs have been revealed to induce cytoprotective autophagy [106]. In addition to surgery, patients with CRC are treated with combination regimens that involve chemotherapy, radiation therapy, and targeted therapy, tailored to specific pathologic staging and genetic status. Exposure to these therapeutic approaches can trigger autophagy that can enable tumor survival via DNA damage response, ER stress response, mTOR and AMPK signaling, and other stress-activated signaling pathways [105]. Indeed, cytoprotective autophagy has been seen as a crucial mechanism underpinning therapeutic resistance in CRC [107]. Moreover, as demonstrated by several preclinical studies and clinical trials, combining autophagy inhibitors with standard conventional therapies can improve the drug response of CRC [107]. Here, we focus on the roles of autophagy under different therapeutic stresses and the mechanism by which it mediates drug resistance in CRC (Figure 6).

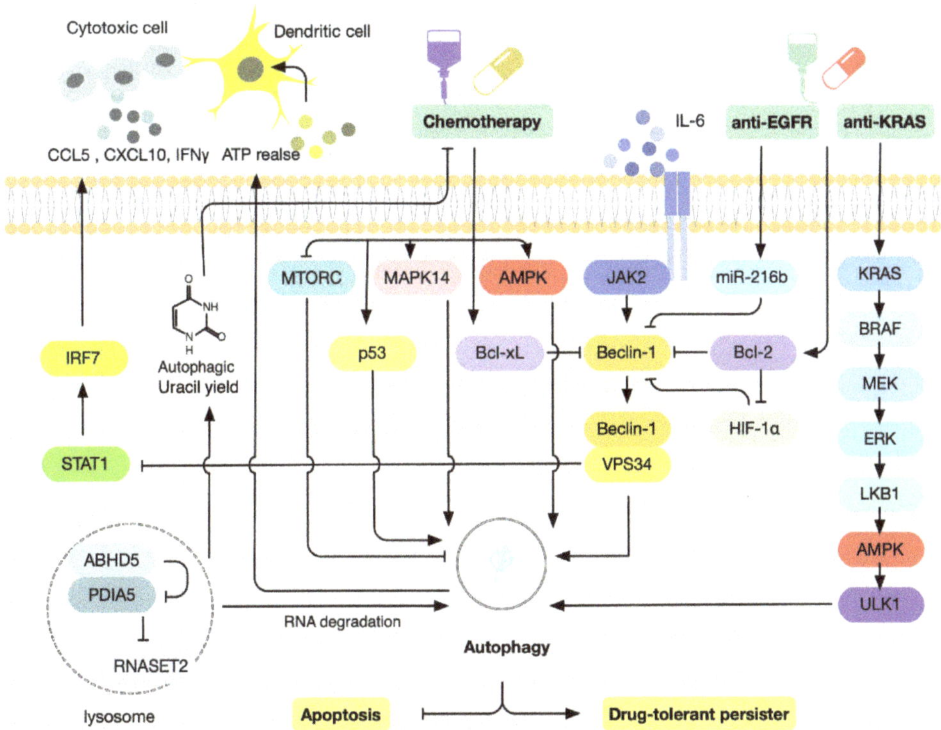

Figure 6. Autophagy under therapeutic stress in CRC. Autophagy activation has been observed during chemotherapies, targeted therapies, and PDT against CRC. In most cases, autophagy serves as a survival mechanism by protecting cells from apoptosis or maintaining cell survival in a DTP state; therefore, autophagy inhibition may be an effective therapeutic strategy in CRC. Paradoxically, autophagy is indispensable in the immune response to chemotherapy in CRC; hence, suppression of autophagy may result in a reduction of immunogenicity of cancer cells and impair antitumor immune immunity. Therefore, whether autophagy inhibitors should be combined with conventional therapies warrants further investigation.

4.1. Autophagy Regulates Cellular Response to Chemotherapy

The commonly used chemotherapeutics for treating CRC include fluorouracil (5-FU), oxaliplatin, and irinotecan, alone or in combination [66]. FU, an analog of uracil, mainly suppresses thymidylate synthase, which prevents the generation of thymidine needed for DNA synthesis, thereby inhibiting the proliferation of CRC cells [108]. In CRC models in vitro and in vivo, autophagy activation has been observed upon 5-FU treatment and protected cells from 5-FU-induced apoptosis [109]. The underlying mechanisms involved the upregulation of Bcl-xL, a key crosstalk factor between autophagy and apoptosis, and activation of the P53-AMPK-mTOR pathway [109]. Abnormal activity of metabolic enzymes also contributes to autophagy-mediated chemoresistance. ABHD5, a lipolytic factor situated in the lysosome, binds to PDIA5 to attenuate its inhibitory effect on ribonuclease RNASET2. In turn, RNASET2 regulates RNA degradation in autophagolysosomes, producing oligonucleotides, including uracil [110]. Treatment with 5-FU triggered metabolic reprogramming in CRC cells, and the expression of ABHD5 enhanced autophagic uracil yield, thus conferring 5-FU resistance due to decreased intake of 5-FU as external uracil [111]. Another report observed the induction of autophagy as a key mechanism of irinotecan resistance in TP53-defective CRC cells through the MAPK14/p38α pathway [112]. In addition, autophagy elicited by extracellular cytokine IL-6 protected CRC cells against the cytotoxic effects of 5-FU and oxaliplatin via the JAK2/BECN1 signaling axis [113]. Similar findings were made in microsatellite instability (MSI) CRCs, where mutation of a key autophagy regulator, UVRAG, led to a significant reduction in functional autophagy and became more responsive to 5-FU, oxaliplatin, and irinotecan [114].

The extensive laboratory studies above supported that autophagy engages in a complex interplay with apoptosis under therapeutic stress. Interestingly, apart from apoptosis, autophagy serves as a key mechanism for maintaining cell survival in a drug-tolerant persister (DTP) state to survive the stressful environment caused by chemotherapy [115,116]. In this context, CRC cells reversibly transition into a largely quiescent or slow-growing state, and after withdrawal of treatment, they exit the DTP state and regain the ability of growth and proliferation [117,118]. Mechanistically, this is achieved by employing an evolutionarily conserved program, diapause, which is adopted by hundreds of mammalian species that can suspend embryonic development under unfavorable environmental conditions [119]. Remarkably, as revealed by analyses of expression signatures of the diapause-like DTP in CRC models, this phenotype was maintained via downregulation of the mTOR pathway and upregulation of the autophagy program [120]. Therefore, combination therapy of chemotherapy and autophagy inhibitors represented an innovative therapeutic strategy to disrupt the survival mechanism and prevent cancer relapse [120].

4.2. Autophagy Regulates Cellular Response to Targeted Therapies

Aberrant activation or upregulation of oncogenes including EGFR, KRAS, NRAS, and BRAF are frequently present in CRC [121]. Biologics targeting EGFR, such as cetuximab and panitumumab, are often incorporated into the chemotherapy regimens based on the mutation status of individual patients [122–125]. Anti-EGFR monoclonal antibodies act by blocking access of ligands to the binding domain of EGFR and promoting its internalization and degradation. The interplay between EGFR and autophagy involves RAS PI3K-AKT-mTOR pathways, which serve as the downstream signaling of EGFR as well as the key regulatory network of autophagy [126]. Given the common mechanisms shared by internalized EGFR and autophagy, it is not surprising that activation of autophagy was demonstrated to underlie the acquired resistance of anti-EGFR therapies. Indeed, it has been reported that treatment with the EGFR antibody cetuximab can elicit autophagy in CRC cells and protect them from therapy-induced apoptosis [127,128]. Mechanistic investigations revealed that cetuximab suppressed the expression of HIF-1α and subsequently Bcl-2, which attenuated the inhibitory effect of Bcl-2 on beclin 1 and enhanced the formation of the beclin 1/hVps34 complex, thus activating autophagy [127]. Moreover, cetuximab downregulated miR-216b, which can impair the translation of Beclin-1 through binding to

3′-UTR of its mRNA, thereby inducing cytoprotective autophagy [128]. Therefore, these studies suggested the potential of autophagy inhibitors to sensitize CRC to anti-EGFR monoclonal antibodies.

BRAF-V600E mutation activates the MEK/ERK pathway, conferring a poor prognosis in CRC patients [129]. Targeted combination therapy with BRAF inhibitor encorafenib plus EGFR inhibitor cetuximab has been shown to extend overall survival and approved for second-line therapy [130]. Intriguingly, targeting MEK/ERK pathway using MEK inhibitor trametinib induced prosurvival autophagy by activating the LKB1/AMPK/ULK1 axis in KRAS-mutated pancreatic ductal adenocarcinoma [131]. This was similarly relevant to CRC, since combination therapy of trametinib with autophagy inhibitor chloroquine demonstrated significant antitumor effects in patient-derived xenografts (PDX) of BRAF-mutated CRC [131]. Another monoclonal antibody, bevacizumab, which targets VEGF and interrupts tumor angiogenesis, has been extensively used in CRC [132]. In mouse xenografts of CRC, bevacizumab elicited autophagy and blockade of autophagy with chloroquine displayed synergistic antiproliferative effects against tumor [133]. Photodynamic therapy (PDT), in which photosensitizers are irradiated and excited by light, leads to ROS generation and accumulation, and eventually cell death [134]. This novel technique has become a complement to traditional cancer treatment. Notably, it has been reported that PDT can activate autophagy in CRC, and pharmacological autophagy inhibitors enhanced therapeutic sensitivity to PDT [135]. Together, autophagy serves as a key survival mechanism in response to chemotherapies, targeted therapies, and PDT against CRC; therefore, autophagy inhibition may be an effective therapeutic strategy in CRC.

4.3. Autophagy Regulates Cellular Response to Immunotherapy

While autophagy was convincingly shown to be hijacked by cancer cells to resist therapeutic challenges, the consensus that combining autophagy inhibitors with chemotherapy should be regarded as a general therapeutic strategy has been challenged. It is important to perceive that conventional chemotherapies exert anticancer effects not only through a direct cytotoxic mechanism, but also partly owing to the re-stimulation of antitumor immune function [136]. Interestingly, evidence has indicated that autophagy has a major role in immunological control in response to immunogenic chemotherapy in CRC [137,138]. In the context of anticancer chemotherapy exposure, autophagy-competent CRC favored ATP secretion from malignant cells, thereby enhancing the recruitment of dendritic cells and T lymphocytes [137]. Moreover, similar findings were revealed in melanoma, where chemotherapy and radiotherapy-induced autophagy has been shown to augment the sensitivity of tumor cells to lysis by cytotoxic T cells [138,139]. Together, these lines of evidence highlighted that suppression of autophagy might, at least in part, result in a reduction in immunogenicity of cancer cells, and hence defective immune response and relapsed disease.

On theoretical grounds, this detrimental side effect exerted by autophagy inhibition that blunts the antitumor immunity in CRC may be circumvented via combined administration with an immune checkpoint inhibitor [105,140]. Indeed, it has been shown that blocking PIK3C3/VPS34 in combination with anti-PD-1/PD-L1 immunotherapy exhibited promising efficacy in CRC [141,142]. However, in this study, autophagy inhibition achieved by targeting PIK3C3/VPS34 promoted the attraction of cytotoxic immune cells via STAT1/IRF7-dependent production of CCL5 and CXCL10 [141,142]. Along similar lines, experimental studies addressing the impact of autophagy on cancer immune landscape have yielded a wealth of controversial results across various cancer types. In mouse models of melanoma and breast cancer, the levels of T cell infiltration and T cell responses remained unchanged upon autophagy inhibition [143], whereas in other studies, loss of autophagy was believed to facilitate recruitment of antitumor immune effector cells to the tumor bed [144–147]. The extrapolation can be made that targeting different autophagy proteins may elicit different impacts on cancer immune response and presumably involves autophagy-independent mechanisms. Hence, there is still a lack of knowledge regarding the interaction between autophagy and antitumor immunity.

5. Clinical Implications and Future Perspectives

Mounting evidence suggested a prominent role of autophagy in the development of cancer, especially in those organs that are constantly challenged by environmental stressors, such as the large intestine [148]. The idea that autophagy serves as a survival mechanism for tumor cells has provided the logical rationale for autophagy inhibition as a therapeutic strategy in CRC [149]. Indeed, autophagy inhibitors, notably chloroquine (CQ) or hydroxychloroquine (HCQ), have been widely adopted in combination with traditional chemotherapy/radiotherapy in clinical trials of multiple tumor types. Other specific inhibitors are also in development and need further investigation in preclinical and clinical trials [150]. Although the safety of these drugs has been demonstrated, the efficacy of autophagy inhibition has varied widely between patients with different types of tumors and at different stages [148]. These reported clinical outcomes, which are not always encouraging, exemplify the underlying limitations of the clinical applications of autophagy inhibition.

It is critical to note that autophagy has multifaceted and opposing roles in the world of oncology. First, it also plays a cytotoxic role under certain circumstances, which is related to its regulation of apoptosis by the degradation of different proapoptotic or antiapoptotic factors. As such, autophagy inhibition is a bad idea since it would protect malignant cells from undergoing programmed cell death. Moreover, in the context of tumor initiation, growth, and therapeutic pressure of CRC, autophagy functions in a context-dependent manner. For different cell types along the course of the adenoma-carcinoma sequence, including normal IECs, hyperproliferative IECs, adenoma cells, and carcinoma cells, autophagy exerts opposing effects in the presence of distinct microenvironmental conditions. For example, autophagic defects predispose normal cells to malignant transformation, whereas tumor cells can exploit autophagy to thrive under the hostile microenvironment and survive anticancer therapy. In the meantime, while accumulating studies support that autophagy operates in a cell-intrinsic fashion, it also has a cell-extrinsic function. A prime example of this is its relevance in immunological control, where autophagy is responsible for the immunostimulatory signal-sending (notably, ATP) and effector immune cell recruitment [151].Thus, based on these observations, autophagy inhibition may be counterproductive in cancer therapy. To address this dilemma, evaluation with appropriate biomarkers of the status of autophagy, that is, prosurvival or prodeath, whether tumorigenic or tumor-suppressive, may aid in selecting patients who will benefit from autophagy inhibition or induction therapy.

Another issue about the clinical implication of autophagy manipulation is drug specificity. Currently, most pharmacological modulators of autophagy do not selectively target autophagy. Various inhibitors that regulate the different steps of autophagy, including those targeting mTORC1, ULK1, Beclin1, and so on, also interfere with other oncogenic signaling cascades. The ubiquitous effects of autophagy on normal tissues may also limit the clinical utility of autophagy regulators, given that deficiency of autophagy can result in neurodegeneration, lysosomal storage diseases, and other organ dysfunction [152]. Thus, with the increasing understanding of the non-autophagic role of autophagy-related proteins, as well as the potential toxicity of global autophagy modulation on non-transformed tissues, specific regulation of autophagy-related functions local to tumor lesions is required to prevent adverse effects.

Overall, it is impossible to achieve long-term remission and cure through a single-agent treatment in cancer; therefore, combination therapy utilizing multiple means holds great potential for optimal management of CRC [105]. Thus, further explorations that shed additional light on the pleiotropic mechanisms of autophagic machinery more accurately will be critical to help enhance the effectiveness of current CRC therapy.

Author Contributions: Conceptualization, H.C. and Z.Z.; investigation, Q.L. and Y.C.; writing original draft preparation, Q.L. and Y.C.; writing—review and editing, Q.L., Y.C. and L.Z.; visualization, Q.L. and Y.C.; supervision, H.C. and Z.Z. All authors have read and agreed to the published version of the manuscript.

Funding: This work was supported by grants from the National Natural Science Foundation of China (82073246) and 1.3.5 project for disciplines of excellence, West China Hospital, Sichuan University (ZYGD20006).

Conflicts of Interest: The authors declare no conflict of interest.

References

1. Foerster, E.G.; Mukherjee, T.; Cabral-Fernandes, L.; Rocha, J.D.B.; Girardin, S.E.; Philpott, D.J. How autophagy controls the intestinal epithelial barrier. *Autophagy* **2022**, *18*, 86–103. [CrossRef] [PubMed]
2. Tjalsma, H.; Boleij, A.; Marchesi, J.R.; Dutilh, B.E. A bacterial driver-passenger model for colorectal cancer: Beyond the usual suspects. *Nat. Rev. Microbiol.* **2012**, *10*, 575–582. [CrossRef]
3. Yang, Z.; Klionsky, D.J. Mammalian autophagy: Core molecular machinery and signaling regulation. *Curr. Opin. Cell Biol.* **2010**, *22*, 124–131. [CrossRef] [PubMed]
4. He, C.; Klionsky, D.J. Regulation mechanisms and signaling pathways of autophagy. *Annu. Rev. Genet.* **2009**, *43*, 67–93. [CrossRef] [PubMed]
5. Sica, V.; Galluzzi, L.; Bravo-San Pedro, J.M.; Izzo, V.; Maiuri, M.C.; Kroemer, G. Organelle-Specific Initiation of Autophagy. *Mol. Cell* **2015**, *59*, 522–539. [CrossRef]
6. Mizushima, N.; Komatsu, M. Autophagy: Renovation of cells and tissues. *Cell* **2011**, *147*, 728–741. [CrossRef]
7. Fearon, E.R. Molecular genetics of colorectal cancer. *Annu. Rev. Pathol.* **2011**, *6*, 479–507. [CrossRef]
8. Vogelstein, B.; Kinzler, K.W. The multistep nature of cancer. *Trends Genet.* **1993**, *9*, 138–141. [CrossRef]
9. Lee-Six, H.; Olafsson, S.; Ellis, P.; Osborne, R.J.; Sanders, M.A.; Moore, L.; Georgakopoulos, N.; Torrente, F.; Noorani, A.; Goddard, M.; et al. The landscape of somatic mutation in normal colorectal epithelial cells. *Nature* **2019**, *574*, 532–537. [CrossRef]
10. Fearon, E.R.; Vogelstein, B. A genetic model for colorectal tumorigenesis. *Cell* **1990**, *61*, 759–767. [CrossRef]
11. Settembre, C.; Di Malta, C.; Polito, V.A.; Garcia Arencibia, M.; Vetrini, F.; Erdin, S.; Erdin, S.U.; Huynh, T.; Medina, D.; Colella, P.; et al. TFEB links autophagy to lysosomal biogenesis. *Science* **2011**, *332*, 1429–1433. [CrossRef] [PubMed]
12. Settembre, C.; De Cegli, R.; Mansueto, G.; Saha, P.K.; Vetrini, F.; Visvikis, O.; Huynh, T.; Carissimo, A.; Palmer, D.; Klisch, T.J.; et al. TFEB controls cellular lipid metabolism through a starvation-induced autoregulatory loop. *Nat. Cell Biol.* **2013**, *15*, 647–658. [CrossRef] [PubMed]
13. Zeng, C.; Matsuda, K.; Jia, W.H.; Chang, J.; Kweon, S.S.; Xiang, Y.B.; Shin, A.; Jee, S.H.; Kim, D.H.; Zhang, B.; et al. Identification of Susceptibility Loci and Genes for Colorectal Cancer Risk. *Gastroenterology* **2016**, *150*, 1633–1645. [CrossRef] [PubMed]
14. Quach, C.; Song, Y.; Guo, H.; Li, S.; Maazi, H.; Fung, M.; Sands, N.; O'Connell, D.; Restrepo-Vassalli, S.; Chai, B.; et al. A truncating mutation in the autophagy gene UVRAG drives inflammation and tumorigenesis in mice. *Nat. Commun.* **2019**, *10*, 5681. [CrossRef]
15. Zhang, P.; Holowatyj, A.N.; Roy, T.; Pronovost, S.M.; Marchetti, M.; Liu, H.; Ulrich, C.M.; Edgar, B.A. An SH3PX1-Dependent Endocytosis-Autophagy Network Restrains Intestinal Stem Cell Proliferation by Counteracting EGFR-ERK Signaling. *Dev. Cell* **2019**, *49*, 574–589.e575. [CrossRef]
16. Stryker, S.J.; Wolff, B.G.; Culp, C.E.; Libbe, S.D.; Ilstrup, D.M.; MacCarty, R.L. Natural history of untreated colonic polyps. *Gastroenterology* **1987**, *93*, 1009–1013. [CrossRef]
17. Sammarco, G.; Gallo, G.; Vescio, G.; Picciariello, A.; De Paola, G.; Trompetto, M.; Curro, G.; Ammendola, M. Mast Cells, microRNAs and Others: The Role of Translational Research on Colorectal Cancer in the Forthcoming Era of Precision Medicine. *J. Clin. Med.* **2020**, *9*, 2852. [CrossRef]
18. Gewirtz, D.A. The four faces of autophagy: Implications for cancer therapy. *Cancer Res.* **2014**, *74*, 647–651. [CrossRef]
19. Chen, L.; He, M.; Zhang, M.; Sun, Q.; Zeng, S.; Zhao, H.; Yang, H.; Liu, M.; Ren, S.; Meng, X.; et al. The Role of non-coding RNAs in colorectal cancer, with a focus on its autophagy. *Pharmacol. Ther.* **2021**, *226*, 107868. [CrossRef]
20. Ballabio, A.; Bonifacino, J.S. Lysosomes as dynamic regulators of cell and organismal homeostasis. *Nat. Rev. Mol. Cell Biol.* **2020**, *21*, 101–118. [CrossRef]
21. Dikic, I.; Elazar, Z. Mechanism and medical implications of mammalian autophagy. *Nat. Rev. Mol. Cell Biol.* **2018**, *19*, 349–364. [CrossRef] [PubMed]
22. Rybstein, M.D.; Bravo-San Pedro, J.M.; Kroemer, G.; Galluzzi, L. The autophagic network and cancer. *Nat. Cell Biol.* **2018**, *20*, 243–251. [CrossRef] [PubMed]
23. Mokarram, P.; Albokashy, M.; Zarghooni, M.; Moosavi, M.A.; Sepehri, Z.; Chen, Q.M.; Hudecki, A.; Sargazi, A.; Alizadeh, J.; Moghadam, A.R.; et al. New frontiers in the treatment of colorectal cancer: Autophagy and the unfolded protein response as promising targets. *Autophagy* **2017**, *13*, 781–819. [CrossRef] [PubMed]
24. Yu, T.; Guo, F.; Yu, Y.; Sun, T.; Ma, D.; Han, J.; Qian, Y.; Kryczek, I.; Sun, D.; Nagarsheth, N.; et al. Fusobacterium nucleatum Promotes Chemoresistance to Colorectal Cancer by Modulating Autophagy. *Cell* **2017**, *170*, 548–563.e516. [CrossRef]

25. Mizushima, N. The ATG conjugation systems in autophagy. *Curr. Opin. Cell Biol.* **2020**, *63*, 1–10. [CrossRef]
26. Kim, J.; Kundu, M.; Viollet, B.; Guan, K.L. AMPK and mTOR regulate autophagy through direct phosphorylation of Ulk1. *Nat. Cell Biol.* **2011**, *13*, 132–141. [CrossRef]
27. Mihaylova, M.M.; Shaw, R.J. The AMPK signalling pathway coordinates cell growth, autophagy and metabolism. *Nat. Cell Biol.* **2011**, *13*, 1016–1023. [CrossRef]
28. Codogno, P.; Mehrpour, M.; Proikas-Cezanne, T. Canonical and non-canonical autophagy: Variations on a common theme of self-eating? *Nat. Rev. Mol. Cell Biol.* **2011**, *13*, 7–12. [CrossRef]
29. Green, D.R.; Levine, B. To be or not to be? How selective autophagy and cell death govern cell fate. *Cell* **2014**, *157*, 65–75. [CrossRef]
30. Galluzzi, L.; Green, D.R. Autophagy-Independent Functions of the Autophagy Machinery. *Cell* **2019**, *177*, 1682–1699. [CrossRef]
31. Feng, Y.; He, D.; Yao, Z.; Klionsky, D.J. The machinery of macroautophagy. *Cell Res.* **2014**, *24*, 24–41. [CrossRef] [PubMed]
32. Boya, P.; Reggiori, F.; Codogno, P. Emerging regulation and functions of autophagy. *Nat. Cell Biol.* **2013**, *15*, 713–720. [CrossRef] [PubMed]
33. Axe, E.L.; Walker, S.A.; Manifava, M.; Chandra, P.; Roderick, H.L.; Habermann, A.; Griffiths, G.; Ktistakis, N.T. Autophagosome formation from membrane compartments enriched in phosphatidylinositol 3-phosphate and dynamically connected to the endoplasmic reticulum. *J. Cell Biol.* **2008**, *182*, 685–701. [CrossRef] [PubMed]
34. Saftig, P.; Haas, A. Turn up the lysosome. *Nat. Cell Biol.* **2016**, *18*, 1025–1027. [CrossRef] [PubMed]
35. Patel, K.K.; Stappenbeck, T.S. Autophagy and intestinal homeostasis. *Annu. Rev. Physiol.* **2013**, *75*, 241–262. [CrossRef]
36. Lassen, K.G.; Xavier, R.J. Mechanisms and function of autophagy in intestinal disease. *Autophagy* **2018**, *14*, 216–220. [CrossRef]
37. Baxt, L.A.; Xavier, R.J. Role of Autophagy in the Maintenance of Intestinal Homeostasis. *Gastroenterology* **2015**, *149*, 553–562. [CrossRef]
38. Blanpain, C.; Horsley, V.; Fuchs, E. Epithelial stem cells: Turning over new leaves. *Cell* **2007**, *128*, 445–458. [CrossRef]
39. Jung, H.; Leal-Ekman, J.S.; Lu, Q.; Stappenbeck, T.S. Atg14 protects the intestinal epithelium from TNF-triggered villus atrophy. *Autophagy* **2019**, *15*, 1990–2001. [CrossRef]
40. Levy, A.; Stedman, A.; Deutsch, E.; Donnadieu, F.; Virgin, H.W.; Sansonetti, P.J.; Nigro, G. Innate immune receptor NOD2 mediates LGR5(+) intestinal stem cell protection against ROS cytotoxicity via mitophagy stimulation. *Proc. Natl. Acad. Sci. USA* **2020**, *117*, 1994–2003. [CrossRef]
41. Trentesaux, C.; Fraudeau, M.; Pitasi, C.L.; Lemarchand, J.; Jacques, S.; Duche, A.; Letourneur, F.; Naser, E.; Bailly, K.; Schmitt, A.; et al. Essential role for autophagy protein ATG7 in the maintenance of intestinal stem cell integrity. *Proc. Natl. Acad. Sci. USA* **2020**, *117*, 11136–11146. [CrossRef] [PubMed]
42. Levy, J.; Cacheux, W.; Bara, M.A.; L'Hermitte, A.; Lepage, P.; Fraudeau, M.; Trentesaux, C.; Lemarchand, J.; Durand, A.; Crain, A.M.; et al. Intestinal inhibition of Atg7 prevents tumour initiation through a microbiome-influenced immune response and suppresses tumour growth. *Nat. Cell Biol.* **2015**, *17*, 1062–1073. [CrossRef] [PubMed]
43. Yang, L.; Liu, C.; Zhao, W.; He, C.; Ding, J.; Dai, R.; Xu, K.; Xiao, L.; Luo, L.; Liu, S.; et al. Impaired Autophagy in Intestinal Epithelial Cells Alters Gut Microbiota and Host Immune Responses. *Appl. Environ. Microbiol.* **2018**, *84*, e00880-18. [CrossRef]
44. Merle, N.S.; Church, S.E.; Fremeaux-Bacchi, V.; Roumenina, L.T. Complement System Part I—Molecular Mechanisms of Activation and Regulation. *Front. Immunol.* **2015**, *6*, 262. [CrossRef]
45. Sorbara, M.T.; Foerster, E.G.; Tsalikis, J.; Abdel-Nour, M.; Mangiapane, J.; Sirluck-Schroeder, I.; Tattoli, I.; van Dalen, R.; Isenman, D.E.; Rohde, J.R.; et al. Complement C3 Drives Autophagy-Dependent Restriction of Cyto-invasive Bacteria. *Cell Host Microbe* **2018**, *23*, 644–652.e645. [CrossRef] [PubMed]
46. Benjamin, J.L.; Sumpter, R., Jr.; Levine, B.; Hooper, L.V. Intestinal epithelial autophagy is essential for host defense against invasive bacteria. *Cell Host Microbe* **2013**, *13*, 723–734. [CrossRef]
47. Conway, K.L.; Kuballa, P.; Song, J.H.; Patel, K.K.; Castoreno, A.B.; Yilmaz, O.H.; Jijon, H.B.; Zhang, M.; Aldrich, L.N.; Villablanca, E.J.; et al. Atg16l1 is required for autophagy in intestinal epithelial cells and protection of mice from Salmonella infection. *Gastroenterology* **2013**, *145*, 1347–1357. [CrossRef]
48. Martin, P.K.; Marchiando, A.; Xu, R.; Rudensky, E.; Yeung, F.; Schuster, S.L.; Kernbauer, E.; Cadwell, K. Autophagy proteins suppress protective type I interferon signalling in response to the murine gut microbiota. *Nat. Microbiol.* **2018**, *3*, 1131–1141. [CrossRef]
49. Snell, L.M.; McGaha, T.L.; Brooks, D.G. Type I Interferon in Chronic Virus Infection and Cancer. *Trends Immunol.* **2017**, *38*, 542–557. [CrossRef]
50. Nighot, P.K.; Hu, C.A.; Ma, T.Y. Autophagy enhances intestinal epithelial tight junction barrier function by targeting claudin-2 protein degradation. *J. Biol. Chem.* **2015**, *290*, 7234–7246. [CrossRef]
51. Ganapathy, A.S.; Saha, K.; Suchanec, E.; Singh, V.; Verma, A.; Yochum, G.; Koltun, W.; Nighot, M.; Ma, T.; Nighot, P. AP2M1 mediates autophagy-induced CLDN2 (claudin 2) degradation through endocytosis and interaction with LC3 and reduces intestinal epithelial tight junction permeability. *Autophagy* **2021**, 1–18. [CrossRef] [PubMed]
52. Zhang, C.; Yan, J.; Xiao, Y.; Shen, Y.; Wang, J.; Ge, W.; Chen, Y. Inhibition of Autophagic Degradation Process Contributes to Claudin-2 Expression Increase and Epithelial Tight Junction Dysfunction in TNF-alpha Treated Cell Monolayers. *Int. J. Mol. Sci.* **2017**, *18*, 157. [CrossRef] [PubMed]

53. Wong, M.; Ganapathy, A.S.; Suchanec, E.; Laidler, L.; Ma, T.; Nighot, P. Intestinal epithelial tight junction barrier regulation by autophagy-related protein ATG6/beclin 1. *Am. J. Physiol. Cell Physiol.* **2019**, *316*, C753–C765. [CrossRef] [PubMed]
54. Vander Heiden, M.G.; Cantley, L.C.; Thompson, C.B. Understanding the Warburg effect: The metabolic requirements of cell proliferation. *Science* **2009**, *324*, 1029–1033. [CrossRef] [PubMed]
55. Kimmelman, A.C.; White, E. Autophagy and Tumor Metabolism. *Cell Metab.* **2017**, *25*, 1037–1043. [CrossRef]
56. Kim, K.H.; Lee, M.S. Autophagy—A key player in cellular and body metabolism. *Nat. Rev. Endocrinol.* **2014**, *10*, 322–337. [CrossRef]
57. White, E.; Mehnert, J.M.; Chan, C.S. Autophagy, Metabolism, and Cancer. *Clin. Cancer Res.* **2015**, *21*, 5037–5046. [CrossRef]
58. Poillet-Perez, L.; White, E. Role of tumor and host autophagy in cancer metabolism. *Genes Dev.* **2019**, *33*, 610–619. [CrossRef]
59. Amaravadi, R.; Kimmelman, A.C.; White, E. Recent insights into the function of autophagy in cancer. *Genes Dev.* **2016**, *30*, 1913–1930. [CrossRef]
60. White, E.; Lattime, E.C.; Guo, J.Y. Autophagy Regulates Stress Responses, Metabolism, and Anticancer Immunity. *Trends Cancer* **2021**, *7*, 778–789. [CrossRef]
61. Barros, J.A.S.; Siqueira, J.A.B.; Cavalcanti, J.H.F.; Araujo, W.L.; Avin-Wittenberg, T. Multifaceted Roles of Plant Autophagy in Lipid and Energy Metabolism. *Trends Plant Sci.* **2020**, *25*, 1141–1153. [CrossRef] [PubMed]
62. Pietrocola, F.; Izzo, V.; Niso-Santano, M.; Vacchelli, E.; Galluzzi, L.; Maiuri, M.C.; Kroemer, G. Regulation of autophagy by stress-responsive transcription factors. *Semin. Cancer Biol.* **2013**, *23*, 310–322. [CrossRef] [PubMed]
63. Lock, R.; Roy, S.; Kenific, C.M.; Su, J.S.; Salas, E.; Ronen, S.M.; Debnath, J. Autophagy facilitates glycolysis during Ras-mediated oncogenic transformation. *Mol. Biol. Cell* **2011**, *22*, 165–178. [CrossRef]
64. Guo, J.Y.; Chen, H.Y.; Mathew, R.; Fan, J.; Strohecker, A.M.; Karsli-Uzunbas, G.; Kamphorst, J.J.; Chen, G.; Lemons, J.M.; Karantza, V.; et al. Activated Ras requires autophagy to maintain oxidative metabolism and tumorigenesis. *Genes Dev.* **2011**, *25*, 460–470. [CrossRef] [PubMed]
65. Yang, S.; Wang, X.; Contino, G.; Liesa, M.; Sahin, E.; Ying, H.; Bause, A.; Li, Y.; Stommel, J.M.; Dell'antonio, G.; et al. Pancreatic cancers require autophagy for tumor growth. *Genes Dev.* **2011**, *25*, 717–729. [CrossRef]
66. Kuipers, E.J.; Grady, W.M.; Lieberman, D.; Seufferlein, T.; Sung, J.J.; Boelens, P.G.; van de Velde, C.J.; Watanabe, T. Colorectal cancer. *Nat. Rev. Dis. Primers* **2015**, *1*, 15065. [CrossRef]
67. Strohecker, A.M.; Guo, J.Y.; Karsli-Uzunbas, G.; Price, S.M.; Chen, G.J.; Mathew, R.; McMahon, M.; White, E. Autophagy sustains mitochondrial glutamine metabolism and growth of BrafV600E-driven lung tumors. *Cancer Discov.* **2013**, *3*, 1272–1285. [CrossRef]
68. Santos, C.R.; Schulze, A. Lipid metabolism in cancer. *FEBS J.* **2012**, *279*, 2610–2623. [CrossRef]
69. Wen, Y.A.; Xing, X.; Harris, J.W.; Zaytseva, Y.Y.; Mitov, M.I.; Napier, D.L.; Weiss, H.L.; Mark Evers, B.; Gao, T. Adipocytes activate mitochondrial fatty acid oxidation and autophagy to promote tumor growth in colon cancer. *Cell Death Dis.* **2017**, *8*, e2593. [CrossRef]
70. Deretic, V. Autophagy in inflammation, infection, and immunometabolism. *Immunity* **2021**, *54*, 437–453. [CrossRef]
71. Wei, J.; Long, L.; Yang, K.; Guy, C.; Shrestha, S.; Chen, Z.; Wu, C.; Vogel, P.; Neale, G.; Green, D.R.; et al. Autophagy enforces functional integrity of regulatory T cells by coupling environmental cues and metabolic homeostasis. *Nat. Immunol.* **2016**, *17*, 277–285. [CrossRef] [PubMed]
72. Bellot, G.; Garcia-Medina, R.; Gounon, P.; Chiche, J.; Roux, D.; Pouyssegur, J.; Mazure, N.M. Hypoxia-induced autophagy is mediated through hypoxia-inducible factor induction of BNIP3 and BNIP3L via their BH3 domains. *Mol. Cell Biol.* **2009**, *29*, 2570–2581. [CrossRef] [PubMed]
73. Brugarolas, J.; Lei, K.; Hurley, R.L.; Manning, B.D.; Reiling, J.H.; Hafen, E.; Witters, L.A.; Ellisen, L.W.; Kaelin, W.G., Jr. Regulation of mTOR function in response to hypoxia by REDD1 and the TSC1/TSC2 tumor suppressor complex. *Genes Dev.* **2004**, *18*, 2893–2904. [CrossRef] [PubMed]
74. Scherz-Shouval, R.; Shvets, E.; Fass, E.; Shorer, H.; Gil, L.; Elazar, Z. Reactive oxygen species are essential for autophagy and specifically regulate the activity of Atg4. *EMBO J.* **2019**, *38*, e101812. [CrossRef]
75. Lucas, C.; Salesse, L.; Hoang, M.H.T.; Bonnet, M.; Sauvanet, P.; Larabi, A.; Godfraind, C.; Gagniere, J.; Pezet, D.; Rosenstiel, P.; et al. Autophagy of Intestinal Epithelial Cells Inhibits Colorectal Carcinogenesis Induced by Colibactin-Producing Escherichia coli in Apc(Min/+) Mice. *Gastroenterology* **2020**, *158*, 1373–1388. [CrossRef]
76. Guo, J.Y.; Karsli-Uzunbas, G.; Mathew, R.; Aisner, S.C.; Kamphorst, J.J.; Strohecker, A.M.; Chen, G.; Price, S.; Lu, W.; Teng, X.; et al. Autophagy suppresses progression of K-ras-induced lung tumors to oncocytomas and maintains lipid homeostasis. *Genes Dev.* **2013**, *27*, 1447–1461. [CrossRef]
77. Youle, R.J.; Narendra, D.P. Mechanisms of mitophagy. *Nat. Rev. Mol. Cell Biol.* **2011**, *12*, 9–14. [CrossRef]
78. Ziegler, P.K.; Bollrath, J.; Pallangyo, C.K.; Matsutani, T.; Canli, O.; De Oliveira, T.; Diamanti, M.A.; Muller, N.; Gamrekelashvili, J.; Putoczki, T.; et al. Mitophagy in Intestinal Epithelial Cells Triggers Adaptive Immunity during Tumorigenesis. *Cell* **2018**, *174*, 88–101.e116. [CrossRef]
79. Liu, M.; Sun, T.; Li, N.; Peng, J.; Fu, D.; Li, W.; Li, L.; Gao, W.Q. BRG1 attenuates colonic inflammation and tumorigenesis through autophagy-dependent oxidative stress sequestration. *Nat. Commun.* **2019**, *10*, 4614. [CrossRef]
80. Qureshi-Baig, K.; Ullmann, P.; Haan, S.; Letellier, E. Tumor-Initiating Cells: A criTICal review of isolation approaches and new challenges in targeting strategies. *Mol. Cancer* **2017**, *16*, 40. [CrossRef]

81. Qureshi-Baig, K.; Kuhn, D.; Viry, E.; Pozdeev, V.I.; Schmitz, M.; Rodriguez, F.; Ullmann, P.; Koncina, E.; Nurmik, M.; Frasquilho, S.; et al. Hypoxia-induced autophagy drives colorectal cancer initiation and progression by activating the PRKC/PKC-EZR (ezrin) pathway. *Autophagy* **2020**, *16*, 1436–1452. [CrossRef] [PubMed]
82. Koukourakis, M.I.; Giatromanolaki, A.; Sivridis, E.; Pitiakoudis, M.; Gatter, K.C.; Harris, A.L. Beclin 1 over- and underexpression in colorectal cancer: Distinct patterns relate to prognosis and tumour hypoxia. *Br. J. Cancer* **2010**, *103*, 1209–1214. [CrossRef]
83. Sun, Y.; Xing, X.; Liu, Q.; Wang, Z.; Xin, Y.; Zhang, P.; Hu, C.; Liu, Y. Hypoxia-induced autophagy reduces radiosensitivity by the HIF-1alpha/miR-210/Bcl-2 pathway in colon cancer cells. *Int. J. Oncol.* **2015**, *46*, 750–756. [CrossRef] [PubMed]
84. Rouschop, K.M.; van den Beucken, T.; Dubois, L.; Niessen, H.; Bussink, J.; Savelkouls, K.; Keulers, T.; Mujcic, H.; Landuyt, W.; Voncken, J.W.; et al. The unfolded protein response protects human tumor cells during hypoxia through regulation of the autophagy genes MAP1LC3B and ATG5. *J. Clin. Investig.* **2010**, *120*, 127–141. [CrossRef] [PubMed]
85. Tam, S.Y.; Wu, V.W.C.; Law, H.K.W. Dynamics of oxygen level-driven regulators in modulating autophagy in colorectal cancer cells. *Biochem. Biophys. Res. Commun.* **2019**, *517*, 193–200. [CrossRef]
86. Park, E.M.; Chelvanambi, M.; Bhutiani, N.; Kroemer, G.; Zitvogel, L.; Wargo, J.A. Targeting the gut and tumor microbiota in cancer. *Nat. Med.* **2022**, *28*, 690–703. [CrossRef]
87. Sepich-Poore, G.D.; Zitvogel, L.; Straussman, R.; Hasty, J.; Wargo, J.A.; Knight, R. The microbiome and human cancer. *Science* **2021**, *371*, eabc4552. [CrossRef]
88. Knippel, R.J.; Drewes, J.L.; Sears, C.L. The Cancer Microbiome: Recent Highlights and Knowledge Gaps. *Cancer Discov.* **2021**, *11*, 2378–2395. [CrossRef]
89. Rossi, T.; Vergara, D.; Fanini, F.; Maffia, M.; Bravaccini, S.; Pirini, F. Microbiota-Derived Metabolites in Tumor Progression and Metastasis. *Int. J. Mol. Sci.* **2020**, *21*, 5786. [CrossRef]
90. Overacre-Delgoffe, A.E.; Bumgarner, H.J.; Cillo, A.R.; Burr, A.H.P.; Tometich, J.T.; Bhattacharjee, A.; Bruno, T.C.; Vignali, D.A.A.; Hand, T.W. Microbiota-specific T follicular helper cells drive tertiary lymphoid structures and anti-tumor immunity against colorectal cancer. *Immunity* **2021**, *54*, 2812–2824.e2814. [CrossRef]
91. Lathrop, S.K.; Bloom, S.M.; Rao, S.M.; Nutsch, K.; Lio, C.W.; Santacruz, N.; Peterson, D.A.; Stappenbeck, T.S.; Hsieh, C.S. Peripheral education of the immune system by colonic commensal microbiota. *Nature* **2011**, *478*, 250–254. [CrossRef] [PubMed]
92. Rubinstein, M.R.; Wang, X.; Liu, W.; Hao, Y.; Cai, G.; Han, Y.W. Fusobacterium nucleatum promotes colorectal carcinogenesis by modulating E-cadherin/beta-catenin signaling via its FadA adhesin. *Cell Host Microbe* **2013**, *14*, 195–206. [CrossRef] [PubMed]
93. Arthur, J.C.; Perez-Chanona, E.; Muhlbauer, M.; Tomkovich, S.; Uronis, J.M.; Fan, T.J.; Campbell, B.J.; Abujamel, T.; Dogan, B.; Rogers, A.B.; et al. Intestinal inflammation targets cancer-inducing activity of the microbiota. *Science* **2012**, *338*, 120–123. [CrossRef]
94. Cougnoux, A.; Dalmasso, G.; Martinez, R.; Buc, E.; Delmas, J.; Gibold, L.; Sauvanet, P.; Darcha, C.; Dechelotte, P.; Bonnet, M.; et al. Bacterial genotoxin colibactin promotes colon tumour growth by inducing a senescence-associated secretory phenotype. *Gut* **2014**, *63*, 1932–1942. [CrossRef] [PubMed]
95. Brennan, C.A.; Garrett, W.S. Fusobacterium nucleatum—Symbiont, opportunist and oncobacterium. *Nat. Rev. Microbiol.* **2019**, *17*, 156–166. [CrossRef] [PubMed]
96. Walker, W.A. Chapter 5—Colorectal cancer and the microbiome: Dysplasia, probiotics, and Fusobacterium nucleatum. In *Colorectal Neoplasia and the Colorectal Microbiome*; Floch, M.H., Ed.; Academic Press: Cambridge, MA, USA, 2020; pp. 79–94.
97. Castellarin, M.; Warren, R.L.; Freeman, J.D.; Dreolini, L.; Krzywinski, M.; Strauss, J.; Barnes, R.; Watson, P.; Allen-Vercoe, E.; Moore, R.A.; et al. Fusobacterium nucleatum infection is prevalent in human colorectal carcinoma. *Genome Res.* **2012**, *22*, 299–306. [CrossRef]
98. Mima, K.; Nishihara, R.; Qian, Z.R.; Cao, Y.; Sukawa, Y.; Nowak, J.A.; Yang, J.; Dou, R.; Masugi, Y.; Song, M.; et al. Fusobacterium nucleatum in colorectal carcinoma tissue and patient prognosis. *Gut* **2016**, *65*, 1973–1980. [CrossRef]
99. Dekker, E.; Tanis, P.J.; Vleugels, J.L.A.; Kasi, P.M.; Wallace, M.B. Colorectal cancer. *Lancet* **2019**, *394*, 1467–1480. [CrossRef]
100. Keller, D.S.; Berho, M.; Perez, R.O.; Wexner, S.D.; Chand, M. The multidisciplinary management of rectal cancer. *Nat. Rev. Gastroenterol. Hepatol.* **2020**, *17*, 414–429. [CrossRef]
101. Chen, Y.; Chen, Y.; Zhang, J.; Cao, P.; Su, W.; Deng, Y.; Zhan, N.; Fu, X.; Huang, Y.; Dong, W. Fusobacterium nucleatum Promotes Metastasis in Colorectal Cancer by Activating Autophagy Signaling via the Upregulation of CARD3 Expression. *Theranostics* **2020**, *10*, 323–339. [CrossRef]
102. Lucas, C.; Barnich, N.; Nguyen, H.T.T. Microbiota, Inflammation and Colorectal Cancer. *Int. J. Mol. Sci.* **2017**, *18*, 1310. [CrossRef] [PubMed]
103. Bonnet, M.; Buc, E.; Sauvanet, P.; Darcha, C.; Dubois, D.; Pereira, B.; Dechelotte, P.; Bonnet, R.; Pezet, D.; Darfeuille-Michaud, A. Colonization of the human gut by E. coli and colorectal cancer risk. *Clin. Cancer Res.* **2014**, *20*, 859–867. [CrossRef] [PubMed]
104. Yu, L.C.; Wei, S.C.; Li, Y.H.; Lin, P.Y.; Chang, X.Y.; Weng, J.P.; Shue, Y.W.; Lai, L.C.; Wang, J.T.; Jeng, Y.M.; et al. Invasive Pathobionts Contribute to Colon Cancer Initiation by Counterbalancing Epithelial Antimicrobial Responses. *Cell Mol. Gastroenterol. Hepatol.* **2022**, *13*, 57–79. [CrossRef] [PubMed]
105. Rebecca, V.W.; Amaravadi, R.K. Emerging strategies to effectively target autophagy in cancer. *Oncogene* **2016**, *35*, 1–11. [CrossRef] [PubMed]
106. Amaravadi, R.K.; Thompson, C.B. The roles of therapy-induced autophagy and necrosis in cancer treatment. *Clin. Cancer Res.* **2007**, *13*, 7271–7279. [CrossRef]

107. Sui, X.; Chen, R.; Wang, Z.; Huang, Z.; Kong, N.; Zhang, M.; Han, W.; Lou, F.; Yang, J.; Zhang, Q.; et al. Autophagy and chemotherapy resistance: A promising therapeutic target for cancer treatment. *Cell Death Dis.* **2013**, *4*, e838. [CrossRef]
108. Wigmore, P.M.; Mustafa, S.; El-Beltagy, M.; Lyons, L.; Umka, J.; Bennett, G. Effects of 5-FU. *Adv. Exp. Med. Biol.* **2010**, *678*, 157–164. [CrossRef]
109. Li, J.; Hou, N.; Faried, A.; Tsutsumi, S.; Kuwano, H. Inhibition of autophagy augments 5-fluorouracil chemotherapy in human colon cancer in vitro and in vivo model. *Eur. J. Cancer* **2010**, *46*, 1900–1909. [CrossRef]
110. Thorn, A.; Steinfeld, R.; Ziegenbein, M.; Grapp, M.; Hsiao, H.H.; Urlaub, H.; Sheldrick, G.M.; Gartner, J.; Kratzner, R. Structure and activity of the only human RNase T2. *Nucleic Acids Res.* **2012**, *40*, 8733–8742. [CrossRef]
111. Ou, J.; Peng, Y.; Yang, W.; Zhang, Y.; Hao, J.; Li, F.; Chen, Y.; Zhao, Y.; Xie, X.; Wu, S.; et al. ABHD5 blunts the sensitivity of colorectal cancer to fluorouracil via promoting autophagic uracil yield. *Nat. Commun.* **2019**, *10*, 1078. [CrossRef]
112. Paillas, S.; Causse, A.; Marzi, L.; de Medina, P.; Poirot, M.; Denis, V.; Vezzio-Vie, N.; Espert, L.; Arzouk, H.; Coquelle, A.; et al. MAPK14/p38alpha confers irinotecan resistance to TP53-defective cells by inducing survival autophagy. *Autophagy* **2012**, *8*, 1098–1112. [CrossRef] [PubMed]
113. Hu, F.; Song, D.; Yan, Y.; Huang, C.; Shen, C.; Lan, J.; Chen, Y.; Liu, A.; Wu, Q.; Sun, L.; et al. IL-6 regulates autophagy and chemotherapy resistance by promoting BECN1 phosphorylation. *Nat. Commun.* **2021**, *12*, 3651. [CrossRef] [PubMed]
114. He, S.; Zhao, Z.; Yang, Y.; O'Connell, D.; Zhang, X.; Oh, S.; Ma, B.; Lee, J.H.; Zhang, T.; Varghese, B.; et al. Truncating mutation in the autophagy gene UVRAG confers oncogenic properties and chemosensitivity in colorectal cancers. *Nat. Commun.* **2015**, *6*, 7839. [CrossRef] [PubMed]
115. Bulut-Karslioglu, A.; Biechele, S.; Jin, H.; Macrae, T.A.; Hejna, M.; Gertsenstein, M.; Song, J.S.; Ramalho-Santos, M. Inhibition of mTOR induces a paused pluripotent state. *Nature* **2016**, *540*, 119–123. [CrossRef]
116. Zhang, Z.; Qin, S.; Chen, Y.; Zhou, L.; Yang, M.; Tang, Y.; Zuo, J.; Zhang, J.; Mizokami, A.; Nice, E.C.; et al. Inhibition of NPC1L1 disrupts adaptive responses of drug-tolerant persister cells to chemotherapy. *EMBO Mol. Med.* **2022**, *14*, e14903. [CrossRef]
117. Liau, B.B.; Sievers, C.; Donohue, L.K.; Gillespie, S.M.; Flavahan, W.A.; Miller, T.E.; Venteicher, A.S.; Hebert, C.H.; Carey, C.D.; Rodig, S.J.; et al. Adaptive Chromatin Remodeling Drives Glioblastoma Stem Cell Plasticity and Drug Tolerance. *Cell Stem. Cell* **2017**, *20*, 233–246.e237. [CrossRef]
118. Sharma, S.V.; Lee, D.Y.; Li, B.; Quinlan, M.P.; Takahashi, F.; Maheswaran, S.; McDermott, U.; Azizian, N.; Zou, L.; Fischbach, M.A.; et al. A chromatin-mediated reversible drug-tolerant state in cancer cell subpopulations. *Cell* **2010**, *141*, 69–80. [CrossRef]
119. Deng, L.; Li, C.; Chen, L.; Liu, Y.; Hou, R.; Zhou, X. Research advances on embryonic diapause in mammals. *Anim. Reprod. Sci.* **2018**, *198*, 1–10. [CrossRef]
120. Rehman, S.K.; Haynes, J.; Collignon, E.; Brown, K.R.; Wang, Y.; Nixon, A.M.L.; Bruce, J.P.; Wintersinger, J.A.; Singh Mer, A.; Lo, E.B.L.; et al. Colorectal Cancer Cells Enter a Diapause-like DTP State to Survive Chemotherapy. *Cell* **2021**, *184*, 226–242.e221. [CrossRef]
121. Guinney, J.; Dienstmann, R.; Wang, X.; de Reynies, A.; Schlicker, A.; Soneson, C.; Marisa, L.; Roepman, P.; Nyamundanda, G.; Angelino, P.; et al. The consensus molecular subtypes of colorectal cancer. *Nat. Med.* **2015**, *21*, 1350–1356. [CrossRef]
122. Bokemeyer, C.; Bondarenko, I.; Makhson, A.; Hartmann, J.T.; Aparicio, J.; de Braud, F.; Donea, S.; Ludwig, H.; Schuch, G.; Stroh, C.; et al. Fluorouracil, leucovorin, and oxaliplatin with and without cetuximab in the first-line treatment of metastatic colorectal cancer. *J. Clin. Oncol.* **2009**, *27*, 663–671. [CrossRef] [PubMed]
123. Van Cutsem, E.; Kohne, C.H.; Hitre, E.; Zaluski, J.; Chang Chien, C.R.; Makhson, A.; D'Haens, G.; Pinter, T.; Lim, R.; Bodoky, G.; et al. Cetuximab and chemotherapy as initial treatment for metastatic colorectal cancer. *N. Engl. J. Med.* **2009**, *360*, 1408–1417. [CrossRef] [PubMed]
124. Bokemeyer, C.; Van Cutsem, E.; Rougier, P.; Ciardiello, F.; Heeger, S.; Schlichting, M.; Celik, I.; Kohne, C.H. Addition of cetuximab to chemotherapy as first-line treatment for KRAS wild-type metastatic colorectal cancer: Pooled analysis of the CRYSTAL and OPUS randomised clinical trials. *Eur. J. Cancer* **2012**, *48*, 1466–1475. [CrossRef]
125. Tong, W.M.; Ellinger, A.; Sheinin, Y.; Cross, H.S. Epidermal growth factor receptor expression in primary cultured human colorectal carcinoma cells. *Br. J. Cancer* **1998**, *77*, 1792–1798. [CrossRef] [PubMed]
126. Tan, X.; Lambert, P.F.; Rapraeger, A.C.; Anderson, R.A. Stress-Induced EGFR Trafficking: Mechanisms, Functions, and Therapeutic Implications. *Trends Cell Biol.* **2016**, *26*, 352–366. [CrossRef]
127. Li, X.; Fan, Z. The epidermal growth factor receptor antibody cetuximab induces autophagy in cancer cells by downregulating HIF-1alpha and Bcl-2 and activating the beclin 1/hVps34 complex. *Cancer Res.* **2010**, *70*, 5942–5952. [CrossRef]
128. Chen, Z.; Gao, S.; Wang, D.; Song, D.; Feng, Y. Colorectal cancer cells are resistant to anti-EGFR monoclonal antibody through adapted autophagy. *Am. J. Transl Res.* **2016**, *8*, 1190–1196.
129. Roth, A.D.; Tejpar, S.; Delorenzi, M.; Yan, P.; Fiocca, R.; Klingbiel, D.; Dietrich, D.; Biesmans, B.; Bodoky, G.; Barone, C.; et al. Prognostic role of KRAS and BRAF in stage II and III resected colon cancer: Results of the translational study on the PETACC-3, EORTC 40993, SAKK 60-00 trial. *J. Clin. Oncol.* **2010**, *28*, 466–474. [CrossRef]
130. Kopetz, S.; Grothey, A.; Yaeger, R.; Van Cutsem, E.; Desai, J.; Yoshino, T.; Wasan, H.; Ciardiello, F.; Loupakis, F.; Hong, Y.S.; et al. Encorafenib, Binimetinib, and Cetuximab in BRAF V600E-Mutated Colorectal Cancer. *N. Engl. J. Med.* **2019**, *381*, 1632–1643. [CrossRef]

131. Kinsey, C.G.; Camolotto, S.A.; Boespflug, A.M.; Guillen, K.P.; Foth, M.; Truong, A.; Schuman, S.S.; Shea, J.E.; Seipp, M.T.; Yap, J.T.; et al. Protective autophagy elicited by RAF–>MEK–>ERK inhibition suggests a treatment strategy for RAS-driven cancers. *Nat. Med.* **2019**, *25*, 620–627. [CrossRef]
132. Baraniskin, A.; Buchberger, B.; Pox, C.; Graeven, U.; Holch, J.W.; Schmiegel, W.; Heinemann, V. Efficacy of bevacizumab in first-line treatment of metastatic colorectal cancer: A systematic review and meta-analysis. *Eur. J. Cancer* **2019**, *106*, 37–44. [CrossRef] [PubMed]
133. Selvakumaran, M.; Amaravadi, R.K.; Vasilevskaya, I.A.; O'Dwyer, P.J. Autophagy inhibition sensitizes colon cancer cells to antiangiogenic and cytotoxic therapy. *Clin. Cancer Res.* **2013**, *19*, 2995–3007. [CrossRef] [PubMed]
134. Piette, J.; Volanti, C.; Vantieghem, A.; Matroule, J.Y.; Habraken, Y.; Agostinis, P. Cell death and growth arrest in response to photodynamic therapy with membrane-bound photosensitizers. *Biochem. Pharmacol.* **2003**, *66*, 1651–1659. [CrossRef]
135. Wei, M.F.; Chen, M.W.; Chen, K.C.; Lou, P.J.; Lin, S.Y.; Hung, S.C.; Hsiao, M.; Yao, C.J.; Shieh, M.J. Autophagy promotes resistance to photodynamic therapy-induced apoptosis selectively in colorectal cancer stem-like cells. *Autophagy* **2014**, *10*, 1179–1192. [CrossRef] [PubMed]
136. Galluzzi, L.; Buque, A.; Kepp, O.; Zitvogel, L.; Kroemer, G. Immunological Effects of Conventional Chemotherapy and Targeted Anticancer Agents. *Cancer Cell* **2015**, *28*, 690–714. [CrossRef]
137. Michaud, M.; Martins, I.; Sukkurwala, A.Q.; Adjemian, S.; Ma, Y.; Pellegatti, P.; Shen, S.; Kepp, O.; Scoazec, M.; Mignot, G.; et al. Autophagy-dependent anticancer immune responses induced by chemotherapeutic agents in mice. *Science* **2011**, *334*, 1573–1577. [CrossRef]
138. Ramakrishnan, R.; Huang, C.; Cho, H.I.; Lloyd, M.; Johnson, J.; Ren, X.; Altiok, S.; Sullivan, D.; Weber, J.; Celis, E.; et al. Autophagy induced by conventional chemotherapy mediates tumor cell sensitivity to immunotherapy. *Cancer Res.* **2012**, *72*, 5483–5493. [CrossRef]
139. Kim, S.; Ramakrishnan, R.; Lavilla-Alonso, S.; Chinnaiyan, P.; Rao, N.; Fowler, E.; Heine, J.; Gabrilovich, D.I. Radiation-induced autophagy potentiates immunotherapy of cancer via up-regulation of mannose 6-phosphate receptor on tumor cells in mice. *Cancer Immunol. Immunother.* **2014**, *63*, 1009–1021. [CrossRef]
140. Robert, C.; Long, G.V.; Brady, B.; Dutriaux, C.; Maio, M.; Mortier, L.; Hassel, J.C.; Rutkowski, P.; McNeil, C.; Kalinka-Warzocha, E.; et al. Nivolumab in previously untreated melanoma without BRAF mutation. *N. Engl. J. Med.* **2015**, *372*, 320–330. [CrossRef]
141. Janji, B.; Hasmim, M.; Parpal, S.; De Milito, A.; Berchem, G.; Noman, M.Z. Lighting up the fire in cold tumors to improve cancer immunotherapy by blocking the activity of the autophagy-related protein PIK3C3/VPS34. *Autophagy* **2020**, *16*, 2110–2111. [CrossRef]
142. Noman, M.Z.; Parpal, S.; Van Moer, K.; Xiao, M.; Yu, Y.; Viklund, J.; De Milito, A.; Hasmim, M.; Andersson, M.; Amaravadi, R.K.; et al. Inhibition of Vps34 reprograms cold into hot inflamed tumors and improves anti-PD-1/PD-L1 immunotherapy. *Sci. Adv.* **2020**, *6*, eaax7881. [CrossRef] [PubMed]
143. Starobinets, H.; Ye, J.; Broz, M.; Barry, K.; Goldsmith, J.; Marsh, T.; Rostker, F.; Krummel, M.; Debnath, J. Antitumor adaptive immunity remains intact following inhibition of autophagy and antimalarial treatment. *J. Clin. Investig.* **2016**, *126*, 4417–4429. [CrossRef] [PubMed]
144. Noman, M.Z.; Janji, B.; Kaminska, B.; Van Moer, K.; Pierson, S.; Przanowski, P.; Buart, S.; Berchem, G.; Romero, P.; Mami-Chouaib, F.; et al. Blocking hypoxia-induced autophagy in tumors restores cytotoxic T-cell activity and promotes regression. *Cancer Res.* **2011**, *71*, 5976–5986. [CrossRef] [PubMed]
145. Mgrditchian, T.; Arakelian, T.; Paggetti, J.; Noman, M.Z.; Viry, E.; Moussay, E.; Van Moer, K.; Kreis, S.; Guerin, C.; Buart, S.; et al. Targeting autophagy inhibits melanoma growth by enhancing NK cells infiltration in a CCL5-dependent manner. *Proc. Natl. Acad. Sci. USA* **2017**, *114*, E9271–E9279. [CrossRef] [PubMed]
146. Yang, A.; Herter-Sprie, G.; Zhang, H.; Lin, E.Y.; Biancur, D.; Wang, X.; Deng, J.; Hai, J.; Yang, S.; Wong, K.K.; et al. Autophagy Sustains Pancreatic Cancer Growth through Both Cell-Autonomous and Nonautonomous Mechanisms. *Cancer Discov.* **2018**, *8*, 276–287. [CrossRef]
147. Wei, H.; Guan, J.L. Blocking tumor growth by targeting autophagy and SQSTM1 in vivo. *Autophagy* **2015**, *11*, 854–855. [CrossRef]
148. Levy, J.M.M.; Towers, C.G.; Thorburn, A. Targeting autophagy in cancer. *Nat. Rev. Cancer* **2017**, *17*, 528–542. [CrossRef]
149. Zhang, Z.; Shi, J.; Nice, E.C.; Huang, C.; Shi, Z. The Multifaceted Role of Flavonoids in Cancer Therapy: Leveraging Autophagy with a Double-Edged Sword. *Antioxidants* **2021**, *10*, 1138. [CrossRef]
150. Li, Z.; Si, W.; Jin, W.; Yuan, Z.; Chen, Y.; Fu, L. Targeting autophagy in colorectal cancer: An update on pharmacological small-molecule compounds. *Drug Discov. Today* **2022**, *27*, 2373–2385. [CrossRef]
151. Galluzzi, L.; Bravo-San Pedro, J.M.; Demaria, S.; Formenti, S.C.; Kroemer, G. Activating autophagy to potentiate immunogenic chemotherapy and radiation therapy. *Nat. Rev. Clin. Oncol.* **2017**, *14*, 247–258. [CrossRef]
152. Gewirtz, D.A. The Challenge of Developing Autophagy Inhibition as a Therapeutic Strategy. *Cancer Res.* **2016**, *76*, 5610–5614. [CrossRef] [PubMed]

antioxidants

MDPI

Brief Report

Melatonin Ameliorates Autophagy Impairment in a Metabolic Syndrome Model

Adrián Santos-Ledo [1], Beatriz de Luxán-Delgado [2], Beatriz Caballero [3,4], Yaiza Potes [3,4], Susana Rodríguez-González [3,4], José Antonio Boga [5], Ana Coto-Montes [3,4,*] and Marina García-Macia [6,7,8,*]

1 Institute of Neurosciences of Castilla y León-INCYL, Institute of Biomedical Research of Salamanca-IBSAL, Cell Biology and Pathology, University of Salamanca, 37007 Salamanca, Spain; santosledo@usal.es
2 Centre for Tumour Biology, Barts Cancer Institute-Queen Mary, University of London, John Vane Science Centre, Charterhouse Square, London EC1M 6BQ, UK; b.luxandelgado@gmail.com
3 Departamento de Morfología y Biología Celular, Área de Biología Celular, Facultad de Medicina, Universidad de Oviedo, Julián Clavería s/n, 33006 Oviedo, Spain; caballerobeatriz@uniovi.es (B.C.); potesyaiza@uniovi.es (Y.P.); gusatan69@gmail.com (S.R.-G.)
4 Instituto de Investigación Sanitaria del Principado de Asturias (ISPA), Av. del Hospital Universitario, s/n, 33011 Oviedo, Spain
5 Servicio de Microbiología, Hospital Universitario Central de Asturias, Celestino Villamil s/n, 33006 Oviedo, Spain; joseantonio.boga@sespa.es
6 Institute of Biomedical Research of Salamanca, University Hospital of Salamanca, 37007 Salamanca, Spain
7 Institute of Functional Biology and Genomics, University of Salamanca, CSIC, 37007 Salamanca, Spain
8 Centro de Investigación Biomédica en Red Sobre Fragilidad y Envejecimiento Saludable (CIBERFES), Instituto de Salud Carlos III, 28029 Madrid, Spain
* Correspondence: acoto@uniovi.es (A.C.-M.); marinagarciamacia@usal.es (M.G.-M.); Tel.: +34-923-294-907 (M.G.-M.)

Citation: Santos-Ledo, A.; Luxán-Delgado, B.d.; Caballero, B.; Potes, Y.; Rodríguez-González, S.; Boga, J.A.; Coto-Montes, A.; García-Macia, M. Melatonin Ameliorates Autophagy Impairment in a Metabolic Syndrome Model. *Antioxidants* **2021**, *10*, 796. https://doi.org/10.3390/antiox10050796

Academic Editors: Angeles Aroca and Cecilia Gotor

Received: 22 April 2021
Accepted: 15 May 2021
Published: 18 May 2021

Abstract: Metabolic syndrome is a global health problem in adults and its prevalence among children and adolescents is rising. It is strongly linked to a lifestyle with high-caloric food, which causes obesity and lipid metabolism anomalies. Molecular damage due to excessive oxidative stress plays a major role during the development of metabolic syndrome complications. Among the different hormones, melatonin presents strong antioxidant properties, and it is used to treat metabolic diseases. However, there is not a consensus about its use as a metabolic syndrome treatment. The aim of this study was to identify melatonin effects in a metabolic syndrome model. Golden hamsters were fed with 60% fructose-enriched food to induce metabolic syndrome and were compared to hamsters fed with regular chow diet. Both groups were also treated with melatonin. Fructose-fed hamsters showed altered blood lipid levels (increased cholesterol and LDL) and phenotypes restored with the melatonin treatment. The Harderian gland (HG), which is an ideal model to study autophagy modulation through oxidative stress, was the organ that was most affected by a fructose diet. Redox balance was altered in fructose-fed HG, inducing autophagic activation. However, since LC3-II was not increased, the impairment must be in the last steps of autophagy. Lipophagy HG markers were also disturbed, contributing to the dyslipidemia. Melatonin treatment improved possible oxidative homeostasis through autophagic induction. All these results point to melatonin as a possible treatment of the metabolic syndrome.

Keywords: melatonin; metabolic syndrome; autophagy; lipophagy; Harderian gland

1. Introduction

Current changes in lifestyle and eating behavior are increasing the prevalence of overweight and obesity, reaching the status of pandemic. These changes include sedentariness and an elevated consumption of high-calorie food and sugary beverages. Although obesity is a known risk factor for metabolic defects, associated problems like insulin resistance and diabetes also affect normal weight people. Metabolic syndrome (MetS), cluster abnormalities including abdominal obesity, insulin resistance, and dyslipidemia, increased blood pressure, pro-inflammatory states, and risk of cerebrovascular accidents [1]. MetS is also characterized by high levels of oxidative stress, which might play a fundamental role in its progression [2], since the accumulation of free radicals is a harmful process that can damage several cellular structures. This state, together with reduced antioxidant defenses [3], suggests that an oxidative imbalance might be very relevant to MetS.

The pineal melatonin is the capital endogenous synchronizer of the circadian rhythms in many organisms [4]. Likewise, melatonin as well as its metabolites, are well-known endogenous antioxidants [5] and they exert anti-inflammatory, antihyperlipidemic, and anti-hypertensive actions and modulates insulin secretion and action [6], mainly fostering the antioxidant system. Melatonin treatment ameliorates metabolic changes associated with obesity in rats fed a high-fat diet [7] or metabolic changes and hypertension associated with obesity in young Zucker diabetic fatty rats [1]. Melatonin was even suggested as a treatment for viral infections like Sars-Cov2 [8]. However, it is still unclear whether melatonin could ameliorate the pathological phenotype during MetS, induced by high-fructose intake in hamsters, which might better emulate the harmful effects of prepared foods and carbonated beverages [9].

The Syrian hamster Harderian gland (HG) is a tubule-alveolar orbital gland secreting lipid that lubricate the cornea [10]. This small gland presents many more functions including the production of pheromones [11], the participation in a pineal–gonadal axis [12], the synthesis of indolamines, as melatonin [13], and an important porphyrin production (which might regulate melatonin production [14]) that is stronger than that in the liver [15]. Moreover, due to the localization of the HG, porphyrins exposed to light produce reactive oxygen species (ROS) through photo-oxidation. Therefore, the gland is an excellent model to comprehend physiological oxidative stress and its control [16]. In previous reports we showed that damage caused by oxidative stress forces the gland to trigger autophagic processes to maintain vital functions, adapting to environmental stress [17]. We wondered if a similar scenario could occur after inducing MetS, and whether melatonin might present protective functions.

Our aim was to identify melatonin effects in a metabolic syndrome model. Particularly, the melatonin role in the autophagic response to oxidative stress. Here, we showed that melatonin ameliorates the fructose-diet induced dyslipidemia. Furthermore, melatonin treatment partially restored impaired autophagy in the most affected organ through a fructose diet.

2. Material and Methods

2.1. Animals

Eight-week-old male Syrian hamsters (*Mesocricetus auratus*) (Harlan Interfauna Ibérica, Barcelona, Spain) were housed 2 per cage, during long days, with a 14:10 light:dark cycle (lights on daily from 07:00 to 21:00) at 22 ± 2 °C (n = 6 per experimental condition). Animals received water and a standard pellet diet, ad libitum. The Oviedo University Local Animal Care and Use Committee approved the experimental protocols. All experiments were carried out according to the Spanish Government Guide and the European Community Guide for Animal Care (Council Directive 86/609/EEC). After 1 month in the animal house, the hamsters were fed either a high-fructose diet (n = 12) (TD.89247, Harlan Interfauna Ibérica, Barcelona, Spain) or a regular chow diet (n = 12) for 4 weeks, following the Kasim-Karakas' protocol to induce Metabolic Syndrome (MetS) [18]. Then, six hamsters of each diet were treated daily with 25 µg of melatonin for 15 days. Melatonin was dissolved in ethanol (final ethanol concentration, 0.5%) and injected subcutaneously (SQ) at Zeitgeber time (ZT) 10 (ZT 0 = onset of light). The controls received the same volume of saline (0.5% final ethanol concentration) under similar conditions, six animals for fifteen days. Weight and food intake was monitored weekly. Blood extraction was performed 1 day before harvest and the blood parameters (glucose, triglycerides, total cholesterol, HDL cholesterol and LDL cholesterol) were analyzed by routine laboratory tests at the Laboratory of Veterinary Analysis of Dr. Barba (Madrid, Spain). Hamsters were sacrificed and the

Harderian glands were immediately removed, weighted, frozen in liquid nitrogen, and stored at −80 °C, until the experiments were performed. The other organs (brain, liver, muscle, heart, and white adipose tissue) were removed, weighted, and stored at −80 °C.

2.2. Isolation of Proteins

HGs (0.1 g) were homogenized using a Polytron homogenizer at 4 °C in 1 mL of lysis buffer (50 mM Tris/HCl, 150 mM NaCl at pH 7.4). The tissue homogenates were then centrifuged for 6 min at 3000 rpm at 4 °C. The supernatants were collected and centrifuged again under the same conditions. The protein concentration of the supernatants was measured by the method of Bradford [19].

2.3. Lipid Peroxidation

A lipid peroxidation kit from Calbiochem (437634, Calbiochem, EMD Biosciences Inc., San Diego, CA, USA) was used to measure the amount of malondialdehyde (MDA) and 4-hydroxy-2(E)-nonenal (4-HNE) as an index of the oxidative destruction of lipids. Data are presented as nmol (MDA+4-HNE) per mg protein.

2.4. Total Antioxidant Activity (TAA)

TAA was determined using the ABTS/H_2O_2/HRP method modified for tissue samples [20,21]. The results are expressed as equivalents of mg Trolox/mg protein.

2.5. Immunoblotting

The protein samples (100 μg) were prepared in Western blotting sample buffer (65.8 mM Tris-HCl, pH 6.8, 2.1% SDS, 26.3% (*w*/*v*) glycerol, 0.01% Bromophenol Blue). The 12% SDS-polyacrylamide gels were run and analyzed, as previously described [21–23]. Primary antibodies applied were—Beclin 1, perilipin (PLIN), p-mTOR, and mTOR from Santa Cruz Biotechnology (Santa Cruz, CA, USA), sequestosome-1 (SQSTM1/p62) from Cell Signaling Technology (Boston, MA, USA), Lysosomal acid lipase (LAL) antibody from Abcam (Cambridge, UK), and LC3 from MBL (Naka-ku Nagoya, Japan). Primary antibodies were diluted 1:1000 in blocking buffer. Goat anti-human β-actin antibody (Santa Cruz Biotechnology, Inc.) diluted at 1:1000 was always assayed as a loading reference. After washing in TBS-T (20 mM Tris-HCl, 150 mM NaCl, pH 7.4 and 0.05% Tween-20), the membranes were then incubated with the corresponding horseradish peroxidase-conjugated secondary antibody (Santa Cruz Biotechnology, Inc.) diluted at 1:2500. Binding of antibodies to their antigens was detected using the Western Blotting Luminol Reagent (sc-2048; Santa Cruz Biotechnology, Inc.), according to the manufacturer's protocol. The results were calculated from at least three separate experiments for each antibody and were normalized to actin. Band intensity was quantified using the Quantity One 1D-analyzes software v. 5.5.1. (Bio-Rad Laboratories Inc., Hercules, CA, USA).

2.6. Statistical Analysis

Data are represented as mean ± standard error of mean (SEM). The Graphpad prism was used to perform ANOVA with a Tukey's post-hoc test. * $p < 0.05$, ** $p < 0.01$, or *** $p < 0.001$ and # $p < 0.05$, ## $p < 0.01$, and ### $p < 0.001$ was considered statistically significant.

3. Results and Discussion

3.1. Melatonin Treatment Ameliorates the Lipid Effect Caused by a Fructose Diet

Syrian hamsters have a reproducible response to dietary manipulation [24]. Their lipid metabolism closely resembles that of humans, and unlike mice and rats, hamsters show cholesterol ester transport protein activity [25]. Fructose is an important nutritional factor in the development of the Metabolic Syndrome (MetS) in humans [2] and high fructose intake in hamsters also leads to the development of MetS-associated symptoms [18]. Syrian hamsters were fed with 60% fructose enriched food for 4 weeks, while control hamsters were fed with a regular chow diet. Then, half of the hamsters of each diet were injected with SQ melatonin for 15 days (25 μg of melatonin per hamster) [22]. Before sacrifice, blood was extracted, and biochemical determinations were performed. Glucose levels showed no significant differences between diets (Figure 1A). As hamster were not fasted before the blood extraction, changes were not expected either [3]. Fructose is a highly lipogenic sugar that is already associated with dyslipidemic symptoms induced by MetS [26]. Accordingly, triglycerides were increased with the fructose diet (Figure 1B, $p < 0.05$) and restored with the melatonin treatment (Figure 1B, $p < 0.05$). Total cholesterol was also increased (Figure 1C, $p < 0.01$), as well as LDL particles (Figure 1E, $p < 0.05$), whilst there was no significant compensation with the HDL cholesterol (Figure 1D). Melatonin restored total and LDL cholesterol levels (Figure 1C,E, $p < 0.05$). These results demonstrates that melatonin improves dyslipidemia in fructose-fed hamsters more efficiently than melatonin treatments in MetS' patients [3]. This might be due to a higher dose in hamsters than in humans [3]. Melatonin's role as a hypolipidemic is related to a decrease in intestinal cholesterol absorption [27] or inhibition of cholesterol biosynthesis and LDL-C accumulation [28] in rats. However, further studies need to be performed to decipher melatonin's hypolipidemic role in hamsters and humans.

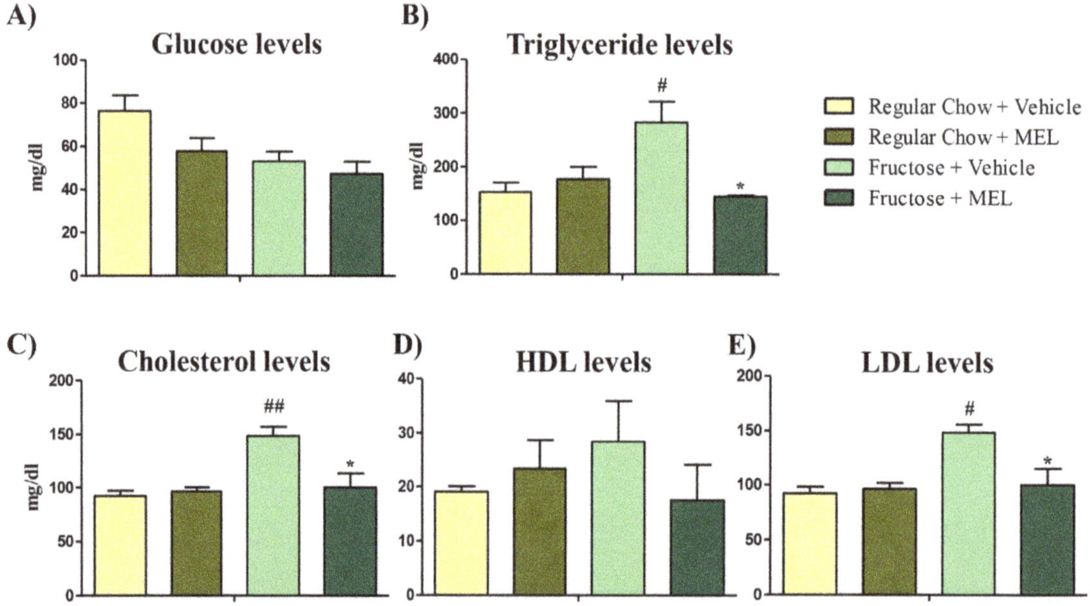

Figure 1. Fructose diet produces an imbalance in serum lipid markers that is ameliorated by melatonin. (**A**) Glucose levels in serum were measured in mg/dl from hamsters fed with regular chow or fructose with or without melatonin treatment (25 μg of melatonin for 15 days). (**B**) Triglycerides, (**C**) Cholesterol, (**D**) HDL, and (**E**) LDL levels in serum were measured in mg/dl from hamsters fed with regular chow or fructose with or without melatonin treatment. Bars are mean ± SEM. * $p < 0.05$ (differences caused by melatonin treatment) and # $p < 0.05$, ## $p < 0.001$ (differences caused by diet).

3.2. The Harderian Gland Is the Organ Most Affected by a Fructose Diet

Weight and food intake was monitored during the experiment. Fructose-fed hamster weight increase tended to be higher (Figure 2A), but they were eating less (Figure 2B, $p < 0.05$), which underlined a metabolic unbalance [29]. At the end of the experiment, experimental groups did not show significant weight differences (Figure 2C). Reports showed that excessive fructose intake induced features of MetS in humans with and without obesity [6], which agreed with our results—fructose-fed hamster showed MetS symptoms like dyslipidemia, but not overweight. Then, we evaluated the different organ weights—brain, liver, muscle, heart, and white adipose tissue (WAT) showed no significative differences in any condition (Figure 2D). Nevertheless, the Harderian gland's (HG) weight increased in the fructose-fed hamster (Figure 2E, $p < 0.05$) and this increase was suppressed with the melatonin treatment (Figure 2E, $p < 0.05$). HG cells accumulated fats in lipid droplets, which were active organelles in this organ [30]. Thus, HG weight increase could be due to an excessive lipid accumulation, impaired lipid metabolism, or defective lipophagy.

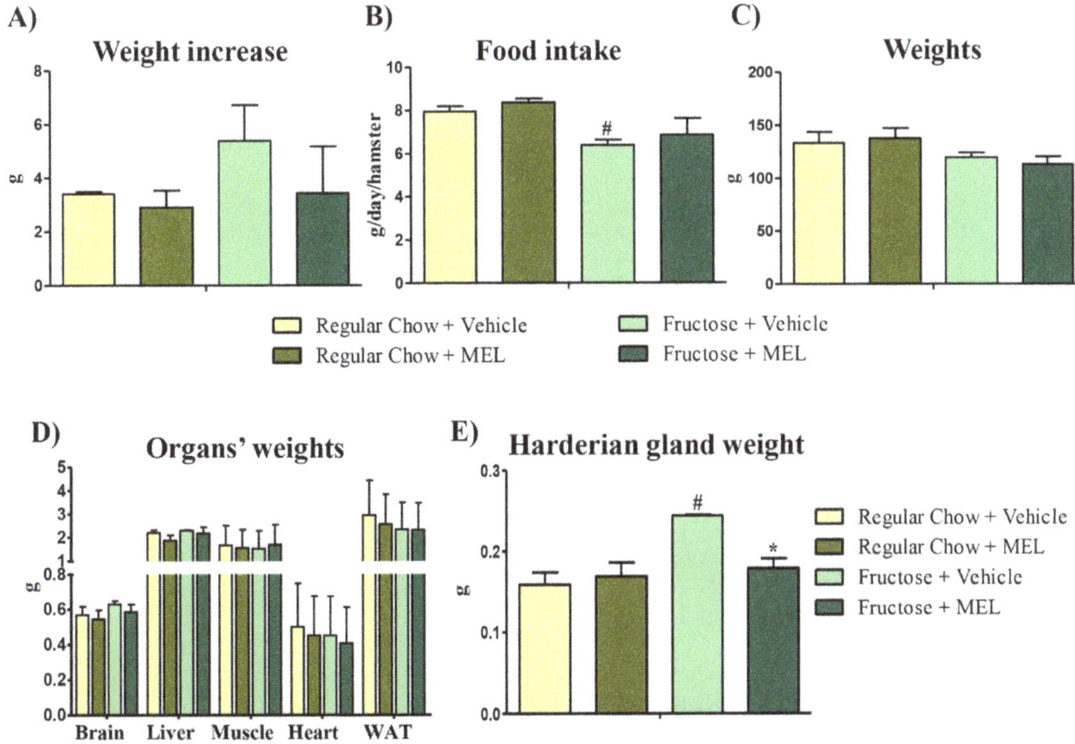

Figure 2. The Harderian gland is the organ most affected by a fructose diet. (**A**) Weight increase in grams of the hamsters fed with regular chow or fructose with or without melatonin treatment (25 µg of melatonin for 15 days). (**B**) Food intake measure in grams per day and hamster from all the above conditions. (**C**) Total weights in grams of all groups. (**D**) Different organs and (**E**) the Harderian gland weights in grams, from all the above conditions. Bars are mean ± SEM. * $p < 0.05$ (differences caused by melatonin treatment) and # $p < 0.05$ (differences caused by diet).

Dyslipidemia and increased lipid oxidation are symptoms of an early MetS [31]. The HG is a well-established model to study oxidative stress [22], and it showed the disease's earliest hallmarks, with respect to other organs. Thus, HG is a good target to study both early stages of MetS and melatonin as a putative therapy. Our results indicated that this antioxidant rescued the weight changes and the dyslipidemia caused by fructose. We then decided to delve into the connections between MetS, oxidative stress, and defects in lipid metabolism and autophagy, using HG as a model.

3.3. Melatonin Ameliorates Autophagic and Lipophagic Impairment in the Harderian Gland Caused by the Metabolic Syndrome.

Our previous results showed that HG was affected the most after fructose feeding. Early MetS is characterized by increased systemic markers of lipid oxidation [6]. Accordingly, the HG from fructose fed hamsters showed the highest levels of oxidated lipids (Figure 3A, $p < 0.01$). Melatonin treatment restored the lipoperoxidation levels (Figure 3A,

$p < 0.01$). Prolonged state of oxidative stress results in reduction of antioxidative enzyme activities [3]. We also observed this effect in animals fed with fructose (Figure 3B, $p < 0.05$), a phenotype once again restored by a melatonin injection (Figure 3B, $p < 0.01$). Interestingly, melatonin had no effect in the regular chow-fed hamsters (Figure 3B).

Autophagy is key to eliminate damaged proteins, lipids, and organelles, favoring cellular survivability under oxidative stress environments or cellular damage in the HG [16,32], and in many other contexts [33,34]. We then wondered whether autophagy was also involved in this MetS scenario. HG from the fructose-fed hamster showed significant raised levels of the autophagy inhibitor mTOR (Figure 3C, $p < 0.05$), evaluated by the ratio of p-mTOR and mTOR expression (representative Westerns showed with their quantification). In a highly-nutrient state, mTOR is activated [35], which would fit with the fructose diet nutrient overload. Melatonin's role in the mTOR pathway was not clarified and both activation and inhibition of the mTOR pathway were described in different contexts [36,37]. Melatonin treatment increased the mTOR activity in the regular chow-fed hamsters (Figure 3C, $p < 0.01$), which might be fostering the mTOR role in storing nutrients [38]. However, in the fructose-fed hamsters displaying an inflammatory environment caused by impaired lipid handling, melatonin treatment inhibited mTOR activity (Figure 3C, $p < 0.001$). This latter result would suggest that melatonin stimulates protective mechanisms when present in a toxic scenario [37]. We also found an increased expression in the autophagy inductor Beclin 1 produced by a fructose diet (Figure 3C, $p < 0.001$) and a further increase after melatonin treatment in both diets (Figure 3C, $p < 0.001$). LC3-II, a marker of autophagosomes, was also elevated (Figure 3C, $p < 0.05$). Likely, fructose diet induces an oxidative and inflammatory state that triggers autophagosome formation as cellular survival response, as previously described (15, 34). However, activated mTOR would keep autophagy inhibited [39]. LC3-II/LC3-I ratio was also measured to better understand the autophagic dynamics. We found that this ratio was increased in the fructose-fed glands when they were treated with melatonin (Figure 3C, $p < 0.05$). Conversely, and according to our experiments, melatonin treatment in the fructose-fed hamster triggers autophagosome formation (LC3-II and Beclin 1 are increased) and at the same time mTOR activity is reduced, which might allow a more efficient autophagy [32].

Autophagy is also involved in a plethora of other functions, such as an alternative energy source through a type of selective autophagy called lipophagy [40]. Moreover, fructose is a highly lipogenic sugar [26] that causes lipid metabolism impairment [25]. To decipher the role of this selective pathway, we analyzed the expression of proteins related with lipophagy in the Harderian gland. We previously described that p62, a key autophagic protein that mediates the selective specific degradation of protein aggregates and cytoplasmic bodies [41], is a key regulator of lipophagy in the HG [30]. Fructose diet reduced p62 expression (Figure 3D, $p < 0.001$), a phenotype partially recovered by melatonin treatment, which also increased p62 expression in a regular diet (Figure 3D, $p < 0.01$). Lipophagic processes were studied by assaying the expression of lysosomal acid lipase (LAL), which was reduced in the HG from fructose-fed hamsters (Figure 3D, $p < 0.05$) and was recovered by melatonin treatment (Figure 3D, $p < 0.05$). Finally, a specific lipid droplet marker, perilipin (Plin), was increased in fructose diet, implying more lipid droplets (Figure 3D, $p < 0.001$), which were reduced by melatonin treatment (Figure 3D, $p < 0.001$). Melatonin seemed to induce a different effect, depending on the diet in both LAL and Plin (Figure 3D, $p < 0.001$), which might depend on the availability of the autophagic machinery [40]. However, deep studies, i.e., through immunofluorescence and differential expression of more perilipins and its posttranslational modifications, would be required to understand how melatonin modifies lipophagy activity.

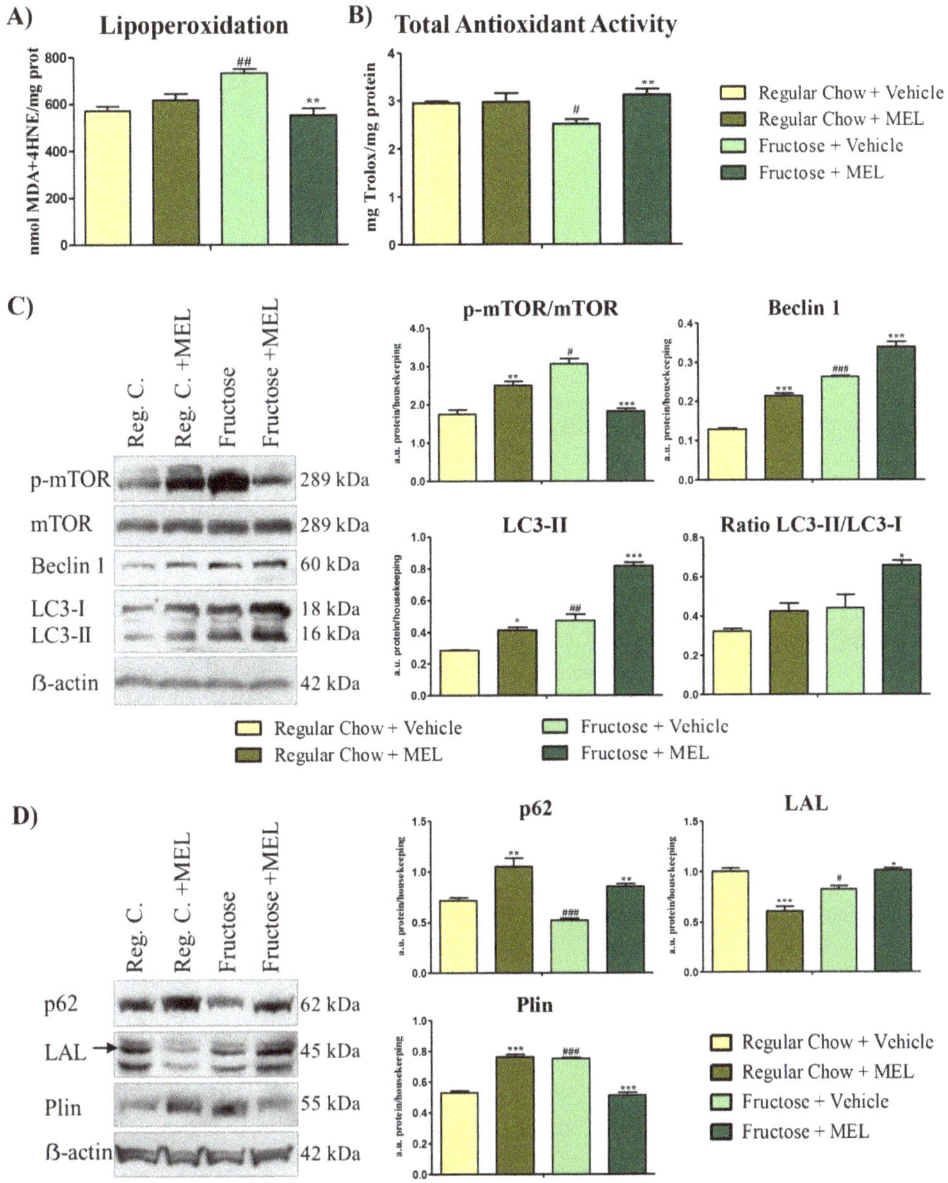

Figure 3. Melatonin ameliorates autophagic and lipophagic impairment in the Harderian gland in animals fed with fructose. (**A**) Harderian gland's lipid peroxidation measured in nmols of 4-HNE+MDA/mg prot from hamsters fed with regular chow or fructose, with or without melatonin treatment (25 µg of melatonin for 15 days). (**B**) Harderian gland's total antioxidant activity measured in mg Trolox/mg protein from hamsters from all the above conditions. (**C**) Autophagy pathway proteins and LC3-II/LC3-I ratio in Harderian gland's homogenates from hamsters from all conditions (quantified in histograms of protein expression/actin expression and mTOR is represent as p-mTOR/mTOR). (**D**) Lipophagy-related proteins expression in Harderian gland's homogenates from hamsters from all conditions (quantified in histograms of protein expression/actin expression). Bars are mean ± SEM. * $p < 0.05$, ** $p < 0.01$, *** $p < 0.001$ (differences caused by melatonin treatment) and # $p < 0.05$, ## $p < 0.01$, ## $p < 0.001$ (differences caused by diet).

According to our results, damage produced in the HG by the fructose diet seem to induce autophagosome formation as an oxidative stress response. However, under nutrient-enrichment, mTOR is strongly activated and autophagy is consequently inhibited. Additionally, selective autophagy markers and lysosome lipase activity are diminished, and lipid droplets are accumulated. Melatonin might have a dual role depending on the cellular situation, when the cells are balanced, the melatonin activates the mTOR, which promotes the nutrient's storage [38] as was observed by the Plin accumulation. However, when homeostasis is broken and oxidative stress levels are high, melatonin seems to activate autophagy and selective autophagy through mTOR inhibition [36], which would have a protective effect (Figure 4).

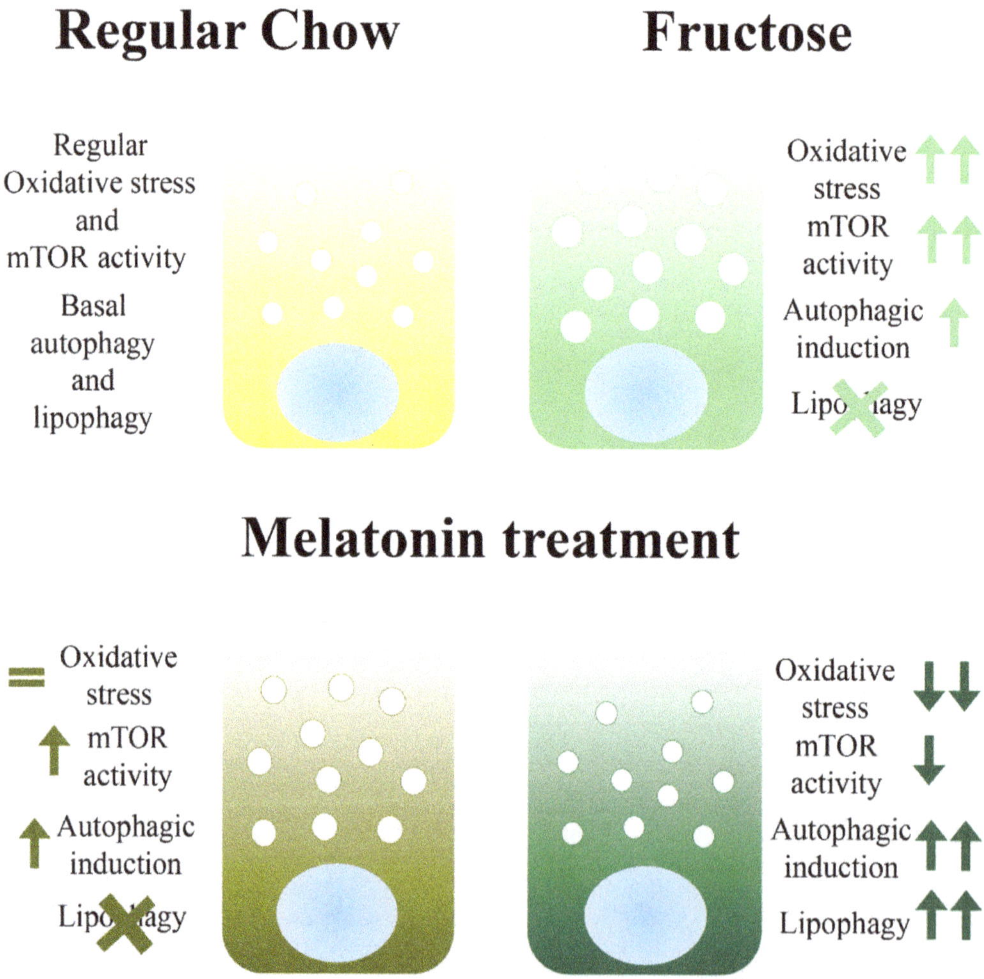

Figure 4. Proposed model—melatonin might have a dual role depending on the cellular situation. When cells are balanced, melatonin activates mTOR that promotes nutrient's storage. When oxidative stress levels are high, melatonin activates autophagy and selective autophagy, through mTOR inhibition, which have a protective effect.

4. Conclusions

The Harderian gland is an ideal model to study early Metabolic Syndrome onset.

Melatonin improves dyslipidemia in a model of Metabolic Syndrome.

Fructose diet activates mTOR and inhibits autophagy in the Harderian glands.

Melatonin seems to activate lipophagy to ameliorate oxidative damage in an early Metabolic Syndrome model.

Author Contributions: M.G.-M., A.S.-L., B.d.L.-D., B.C., Y.P., S.R.-G. performed experiments and analysed the data; M.G.-M. and A.S.-L. performed statistical analysis; M.G.-M., J.A.B. and A.C.-M. designed the study design and supervised the project; M.G.-M., A.S.-L. and A.C.-M. wrote the paper. All authors reviewed the manuscript. All authors have read and agreed to the published version of the manuscript.

Funding: This work was supported by Instituto de Salud Carlos III (ISCIII), Ministerio de Ciencia, and Innovación y Universidades, Spain (FISS-18-PI17/02009); and Newcastle University, United Kingdom (C0120R3166, C0245R4032, and BH182173). M.G.-M. has a postdoctoral fellowship from the Instituto de Salud Carlos III (ISCIII), Ministerio de Ciencia, and Innovación y Universidades. Financial support from the University of Oviedo, Spain, is also acknowledged.

Institutional Review Board Statement: The Oviedo University Local Animal Care and Use Committee approved the experimental protocols (reference 33443591). All experiments were carried out according to the Spanish Government Guide and the European Community Guide for Animal Care (Council Directive 86/609/EEC).

Informed Consent Statement: Not applicable.

Data Availability Statement: The authors declare that the data supporting the findings of this study are available within the paper.

Acknowledgments: We are members of the Women in Autophagy (WIA), INPROTEOLYS, SEBBM, SEFAGIA, and INEUROPA network.

Conflicts of Interest: The authors declare no conflict of interest.

References

1. Agil, A.; Navarro-Alarcon, M.; Ruiz, R.; Abuhamadah, S.; El-Mir, M.Y.; Vazquez, G.F. Beneficial effects of melatonin on obesity and lipid profile in young Zucker diabetic fatty rats. *J. Pineal Res.* **2011**, *50*, 207–212. [CrossRef] [PubMed]
2. Demirtas, C.Y.; Pasaoglu, O.T.; Bircan, F.S.; Kantar, S.; Turkozkan, N. The investigation of melatonin effect on liver antioxidant and oxidant levels in fructose-mediated metabolic syndrome model. *Eur. Rev. Med. Pharmacol. Sci.* **2015**, *19*, 1915–1921.
3. Kozirog, M.; Poliwczak, A.R.; Duchnowicz, P.; Koter-Michalak, M.; Sikora, J.; Broncel, M. Melatonin treatment improves blood pressure, lipid profile, and parameters of oxidative stress in patients with metabolic syndrome. *J. Pineal Res.* **2011**, *50*, 261–266. [CrossRef] [PubMed]
4. Reiter, R.J. The pineal and its hormones in the control of reproduction in mammals. *Endocr. Rev.* **1980**, *1*, 109–131. [CrossRef] [PubMed]
5. Negi, G.; Kumar, A.; Sharma, S.S. Melatonin modulates neuroinflammation and oxidative stress in experimental diabetic neuropathy: Effects on NF-kappaB and Nrf2 cascades. *J. Pineal Res.* **2011**, *50*, 124–131. [CrossRef] [PubMed]
6. Kitagawa, A.; Ohta, Y.; Ohashi, K. Melatonin improves metabolic syndrome induced by high fructose intake in rats. *J. Pineal Res.* **2012**, *52*, 403–413. [CrossRef] [PubMed]
7. Rios-Lugo, M.J.; Cano, P.; Jimenez-Ortega, V.; Fernandez-Mateos, M.P.; Scacchi, P.A.; Cardinali, D.P.; Esquifino, A.I. Melatonin effect on plasma adiponectin, leptin, insulin, glucose, triglycerides and cholesterol in normal and high fat-fed rats. *J. Pineal Res.* **2010**, *49*, 342–348. [CrossRef]
8. Anderson, G.; Reiter, R.J. Melatonin: Roles in influenza, Covid-19, and other viral infections. *Rev. Med. Virol.* **2020**, *30*, e2109. [CrossRef] [PubMed]
9. Balakumar, M.; Raji, L.; Prabhu, D.; Sathishkumar, C.; Prabu, P.; Mohan, V.; Balasubramanyam, M. High-fructose diet is as detrimental as high-fat diet in the induction of insulin resistance and diabetes mediated by hepatic/pancreatic endoplasmic reticulum (ER) stress. *Mol. Cell. Biochem.* **2016**, *423*, 93–104. [CrossRef]
10. Sakai, T. The mammalian Harderian gland: Morphology, biochemistry, function and phylogeny. *Arch. Histol. Jpn.* **1981**, *44*, 299–333. [CrossRef]
11. Payne, A.P.; McGadey, J.; Moore, M.H.; Thompson, G.G. Changes in Harderian gland activity in the female golden hamster during the oestrous cycle, pregnancy and lactation. *Biochem. J.* **1979**, *178*, 597–604. [CrossRef] [PubMed]

12. Hoffman, R.A.; Johnson, L.B.; Reiter, R.J. Harderian glands of golden hamsters: Temporal and sexual differences in immunoreactive melatonin. *J. Pineal Res.* **1985**, *2*, 161–168. [CrossRef] [PubMed]
13. Menendez-Pelaez, A.; Lopez-Gonzalez, M.A.; Guerrero, J.M. Melatonin binding sites in the Harderian gland of Syrian hamsters: Sexual differences and effect of castration. *J. Pineal Res.* **1993**, *14*, 34–38. [CrossRef] [PubMed]
14. Santillo, A.; Chieffi Baccari, G.; Minucci, S.; Falvo, S.; Venditti, M.; Di Matteo, L. The Harderian gland: Endocrine function and hormonal control. *Gen. Comp. Endocrinol.* **2020**, *297*, 113548. [CrossRef]
15. Coto-Montes, A.; Boga, J.A.; Tomas-Zapico, C.; Rodriguez-Colunga, M.J.; Martinez-Fraga, J.; Tolivia-Cadrecha, D.; Menendez, G.; Hardeland, R.; Tolivia, D. Porphyric enzymes in hamster Harderian gland, a model of damage by porphyrins and their precursors. A chronobiological study on the role of sex differences. *Chem. Biol. Interact.* **2001**, *134*, 135–149. [CrossRef]
16. Vega-Naredo, I.; Caballero, B.; Sierra, V.; Huidobro-Fernandez, C.; de Gonzalo-Calvo, D.; Garcia-Macia, M.; Tolivia, D.; Rodriguez-Colunga, M.J.; Coto-Montes, A. Sexual dimorphism of autophagy in Syrian hamster Harderian gland culminates in a holocrine secretion in female glands. *Autophagy* **2009**, *5*, 1004–1017. [CrossRef]
17. Vega-Naredo, I.; Coto-Montes, A. Physiological autophagy in the Syrian hamster Harderian gland. *Methods Enzymol.* **2009**, *452*, 457–476.
18. Kasim-Karakas, S.E.; Vriend, H.; Almario, R.; Chow, L.C.; Goodman, M.N. Effects of dietary carbohydrates on glucose and lipid metabolism in golden Syrian hamsters. *J. Lab. Clin. Med.* **1996**, *128*, 208–213. [CrossRef]
19. Bradford, M.M. A rapid and sensitive method for the quantitation of microgram quantities of protein utilizing the principle of protein-dye binding. *Anal. Biochem.* **1976**, *72*, 248–254. [CrossRef]
20. Arnao, M.B.; Cano, A.; Acosta, M. The hydrophilic and lipophilic contribution to total antioxidant activity. *Food Chem.* **2001**, *73*, 239–244. [CrossRef]
21. Garcia-Macia, M.; Vega-Naredo, I.; De Gonzalo-Calvo, D.; Rodriguez-Gonzalez, S.M.; Camello, P.J.; Camello-Almaraz, C.; Martin-Cano, F.E.; Rodriguez-Colunga, M.J.; Pozo, M.J.; Coto-Montes, A.M. Melatonin induces neural SOD2 expression independent of the NF-kappaB pathway and improves the mitochondrial population and function in old mice. *J. Pineal Res.* **2011**, *50*, 54–63. [CrossRef]
22. Vega-Naredo, I.; Caballero, B.; Sierra, V.; Garcia-Macia, M.; de Gonzalo-Calvo, D.; Oliveira, P.J.; Rodriguez-Colunga, M.J.; Coto-Montes, A. Melatonin modulates autophagy through a redox-mediated action in female Syrian hamster Harderian gland controlling cell types and gland activity. *J. Pineal Res.* **2012**, *52*, 80–92. [CrossRef]
23. Vega-Naredo, I.; Poeggeler, B.; Sierra-Sanchez, V.; Caballero, B.; Tomas-Zapico, C.; Alvarez-Garcia, O.; Tolivia, D.; Rodriguez-Colunga, M.J.; Coto-Montes, A. Melatonin neutralizes neurotoxicity induced by quinolinic acid in brain tissue culture. *J. Pineal Res.* **2005**, *39*, 266–275. [CrossRef] [PubMed]
24. Dalboge, L.S.; Pedersen, P.J.; Hansen, G.; Fabricius, K.; Hansen, H.B.; Jelsing, J.; Vrang, N. A Hamster Model of Diet-Induced Obesity for Preclinical Evaluation of Anti-Obesity, Anti-Diabetic and Lipid Modulating Agents. *PLoS ONE* **2015**, *10*, e0135634. [CrossRef]
25. Horne, R.G.; Yu, Y.; Zhang, R.; Abdalqadir, N.; Rossi, L.; Surette, M.; Sherman, P.M.; Adeli, K. High Fat-High Fructose Diet-Induced Changes in the Gut Microbiota Associated with Dyslipidemia in Syrian Hamsters. *Nutrients* **2020**, *12*, 3557. [CrossRef] [PubMed]
26. Malo, E.; Saukko, M.; Santaniemi, M.; Hietaniemi, M.; Lammentausta, E.; Blanco Sequeiros, R.; Ukkola, O.; Kesaniemi, Y.A. Plasma lipid levels and body weight altered by intrauterine growth restriction and postnatal fructose diet in adult rats. *Pediatr. Res.* **2013**, *73*, 155–162. [CrossRef]
27. Hussain, S.A. Effect of melatonin on cholesterol absorption in rats. *J. Pineal Res.* **2007**, *42*, 267–271. [CrossRef] [PubMed]
28. Chan, T.Y.; Tang, P.L. Effect of melatonin on the maintenance of cholesterol homeostasis in the rat. *Endocr. Res.* **1995**, *21*, 681–696. [CrossRef]
29. Martinez-Lopez, N.; Garcia-Macia, M.; Sahu, S.; Athonvarangkul, D.; Liebling, E.; Merlo, P.; Cecconi, F.; Schwartz, G.J.; Singh, R. Autophagy in the CNS and Periphery Coordinate Lipophagy and Lipolysis in the Brown Adipose Tissue and Liver. *Cell Metab.* **2016**, *23*, 113–127. [CrossRef]
30. Garcia-Macia, M.; Santos-Ledo, A.; Caballero, B.; Rubio-Gonzalez, A.; de Luxan-Delgado, B.; Potes, Y.; Rodriguez-Gonzalez, S.M.; Boga, J.A.; Coto-Montes, A. Selective autophagy, lipophagy and mitophagy, in the Harderian gland along the oestrous cycle: A potential retrieval effect of melatonin. *Sci. Rep.* **2019**, *9*, 18597. [CrossRef]
31. Cardinali, D.P.; Vigo, D.E. Melatonin, mitochondria, and the metabolic syndrome. *Cell. Mol. Life Sci.* **2017**, *74*, 3941–3954. [CrossRef] [PubMed]
32. Garcia-Macia, M.; Rubio-Gonzalez, A.; de Luxan-Delgado, B.; Potes, Y.; Rodriguez-Gonzalez, S.; de Gonzalo-Calvo, D.; Boga, J.A.; Coto-Montes, A. Autophagic and proteolytic processes in the Harderian gland are modulated during the estrous cycle. *Histochem. Cell Biol.* **2014**, *141*, 519–529. [CrossRef] [PubMed]
33. Yang, Y.; Karsli-Uzunbas, G.; Poillet-Perez, L.; Sawant, A.; Hu, Z.S.; Zhao, Y.; Moore, D.; Hu, W.; White, E. Autophagy promotes mammalian survival by suppressing oxidative stress and p53. *Genes Dev.* **2020**, *34*, 688–700. [CrossRef] [PubMed]
34. Hosseini, A.; Amiri, F.; Khalighi, F.; Mohammadi Roushandeh, A.; Kuwahara, Y.; Bashiri, H.; Habibi Roudkenar, M. Cell Survival Effects of Autophagy Regulation on Umbilical Cord-Derived Mesenchymal Stem Cells Following Exposure to Oxidative Stress. *Iran. J. Med. Sci.* **2019**, *44*, 493–500. [CrossRef] [PubMed]

35. Rabanal-Ruiz, Y.; Korolchuk, V.I. mTORC1 and Nutrient Homeostasis: The Central Role of the Lysosome. *Int. J. Mol. Sci.* **2018**, *19*, 818. [CrossRef]
36. Mohamed, D.Z.; El-Sisi, A.E.E.; Sokar, S.S.; Shebl, A.M.; Abu-Risha, S.E. Targeting autophagy to modulate hepatic ischemia/reperfusion injury: A comparative study between octreotide and melatonin as autophagy modulators through AMPK/PI3K/AKT/mTOR/ULK1 and Keap1/Nrf2 signaling pathways in rats. *Eur. J. Pharmacol.* **2021**, *897*, 173920. [CrossRef] [PubMed]
37. Cui, Y.; Yang, M.; Wang, Y.; Ren, J.; Lin, P.; Cui, C.; Song, J.; He, Q.; Hu, H.; Wang, K.; et al. Melatonin prevents diabetes-associated cognitive dysfunction from microglia-mediated neuroinflammation by activating autophagy via TLR4/Akt/mTOR pathway. *FASEB J.* **2021**, *35*, e21485. [CrossRef]
38. Zoncu, R.; Efeyan, A.; Sabatini, D.M. mTOR: From growth signal integration to cancer, diabetes and ageing. *Nat. Rev. Mol. Cell Biol.* **2011**, *12*, 21–35. [CrossRef]
39. Wullschleger, S.; Loewith, R.; Hall, M.N. TOR signaling in growth and metabolism. *Cell* **2006**, *124*, 471–484. [CrossRef]
40. Singh, R.; Kaushik, S.; Wang, Y.; Xiang, Y.; Novak, I.; Komatsu, M.; Tanaka, K.; Cuervo, A.M.; Czaja, M.J. Autophagy regulates lipid metabolism. *Nature* **2009**, *458*, 1131–1135. [CrossRef]
41. Klionsky, D.J.; Abdel-Aziz, A.K.; Abdelfatah, S.; Abdellatif, M.; Abdoli, A.; Abel, S.; Abeliovich, H.; Abildgaard, M.H.; Abudu, Y.P.; Acevedo-Arozena, A.; et al. Guidelines for the use and interpretation of assays for monitoring autophagy (4th edition). *Autophagy* **2021**, 1–382. [CrossRef] [PubMed]

MDPI
St. Alban-Anlage 66
4052 Basel
Switzerland
www.mdpi.com

Antioxidants Editorial Office
E-mail: antioxidants@mdpi.com
www.mdpi.com/journal/antioxidants

www.ingramcontent.com/pod-product-compliance
Lightning Source LLC
Chambersburg PA
CBHW040744250926
51850CB00007BB/1403

* 9 7 8 3 0 3 6 5 8 8 9 4 0 *